Parasite-Faunas of Jammu, Kashmir and Ladakh (Indian Himalayas): *Diversity, Impact and Importance*

THE AUTHOR

Prof. (Dr.) Ramesh Chander Bhagat, earlier served as Asstt. Research Officer, Sericulture Research Centre of Kashmir. Thereafter, he joined Deptt. of Zoology, Kashmir University and served at various positions, including Head of the Deptt. He also held the position of Dean, Faculty of Science (multidisciplinary and research) of Kashmir University. Prof. Bhagat has specialized in the fields of entomology, insect parasites, biological control agents and faunal biodiversity. The author of this book possesses more than 30-year of experience in teaching general and applied Zoology, and Entomology, conducting research, guiding research (M.Phil. /Ph.D.) and as Principal Investigator of research Projects (UGC, CSIR). There are more than 185 research and scientific publications to the credit of the author, including 05 reference/ research books on Zoological studies and Biodiversity of Jammu & Kashmir and Ladakh Himalayas. He has also received many awards / honours and remained active member of academic / expert committees and of scientific and statutory bodies of the Universities.

Parasite-Faunas of Jammu, Kashmir and Ladakh (Indian Himalayas): *Diversity, Impact and Importance*

– Author –

Ramesh Chander Bhagat

2022

Daya Publishing House®

A Division of

Astral International Pvt. Ltd.

New Delhi – 110 002

ISBN: 9789354615108

Published by : **Daya Publishing House®**
A Division of
Astral International Pvt. Ltd.
– ISO 9001:2015 Certified Company –
4736/23, Ansari Road, Darya Ganj
New Delhi-110 002
Ph. 011-43549197, 23278134
e-mail: info@astralint.com
Website: www.astralint.com

Preface

The Parasite- Faunas, included in the present book, are animal parasites (Invertebrates), parasitic on Invertebrates, Vertebrates, including Humans, pertaining to different Phyla of Animal Kingdom, such as : Protozoans, Apicomplexans and Cnidarians (microparasites); Metazoans (macroparasites)- Helminths (trematodes, cestodes, nematodes, acanthocephalans), Annelids (leeches), Pentastomida and Arthropods (crustaceans, insects, ticks, mites). Apart from Invertebrate and Vertebrate hosts, parasitic nematodes that feed on different parts of plants, known as plant- parasitic nematodes, have also been incorporated in this book The parasites of diverse groups, have been found to be parasitizing Invertebrates (arthropods), Vertebrates (fishes, amphibians, reptiles, birds, mammals), besides humans as well as plants (wild and cultivated/crops), have lucidly been dealt with in this book. The umpteen animal parasite species, associated with diverse groups of host animals, humans, plants, are known to be occurring in vast and wide areas, and localities of Jammu, Kashmir and Ladakh region, as Zoogeographically important regions and biodiversity hot spots in the Himalayas of Indian sub-continent.

The Jammu and Kashmir, and Ladakh, are Union Territories (UTs) of Indian sub-continent. Jammu and Kashmir (UT), consisting of two Divisions: Jammu and Kashmir Division, with each divided into 10 Districts. Ladakh, a part of erstwhile J&K State, is also a separate UT, with 2 Districts (Kargil, Leh). Jammu, Kashmir and Ladakh Himalayan regions, are geographically as well as climatically different: Jammu (sub-tropical), situated in between the valley of Kashmir to Shiwalik (= Siwalik hills), outer Himalayas (north-west India); Kashmir (temperate), including main Himalyan range of Kashmir valley, Pir Panjal Himalayan range; Ladakh (cold desert), highest altitude Plateau region in India, including parts of the Himalayan and Karakorum mountain ranges and the upper Indus River Valley. Due to high-altitude environment, the climate of Ladakh is extremely cold and dry.

The parasites living on or in a different kinds of organisms (Hosts), from which these get some or all their nourishment. Parasites are responsible for causing harms and diseases to their hosts. The parasites are mainly grouped as ectoparasites and endoparasites. The parasites that live on the host body, are called ectoparasites (mosquitoes, fleas, lice, leeches, *etc.*) and those that live inside hosts are called endoparasites (protozoans, apicomplexans, helminths, *etc.*).The parasites have also been categorized as : temporary parasites, visit their host for short period; permanent parasites, lead a permanent condition throughout whole period of life;facultative parasite, live either a parasitic or non-parasitic existence; obligatory parasite cannot live without or parasitic life. In addition to the above mentioned groups of parasites, parasitoids are also prevalent in J&K and Ladakh Himalayan regions. Parasitoids are entomophagous parasitic insects that live in or on, and finally kill a large insect hosts. Besides, hyperparasites show existence, are parasites whose hosts often insects,are also parasites, often specifically parasitoids.

Microparasites as well as macroparasites are responsible for causing harms and diseases to different animals, humans and plants/crops. Helminths especially nematodes (Strongylids) cause huge economic loss globally by affecting food animals (sheep, goats, cattle, pigs). The plant-parasitic nematodes, particularty root-knot nematodes (*Meloidogyne, etc.*), cyst nematodes (*Heterodera, etc.*), Lesion nematode (*Pratylenchus.*) and burrowing nematode (*Radopholus*), pose threat to successful crop cultivation, attack crops and plants of economic importance, leading to crop losses.

The parasitic arthropods (insects, ticks, mites) are potential vectors of viral, spirochaetal, rickettesial, bacterial, apicomplexan and helminth diseases (*Dirofilaria, Thelazia, Acanthocheilema, etc.*) in animals and humans. The major diseases of public health and animal husbandry importance through arthropod vectors are : dengue, malaria (human, bird), human leishmaniasis (cutaneous, visceral), human plague, relapsing fever, tick typhus, Crimean- Congo- haemorrahagic fever, of medical importance; trypanosomiasis or ' Surra disease ', babesiosis and theileriosis in domestic animals.The Zoogeographical regions of J&K and Ladakh, have also shown the high prevalence of zoonotic parasites of medical and veterinary importance. A zoonotic disease or zoonosis, through parasites, spread between animals and humans. The major zoonotic parasites of humans are : protozoans (*Cryptosporidium, Babesia, Endolimax, Entamoeba, Giardia, Toxoplasma*); helminths, (*Fasciola, Taenia saginatta, Ancylostoma caninum, Trichuris*). Besides these, anthroponotic parasites, cause antroponosis or zooanthroponosis (reverse zoonosis). The anthroponotic parasites can be transmitted from humans to animals. The important anthroponotic parasites are: protozoans (*Giardia lambi, Entamoeba histilytica*); nematodes (*Ancylostoma duodenale, Ascaris lumbricoides and Trichuris trichiura*).

Parasites are not exclusively harmful, however, some groups possess usefulness to humans and other creatures. These benefits are :

(a) Playng an important role in regulating population of their hosts and balance overall ecosystem.

(b) Acting as bio-indicator of environmental degradation

(c) For the treatment of medical condition or diseases, such as : parasitic leeches are beneficial in many ways in curing human diseases; maggot (insect) debridement therapy or bio-surgery for trearment of non-healing wounds by calliphorid (flies) maggots.

(d) Parasitic insects like necrophagous dipterans are forensic indicators, as most important ecological evidence associated with decomposing corps, provide insight into estimating post-mortem intervals (PMI).

(e As biological control agents like hymenopteran and dipteran entomophagous parasitic insects, are being utilized in different parts of the world for the suppression or control of insect pests of agricultural/ horticultural and forestry importance, *etc.*

In the present book, systematic survey, distribution, prevalence, species richness and host range of different groups of Parasite- faunas, with special focuss on their importance, in Jammu, Kashmir and Ladakh Himalayan region, have comprehensively been dealt with. The book embodies as many as 10 separate Chapters. Chapter 1, encompasses parasites, belonging to groups *viz.*, Microsporidians, Amoebozans, Ciliophorans, Metamonads (=Protozoans), Euglenozoans, Apicomplexans, Protozoans and Cnidarians, parasitic on insects, fishes, birds, mammals and humans, with special reference to parasites of medical and veterinary importance. The Protozoans and Apicomplexan parasites of medical importance, causing diseases include amoebiasis, leishmaniasis (cutaneous, visceral), giardiasis, trichomoniasis, malaria, cryptosporidiosis and toxoplasmosis. The protozoan and apicomplexan parasites of veterinary and animal husbandry importance, responsible for causing diseases, discussed in the present work are: animal trypanosomiasis, giardiasis, trichomoniasis, histomoniasis, babesiosis, theileriosis, cryptosporidiosis, *Toxoplasma* sp., *Cystoisospora, Heptozoon canis, Sarcocystis, Eimeria* spp. and, animal and poultry coccidiosis.

Helminth parasites : trematodes, cestodes and nematodes (animal, human), are spread over two separate Chapters (2, 3), providing information on diversity, host range and importance of these parasites. In addition to these, widely and highly prevalent trematodes, cestodes and nematodes of medical and veterinary importance, have been highlighted. Chapter 4, is soley devoted to up-to-date Catalogues on distribution, prevalence and, high infection in age group and gender due to nematode, cestode and trematode of of medical and veterinary importance.

Acanthocephalan parasites and plant-parasitic nematodes find place in Chapter 5, in which the systematic survey of acanthocephalans, infecting fishes have been discussed. The systematic survey, diversity, and species richness of plant parasitic nematodes affecting cereals (paddy, maize, wheat), vegetables (brassicas,cucurbits, solanaceous and other vegetables), fruit plantations (temperate, tropical), medicinal/ aromatic and ornamental plants, besides forest range plantation and grasses, have been incorporated in this Chapter.

Chapter 6 pertains to parasitic leeches, crustaceans, pentastomida, ticks and mites. Brief description is provided in this Chapter on : diversity of leeches, with reference to medicinal leeches; copepods; branchiurans and *Linguatula serrata*. The faunistic survey, host range of ticks and mites, causing and transmitting diseases in animals and humans, have been given in this Chapter. The ticks of veterinary importance include diseases like tick-borne spirochaetal, rickettsial, bacterial and apicomplexan diseases. The tick-borne diseases of medical importance are : tick typhus, relapsing fever and Crimean-Congo haemorrhagic fever. The mites, parasitic on insects, birds (poultry), sheep, small mammals, besides mites of medical importance, are also included in this Chapter.

A detailed account on systematic survey, diversity and host range of fleas and lice (sucking and chewing), with their medical and veterinary importance, have been provided in Chapter 7. The fleas (*Xenopsylla* spp., *Pulex* spp.), are of medical importance. The flea, *Xenopsylla cheopis* (Oriental rat flea) is an important vector of diseases, *viz.* human plague and murine typhus. An elaborate information on mosquito-fauna and parasitic dipteran flies, with their importance, has been presented in Chapter 8. In this Chapter, an up-to-date annotated checklist of mosquitoes, besides systematic survey, species richness and mosquito-borne diseases (malaria, dengue), have been presented. In addition to this, Chapter 8, includes other parasitic dipteran flies, *viz.* sandflies, blackflies, deerflies, clegs, muscids, pupipara, ceratopogonids and dipteran flies of myiasis-causing importance in animals and humans, have been dealt with.

The Chapter 9 and 10, are concerned with entomophagous parasitic insects (parasites and parasitoids). Chapter 9, focuses on systematic survey, distribution, diversity, host range and importance of these parasites/parasitoids, belonging to as many as 6 families, under 4 superfamilies of order Diptera and 13 families of 5 superfamilies of order Hymenoptera. The Dipteran super-families are : Bombyloidea; Conopoidea; Muscoidea and Syrphoidea. The Hymenopteran superfamilies include Ceraphronoidea, Chalcidoidea, Cynipoidea, Ichneumonoidea and Platygastroidea. Chapter 10 depicts a detailed Catalogue of Host plant species/family, Host insect species/family and Entomophagous parasitic insect species/family. A total of more than 152 species of plants, belonging to 43 families, has been found to be associated with number of insects species as hosts of parasites and parasitoids. The host plant species are not only wild but also cultivated, including agriculture crops and other economically important plant species.

The valuable information and databases on Parasite-faunas, contained in the present book, have been updated for the first time, in the light of recent researches, accomplished by number of research workers in different fields of animal and human parasites besides, plant-parasitic nematodes of Jammu, Kashmir and Ladakh Himalayan regions. Special emphasis has been laid on updation with respect to systematic and nomenclatural changes of parasites. The valid or most generally accepted names and synonyms are cited in the text of the book. The clinical aspects of parasitic diseases of humans and animals, have not been discussed.

The book is laced with number of tables (more than 60), based on copious amount of data, spread under different Chapters. The sources of information and references in the form of publications (research journals, books, *etc.*), perused by the author are given at the end of the book, under separate References and Bibliography section. In this section, number of references included are marked with astrick (*), not finding place in the main text of the book. These references are added for the purpose of as additional/further reading. An exhaustive Subject Index, has also been added in the book.

In preparation of this reference book, pioneering efforts have been made, leading to an up-to-date work, in light of latest research findings on different parasite faunas, related to diverse fields of human activities. The databases developed,will serve the purpose of scientific communities not only in India but also elsewhere in the world. It is hoped that this book undoubtedly, will cater to the growing needs of Science Professionals, Experts, Researchers and Students in different fields of biosciences, zoology, parasitology, entomology, besides agriculture, veterinary and medical. The author will welcome suggestions and criticisms for further improvement of the book.

Lastly, I express my sincere gratitude to my wife, Dr. (Mrs.) Poonam Bala, Former Lecturer, for her constant support and encouragement besides valuable help rendered in: Compilation of data on parasite taxas; preparation of Bibliography and subject index. Also, I am highly indebted to my son Mr. Rishabh Bhagat and daughter-in-law, Ms. Aditi for providing valuable assistance in processing and computerization of the manuscript of the present book.

Ramesh Chander Bhagat

Contents

Chapter 1

Parasitic Protozoans, Apicomplexans and Cnidarians

The parasitic protozoans under earlier classification, have now been grouped under a separate Kingdom of Protozoa, including various Phyla as Euglenozoa, Amoebozoa and Metamonda, *etc.* (Cavalier- Smith, 2010; Ruggiero *et al.,* 2015). The former ciliate protozonas are also ranked as a Phylum in most of system of taxonomy, " Ciliophora" (ITIS Report, 2014; Adl *et al.*, 2012), as a phylum either in Kingdom Protista or Protozoa. Apicomplexa form a large Phylum, formerly known as Sporozoa. Myxozoan parasites were previously considered as protozoans (Sporozoa), however, on the basis of molecular analysis, find place in Cnidaria (Cairns *et al.*, 2009).

In Jammu and Kashmir, and Ladakh Union territories (UTs) of India, the parasitic Protozoans (Amoebozoa, Ciliophora, Euglenozoa, Metamonda), Apicomplexans and Cnidarians show wide diversity, and are of great importance. These parasitic groups have incorporated as many as 93 species, affecting diverse range of invertebrate and vertebrate hosts, including humans, of these regions of paramount zoogeographical importance (**Table 1.1**). The parasitic Amoebozoans are small in size, with single nucleus and lobose pseudopodia. A total of 5 species of parasitic Amoebozoans are known to occur in UTs, infecting human population, belong to three genera, *viz. Endolimax, Entamoeba* and *Iodamoeba* (**Table 1.1**). *Entameba histolytica* casuses human amoebic dysentery or amoebiasis.

The parasitic Euglenozoans are flagellates (single cell), infect humans as well as animals (insects, fishes, birds, rodents, domestic animals, livestock) and some members like Crithidia (*C.bombi, C. expecki*), *Trypanosoma* (*T. avium, T. evansi, T. lewisi, T. kashmirensis, T. mukasi, Trypanosoma* sp.), *Latmania passim, Leishmania* (*L. donvani, L. tropica*), are responsible for cauising severe diseases in these hosts. The parasitic Euglenzoans of this region include a total of 10 species, under 4 genera (**Table 1.1**). The Metamonads have multiflagellate cells. These are parasitic on humans and other animals of UTs, covering a total of 7 species, belonging to 6 genera of great medical and veterinary importance as *Chilomastix, Enteromonas, Giardia, Histomonas, Retortamonas* and *Trichomonas* (**Table 1.1**).

Table 1.1: Systematic and Annotated Checklist of Parasitic Protozoans, Apicomplexans and Cnidarians of Invertebrate and Vertebrate, including Humans of Jammu, Kashmir and Ladakh (Key to the numerals in long brackets, given at the end of Checklist)

Phylum : Amoebozoa

Subphylum : Conosa

Class: Archamoeba

Family: Endolimacidae

1. *Endolimax nana*: Hosts- Humans [5]

Family : Entamoebidae

2. *Entamoeba coli* : Hosts- Humans [5]
3. *Entamoeba hartmani* : Hosts- Humans [5]
4. *Entamoeba histolytica*: Hosts- Himans [1,5]

Family : Endamoebidae

5. *Iodamoeba bustschili*: Hosts- Humans [5]

Phylum : Ciliophora

Class: Kinetofragminophora

Order : Trichostromatida

Family : Pycnotrichidae

6. *Buxtonella sulcata* : Hosts- Bovines [13]

Class: Litastomatea

Order: Vestibulifevida

Family: Balantiidae

7. *Balantidium coli* : Hosts : Zoma, Cattle, Yak [65]

Class: Oligohymenophorea

Family: Ichthyophiridae

8. *Ichthyophthirus multifiliis*: Host- Fishes (*Aspidiopara murar*, *Triplophysa kashmirensis, Schizothorax niger* [9,10, 36]

Subclass: Peritricha

Order: Mobilida

Family : Trichodinidae

9. *Trichodina heterodenta* : Hosts- Fishes (*Catla catla*, *Cyprinus carpio*, *Labeo rohita, Puntius ticto, Schizothorax niger* [9,10, 57]

Class: Phyllopharyngea

Order: Crytophorida

Family : Chilodonellidae

10. *Chilodonella uncinata* : Hosts - Fish (*S.niger*) [10]
11. *Chilodonella* sp. : Hosts- Fishes (*Labeo rohita*, *S. niger*) [4, 56]

Phylum : Euglenozoa

Class: Kinetoplastida

Subclass: Metakinetoplastina

Family: Trypanosomatidae

12. *Trypanosoma avium* : Host- *Glaucidium cuculoides* (Asian Barred owlet)[32, 34,35]
13. *Trypanosoma evansi* : Hosts- Cattle, Camel, Buffaloes, Ponies, Horses [23, 39, 46, 51]
14. *Trypanosoma lewisi*: Hosts- Rodents, *Rattus* sp. [34]
15. *Trypanosoma kashmirensis*: Host- Fish (*Schizothorax plagiostomus*)[26]
16. *Trypanosoma mukasi*: Hosts- Fishes (*Carassius carassius, Schizothorax curvifrons, Triplophysa marmorata* [45]
17. *Trypanosoma* sp.: Host- Dog [6]

Subfamily: Leishmaniinae

Infrafamily: Leishmaniatae

18. *Leishmania donavani*: Hosts- Humans [19]
19. *Leisihmania tropica*: Hosts- Humans [24]

Infra-family: Crithidiatae

20. *Crithidia bombi*: Hosts- Insects (*Bombus simillimus, Bombus asciatus, Bombus trifasciatus* [63]
21. *Crithidia expecki*: Hosts- Insects (*B. simillimus, B.asciatus, B. trifasciatus*) [63]
22. *Lotmania passim*: Hosts- Insects, Honeybees (*Apis cerana, Apis mellifera, Apis dorsata* [63]

Phylum: Metamonada

Class Eopharyngia

Order: Diplomonadida

Family: Hexamitidae

Sub-family: Giardiinae

23. *Giardia intestinalis* (= *G. lamblia, G.duodenalis*): Hosts- Humans [1, 5, 22, 28]
24. *Giardia* sp.: Host- dogs [47]

Suborder: Distomatina

Family: Enteromonadidae

25. *Enteromonas hominis*: Hosts- Humans [5]

Class: Retortamonadea

Order: Retortamonadida

Family: Retortamonadidae

26. *Chilomastix mesnile*: Host- Humans [5]
27. *Retortamonas intestinalis* (= *Enbadomonas intestinalis*): Host- Humans [5]

Class: Parabasalia

Order : Trichomonadida

Family: Monocercomonadidae

28. *Histomonas meleagridis* : Host- Rock Patridge (*Alectoris gracea* = *Gracia electoris*) [20, 29, 34]

Family: Trichomonadidae

29. *Trichomonas vaginalis* : Hosts- Humans [25]

30. *Trichomonas foetus* : Host - Bulls [59]
31. *Trichomonas gallinae* : Host-Pigeon [Mehmood *et al.*, 2019]

Phylum: Cnidaria

Class: Myxozoa

Sub-Class Myxosporea

Order: Bivalvulida

Family: Myxobolidae

32. *Myxoboluis chushin* : Host – Fish (*Schizothorax niger*) [11]

Suborder: Variisporina

Family: Myxidiidae

33. *Myxidium cholai* : Host – Fish (*Puntius chola*) [36]
34. *Myxidium himalayensis* : Host – Fish (*Schizothorax richardsonii*) [Ahmad *et al.*, 2019]

Phylum : Apicomplexa

Class: Aconoidasida

Sub-class: Haemosporidiasina

Order: Chromatarida

Sub-order: Laveraniinae

Family: Plasmodiidae

35. *Plasmodium faliciparum*: Hosts- Humans [30, 64]
36. *Plasmodium malariae*: Hosts- Humans [30, 14]
37. *Plasmodium relictum* : Hosts- *Muscicape thalassina thalassina*) (Indian Verditer flycatcher [35]
38. *Plasmodium vivax*: Hosts- Humans [14, 30, 64]
39. *Plasmodium* spp. : Hosts- Birds, Long-tailed shrike (*Lanius schach*), *Zosterops palpebrosus* [34]

Family: Haemoproteidae

40. *Haemoproteus columbae*: Host- Pigeon [7]
41. *Haemoproteus borgesi*: Host- *Jynx torquilla Himalayans* (Eurasian Wryneck) [35]
42. *Haemoproteus celli*: Hosts- *Glaucidium cuculoides*(Asian barred owlet), *Strix aluco* (Tawny owl) [35]
43. *Haemoproteus dicrurid* : Host- *Dicrurus adsimilis* (Forked-tailed drongo)[35]
44. *Haemoproteus oryzivorae*: Host - *Myophonus caeruleus* (Blue whistling thrush) [35]
45. *Haemoproteus pastoris*: Host – *Acridotheres tristis tristis* (Indian Mynah) [35]
46. *Haemoproteus passeris* : Host- Bird *Passer domesticus* (The house sparrow) [35]
47. *Haemoproteus* spp.: Hosts: Birds- *Accipiter nisus* (Eurasian sparrow), *Butea vulpinus vulpinus* (Common Buzzard), *Carpodaeus erythrinus* (Common Rosefinch), *Erithaeus brunneus*(Indian Blue Robin), *Garrulus* lanceolatus(Black-headed Jay), *Lanius schach* (The long-tailed shrike), *Monticola cinchorhyncha* (The blue-capped rock thrush), *Motacilla cinereal* (Wagtail), *Passer domesticus* (Sparrow), *Pericrocotus brevirostris* (Short-billed minivet), *Phylloscopus occipitalis* (Western crowned warbler), *Pycnonotus leucogenys* (The Himalayan Bulbul), *Pycnonotus luteolus* (White-browed bulbul), *Sturnus vulgaris* (Common starling), *Sylvia curruca* (Lesser white throat), *Zosterops palpebrosus*, (The oriental white-eye) [32, 33, 34,35]; *Parahaemoproteus velans* (= *Haemoproteus velans*) : Hosts- *Dendrocopus mahrattensi* (Yellow-crowned wood-pecker), *Jynx torquilla* (Eurasian wryneck) [34]

Class: Aconoidasida

Sub-class: Coccidia

Order: Eucoccidiorida

Sub-order: Adeleorina

Family: Dactylosomatidae

48. *Babesioma* sp. : Host- Fish (*Triplophysa marmorata*)[45]

Order: Achromatorida

Sub-order : Babesiina

Family: Babesiidae

49. *Babesia bigemina*: Host- Cattle [12, 51, 52, 54, 55]
50. *Babesia bovis* (=*Babesia argentina*): Hosts- Cattle [12, 40, 51, 52, 60]
51. *Babesia motasi*: Hosts- Sheep and goat [43, 61]
52. *Babesia ovis* : Hosts- Sheep and goat [43, 61]
53. *Babesia* sp.: Hosts- Sheep and goats [15,16]

Sub-order 2: Theileriina

Family 1: Theileriinae

54. *Theileria annulata* : Host- Cattle [46, 52, 52, 53]
55. *Theileria parva* : Host- Cattle[40, 43]

Family 2: Leucocytozoidae

56. *Leucocytozoon dubreuili*: Host- bird, *Pycnonotus leucogenys* (The Himalayan bulbul) [33, 34, 35]
57. *Leucocytozoon fringillinarum* : Host- *Carpodaeus erythrinus* (Common Rose finch) [33, 34, 35]
58. *Leucocytozoon majoris*: Host – *P. leucogenys* [31, 34]
59. *Leucocytozoon squamatus*: Host- bird, *Picus squamatus* (Scaly-bellied woodpecker) [31, 34]
60. *Leucocytozoon* spp. : Host- birds, *Carpodacus erythrinus*, *Coccothraustes esteroides* (Hanfinch), *Emberiza cia* (Rock bunting), *Ficedula superciliaris* (Ultramarine flycatcher), *Trochalopteron variagatum* (=*Garrulas variegatus*), (Vriegated laughingthrush) *Hypsipetes madagascariensis* (Malagasy bulbul), *Mycerobas affinis* (Collared Grosbeak), *Oriolus oriolus* (The Eurasian golden oriole), *Phoenicurus leucocephalus* (White capped water redstart), *Picus squamatus squamatus* (Scaly-bellied Woodpecker), *Pyrrhula aurantiaca* (Oriental Bullfinch), *Parus major* (The great tit),*Parus monticollus* (Green-backed tit), *Tarsiger cyanurus* (Red-flanked bluetail), *Turdus ruficollis* (Red-throated thrush), *Turdus unicolor* (Tickell's thrush) [32, 33, 34, 35]

Class 2: Conoidasida

Sub-class: Gregarinasina

Order: Eugregarinoida

Sub-order: Aseptoria

Family: Diptocystidae

61. *Diptocystis kashmirensis*: Host- *Blatta orientalis* (Cockroach) [44]

Sub-class : Coccidiasina

Order: Eucoccidiorida

Family : Cryptosporidiidae

62. *Cryptosporidium*: Humans, Buffaloes and Lambs [1, 3, 27, 67]

Family : Eimeriidae

63. *Eimeria acervulina*: Hosts- Chicken, Poultry [2,48,50]
64. *Eimeria alijevi*: Hosts- Small ruminants [37]
65. *Eimeria arloingi* : Hosts – Ruminants(Caprine) [17,18, 37]
66. *Eimeria auburnensis* : Hosts – Cattle (calves)[38]
67. *Eimeria bovis* :Hosts- Cattle (calves) [38]
68. *Eimeria brasiliensis*: Host- *Bos grunniens* (Yak)41]
69. *Eimeria bukidnonensis*: Host- cattle (calves)[38]
70. *Eimeria canadensis* : Cattle(calves) [38]
71. *Eimeria caprina* : Host- Goat [17, 18, 66]
72. *Emeria christenseni*: Host- Goat) [17, 18, 37, 66]
73. *Eimeria crandallis*: Hosts- *Capra hircus hircus*, *Ovies aries* (JK) [34]
74. *Eimeria cylindrica*: Hosts- Cattle(calves) [38]
75. *Eimeria ellipsoidalis*: Hosts- Cattle(calves) [38]
76. *Eimeria faurei*: Host- Goat [17, 18]
77. *Eimeria hirci*: Hosts- Sheep and goat [37]
78. *Eimeria kocharii*: Host- Goat [17,18, 66]
79. *Eimeria leuckarti* : Host- Horses [Pandit *et al.,* 2008]
80. *Eimeria mitis*: Hosts- Chicken, Poultry [2, 48, 49, 50]
81. *Eimeria maxima*; Hosts- Chicken, Poultry) [2, 48, 49,50]
82. *Eimeria necatrix*: Hosts- Chicken, Poultry [2, 48, 49, 50]
83. *Eimeria ninakohlyakimovae*: Hosts- Sheep, goat [17, 18, 37]
84. *Eimeria parva*: Host- Goat [17, 18]
85. *Eimeria stiedae*: Host- *Oryctolagus cuniculus* (The European rabbit)[34]
86. *Eimeria subspherica*: Host- Cattle (calves) [38]
87. *Eimeria tenella*: Hosts- Chicken, Poultry [2, 48, 49, 50]
88. *Eimeria zuernii*: Host- Cattle(Calves), Indian Yak [38]
89. *Eimeria* spp.: Hosts- Domestic geese[Hamdani *et al.,* 2017]; Pig [21], Zoma, Ass, Donkey [65]; Chiru (Tibetan antelope)[Ganai *et al.,* 2008]; Live stock in Zanskar (Ladakh) [Wani *et al.,* 2013]
90. *Cystoisospora* (= *Isospora*) sp. : Hosts- Cat [8], dogs [69]

Sub-order 1: Eimeriorina

Family: Sarcocystidae

Subfamily: Sarcocystinae

91. *Sarcocystis* sp. : Hosts- Yalk [41]; dogs [69]

Subfamily: Toxoplasmatinae

92. *Toxoplasma gondii*: Hosts- Human, Cat [58,68]

Sub-order 2: Adeleorina

Family: Hepatozoidae

93. *Hepatozoon canis*: Host- Dog [42, 62]

Key to code numbers given in long brackets, infront of the listed texa:

1 = Abdullah *et al.* (2016); 2 = Ahad *et al.* (2015); 3 = Ahmad *et al.* (2013); 4 = Ambani (2014); 5 = Arora *et al.* (1976); 6 = Bhardwaj *et al.* (2015); 7 = Borkataki *et al.* (2013 a); 8 = Borkataki *et al.* (2013b); 9 = Dar *et al.* (2016a); 10 = Dar *et al.* (2016 b); 11 = Dar *et al.* (2017); 12 = Das and Shaw (1968); 13 = Ganai *et al.* (2015); 14 = Harish and Gupta (2009); 15 = Haq *et al.* (2017a); 16 = Haq *et al.* (2017b); 17 = Iqbal *et al.* (2012); 18 = Iqbal *et al.* (2013b); 19 = Jacob and Kalra (1951); 20 = Kamil *et al.* (2006); 21' = Khajuria *et al.* (2010); 22 = Khan *et al.* (2002); 23 = Kumar *et al.* (2013); 24 = Leherwal *et al.* (2004); 25 = Malla *et al.* (1989); 26 = Maqbool and Ahmmed (2016); 27 = Masarat *et al.* (2012); 28 = Mir *et al.* (1979); 29 = Mir *et al.* (1996); 30 = Nair (1973); 31 = Nandi (1986); 32 = Nandi (1989); 33 = Nandi (1999); 34 = Nandi and Das (2010); 35 = Nandi and Mandal (1978); 36 = Narian and Raina (1992); 37 = Pandit (2006); 38 = Pandit (2009); 39 = Raina and Bhadwal (2000); 40 = Raina *et al.* (2011); 41 = Ranga-Rao *et al.* (1994); 42 = Rani-Puteri *et al.* (2011); 43 = Rather *et al.* (2015); 44 = Rehana *et al.* (2016); 45 = Shahi *et al.* (2013); 46 = Sharma *et al.* (1993); 47 = Sharma *et al.* (2013a); 48 = Sharma *et al.* (2013b); 49 = Sharma *et al.* (2015a); 50 = Sharma *et al.* (2015b); 51 = Shaw (1980); 52 = Shaw (1989); 53 = Shaw and Das (1970); 54 = Singh *et al.* (2014); 55 = Singh *et al.* (2014); 56 = Tak *et al.* (2014); 57 = Tantary *et al.* (2016); 58 = Thokar (1986); 59 = Tikku (1980); 60 = Tufani *et al.* (2009); 61 = Tufani *et al.* (2017); 62 = Upadhaya *et al.* (2013); 63 = Varilova *et al.* (2017); 64 = Verma and Magotra; 65 = Wani *et al.* (2013); 66 = Yadav *et al.* (2007); 67 = Yadav *et al.* (2012); 68 = Zargar *et al.* (1998); 69 = Rebecca *et al.* (2014).

Addendum= Pandit *et al.* (2008); Hamdani *et al.* (2017); Ganai *et al.* (2008); Wani *et al.* (2001).

The Ciliophora group of protozoan parasites are most commonly known as ciliates because of the presence of hair like structures in large number at the surface of parasites. 6 species, under 5 genera (*Buxtonella, Balantidium, Ichthyophthirus, Trichodena* and *Chilodonella*), of parasitic Ciliophorans are known to be prevalent in the UTs, reported to be parasitizing domestic animals/livestock and fish-fauna. The Myxozoans are believed to be highly reduced Cnidarians, represented by two species, *viz. Myxobolus chushin* and *Myxidium chola*, parasitizing fishes of Jammu and Kashmir regions (**Table 1.1**).

The Apicomplexans are obligate interacellular parasites, representing a total of 59 species, belonging to 13 genera in UTs. Various genera and their number of species are: *Plasmodium* (5 spp.), *Haemoproteus* (8), *Babesia* (5), *Babesiosma* (1), *Theileria* (2), *Leucocytozoon* (5), *Diplocystis* (1), *Cryptosporidium* (1), *Eimeria* (27) *Cystoisospora* (= *Isospora*) (1), *Toxoplasma* (1), *Sarcocysts* (1) and *Hepatozoon* (1) (**Table, 1.1**). The Apicomplexans are found to be having wide host range of invertebrates as well as vertebrates, including humans and economically important livestock, causing several kinds of infections and diseases in these hosts.

1.1 Insect Parasites (Microsporidians, Euglenozoans and Apicomplexans)

In J&K, and Ladakh UTs, Microsporidian, Euglenozoans and Apicomplexans parasites are found to infect insects like honey bees, bumble bees, silkworm and cockroach in different regions and localities.The Microsporidian parasites, *viz. Nosema bombi, N.ceranae* and *Nosema D*, are reported to infect honey bees belonging to genus *Apis* (*Apis cerana, A. mellifera, A. dorsata*). *Nosema* D is also reported to parasitize bumble bee, *Bombus trifasciatus*. *Nosema* spp. have recently been placed in Kingdom Fungi, Phylum Microsporidia, Class Microsporea, Family Nosematidae. Earlier, Microsporodians were considered as Protozoans or Protists, however, now known to be fungi (Hibbett *et al.*, 2007). These are spore-forming uni-cellular obligate parasites of beneficial insects like honey bees, bumble bees and silk worm in the UTs.

N. bombycis is a very economically important parasite, it occurs in all tissues of the embryo, larvae, pupae and adult of *Bombyx mori* (Silkworm). The infested silkworm shows numerous minute brownish-black spots, scattered over their bodies. The infected larvae is unable to spin cocoon and dies without pupation. Transovarian infection also occurs in this beneficial insect. *N. apis*, causes great loss to workers frequently and, drones and queens also become suscepitible to nosema disease. *N. apis* is found almost exclusively in the epithelial cells lining of host's gut, unlike *N.bombycis* does not occur transovarially.

Nosema apis and *N. ceranae* cause virulent disease in honey bees, known as nosemosis, resulting in colony loss. *N. ceranae* is believed to be more virulent than *N. apis*, able to infect different *Apis* spp. and are widely distributed in the world, with capacity to infect other hymenopterans (Brown, 2017). *Nosema D.* has newly been recorded from bumble bee (*Bombus trifasiatus*) populations in Jammu and Kashmir (Vavilova *et al.*, 2017). *Nosema bombycis* causes microsporidiasis in silk worm (*Bombyx mori*), is one of the deadliest diseases and can result complete loss of Sericulture Industry. In Jammu and Kashmir, *Nosema* spp., have shown wide prevalence, affecting *Apis* spp. (Honey bees) and Silk worm. These parasites are considered to be economically very important in this region and showed prevalence among silkworm populations. (Sahaf, 2002; Ganie *et al.*, 2008) and severely affecting honey bees (Shah and Shah,1988; Abrol, 2013; Dar *et al.*, 2013).

The prevalence of *Nosema* spp. and parasites like *Crithidia* and *Lotmania* (Euglenozoa: Trypanosomatidae) of honey bees and bumble bees in UTs, has been studied by using PCR with primers specific for the ribosomal RNA genes cluster of *Nosema*, *Crithidia* and *Lotmania* species by Vivilova *et al.* (2017). These authors have reported the prevalence of *Nosema bombi* and *Nosema ceranae* from honey bees (*Apis cerana, A.mellifera, A. dorsata*). *Nosema D.*, is reported only from bumble bee population, *viz. Bombus trifasciatus*

The Trypanosomatids *viz.*, *Crithidia expocki* and *C. bombi* are known to infect *Bombus* spp. of this region besides, *Lotmania passim*, parasitizing *Apis* spp. However, no *Crithidia* has been found in case of honey bees and *Lotmania passim* in bumble bee population (**Table 1.1**). The prevalence of *Nosema* spp. in honey bee population is reported to be 25 per cent and 10 per cent in bumble bees and, trypanosomatid parasites 16 per cent and 31 per cent in honey bees and bumble bees respectively (Vavilova *et al.*, 2017).

Host sharing of Trypanosomatid parasites *Lotmania passim* and *Nosema ceranae* of honey bees (*A. mellifera*) is kown to exist, however, nature of interaction between the two is not fully understood (Tritscher *et al.*, 2016). Co-infection by Microsporidian and Trypanosomatids, has been identified in several honey bee and bumble bee species of Jammu and Kashmir. Co-infection of *Nosema* spp. and Trypanosomatid parasite, *Lotmania passim*, detected in honey bees (*A. cerana, A. mellifera*), and *Nosema* D. with *Crithidia expoeki* in *Bombus trifasciatus* of this region (Vivilova *et al.*, 2017). A new apicomplexan parasite, *viz. Diplocystis kashmirensis* (Conoidasida:

Gregarinasina: Diplocystidae), has been found in haemocoel and gut of Cockroach, *Blatta orientalis* in Kashmir Valley (Rehana *et al.*, 2016).

1.2 Fish Parasites

Various protozoan parasites (Ciliophora, Euglenozoa), Apicomplexan and Cnidarian parasites of fish-fauna, inhabiting different water bodies in diverse areas and localities of UTs, have been reported. These are responsible for causing diseases in local (endemic) and exotic fish species. A total of 10 species of fish parasites, belonging to Ciliophora, Euglenozoa, Apicomplexa and Cnidaria (Myxozoa) are known to occur these regions (**Table 1.1**).

1.2.1 Protozoan Parasites

Earlier, in Jammu and Kashmir regions, the parasitic protozoan, *viz. Ichthophthirus multifiliis* (Ciliophora), was noticed to infect gills and skin of fish, *viz. Cabdio morar* (= *Aspidoparia morar*) and *Triplophysa kashmirensis*, from Sethi Nalla (Jammu) and Telbal Nalla (Kashmir) respectively (Narian and Raina,1992). These authors reported heavy parasitic infection in these fishes, with gills in various stages of degeneration, hyperplasis of connective tissues and deformity of repiratory folds. *I. multifiliis* is responsible for causing a very persistent disease in the fishes, known as White Spot disease or " Ich". This disease has damaging affect on the fishes and can cause breathing and mobility problems, and even death. In Kashmir Valley, the cold water cyprinid fish, *Schizothoraz niger* in Dal and Wullar Lake has been found to be infected with *I. multifiliis*, affecting severely the gills of the fish (Dar *et al.*, 2016 a, b), with infection rate 60 per cent (Dar *et al.*, 2016 b).

Ciliophoran parasite, *Trichodina heterodenta* has been found to infect fishes (*Cyprinus carpio*, *Labeo rohita*, *Catala catla* and *Punitus ticto*) in fish ponds of Jammu region (Tantary *et al.*, 2016) and *Schizothorax niger* in Dal Lake and Wullar Lake of Kashmr valley, showed the prevalence rate of parasite infection as 21.7 per cent (Dar *et al.*, 2016 b). *T. heterodenta* causes a disease known as Trichodiniasis. Due to this disease the fishes show sluggish movement, loss of appetite, black colouration, necrosis and ulcer in different parts of the body.

Third species of Ciliophora, *viz. Chilodinella uncinata* has been reported from Kashmir valley, infecting cyprinid fish, *Schizorthorax niger* in Dal and Wullar Lake, indicated prevalence rate 34.92 per cent (Dar *et al.*, 2016 b). *Chilodinella* sp. has also been found to infect the gills of fish (*S. niger*) in River Jehlum and Dal Lake of Kashmir, with highest prevalence rate as 60 per cent during December to Febuary, followed by 12 per cent during September to November and no infection recorded from June to August (Ambani, 2014; Tak *et al.*, 2014).

The Euglenozoan parasites infecting fishes of Kashmir region included *Trypanosoma kashmirensis* and *Trypanosoma mukasai*. *T. kashmirenis*, described as a new species, parasitizing fish, *Schizothorax plagiostomus* from River Jehlum, showed 61.1 per cent of infection rate, with significant decrease ($p < 0.05$) in red blood cell count (Maqbool and Ahmed, 2016). *T. mukasai*, found to be infecting fish-fauna, such as: *Carassius carassius* (overall prevalence 2.32 per cent); *Schizothorax curvifrons* (7.6 per cent) and *Triplophysa marmorata* (41.6 per cent), from Anchar Lake and River

Jehlum, indicated significant ($p < 0.01$) decrease in red bloos cell count in fishes (Shahi *et al.*, 2013).

1.2.2 Apicomplexan and Cnidarian Parasites

An Apicomplexan parasite, *viz. Babesiosma* sp., has been found to infect fish (*Triplophysa marmorata*), inhabiting Anchal Lake in Kashmir Valley, with overall prevalence as 16.6 per cent and significant reduction in red blood cell count in the fish (Shahi *et al.*, 2013). The Cnidarian parasites (Myxozoans) of fishes of Jammu and Kashmi are represented by three species, *viz.*, *Myxobolus chushin, Myxobolus himalayensis* and *Myxidium chola*. *M. chushin* has been described as new species, parasitic on *Schizothorax niger* from Wullar lake of Kashmir valley (Dar *et al.*, 2017). These authors have also utilized partial 18 Sr DNA sequencing and studied pathogenicity, indicating function loss of gill lamellae in *S. niger*.

Another Cnidarian parasite, *Myxidium chola* has been found to be parasitizing fish, *Puntius chola,* inhabiting Ghoomansara Nallah (Jammu), with infection in gall bladder, showed slight enlargement and brownish bile (Raina and Narain, 1992). The morphological description and molecular data of a new species of *Myxobolus himalayensis,* infecting gills of cyprinid fish (*Scizothorax richardsonii*) from the River Poonch, Jammu, have been given by Ahmad *et al.* (2019).

1.3 Bird (Wild) Parasites

The earliest detailed investigations on rthe blood parasites of wild birds of Jammu and Kashmir region were conducted by Nandi and Mandal (1978) and Nandi (1986). The various areas covered during the course of survey included Anantnag, Ducksum, Doda, Jajjar Kotli, Khrew, Patnitop, Ramban and Udhampur. The blood examination of 39 species of birds, belonging to 11 families in these localities, revealed parasitaemia in 17 species/subspecies of birds. The bird species of different families showed infection due to 16 species, under 5 genera of parasites, pertaining to Phylum Euglenozoa and Apicomplexa. The parasite showed not only the species richness but also wide diversity. The higher incidence (26.8 per cent) was reported to be existing in Jammu and Kashmir in comparison to to that of Nasik and Ahmadnagar (Maharashtra) (13.3 per cent per cent) and Orissa (12.7 per cent) (Nandi and Mandal, 1978).

Through the publications (Nandi and Mandal,1978; Nandi,1986; Nandi,1999; Nandi and Das, 2010), it is evident that a total of 17 spp. parasites (Euglenozoans and Apicomplexans), are known to be prevalent in 48 species of wild bird in Jammu and Kashmir region. The number of parasite species, belonging to different genera affecting number of bird species are : 8 spp. of *Haemoprotreus,* infected 25 spp. of bird; 4 spp. of *Leucocytozoon*, parasitic on 18 spp. of bird; *Plasmodium relictum* on 2 spp. of birds- Indian verditer flycatcher, Long-tailed Shrika; *Parahaemoproteus velans* (= *Haemoproteus velans*) on Yellow- crowned woodpecker and Eurasian Wryneck; *Trypanosoma avium* on Asian barred owlet (**Table 1.1**).

8 spp. of *Haemoproteus* (*H. borges, H. columbae, H. celli, H. dicruri, H. oryzivorae, H. passeris, H. pastoris, Haemoproteus* sp.), have been found to cause blood infection in 24 spp. of wild birds in diverse localities of Jammu and Kashmir. These bird species

are: Asian barred owlet, Black headed Jay, Blue- capped rock thrush, Blue whistling thrush, Common buzzard, common rose finch, Common starling, Eurasian sparrow, Eurasian wryneck, Forked-tailed drongo, Himalayan bulbul, House sparrow, Indian blue robin, Indian mynah, Indian verditer flycatcher, Lesser white throat, Long-tailed shrike, Pigeon, Oriental white- eye, Short -billed minivet, Tawny owl, Wagtail, Western crowned warbler and White-browed bulbul (**Table 1.1**).

4 species of *Leucocytozoon* (*L. dubreuili, L.fringillinarum, L. majoris, L. squamatus, Leucocytozoon* sp.), are known to be parasitic on 17 species of wild birds of Jammu and Kashmir region. The host birds are *viz.*, Collared grosbeak, Common rose finch, Eurasian golden oriole, Great tit, Green-backed tit, Hanfinch, Himalayan bulbul, Malagasy bulbul, Oriental bullfinch, Red-throated thrush, Red-flanked blue-tail,Rock bunting, Scaly-bellied woodpecker,Tickell's thrush, Ultramarine flycatcher, Variegated laughing thrush, White-capped water redstart (**Table 1.1**).

Haemoproteus parasites are the common blood parasites of birds, especially wild. These are generally believed to be less pathogenic.The infection in birds is transmitted by blood sucking insects (mosquitoes, biting midges, tabanid flies and louseflies). *Haemoproteus columbae* causes major parasitic infection in pigeons, both domestic as well as wild pigeons. *H. columbae* causes a disease known as pigeon malaria. This disease can remain in the body for life, becomes active under stress. The parasite attacks the red blood cells of pigeons. Hippoboscid fly, *Pseudolynchia canariensis* acts as a vector of pigeon malaria. The incidence of *H. columbae* in pigeons reared in purely domesticated and in free-ranging/wild habitation, in Jammu region has been studied by Borkataki *et al.* (2013). These authors found higher incidence rate (78.28 per cent) of infection in domestic pigeons than in wild, having 5 per cent besides, the occurrence of insect vector of pigeon malaria, *P. canariensis* from underneath of feathers has been reported.

Leucocytozoon parasites are also blood parasites of birds and infect the reticulo-endothelial tissues and blood cells of birds. The infection is mainly transmitted by blackflies (*Simulium*). *Leucocytozoon* causes a disease known as Leucocytozoonosis. This disease is usually asymptomatic and characterized by listlessness, lethargy, anaemia, anorexia, leading to weight loss. In water fowl, there can be morbidity and mortality.

Plasmodium relictum causes malaria in birds and may be fatal to bird species which have not developed resistance to the disease.The avian malaria disease causing agent is carried and transmitted by *Culex quinquefasciatus* (Common house mosquito).The clinical signs of avian malaria are tissue damage and there may not be enough destruction within the red blood cells to cause clinical anemia (Global Invasive Species Database, 2018).*Trypanosoma avium*, is causative agent of avian trypanosomiasis disease and in this disease, normally there is no pathogenicity (Peirce, 2003). However, in bird of prey, Falcons from Kuwait, clinical signs have been reported, including incapacity of flying high, poor appetite, lethargy, weight loss, weakness, dyspnea and death (Tarello, 2005). The transmission of avian trypanosomiasis is not clear and various blood sucking insects like black flies are believed to be as possible vectors of this disease.

1.4 Protozoan and Apicomplexan Parasites of Medical Importance

A detailed survey for the human intestinal protozoans (Amoebozoa, Metamonada), based on stool samples of rural community in Lamberi village of district Rajouri (Jammu) was under taken by the team from National Institute of Communicable Diseases (NICD), during the month of March, 1974. A total of nine protozoan parasites were reported, including 5 species of Amoebozoans (*Endolimax nana, Entamoeba coli, Entamoeba hartmani, Entamoeba histolytica* and *Iodamoeba bustschlii*) and 4 species of Metamonad (*Giardia intestinalis, Enteromonas hominis, Chilomastix mesnili* and *Retortamonas intestinalis*). The findings of NICD (Arora *et al.*, 1976), revealed that 72.1 per cent of the population of Lamberi Village was found to be infected with one or more parasite species. *Entamoeba coli* showed highest percentage (48.6 per cent) of existence in the community. The lowest percentage of intestinal protozoas was 16.5 per cent and 8.3 per cent of *G. intestinalis* (= G. *lamblia*) and *E. histolytica* respectively.

In addition to four species of Metamonada, Malla *et al.* (1989) reported another species of Metamonad, *viz. Trichomonas vaginalis* from Kashmir region. A total of 5 species each of Amoebozoan and Metamonad is known to be prevalent of human population of Jammu and Kashmir. The Euglenozoans (Trypanosomatidae: Leishmaniinae) parasitic on human population of this region are represented by *Leishmania donvani* and *Leishmania tropica*. The Apicomplexan parasites showing their prevalence in these region are: *Plasmodium* (*P. falciparum, P. malariae, P. vivax*), *Cryptosporidium* and *Toxoplasma gondi* (**Table 1.1**).

The parasite species *viz., E. nana, E. coli, E. hartmani, I. bustschlii, C. mesnili, R. intestinalis* and *E. hominis*, are non- pathogenic or mildly pathogenic. The pathogenic human parasites of this region, which are responsible for causing diseases are: *Entamoeba histolytica* (causing amoebiosis disease); *Leishmania donvani* (visceral leishmaniasius or Kala azar); *Leishmania tropica* (cutaneous leishmaniasis or oriental sore); *Trichomonas vaginalis* (Trichomoniasis); *Plasmodium falciparum., P. malariae, P. vivax* (Human malaria); Cryptosporidia (Cryptosporidiasis) and *Toxoplasma gondi* (Toxoplasmosis). For details on vector borne diseases such as Leishmaniasis (visceral and cutaneous) and human malaria, *see* Chapter 8 and 1 respectively.

1.4.1 Amoebiasis

The *Entamoeba histolytica* is a pathogenic parasite, causing disease known as amoebiasis, including intestinal as well as extra-intestinal infection. In case of extra-instestinal amoebiasis, the parasite spread to other organs like liver, lungs, *etc.* and in liver causing amoebic liver abscess. The infection of amoebosis is caused when a person ingests cysts of *E. histolytica* through food or water, besides cysts entering the body through direct contact with faecal matter.

In Kashmir region, during the year Dec., 1982 to Sept., 1983, patients attending hospital (SKIMS) in Srinagar, were detected to be suffering from amebiasis, with prevalence rate 2 per cent and clinical symptoms like amoebic colitis to fulminating dysentery (Khan *et al.*, 1985). In a sero-epidemiological survey of human cystic echinococcosis in Kashmiri population, 15 samples were found positive for the prevalence of amoebiasis (Fomda *et al.*, 2015). In a detailed investigation on clinical profile of liver abscess in children of Kashmir, 40 per cent amoebic liver abscess cases have been detected (Sheikh *et al.*, 2007a).

In Jammu region, the prevalence of amoebiosis among patients, attended Hospital (tertiary care), during the period June, 2013 to June, 2015, has indicated the prevalence rate of amoebiasis as 21 (17.35 per cent) (Jad *et al.*, 2015). These authors are of the opinion that there has been decline in the prevalence of amoebiosis causing parasite as compared to observations made in the hospital 35 years ago. Liver abscess in amoebic dysentery due to *E. histolytica* has also been reported from Jammu region, with prevalence rate 73.33 per cent (patients of all age group and both genders) (Naveed *et al.*, 2015).

1.4.2 Human Leishmaniasis (Visceral and Cutaneous)

In Jammu and Kashmir regions, human leishmanisis caused by Euglenozoan parasites, *Leishmania donovani* and *Leishmania tropica* (Trypanosomatidae : Leishmaniinae). Visceral leishmaniasis (VL) or Kala azar disease in human is caused by the protozoan parasite, *Leishmania donovani*, is transmitted by the bite of infected sandfly, *Phlebatomus argentipes*, the only known vector of this disease in India (Anon, 2016). The synonyms of Kala azar are Black disease, Dumdum fever and Tropical splenomegaly. This disease is mainly characterized by weight loss, hepatosphenomegaly and anaemia. Death can occur in untreated cases of visceral leishmaniasis. Cutaneous leishmaniasis (CL), also known as Oriental sore or tropical sore, Chiclero ulcer or Aleppo boils, affecting human. CL usually produces ulcers on the exposed parts of the body like face, arms and legs. This disease is caused by protozoan parasite, *Leishmania tropica*, transmitted by the bite of an infected sandfly, *Phelabotomus major*. The earliest evidence of human visceral leishmaniasis cases in Kashmir and Jammu regions, has been given by Jacob and Kalra (1951) and Khan *et al.* (1975) respectively.

(a) Visceral Leishmaniasis

The earlier definite cases of Kala- azar patients in the erstwhile Jammu and Kashmir State, have been detected in local inhabitants of Kashmir Valley, who had acquired infection inside the State. Jacob and Kalra (1951) confirmed this disease in two Kashmiris, one belonging to the district Anantnag and other to district Srinagar, who had never gone outside the State. According to these authors, 12 cases were reported from J&K and these included 7 Gurkhas, 2 Kashmiris and 3 Dogras. The 2 Kashmiri had never been outside State and 3 Dogras (2 from Samba and 1 from Poonch) had also never left the State in their life. The village of Budil was incriminated epidemiologically, as likely endemic area, which had only one species of sandfly *i.e. P. major*, as the most abundant species. in whole of Jammu and Kashmir. Further, the authors did not find parasites in tissues of the sandflies. The indigenous cases of visceral leishmaniasis have been reported during 1975 from Jammu region: village Dugan in Kathu district and in Kishtwar region of district Doda.

In the recent yeas, sporadic cases of VL have been reported not only from Jammu but also from Kashmir (Naik *et al.*, 1979; Anon, 2016). The occurrence of this disease in Jammu and Kahmir is posing threat to the public health and there is urgent need of continuous surveillance, detailed epidemiological investigation and control measures. A case of Kala azar was found in 44 year male from Anantnag (South Kashmir), suffered hepatomegaly and massive splenomegaly (Nabi *et al.*, 2015). From Uri area in North Kashmir, Baramula District, a five year old child was reported to be suffered from VL, diagonsed on the basis of bone marrow examination (Mahajan *et al.*, 2009). In Bani region of District Kathua of Jammu, Bhat *et al.* (2014), reported a rare indigenous case of VL.

(b) Cutaneous Leishmaniasis

During recent past years, there have been increasing number of cases of human cutaneous leishmaniasis (CL), reported from Jammu as well as from Kashmir region. The reports of cases of CL, were from different non-endemic geographical regions Kashmir: Karnah and Uri belt in South-west Kashmir (Leherwal, 2004; Masood *et al.*, 2009; Wani *et al.*, 2015); Kupwara and Barmulla (Northern-western frontiers) (Rather and Yaseen, 2017; Rather *et al.*, 2019; Sajad *et al.*, 2017).The CL patients from these areas showed the symptoms in the form of lesions of noduloulceration kind on the face, including lip and nose. According to the observation made by Rather and Yaseen (2017), the patients with CL from two districts, *viz.* Kupwara and Baramulla, had "Tin Tack" sign in localized lesion, reported for the first time from these areas of Kashmir Valley.

A new focus of CL has been identified in Jammu Division. The disease affected more than hundred patients (67 females, 53 males), from November, 2012 to October, 2013 and the patients belonged to three Districts of Jammu, *viz.* Doda, Kishtwar and Poonch (Kaul *et al.*, 2016). Recently, more cases of CL have been reported from Jammu, with CL in lower eye lid of 6-year old girl, a rare entity (Arora *et al.*, 2017; Gupta,2017). The emergence of CL in various areas of Jammu and Kashmir in the recent years is of great concern and the need of the hour is to conduct detailed epidemiological and molecular investigations to ascertain the vector, reservoir of infection and causative agent disease in J&K (UT)

1.4.3 Human Giardiasis

Giardiasis disease in humans is caused by intestinal protozoan parasite, *Giardia intestinalis* (= *Giardia lamblia*). The main clinical symptoms of this disease include diarrhea and malabsorption or nutritional disorder. The infection of giardiasis is transmitted directly by ingestion of contaminated water or food or indirectly by faeco-oral route.

Ahmad (1978), for the first time studied the incidence of *G. intestinalis* (= *Giardia lamblia* in rural population of Kashmir. This author found that among other parasites (*Ascaris lumbricoides* and *Taenia saginata*) in rural population of Kashmir, *G. intestinalis* accounted for 9.4 per cent of total cases. Its prevalence was shown lower in males (8.1 per cent) and higher in females (10.8 per cent). The highest parasitic infection was observed in school going children, age groups (5-15 years). The faeco-oral contact of poor hygiene and extensive contamination of soil and water were believed to be responsible for highest incidence of parasitic infection.

The incidence of giardiasis in urban areas (Srinagar) and surrounding areas, has been studied by Mir *et al.* (1979). These authors made detailed investigation on the incidence of *G. intestinalis* parasite in Kashmiri children. For this purpose, 500 children from Srinagar and surrounding areas were selected for a period of one year. These children belonged to age group 1-10 years. This study revealed total incidence of giardiasis 9.2 per cent (46 cases) and giardiasis with ascariasis was 2.60 per cent (13 cases). The maximum incidence was reported in children of age group 3- 4 years. The personal hygiene of children and in sanitary condition were responsible for higher incidence of giadiasis in Srinagar. Khan *et al.* (2002) reported the prevalence of giardiasis in Kindergarten school children of urban Kashmir, covering 142 school of Srinagar city, indicated giardiasis in a total of 184 children (12.28 per cent).

In recent studies on the prevalence of giardiasis in school childrens of Kashmir, 7 per cent of prevalence of this disease, has been reported, with higher prevalence of parasites anong rural children (Singh *et al.*, 2012). The prevalence of giardiasis among childerns in District Anantnag of Kashmir valley was found to be 8.2 per cent, with highest rate in age group 5-8 years and childrens of rural areas were more infected than in urban are (Abdullah *et al.*, 2016 b). From Jammu region, the prevalence rate of giardiasis in patients who visited Jammu hospital during the period from June, 2013- June, 2015, was reported to be 73.5 per cent (Jad *et al.*, 2015). The investigation on impact of giardiasis disease on haematological parameters of children (268), both gender in district Anantnag of Kashmir has been done by Abdullah *et al.* (2017), showed decrease in haemoglobin and hematocrit values and increase in leukocyte count in infected childrens.

1.4.4 Human Trichomoniasis

Metamonadan parasite, *viz. Trichomonas vaginalis* is responsible for the cause of human trichomoniasis in Kashmir region. This disease is sexually transmitted disease (STD), both in females as well as males, with high morbidity. *T. vaginalis* infects the lower genital tract in females and males, the most infected part is the urethra. In females the infection can progress to vaginal spotting or bleeding, whereas in affected males, there is burning sensation after urination or ejaculation and discharge from the penis.

The earlier clinical and laboratory study on trichomoniasis in Kashmiri population was conducted by Malla *et al.* (1989). These authors have shown association of *T. vaginalis* in female genital tract and urinary tract of 66 per cent of patients. The recent findings, based on the examination of case (Hospital) record of five years (2009-2013), 42 female patients' vaginal discharge, 10 (23.81 per cent) had trichomoniasis in Kashmir valley (Anwar *et al.*, 2015).

1.4.5 Human Malaria

Human malaria is caused Apicomplexan parasites belonging to genus *Plasmodium* (Aconoidasida : Chromatorida : Plasmodiidae) and the infection is due to three species of parasite in Jammu and Kashmir regions, *viz. Plasmodium falciparum, Plasmodium malariae* and *Plasmodium vivax* (Nair, 1973; Verma and Mangotra, 1976; Harish and Gupta, 2009). Malaria disease is transmitted by the bite of sporolzoites-bearing anopheline mosquitoes. The mosquito vectors of human malaria in J&K region include *Anopheles annularis, An. culicifacies, An. fluviatilis, An. maculatus* and *An. stepheni* (See Chapter 8,pp.188-189)

P. falciparum can cause sever type of malaria, this parasite multiplies rapidly in the blood. In this malaria, there is anaemia and clogging of small blood vessels, when it occurs in the brain, leading to severe complication and even death. *P. vivax* is the most prevalent type of malaria, having dormant liver stages (hypnozoites) that can activate and invade blood (relapse), several months or years after the infecting bite by mosquitoe. *P. malariae* has quartan cycle (3- day cycle). If untreated, *P.malariae* causes a long lasting chronic infection and in some cases causing serious complications. The details on mosquitoes as vectors of malaria disease vectors in Jammu and Kashmir are given in Chapter, 8.1.2, pp. 188-189.

The earliest report regarding the prevalence of malaria disease from Kashmir was from Baramulla, Uri and Domel areas, and *An.* (*C.*) *willmorei* shown to be as a vector of indigenous malarial disease in these areas (Gill,1920). In Kashmir, the incidence of human malaria, existed in many villages at higher-altitudes (above 1,524 mts). This was for the first time shown by Jacob (1950), during his survey between April and October. Later, in sixties, the high incidence of fever among the inhabitants of some villages in and around Uri, Tetwal and Keran region of district Baramulla of Kashmir, led National Malaria Eradication Programme (NMEP) Organisation, to collect blood samples from these fever cases and also to conduct entomological survey for Anopheline as vector of malaria disease in Kashmir Valley. During August – September (1963), C.P. Nair of National Institute of Communicable Diseases, Delhi, conducted an epidemiological study in district Baramulla (Kashmir) to establish local transmission of malaria, incriminate vector and to determine the degree of endemicity. The various findings of Nair (1973) are briefly given as under:-

The malaria survey incorporated about 30 villages (15 in Karnah, 8 in Uri and 7 in Handwara) of district Baramulla of Kashmir. The percentage of person either suffering from fever or gave very recent history fever, were 10.2 per cent in Handwara, 11 per cent in Karnah and 12.5 per cent in Uri. The most widely prevalent malarial parasite species was found to be as *Plasmodium vivax* (88.4 per cent) and *P. faliciparum* (11.6 per cent). The inhabitants of the villages, situated at an elevation from 1,524 to 1, 8289 mts, showed the cases of *P. faliciparum*. Two individuals were also found to have mixed infection. Apart from this, in one case, it was *P. vivax* and *P. faliciparum* in the other *P. vivax, P. malariae*. The percentage of heavy infection to total infection was shown to be 74.7 per cent. The gametocytes carrier among the positive ranged from 77.4 per cent in case of *P. falciparum* and 88.9 per cent in case of *P. vivax*. In both the infections, the gametocyte density was also shown to be high.

During the course of entomological survey for anopheline as vector of malaria, *Anopheles fluviatilis* was identified as a vector of malaria, distributed near an elevation of 8,129 mts. The mosquito species was collected at Pringle Haridal village in Karnah and one individual of A. fluviatilis, out of 56 dissected showed the presence of sporozoites of malarial parasites. The density of this mosquito was.33 to 5.3 per man hour. Apart from this, *A. culicifacies* and *A. leucosphyrus* were also present in various villages surveyed, however, in very low density at high-altitude.

Keeping in mind, the various epidemiological findings, Nair (1973) has inferred that an unstable endemic malaria existed in Baramulla district of Kashmir Valley, at all the elevation from 1,676 – 2,134 mts, in different localities surveyed by the author and transmission of this disease was seasonal, with prevalence mostly during warmer months. This malaria had characteristics of that of subtropical zone malaria.

In the recent years, number of malaria cases have been reported from different Districts of Jammu region and this disease is posing threat to the public health of the region (National Family Health Survey, India, 1999; National Rural Health Mission, 2007). Malaria is more restricted to Jammu region rather than Kashmir region. Investigations on malaria and its control in Jammu, has been made by Dwivedi *et al.* (1979) and Kaul *et al.* (1995). *Plasmodium vivax* malaria cases, with severe complications like cerebral malaria, caused by parasite, *Plasmodium falciparum*, have been reported from Jammu (Verma and Magotra, 1976; Harish and Gupta, 2009).

1.4.6 Human Cryptosporidiosis

Human crypotosporidiosis is caused by the apicomplexan parasite of genus *Cryptosporidium*. This disease is a major cause of diarrhoea in developing countries, mainly affecting childrens and HIV infected individuals or AIDS patients. In case of childrens, it causes persistent diarrhoea and, severe and persistent in case of immunocompromised and malnourished persons. The *Cryptosporidium* species reported to infect HIV- infected adults and childrens in India are *viz., Cryptosporidium hominis* and *C. parva* (Ajjampur *et al.,* 2008). The humans become infected with cryptosporidiosis disease by accidently ingesting environmentally resistant spores through parasite contaminated water, food or by contact with faeces or unwashed hands (formites). This disease is characterized by watery diarrhoea, stomach cramp, nauesia and pooe appetite, besides fever and muscle aches may occur

The cryptosporidiosis due to *Cryptosporidium* sp., has been reported from Kashmir region. The analysis of 45 stool samples pertaining to the patients in this region revealed : symptomatic (diarrhoeic), n= 9 and asymptomatic (non-diarrhoeic), n=36, patients, infected with HIV, 80 per cent patients (non-Kashmiris and travellers), but carrier of *Cryptosporidium* and 20 per cent patients – HIV positive (local Kashmiris), who visited different States of India), were diarrhoeic (symptomatic) and also carrier of *Cryptosporidium* (Masarat *et al.,* 2012)

1.4.7 Human Toxoplasmosis

In Kashmir region, the incidence of toxoplasmosis has been reported among human population. Toxoplasmosis is caused by an Apicomplexan parasite, *viz. Toxoplasma gondii.*

The toxoplasma infection is transmitted to humans either by ingestion of oocytes shed in faeces of cat or other Felidae hosts or by the consumption of tissue cysts present in the raw or uncooked meat (lamb, pork). Toxoplasmosis can also occur by drinking water contaminated with *T. gondii*, accidentally ingesting contaminated soil or eating unwashed fruits or vegetables, besides mother-to-child (congenital) transmission, infected organ transplant or rarely through blood transfusion. The final host of *T. gondii* is domestic cat, flies and cockroaches act as transport hosts.

The symptoms due to toxoplasmosis disease vary from swollen lymph glands,or muscle aches and pain, to severe toxoplasmosis, causing damage to the brain, eyes (ocular toxoplasmosis) or other organs. Toxoplasma infection present in pregnant women may lead to spontaneous abortion, still birth or premature delivery, besides various serious fetal defects. Most of the toxoplasma infection are detected with the help of serological tests. In Kashmir region, the earlier serological study of patients clinically suspected to have toxoplasmosis, was conducted by Thokar (1986) and Thokar *et al.* (1988). The prevalence rate of toxoplasmosis in this region was reported as 26.9 per cent by the use of various serological tests, like IHA(Indirect haemogglutination), IFA (Indirect florescent antibody) and ELISA (Enzyme linked immunosorbent assay) (Thokar *et al.*, 1988). The investigations on sero-prevalence of toxoplasmosis in women(285), with repeated abortions, 141 (49.47 per cent) serum samples were found positive for IgM toxoplasma antibody (Zargar *et al.*, 1998).

During the course of sero-epidemological survey of human cystic Echinococcosis in Kashmiri population, Fomda *et al.* (2015), have found toxoplasmosis in five patients. The cases of maternal and congenital toxoplasmosis have also been reported in Kashmir valley. In this direction, a hospital based study has indicated 27 per cent serological prevalence of toxoplamosis in pregnant women, 1.1 per cent of incidence of primary toxoplasmosis during pregnancy, 43.2 per cent rate of vertical transmission and 0.5 per cent incidence of congenital toxoplasmosis (Ahmad *et al.*, 2010). In recent seroprevalence of *T. gondi* infection in healthy human population in Kashmir valley, an Ig G antibody prevalence of 24.2 per cent, has been found and there was significant increase in toxoplasmosis with increase age (Rather *et al.*, 2019)

1.5 Protozoans of Veterinary Importance

A total of 6 species of parasitic protozoans of veterinary importance have been reported from Jammu and Kashmir, and Ladkah UTs, causing diseases in domestic animals, live stock and poultary, *etc.* These are: *Balantidium coli* and *Buxtonella sulcata*, belonging to Phylum Ciliophora; *Trypanosoma evansi* (Phylum Euglenozoa); *Giardia* sp., *Histomonas meleagridis* and *Trichomonas foetus* (Phylum Metamonada) (**Table 1.1**). The Ciliophorans, *B. coli*, has been found to infect domestic animals in Ladakh region, whereas *Buxtonella sulcata* affecting bovines in Jammu region. *B. coli* causes infection in large intestine of animals and the disease caused is known as Balantidiasis or Balantidiosis. *B. coli* is largest of parasitic ciliate protozoan, commensal in several animals, however, can cause chronic diarrhea in cattle and pigs, are reservoir hosts of *B. coli*. The transmission of infection is through ingestion of cyst in contaminated water or food. Balantidiasis is also zoonotic disease, acquired by humans by faeco-oral route. The prevalence of *B. coli*, has been found in live stock, including zomo, cattle, and yak in Zanskar region of Kargil (Ladakh), showing infection rate as 11.42 per cent (Wani *et al.*, 2013).

Buxtonella sulcata, is an opportunistic ciliate protozoan, inhabiting colon of bovines. Ganai *et al.* (2015) reported the prevalence of *B. sulcata* in bovines of R. S. Pora (Jammu), suffered diarrhoea. These authors recorded 20.9 per cent of *B. sulcata* incidence in bovines, with cattle showing infection rate 23.6 per cent and the buffaloes (18.5 per cent). Normally, *Buxtonella* is considered as commensals of intestinal tract of ruminants and helps in digestion of plant material.

1.5.1 Animal Trypanosomiasis

Euglenozoan parasitic protozoan, *viz. Trypanosoma evansi,* is responsible for causing serious disease in animals known as animal trypanosomiasis or 'Surra' disease, affecting wide range of domestic and wild animals, including camels, equines, cattle, buffaloes, sheep, goats, dogs and other carnivores. This disease is transmitted by haematophagous insects like tabanids, stomoxys and *Musca* spp. *T. evansi* can exhibit very high parasitaemia especially in camels, horses and dogs. It is a both blood as well as tissue parasite. 'Surra' is a major disease in camels, equines and dogs, in which it can often be fatal, in the absence of proper treatment, showing symptoms like anaemia, loss of weight, abortion and death (Desquesnes *et al.*, 2013).

Shaw (1980), for the first time reported the prevalence of Animal trypanosomiasis (Surra) disease for the first time in Jammu Division and its absence in Kashmir. The prevalence of *T.evansi,* causing high incidence of trypanosomiasis in cross bread and local cows, also in cattle and horses, is reported from high pasture areas of Jammu (Poonch, Rajouri, Udhumpur), and the biological vector of this disease, were confirmed as *Tabanus rubidus* (horse fly), *Stomoxys calcitrans* (Stable fly), and *Haematobia exigua* (= *Lyperosia exigua*) (buffalo fly) (Sharma 1993 a)

In Jammu and Kashmir region, the serum samples of equines (32/221) were found to be positive for the presence of anibodies against *T. evansi,* through antibody ELISA method, confirming the 'Surra'disease prevalence in these regions (Kumar *et al.*, 2013). The occurrence of trypanosomiasis in buffaloes and ponies, has been reported from R.S. Pora and adjoining Samba Tehsil of Jammu and the outbreak of disease in bovines was shown to be directly proporional to onset of mansoon to post monsoon (Raina *et al.*, 2000).The canine trypanosomiasis in bakerwali dog in Jammu Region, has been reported by Bhardwaj *et al.* (2015). In canine Trypanosomiasis, the symptoms of this disease include facial swelling, involving eyelids, lips and skin beneath the lower jaw, besides weakness, lethargy and keratitis.

1.5.2 Animal Giardiasis, Trichomoniasis and Histomoniasis

The Metamonad parasites of veterinary importance in J&K, UT are : *Giardia* sp., causing giardiasis in dogs in Jammu; *Trichomonas foetus* (Trichomoniasis) in breeding bulls in Jammu and *Histomonas melegridis* (Histomoniasis/enterohepatitis) of partridge and poultry in Kashmir Valley. Giardiasis results in major diarrhoea and other complications. The prevalence of *Giardia* spp. in pet dog population in Jammu region has been investigated by Sharma *et al.* (2015). The prevalence rate of *Giardia* spp. was found by these authors to be 40.83 per cent, highest in mansoon season (66.67 per cent), lowest in winter season (13.33 per cent) and high in female dogs (46 per cent). Prevalence (32.5 per cent) of giardiasis in bovine calves in district Jammu, has been found (Sharma *et al.*, 2014).

The Trichomoniasis disease in cattle is a venereal disease. Bovine trichomoniasis is also known as ' Trich'. *T. foetus* harbour the internal sheath and prepuce of the bulls and cows, multiplies in the internal reproductive tract. Trichomoniasis is transmitted from cow to cow by bulls during breeding season. Bulls show no clinical signs and cows can clear the infection with in a few months. *T. foetus* has been found in preputial washings of 3 bulls in Jammu (Ticku, 1980). The Avian trichomoniasis, with special reference to pigeon in Kashmir has been reported, affecting upper digestive tract, crop and oesophagous of pigeons (Bulbul *et al.*, 2018). The infection of trichomoniasis due to *Trichomonas gallinae*, with prevalence of 40 per cent has been observed in domestic pigeon (*Columba livia*) in Jammu region (Mehmood *et al.*, 2019)

Histomoniasis also known as blackhead disease of poultry and turkeys. This disease causes necrotizing typhlitis and hepatitis in gallinaceous birds, including turkeys and partridge. Also, this disease affects intestinal caeca, liver, causing tissue destruction.The parasite of this disease (*Histomonas meleagridis*), is transmitted by nematode, *Heterakis gallinarum.* There was an outbreak of histomoniasis in hen reared partridge *Gracia eletoris* (=*Alectoris graeca*) in Kashmir (Mir *et al.*, 1996) and in flock of broiler breeder chicken, reared under deep litter system, with possible transmission by nematode, *H. gallinarum* (Kamil *et al.*2006).

1.6 Apicomplexan of Veterinary Importance

The Apicomplexan parasites of veterinary importance J&K, and Ladakh UTs include 4 species of *Bebesia*, 2 species of *Theilaria*, *Cryptosporidium* sp., 25 species of *Eimeria*, *Heptozoon canis*, *Isospora* sp. *Sarcocyst* sp. and *Toxoplasma* sp (**Table 1.1**).

1.6.1 *Babesia* spp. and *Theileria* spp.

Babesia (Order Achromatorida, sub-order Babesiina, family Babesiidae) of the Union Territories, is represented by 4 spp., *viz. B. bigemina*, *B.bovis*, causing tick-borne babesiosis disease in cattle; *Babesia motasi* and *B. ovis*, causing tick-borne babesiosis disease in sheep and goats. *Theileria* (Achromatorida, Theileriina,Theileriidae), is having 2 species in this region, *viz. Theileria annulata* and *T. parva*, responsible for causing tick-borne disease in cattle, known as theileriosis.

Babesia spp. are primarily parasites of R B Cs, whereas *Theileria* spp. use successively, W B Cs and R B Cs for completion of their life cycle in mammalian hosts. Both babesiosis and theileriosis diseases are of great economic importance of domestic animals (cattle, sheep and goats) of Union Terrotories and are known to cause much economic loss. The infection in babesiosis disease is acquired by female tick, which transmits the parasite transovarially to larval stages. The larval ticks inject sporozoites into blood of cattle, which invade erythrocytes. The infected erythrocytes clump together, resulting in blocking of capillaries, leading to brain damage and anoxia of internal organs. In Kashmir valley, Das and Shaw (1968), reported the incidence of babesiosis in imported Australian Jersy cows and bulls. In theileriosis disease, there is hypertrophy of lymphoid tissues, fever, weight loss and haemolytic anaemia. Theileriosis in cross bred Jersy cows in Kashmir was reported by Shaw and Das (1970). In the later years, detailed investigations on prevalence, distribution, clinical profile and therapeutic management of these important diseases of veterinary importance, have been accomplished by : Shaw (1989); Gupta *et al.* (1994); Raina *et al.* (2005); Rather *et al.* (2015); Haq *et al.* (2017 a, b) and Tufani *et al.* (2009, 2015, 2017).

1.6.2 Animal Cryptosporidiosis

The infection due to *Cryptosporidium* (Cryptospridiidae) is known as cryptosporidiosis, associated with diarrhoea. This disease affects most animals, however, can be severe in young and weak animals. This infection is transmitted by ingesting organisms contaminated water, food or by contacting or licking of living objects (formites), also in rare case, possible to inhale (aerosol) the organisms. In Kashmir region, for the first time the prevalence of cryptosporidiosis was found in diary calves, age group 1 day to 6 months, with infection rate 29.37 per cent (Sheikh *et al.*, 2007).

Cryptosporidiosis in calves of buffaloes in Jammu region has been reported by Yadav *et al.* (2012). The cryptosporidiosis in young ruminants (lambs) in Jammu and the relationship between intensity of infection and various epidemiological factors indicated highest of infection in 0-1 month old lambs in winter season (Ahmad *et al.*, 2015). Oocysts of *Cryptosporidium* (4 per cent), have been reported from the stary cats of Jammu region (Borkataki *et al.*, 2013 a, 2013 b). The first report of *Cryptosporidium* infection in local Changthangi and Russian Merino sheep of Leh (Ladakh), with 54.05 per cent of infection rate, has been given by Bhat *et al.* (2019). In domestic pigeons (*Columba livia*) in Jammu region, Mehmood *et al.* (2019), found 50 per cent prevalence of cryptosporidiosis.

1.6.3 *Toxoplasma* sp., *Cystoisopora* sp., *Sarcocystis* sp. and *Hepatozoon canis*

The Prevalence of oocysts of *Toxoplasma* sp. (Sarcocystidae: Toxoplasmatiinae) and *Cystoisospora* (= *Isospora*) sp. (Eimeriidae), (8 per cent), have been found in stary cats of Jammu region (Borkataki *et al.*, 2013). Cats can get the toxoplasma infection by killing and eating of prey. The disease (toxoplasmosis) can also be passed on from cats to human. Cats are usually immune to toxoplamosis. The prevalence of *Cystoisospora* has been reported in stray and refuge dogs in Ladakh region (Rebecca *et al.*, 2014). *Cystoisospora* infection causes a pathological condition, known as Cystoisoporiasis, leading to watery mucous like diarrhoea. This disease is transmitted through contaminated water and food.

Ranga-Rao *et al.* (1994), for the first time reported the occurrence of Sarcocystis (Sarcocystidae : Sarcocystinae) in Indian Yak (*Bos grunniens*) in Ladakh region. *Sarcocystis* spp. have also been reported from Ladakh region, infecting stray and refuge dogs of this region (Rebecca *et al.*, 2014). *Sarcocystis* causes a disease known as sarcocystosis. In this disease, the endothelium and muscles and other tissues are invaded by this parasite.

The tick-borne apicomplexan parasite, *Hepatozoon canis* (Hepatozoidae), with infection rate 24 per cent, has been reported in dogs of Ladakh (Rani-Puteri *et al.*, 2011) and concurrent infection due to *H. canis* in German Shephard dog in Jammu region (Upadhyay *et al.*, 2013). *H. canis* infects canine blood cells (leucocytes) and parenchymal tissues, causing disease known as hepatozoonosis, transmitted in dogs by ingestion of ticks harbouring oocysts. The infection with *H.canis* could be asymptomic in dogs with low level of parasitaemia or manifested as a severe life threatening disease with fever, lethargy, anaemia, cachexia in dogs, with high parasitaemia (Ivanov and Tsachev, 2008).

1.6.4 *Eimeria* spp., Animal and Poultry Coccidiosis

In Union territories J&K and Ladakh, a total of 26 species of *Eimeria* (Eucoccidiorida : Eimeriidae) of veterinary importance have been found. The highest number of species of *Eimeria i.e.*, 10, parasitize ruminants (sheep and goats), followed by 8 spp. parasitic on cattle, 5 spp. on poultry and 1 sp. each on rabbit, Indian yak, horse and pig. Besides these, one undetermined species each of *Eimeria* is reported to parasitize zoma, ass and donkey in Ladakh region (**Table 1.1**).

Coccidiosis (Eimeriosis *sensu stricto*), infection caused by several species of genus *Eimeria* in cattle, sheep, goats, pigs, rabbits and poultry. Coccidiosis is of great economic importance, causing losses due to clinical diseases like diarrhoea, fever, inappetence,weight loss, emaciation, and in extreme cases death may occur. The transmission of coccidiosis in domestic animals is through faecal -oral route and by ingestion of contaminated water, faeces or by licking itself. Coccidiosis in chickens and poultry, is also responsible for causing high losses. Coccidiosis in chicken and poultry mainly causes droppiness, listlessness and loss of appetite. The affected birds also shed blood or mucus in faeces, also diarrhea and even death can occur. The transmission of coccidiosis in poultry is by eating of oocysts that have been shed in the droppings of infected chickens.

(a) Coccidiosis in Domestic Animals

The prevalence of coccidiosis in sheep of Jammu and Kashmir, for the first time was investigated by Pandit and Mir (1988). These authors reported overall prevalence of 62.83 per cent, with Eimerian infection, highest incidence 73.98 per cent in lambs, followed by weaners (66.14 per cent) and adult(47.98 per cent). Bhat *et al.* (2012) showed the prevalence (9.8 per cent) Eimerian infection (*Eimeria* spp.) in sheep of Kashmir region. The non-local or exotic sheep breed were found to be with highest percentage of prevalence(68.18 per cent) than local breed as 36.5 per cent, and more infection in large flock reared on non-cemented floor pattern (Rishi *et al.*, 2014).

Caprine coccidiosis in Kashmir valley is reported to be due to *Eimeria* spp., as *E. arloingi*, *E. ninekohlayakimovae*, *E. parva*, *E. hirei*, *E. christenseni* and *E. alijevi*, with overall prevalwence 72.56 per cent, highest in summer (80.98 per cent), followed by spring (69.75 per cent) and, in autumn and summer (68.51 per cent) (Pandit, 2006). Caprine coccidiosis in Jammu region, is found to be due to seven species of *Eimeria*, *viz. E. arloingi*, *E. christenseni*, *E. ninekohlayakimovae*, *E. parva*, *E. flauri*, *E. caprina* and *E. kochari*, *E.arloingi* as predominant species (Yadav *et al.*, 2007; Iqbal *et al.*, 2012, 2013b). Coccidiosis in goats in Unorganized Farm of Jammu region was found to be highest in monsoon as 73.58 per cent and lowest 37.27 per cent in summer (Iqbal *et al.*, 2013).

The prevalence of coccidiosis in cattle (calves) in Kashmir was reported to be due to seven species, with *E. zuernii* and *E. bovis* are most dominant species. Besides this, the Eimerian infection rate was reported as 70.7 per cent and 75.8 per cent from organized and unorganized managed Farms respectively (Pandit *et al.*, 2009). Coccidial infection has also been found in the three breeds of Rabbits, *viz.* Angora, Newzeland white and Chinchilla at Wussan Rabbit Farm in Kashmir, with overall infection 36.6 per cent, highest in Newzealand white (54 per cent), followed by Chinchilla (38 per cent) and Angora (18 per cent)(Bihaqi *et al.*, 2013). *Eimeria leuckarti* oocysts (0.34 per cent), have been detected in horses in Kashmir (Pandit *et al.*, 2008) In Jammu region, prevalence of coccidian oocyst (34.83 per cent), in pigs has been reported by Khajuria *et al.* (2010). In addition to these, the prevalence of *Eimeria* sp. in Chiru (Tibetan antelope) of Ladakh, has been reported (Ganai *et al.*, 2008).

(b) Coccidiosis in Chicken and Poultry Birds, Rabbit

A total of 5 species of *Eimeria* are known to affect the chickens and Poultry of Jammu, Kashmir and Ladakh (**Table 1.1**). The most prevalent species of this region of India is *Eimeria tenella*, causing caecal or bloody type of coccidiosis. *E. necatrix*, is responsible for bloody intestinal coccidiosis. *E. acerulina* and *E. maxima* cause chronic intestinal coccidiosis. The stages of coccidia in chickens appear both within the host as well as outside. In South Kashmir, the prevalence of coccidiosis in domestic and free range chicken was found all round the year, with highest incidence (68.75 per cent) during winter season and lowest in summer season (Yousuf *et al.*, 2013). However, Ahad *et al.* (2015), have shown highest prevalence rate (45.12 per cent) of coccidian infection during autumn and minimum (20.29 per cent) during winter, in broiler chicken in poultry farms in Kashmir valley. These authors reported an overall prevalence rate (29.87 per cent) of coccidiasis throught out the year, caused by five species of *Eimeria, viz. E. tenella, E. acervuline, E. maxima, E. nectrix, E. mitis.* The dominant species being *E. tenella*, with highest prevalence rate (18.13 per cent) and *E. mitis* showed loswest prevance rate (1.86 per cent).

In Jammu region, investigation on coccidiosis in organized and backyard poultry farms, have shown overall prevalence of infection as 33.33 per cent (Sood *et al.*, 2009). Five species of *Eimeria*, with *E. tenella* being dominant, also reported from Kashmir region, causing coccidiosis in organized and unorganized Farms of Jammu region (**Table 1.1**). In this region (Jammu), the highest prevalence of infection was 53.61 per cent in un-organized Farms (backyard poultry birds) as compared to organized poultry birds as 25.55 per cent, and maximum infection found in mansoon season (July, 2011), 68.9 per cent and lowest in May, 2011 as 12.5 per cent (Sharma *et al.*, 2013 b; Sharma *et al.*, 2015 b). Coccidian parasitizing domestic pigeons in Jammu region, has shown prevalence of 58.3 per cent (Mehmood *et al.*, 2019).

The pathomorphological changes due to coccidiosis by five reported species of *Eimeria* in chicken in Jammu region, have been studied by Sharma *et al.* (2015 a). These authors have found in the affected chickens the clinical signs and changes like extremely ballooned intestine and caeca, haemorrhage in intestinal mucosa, watery ingesta mixed with mucosa and blood, and haemorrhagic enteritis in whole intestine.

An outbreak of coccidiosis inbroilers in Leh (Ladkah), cold desert region has been reported by Kumar *et al.* (2008). An outbreak of hepatic coccidioisis disease in Rabbit Farm, caused by coccidian, *viz. Eimeria stiedae*, with 58.14 per cent morbidity and 62 per cent of mortality rate, and sudden death in rabbit kits (Ahmad *et al.* (2015).

Chapter 2

Platyhelminthes (Trematodes and Cestodes)

2.1 Trematode Parasites

2.1.1 General Account and Species Diversity

Trematode parasites, commonly referred to as flukes, belong to Phylum Platyhelminthes (flatworm) and class Trematoda, characterized by the presence of organs of attachments (suckers). These are permanent parasites of internal organs of mollucs, harbouring larval asexual or occasionally sexual adult stage and of vertebrates, harbouring adult sexual stage, infecting different regions in the body cavity, organs and tissues.

In Union territories Jammu and Kashmir, and Ladakh, trematode parasites are represented by two important Orders, *viz.*, Digenea and Monogenea. The monogenetic trematodes are usually ectoparasites of fish hosts, usually attached to the gill, scales and fins, *etc.*, for example *Diplozoon* spp. Besides, one species, *Eupolystoma rajai*, has been reported to be true endoparasite in the urinary bladder/ rectum of frog (*Rana* sp.). The monogeneids remain attached to the host surface by a large attachment organ, known as opisthaptor, species specific, provided with hooks and is present at posterior end of trematode, usually, a pair of adhesive structures are located at the anterior end.

The digenetic trematodes of this region are very common and widespread parasites of various classes of vertebrates (fishes, amphibians, birds and mammals), occurring in adult as well as juvenile form, in every organ of their hosts. The digenetic trematodes are characterized by the presence of well developed sucker around the mouth, known as oral sucker. Besides, second ventral sucker or acetabulum is present in these types of trematodes. The suckers are involved in the attachment to the host as well as in locomotion. The organs of attachment in digenetic trematodes are poorly developed than those in monogenetic trematodes. Digenean parasites' development include at least two different hosts and several generations.

The first classified checklist of trematode parasites and their invertebrate host (snail) and vertebrate hosts (fishes, amphibians and birds) of Kashmir valley, has been given by Gupta (1967). This author has listed a total of 31 Species of trematodes, based on earlier reports from this region by various authors like Faust (1927), Jordon (1930), Mehra (1937), Kaw (1950), Fotedar (1964 a, 65) and Fotedar and Raina (1964, 65). The listed species of trematode parasites included 9 spp. of trematodes parasitizing 9 spp.of fishes, 17 spp. parasitic on 5 spp. of amphibians, 4 spp. parasitizing 4 spp. of birds. Besides these, 1 sp. of larval trematode infecting land snail, has been added in this checklist.

During the recent years, a great deal of research work has been done on taxonomy and ecology of trematode parasites of vertebrates- fishes amphibians, birds and mammals, besides invertebrate (molluscan host) of Jammu, Kashmir and Ladakh region. In this direction, an indepth study by the present author, of vast and scattered research publications, has revealed that till date, a total of 105 species of trematode parasites belonging to 56 genera, spread over more than 18 families, under order Monogenea and Digenea, is known to occur in these regions. The species diversity of trematode parasites infecting different host groups, has been found to be highest in case of birds, with 49 trematode species, followed by mammalian hosts, with 23 spp., amphibians with 22 spp. and fishes with 18 spp., belonging to diverse families (**Table 2.1**).

Table 2.1: Species Diversity of Trematode Parasites on different Vertebrate Host Groups of Jammu, Kashmir and Ladakh Region

Host Group	*Genera*	*Number of Species*	*Families*	*Super Family*
Fishes	10	18	07	01
Amphibians (Frogs and Toads)	16	22	06	01
Birds (Wild and Poultry)	19	49	09	—
Mammals (Bat, domestic and Yak)	12	23	05	—

Note: Two trematode, *viz. Neascus vetastai* from fish host and *Tetracotyle ranae* from frog, have only been assigned super family Diplostomomidea (after Niewiadomska, 2002), reported from this region as larval trematodes.

2.1.2 Molluscan and Fish Trematodes

The larval stages of trematode genus *Brachylaima*, belonging to the family Brachylaimidae (Digenea), have been reported for the first time from Kashmir Valley, parasitic on various organs of land snail, *Buliminus dextrosinister* by Fotedar (1965). He also described the various larval stages of *Brachylaima*, such as sporocyst, cercaria and metacercaria. The adults of Brachylaimid trematodes have been found in alimentary tracts of birds and mammals, rarely amphibians. The life-cycle of these trematodes include two intermediate hosts, both terrestrial molluscs. The larval stages: daughter-sporocyst are branched, cercaria with rudimentary, stumpy tail, leaves first intermediate host; metacercaria encysted or not in kidney or pericardium of second intermediate host (Pojmanska, 2002). Sharma *et al.* (1990) studied the post dormancy survival of molluscan host, *Lymnea auricularia* and with trematode, *viz. Fasiola gigantica. F. gigantica* is a liver fluke of larger size and is found in sheep, goat and cattle in all parts of the world, including Jammu and Kashmir region.

Table 2.2: Trematode-Fauna of Fishes, with their Location and Distribution in J&K, UT

Sl.No.	*Trematode (Location in Host)*	*Distribution*	*Reference (s)*
1	*Allocreadium fotedari* (intestine)	Kashmir	Dhar and Kharoo (1985
2	*Allocreadium Nemacheilus* (Intestine)	Bandipore, Ganderbal	Kaw (1950)
3	*Allocreadium schizothoracis* (Intestine)	Dal Lake, Manasbal lake	Pande (1938), Kaw (1944), Dar *et al.* (2013)
4	*Allocreadium transversale* (Intestine)	Dal and Manasbal Lake	Chishti *et al.* (2002)
5	*Astiotrema fotedari* (Intestine)	Poonch	Dhar (1978)
6	*Clinostomum schizothoraxi* (Gills, muscles, caudal fin, eyes)	Jehlum river, Dal and Manasbal lakes	Kaw (1950)
7	*Clinostomum shafi* (Intestine, gills)	Anchar and Manasbal lakes	Ahmad and Chishti (2000)
8	*Crepidostomum indicum* (Intestine)	Dal Lake	Kaw (1944)
9	*Dactylogyrus rectotrabus (Gills)*	Poonch river/tributaries	Ahmad and Sharma (2016)
10	*Diplozoon aegytensis* (Gills)	Dal Lake	Ahma and Chishti (1999), Dar *et al.* (2013)
11	*Diplozoon cauveryi* (Gills)	Dal and Mansbal Lake	Chishti *et al.* (2002)
12	*Diplozoon guptai* (Gills, intestine)	Anchar and Dal	Ahmad and Chishti (1999)
13	*Diplozoon kashmirensis* (Gills)	Dal, Manasbal, Anchar and Wular Lakes	Kaw (1950),Dar *et al.* (2015)
14	*Diplozoon nipponicum* (Gills)	Kashmir	Fotedar and Parveen (1987)
15	Diplozoon poonchensis (Gills)	Poonch river	Gupta *et al.* (2014)
16	**Neascus vetastai* (Fin, gill, skin, viscera)	Jehlum river, Dal and Manasbal Lake	Kaw (1950)
17	*Gyrodactylus elegans indicus* (not known)	Jehlum river, Dal and Manasbal lake	Dar *et al.* (2015)
18	*Neodiplostomum kashmirensis* (Muscle)	Jehlum river and Anchar lake	Faust (1927) (*cf.* Gupta, 1967)
19	*Phyllodistomum loosi* (Body cavity)	Dal and Manasbal lake, Jehlum (Shadipora)	Kaw (1950
20	*Phyllodistomum megacotyle* (Intestine)	Poonch	Fotedar (1969)
21	*Strigea annadalai* (Intestine)	Kashmir	Faust (1927) (*cf.*Gupta,1967)

The systematic position of trematode species, cited in above table, under Serial nos.: 1-15, 17-21

Class: Trematoda

Subclass: Monogea

Family: Dactylogyridae

Subfamily : Dactylogyrinae (Serial no. 9)

Family: Diplozoidae

Subfamily: Diplozoinae (Serial nos.10-15)

Family : Gyrodactylidae

Subfamily: Gyrodactylinae (Serial no.17)

Subclass: Digenea

Family: Allcreadiidae

Subfamily: Allocreadidae (Serial nos. 1-4 and 8)

Family: Plagiorchiidae

Subfamily: Astiotrematinae (Serial nos.5)

Family: Clinostomidae

Subfamily: Clinostominae (Serial nos.6-7)

Family : Diplostomidae

Subfamily : Diplostominae Serial no. (18)

Family : Goroderidae

Subfamily: Gorgoderinae (Serial nos. 19-20)

Family : Strigeidae

Subfamily: Strigeinae (Serial no.21)

The trematode species, *viz. Neascus vetastai*, cited at serial no. 16 in above table was not assigned systematic position to this larval trematode (Yamaguti, 1971) and adult of this species is unknown in other places. However, according to system of classification of Trematode, Niewiadomska (2002) has placed genus *Neascus*, under Superfamily Diplostomoida of sub-class Digenea.

Till date, a total of 21 species of trematode parasites, belonging to 10 genera under 6 families of sub-order Prosostomata (order Digenea) and of sub-order Polyopisthocotylea (order Monogenea), besides larval from Neascus vetastai, under super-family Diplostomoidea of Digenca (after Niewiadomska, 2002), has been reported from Jammu and Kashmir region, infecting 18 species of fishes, distributed over 10 genera (**Table 2.3**). The Allocreadiids are represented by *Allocreadium fotedari, A. Nemacheilus, A. schizothoraxi, A. transversale* and *Crepidostomum indicum*. Dactylogyrid includes single species, *viz. Dactylogyrus rectotrabus*. The species of Diplozoids include *Diplozoon aegyptensis, D. cauveryi, D. guptai, D. kashmirensis* and *D. nipponicum*. The Plagiorchiid includes only one species, *viz. Astiotrema fotedari*.

The Clinostomidids reported from this region are: *Clinostomum schizothoraxi* and *C. shafi*. The Gorgoderids are represented by two species,*viz. Phyllodistomum loosi* and *P. megacotyle*. The family Diplostomidae and Strigeidae, is represented by *Neodiplostomum kashmirensis* and *Strigea annandalai* respectively.

Table 2.3: Host-Fish Species Diversity of Trematode Parasites of J&K, and Ladakh UTs

(1) *Carassius carassius*

Clinostomum shafi, C. schizothoraxi, Diplozoon aegyptensis, D. cauveryi, D. guptai and *D. kashmirensis*

(2) *Crossocheilus latius latius* (= *Crossochcilus latius*)

Neodiplostomum kashmirensis

(3) *Cyprinus carpio specularis*

C. schizothoraxi, D.kashmirensis, D. nipponicum

(4) *Cyprinus* sp.

Diplozoon cauveryi

(5) *Garra gotyla*

Dactylogyrus ractotrabus

(6) *Garra mullya*

Phyllodistomum megacotyla

(7) *Glyptothorax* sp.

Clinostomum schizothoraxi

(8) *Labeo dero*

Astiotrema fotedari

(9) *Labeo* sp.

Clinostomum schizothoraxi, Diplozoon aegyplensis, D. cauveryi

(10) *Nemacheilus rupeeola*

Strigea anandalai

(11) *Nemacheilus* sp.

Clinostomum schizothoraxi, Allocreadium nemacheilus

(12) *Oreinus* sp.

Diplozoon aegyptensis, D. cauveryi

(13) *Schizothorax curvifrons (= Schizothorax hugelii) (= Schizothorax micropogon)*

Clinostomum schizothoraxi, Diplozoon kashmirensis, Neodiplostomum kashmirensis, Neaseus vetastai

(14) *Schizothorax esocinus*

Gyrodactylus elegans indicus, Neascus vetastai, Clinostomum schizothoraxi, Diplozoon kashmirensis and *Phyllodistonum loossi*

(15) *Schizothorax niger*

Clinostomuim shafi, Diplozoon kashmirensis, C. schizothoraxi, D. aegyptensis, D. guptai, Allocreadium fotedari, A. schizothoracis, A. transversale, Crepidostomum indicum

(16) *Schizothorax plagiostomum*

Allocreadium schzothoracis, Diplozoon aegyptensis

(17) *Schizothorax progastrus*

Allocreadium schizothoracis, Diplozoon aegyptenis

(18) *Schizothorax richardsonii (= Orienus sinuatus)*

Clinostomum schizothoraxi, Diplozoon kashmirensis, Diplozoon poonchensis, Neascus vetastai

(19) *Schizothorax* sp.

Allocreadium schizothoraxi, Diplozoon kashmirensis and *Phyllodistonum loosi*

(21) *Triplophysa marmorata (= Nemacheilus kashmirensis)*

Allocreadium nemacheilus, Clinostomum schizothoraxi and *Diplozoon kashmirensis*

Source: Ahmad Chishti (1994, 1999, 2000); Ahmad *et al.* (2015); Ahmad and Sharma (2016); Chishti *et al.* (2002); Dar *et al.*, 2013, 2015), Dhar (1978); Dhar and Kharoo (1985, 87); Dhar and Majidah (1987); Fotedar (1969); Fotedar and Parveen (1987); Faust (1927); Gupta (1967); Gupta *et al.* (2014); Kaw (1944, 50); Khan and Majidah (1994); Majidah and Khan (1998) and Pande (1938).

In addition to the above trematode species, *Neascus vetastai*, reported by Kaw (1950) as larval form, has been included in the system of classification provided by Yamaguti (1971), in his monograph. However, in accordance with the recent classification, provided by Niewaidomska (2002), *N. vetastai* has been kept under super-family Diplostomoidea Poirier, 1886 of sub-class Digenea.

The trematode parasites have been found to infect various organs and different parts of body of fish hosts in Jammu and Kashmir region. These are: Intestine, infected by *Allocreadium* spp., *Clinostomum shafi, Crepidostomum indicum, Diplozoom guptai, Phyllodistomum megacotyle* and *Strigea annadalai*; gills and gill cover by *Clinostomum schizotharaxi, C. shafi, Diplozoon* spp., *Neascus vetastai*; fins by *C. schizothoraxi* and *N. vetastai*; muscles by *C. schizothoraxi* and *Neodiplostomum kashmirensis*; eyes by *C. schizothoraxi*; viscera by *N. vetastai* and body cavity by *C. schizothoraxi* (**Table 2.2**).

From the Table 2.3, it is evident that maximum number of parasites *i.e.*, 9 different species, has been found to infect the fish, *viz. Schizothorax niger*. It is followed by 6 spp. of parasites in fish, *Carassius carassius*, 5 spp. in *S. esocinus*, 4 spp. each in *Schizothorax curvifrons* and *S richardsonii*, 3 spp. each in *Schizothorax* spp., *Triplophysa marmorata, Cyprinus, Carpio specularis, Labeo* spp., 2 spp. in *Oreinus* spp. Apart from these, 1 sp. of each parasite, is known to infect fishes, *viz. Crossochelus latius, Cyprinus* spp., *Garra mullya, Glyptothorax, Labeo dero, Nemacheilus rupecola, Nemacheilus* sp.

Dhar and Kharoo (1986) were first to made a detailed field study on trematode parasite, *viz. Clinostomum schizothoraxi* of fresh water fishes of Kashmir. These authors reported the existence of large number of metacercariae form of *C. schizothoraxi*, from gills, gill cover, fins and body surface of number of fresh water fishes, *viz. C. carpio specularis, S. niger, S. esocinus, S. curvifrons, S. richardsonii* (=*N. kashmirensis*) and *T. kashmirensis* (=*N. kashmirensis*).

Fish parasitization in detailed, covering helminthes, including Trematodes, *viz. C. schizothoraxi* and *D. kashmirensis*, from Wular Lake of Kashmir, was studied by Dhar and Majidah (1987). *C. schizothoraxi* was found from the skin and base of tail of *S. niger*, *S. esocinus* and *Nemacheilus* sp. whereas *D. kashmirensis* infected the gills of *S. niger*, *S. esocinus* and *C.c. specularis*.

Majidah and Khan (1998) studied the seasonal population dynamics of two trematode species of fishes, *viz. Diplozoon kashmirensis* and *Clinostomum schizothoraxi* in Wular Lake of Kashmir. These parasites were found to infect fishes like *S. niger*, *S. curvifrons*, *S. esocinus*, *Oreinus plagiostomus*, *Cyprinus carpio specularis* and *Nemacheilus kashmirensis*. The infra-population of parasite was shown to be relatively more during summer. The maximum infection was noticed in *S. esocinus*, *S. curvifrons* and *S. niger*.

In a comprehensive study of fish trematode parasite diversity in fresh water lakes of Kashmir, Chishti *et al.* (2002) found the occurrence of 5 spp. of trematodes *i.e.*, 3 spp.of Diplozoon and 1 spp. each of *Clinostomum* and *Allocreadium* in fishes, *viz. Schizothorax*, *Oreinus*, *Labeo*, *Cyprinus* and *Carassius*. These authors found that digenetic trematode (*Clinostomum*) having a degree of host specificity and marked seasonal occurrences, whereas endoparasite trematode, *Allocreadium* showed occurrences in mid spring to late summer. The monogenean infections in *Schizothorax esocinus* in water bodies, *viz.* Dal Lake, Manasbal Lake and River Jehlum, have revealed 12.64 per cent prevalence and 4.59 per cent prevalence of *Diplozoon kashmirensis* and *Gyrodactylus elegans indicus* respectively (Dar *et al.*, 2015). These authors also showed highest infestation due to these fish parasites in adult specimens. In fish, *Schizothorax richardsonii*, inhabiting Poonch river, Jammu, more than 20 per cent prevalence of trematode (*Diplozoon poonchensis*) infection has been reported in this fish (Gupta *et al.*, 2014). Monogenetic trematode *viz.*, *Dactylogyrus rectotrabus*, has been reported from Poonch river and its tributaries in Poonch area of Jammu, infecting gills of fish, *Garra gotyla*

2.1.3 Amphibian (Frog and toads) Trematode Parasites

In Jammu and Kashmir region, so far 21 spp. of trematode parasites, belonging to 14 genera, distributed over 6 families, under Order Monogenea and Digenea of class Trematoda, infecting frogs and toads, have been recorded. The Monogenea trematode includes a single species, *viz. Eupolystoma rajai*, belonging to sub-family Eupolystomatinae and family Potystomatidae. The Digenea includes rest of the trematode species *i.e.* 20.

Toads (*Bufo* spp.) have been found to be parasitized by 9 different species of trematodes, *B. viridis* by 9 spp. and *Bufo* sp. by 1 sp., infecting intestine, rectum and urinary bladder of the host species. Of these Species of trematode, 9 spp. are known from Kashmir Himalayan region and 1 sp. from Jammu region (**Table 2.4**).

Frog, *viz. Rana cyanophlyctis* has been reported to be infected with a total of 13 spp. of trematode parasites, from J&K with 8 spp. from Kashmir region and 7 spp. from Jammu. These trematode species infect urinary bladder, instestine and lungs of *R. cyanophlyctis*. *Rana tigrina* from Jammu has been found to be parasitized by 3 spp. of trematodes, infecting intestine and urinary bladder of the host amphibian.

Apart from the above mentioned species of toads and frogs; *Rana* sp. from Poonch (Jammu) and *Bufo* sp. from Srinagar (Kashmir) have been found to be infected by trematodes, *E. rajai* and *Ganeo bufonis* respectively.The most widely prevalent family of trematode parasite of amphibians in Jammu and Kashmir is Lecithodendriidae, incorporating 11 spp. of trematodes, distributed over 8 genera, followed Gorgoderidae, having 5 spp., Plagiorchiidae, having 2 spp. and rest of the families, *viz.* Polystomatidae (Monogenea), Paramphistomidae, Diplostomidae having one species each. Apart from these, 1 spp., *viz. Tetracotyle ranae*, has not been placed by Yamaguti (1958, 71) in any family, however, according to recent system of classification given by Niewiadomska (2002), genus Tetracotyle has been placed under Super-family Diplostomoidea (**Table 2.4**).

Different amphibian species of toad and frog, parasitized by different trematode parasite genera, with total number of species belonging to these are: *Bufo viridis*, infected with genus *Diplodiscus* (1 sp.), *Diplostomum* (1), *Gorgoderina* (1), *Loxogenes* (1), *Loxogenoides* (1), *Phyllodistomum* (1), *Plagiorchis* (1), *Pleurogenoides* (1), *Pleurorchis; Rana cyanophlyctis*, with *Diplodiscus* (1), Ganeo (3 spp.), *Gorgoderina* (2 spp.) *Haematoloechus* (1), *Indopleurogenes* (1), *Loxogenes* (2), *Loxogenoides* (1), *Mehraorchis* (1), *Phyllodistomum* (1), *Prosotocus* (1), *Tetracotyle* (1); *Rana tigrina*, with *Diplodiscus* (1), *Ganeo* (1) and *Phyllodistomum* (1); *Rana* sp. with *Eupolystoma* (1); *Bufo* sp. with *Ganeo* (1) (**Table 2.4**).

Table 2.4: Systematic Checklist of Trematode-Fauna, Parasitizing Amphibians (Frogs and Toad) of Jammu and Kashmir (UT)

Order: Monogenea

Family: Polystomatidae

Subfamily: Eupolystomatinae

1. *Eupolystoma rajai*

Host: *Rana* sp.; Distribution: Poonch (Jammu)

Superfamily :Paramphistomatoidea

Family: Paramphistomidae

Subfamily: Diplodiscinae

2. *Diplodiscus amphichrus* (= *Diplodiscus mehari*)

Hosts: *Bufo viridis*, *Rana cyanophlyctis*, *R. tigrina*; Distribution: Srinagar, Poonch, Jammu

Superfamily: Diplostomoidea

Family: Diplostomidae

Subfamily: Diplostominae

3. *Diplostomum bufonis*

Host: *B.viridis*; Distribution: Srinagar

Superfamily: Gorgoderoidea

Family: Gorgoderidae

Subfamily: Gorgoderinae

4. *Gorgoderina elliptica*

Host: *R. cyanophlyctis*; Distribution: Jammu

5. *Gorgoderina vitelliloba* (= *Phyllodistomum frequentum*)
   Hosts: *R.cyanophlyctis*, *B.viridis*; Distribution: Kashmir
6. *Phyllodistomum almorai*
   Host : *R. cyanophlyctis*; Distribution : Kashmir
7. *Phyllodistomum shandrai*
   Host: *R. tigrina*; Distribution : Jammu
8. *Phyllodistomum* sp.
   Host: *B.viridis*; Distribution: Srinagar

Superfamily: Plagiorchoidea

Family: Plagiorchiidae

Subfamily: Plagiorchiinae

9. *Plagiorchis himalayai* (= *Lepoderma himalayai*)
   Host: *B.viridis*: Distribution: Srinagar

Family : Haematoloechidae

Subfamily: Haematoloechinae

10. *Haematoloechus almorai* (= *Pneumoneces almorai*)
    Host: *R. cyanophlyctis*; Distribution : Kashmir

Superfamily : Microphalloidea

Family: Lecithodendriidae

Subfamily: Ganeoninae

11. *Ganeo bufonis*
    Host: *Bufo* sp.; Distribution: Srinagar
12. *Ganeo srinagarensis*
    Host: *R. cyanophlyctis;* Distribution : Srinagar, Kashmir
13. *Ganeo tigrinus*
    Hosts: *R. cyanophlyctis*, *R. tigrina;* Distribution : Jammu

Subfamily: Pleurogeninae (= Pleurogenetiinae)

14. *Indopleurogenes yamaguti*
    Host: *R. cyanophlyctis*; Distribution :Jammu
15. *Loxogenes jammuensis*
    Host: *Rana cyanophlyctis;* Distribution : Jammu
16. *Loxogenes kashmirensis* (= *Nenimandijea kashmirensis*)
    Hosts: *R. cyanophlyctis*, *B.viridis;* Distribution : Kashmir
17. *Loxogenoides bicolor* (= *Loxogenes bicolor*, *L. arcanum*)
    Hosts: *R. cyanophlyctis*, *B. viridis;* Distribution : Kashmir
18. *Pleurogeneoides bufonis*
    Host: *B.viridis*; Distribution: Srinagar
19. *Pleurorchis nenei*
    Host : *B.viridis*; Distribution : Kashmir

Subfamily: Prostocinae

20. *Mehraorchis ranarum*

Host : *R. cyanophlyctis*; Distribution : Jammu

21. *Prosotocus himalayai* (= *Prosotocus partapus*, *P. kashabia*)

Hosts : *R. cyanophlyctis*; Distribution : Jammu and Kashmir

The location of trematode parasite species in frogs and toad, listed (serial 1-21) in above given checklist

Serial No./s	*Location of Parasite in Host Body*
1 and 2	Rectum
1, 5 and 6-8	Urinary bladder
10	Lungs
1-4, 9, 11- 21	Intestine

Source: Bhaleroa (1937); Fotedar (1959, 1967, 1970,1971); Gupta (1967); Jordan (1930); Kaw (1943, 1945, 1950); Singh *et al.* (2001); Tickoo (1970).

In addition to the above listed trematode species (21), *Tetracotyle ranae*, infecting intestine of *Rana cyanphlyctis* in Kashmir region has been reported (Kaw, 1950). This species of trematode was not assigned any family/subfamily by Yamaguti (1958,1971) because of being a larval trematode. In the latest version of systematics of trematodes, given by Niewiadomska (2002), has placed genus *Tetracotyle* in the Superfamily Diplostomoidea under Sub-class Digenea of class Trematoda.

2.1.4 Bird (Wild and Poultry) Trematode Parasites

Till-to-date, a total of 43 species of trematode have been reported to parasitize 18 different species of birds (15 spp. of wild species and 3 spp. domesticated/poultry), occurring in various regions and localities of Kashmir Valley. The majority of the species of parasites, have been found in the birds of wetland habitats (Anchar, Hokersar, Dal Lake), and other localities of the valley like Litter, Dangerpora, Shalteng, Qamarwari, Nowgam, Bijbehara, Zoorpora, Bachpora, Ganderbal, and other regions and localities in district Srinagar. The 43 spp. of trematodes, are belonging to 19 different genera, covering 9 families, under subclass Digenea, are listed in **Table 2.5**.

The trematode parasites have been found to infect various regions in hosts' body such as: intestine (*e.g. Echinochasmus* spp., *Echinoparyphium* spp.); bile passage (*Lyperosomum kavini*); intestine and bile duct (*Pseudoechinostomum indicum*); cloaca (*Prosthogonimus cuneatus*); Caeca (*Paramonostomum* spp.). Besides these sites of infection, trematode, *viz. Prosthogonimus putsehkowskii*, has been reported from the egg of domestic hen.

The various families of trematode parasites, incorporating different genera, with their total number of species are: Cyclocoelidae, including genus *Typhlocoelum* (1 sp.); Dicrocoeliidae, genus *Lyperosomum* (1 sp.); Diplostomidae,genus– *Posthodiplostomum* (1); Prosthogonimidae, genus *Prosthogonimus* (2); Echinostomatidae, 7 genera, *viz. Echinochasmus* (2 spp.), *Echinoparyphium* (8), *Echinostoma* (5), *Euparyphium* (1), *Hypoderacum* (1), *Paryphostomum* (4), *Pseudoechinostomum* (2); Leucochloridiidae, genus *Leucochloridium;*(1), Opisthorchiidae, genus *Opisthorchis* (1); Notocotylidae, 3 genera – *Catatropis* (1), *Notocotylus* (4), *Paramonostomum* (4); Plagiorchiidae, genus *Plagiorchis* (2); Strigeidae, genus – *Apharyngostridea* (1), and genus- *Cotylurus* (1), The checklist of trematode parasites recorded from various bird species (wild, poultry / domesticated), is incorporated in the **Table 2.5.**

Table 2.5: Annotated Checklist of Trematode-Fauna, Parasitizing Birds (Wild and Poultry) in Kashmir

Sl.No.	Trematode Species	Author(s) First Reporting/ References
1	*Aphayngostridea garciai*	Dhar and Kharoo (1991)
2	*Catatropis kashmirensis*	Khan and Chishti (1989)
3	*Cotylurus cornutus* Rudolphi	Dhar and Ahmad (1992)
4	*Echinochasmus coaxatus* Dietz	Fotedar and Khateeb (1987)
5	*Echinochasmus (E.) fotedari*	Chishti and Mir (1990)
6	*Echinoparyphium angatum* Dietz	Ahmad and Chishti (1997)
7	*Echinoparyphium chinense* Lee *et al.*	Khan and Ahmad (1992)
8	*Echinoparyphium cleric* Skrybin	Dhar and Ahmad (1992)
9	*Echinoparyphium distinctum* Dietz	Ahmad and Chishti (1997)
10	*Echinoparyphium haneyanum* Johnston	Ahmad and Chishti (1997)
11	*Echinoparyphium kashmirensis* Chishti and Tanveer	Chishti and Tanveer (1992)
12	*Echinopanyphium pindchi* Dhar and Kharoo	Dhar and Kharoo (1991)
13	*Echinioparyphium recurvatum* (Linstow)	Ahmad and Chishti (1997)
14	*Echinostoma anseris casarcii* Dhar and Kharoo	Dhar and Kharoo (1991)
15	*Echinostoma azerbaidjanicum* Kasimov	Fotedar and Khateeb (1987)
16	*Echinostoma kashmirensis* Chishti and Ahmad	Chishti and Ahmad (1992)
17	*Echinostoma revolutum* (Froelich)	Khan and Chishti (1982a)
18	*Echinostoma turkestanicus* Kurova	Fotedar and Khateeb (1987)
19	*Euparyphium pindchi* Khan and Chishti	Khan and Chishti (1984)
20	*Hypoderacum conoideum* (Block)	Khan and Ahmad (1992)
21	*Leucochloridium gallinuli* Kharoo and Dhar	Kharoo and Dhar (1982)
22	*Lyperosomum kavini* Fotedar and Raina	Fotedar and Raina (1965)
23	*Notocotylus attenuatus* Rudolphi	Ahmad and Chishti (1995)
24	*Notocotylus babai* Bhalerao	Ahmad and Chishti (1995)
25	*Notocolyus imbricatus* (Looss)	Ahmad and Chiushti (1995)
26	*Notocotylus intestinalis*	Kharoo (2011)
27	*Opisthorchis anasi* Khan	Khan (1987)
28	*Paramonostomum elongatum* Yamaguti	Tanveer and Chishti (2001)
29	*Paramonostomum fulicai* Bangh	Tanveer and Chishti (2001)
30	*Paramonostomum galli*	Tanveer and Chishti (2001)

Sl.No.	*Trematode Species*	*Author(s) First Reporting/ References*
31	*Paramonostomum microstomus*	Kharoo (2011)
32	*Paryphostomum mukhtarensis* Ahmad	Ahmad (2000)
33	*Paryphostomum radiatum* Dietz	Ahmad (2000)
34	*Paryphostomum srivastavi* Chisti and Tanveer	Chishti and Tanveer (1993)
35	*Paryphostomum* sp.	Ahmad (2000)
36	*Plagiorchis bulbuli* (Mehva)(= *Lepoderma bulbuli* Mehva)	Mehra (1937) (cf. Yamaguti 1958, 71)
37	*Plagiorchis (P). tehri* Khan and Chishti	Khan and Chishti (1983)
38	*Posthodiplostomum milvi* Fotedar and Raina	Fotedar and Raina (1964)
39	*Prosthogonimus (M.) cuneatus* (Rudolphi)	Fotedar et al. (1970)
40	*Prosthogonimus putschkowskii* Skrjabin	Raina (1971)
41	*Pseudoechinostomum indicum* Mehva	Chishti and Ahmad (1992)
42	*Pseudoechinostomum srinagarensis* Dhar and Ahmad	Dhar and Ahmad (1992)
43	*Typhlocoelium indicum* Fotedar	Fotedar (1964a)

The location of tematode parasite species (serial 1- 43) in different host birds (wild and domestic), is given as under:

Serial No./s	*Location of Parasite in Host Body*
1–21, 23–26, 28–36, 38, 42, 43	Intestine
22	Bile passage
27	Caccum
37	Cloaca
40	Egg
41	Intestine and bile duct

Table 2.6: Classification of Bird Trematode Species (Class: Trematoda), Listed in the Table 2.5, Based on Classification Provided by Yamaguti (1958, 1971) and Jones *et al.* (2002)

Serial No./s (Refer to Listed spp. in Table 2.5	*Sub-Class*	*Family*	*Sub-family*
1, 3	Digenea	Strigeidae	Strigeinae
38	Digenea	Diplostomidae	Crassiphialiinae
2, 23–26, 28–31	Digenea	Notocotylidae	Notocotylinae
4–13	Digenea	Echinostomatidae	Echinochasminae
14–19, 32–35	Digenea	Echinostomatidae	Echinostomatinae
41–42	Digenea	Echinostomatidae	NA
20	Digenea	Echinostomatidae	NA
36–37	Digenea	Plagiorchiidae	Plagiorchiinae

Serial No./s (Refer to Listed spp. in Table 2.5	Sub-Class	Family	Sub-family
21	Digenea	Leucochloridiidae	Leucochloridiinae
22	Digenea	Dicrocoeliidae	Dicrocoeliinae
27	Digenea	Opisthorchiidae	Ophisthorchiinae
43	Digenea	Typhlocoelidae	Typhlocoelinae
39–40	Digenea	Prosthogonimidae	Prosthogonominae

NA: Not available.

Table 2.7: Host-Bird Species-Trematode Species Catalogue

1. *Anas penelope* (Wigeon)

   Notocotylus attenuatus

2. *Anas platyrhynchos domesticus* (Domestic Duck)

   Cotylurus cornutus, Echinoparuphium agnatum, E. recurvatum, Echinostoma kashmirensis, E. revolutum, Hypoderatum conoideum and *Pseudoechinostomum srinagarensis*

3. *Anas platyrhynchos* (= *A. platyrhynchos platyrhynchos*) (Mallard/Wild Duck)

   Opisthorchis anasi, Pseudoechinostomum indicum

4. *Anser anser domesticus* (Goose)

   Notocolylus babai, Notocotylus sp., *N. imbricatus*

5. *Anser anser* (Greyleg Goose)

   Notocotylus intestinalis

6. *Aythya rufa* (White-eyed pochard)

   Notocotylus babai

7. *Casarca ferruginea* (Ruddy sheldrake)

   Echinostoma anseris casarcii

8. *Corvus macrorhynchos* (Jungle Crow)

   Echinostoma revolutum

9. *Corvus monedula monedula*)(= *Corvus monedula soemmeringi*) (Eastern Jackdaw)

   Lyperosomum kavini

10. *Fulica atra atra* (Coot/kotru)

    Paramonostomum fulicai

11. *Gallinula chloropus indica* (Indian Moorhen)

    Echinoparyphium pindchi, Enparyphium pindchi and *Leucochloridium gallinuli*

12. *Gallus gallus domesticus* (Domestic fowl)

    Catatropis kashmirensis, Echinochasmus coaxatus, Echinoparyphium kashmirensis, Echinostoma azerbaidjanicum, E.revolutum, E. turkestanicus, Paramonostomum elongatum, P. galli, Paryphostomum srivastavi, Prosthogonimus putschkowskii (egg parasite).

13. *Grus grus liffordi* (Crane)

    Apharyngostridea garciai

14. *Malpastes haemorhous intermediates*

Plagiorchis bulbuli

15. *Milvus migrans lineatus* (Kite)

Posthodiplostomum milvi

16. *Passer domesticus* (House sparrow)

Plagiorchis tehri, Prosthogonumus cuneatus

17. *Phalocorocavax carbo sinensis* (Indian large Carmorant)

Notocotulus attenuatus, Paryphostonium mukhtarensis, P. radiatum and *Paruphostomum* sp.

18. *Podiceps ruficollis capensis* (Indian little Grebe/Dabchick)

Echinochasmus fotedari and *Euparyphium pindchi*

19. *Tadorna ferruginea* (The ruddy Shelduck or Brahiny duck)

Paramonostomum microstomus

20. *Tringa glareola* (Wood sandpiper)

Echinostoma revolutum

Source: Ahmad (2000); Ahmad and Chishti (1995, 97, 2000, 02); Chishti and Ahmad (1992); Chishti and Mir (1990); Chishti and Tanveer (1993); Dhar and Ahmad (1992); Dhar and Kharoo (1991); Fotedar (1964); Fotedar *et al.* (1970); Fotedar and Khateeb (1987); Fotedari and Raina (1964, 65); Hamdani *et al.* (2017); Khan (1987); Khan and Chishti (1982, 83, 84, 89); Kharoo and Dhar (1982); Khan and Ahmad (1992); Mehra (1937); Pandit *et al.* (1991); Raina (1971); Tanveer and Chishti (2001).

From the Table 2.7, it is evident that single trematode species parasitization, has been reported from the wild bird species, *viz. A. anser, A. rufa, A. clypeata, A. penelope, C. ferruginea, C. macrorhynchos, C. m. monedula, F. a. atra, G. g. liffordi, M. h. intermedictus, M. m. lineatus, T. glaeola.* Two different species of trematode parasites, have been reported to infect birds like *A. platyrhynchos, Passer domesticus, Podiceps domesticus* and *P. r. capensis.* Three and four different species of trematode parasites in Kashmir valley, have been found in bird like *Gallinula indica* and *Phalocorocarax carbo sinensis* respectively. Birds of poultry importance, *viz. A.p. domesticus* and *G. g. domesticus,* are affected by eleven and ten different species of parasites respectively.

The trematodes of this region, having wide host-spectrum are: *Notocotylus attenuatus,* parasitic on birds like Wigeon and Indian large Carmorant. *Euparyphium pindchi* is also parasitic on two different bird hosts, *viz.* Indian Moorhen and Indian little Greb. *Echinostoma revolutum* has been found in different host birds– domestic duck, domestic fowl, jungle crow and *Tringa glaeola.*

(a) Trematode Parasites of Domestic Fowl in Kashmir

In Kashmir Valley, the detailed studies on the prevalence, occurrence and seasonal variations of trematode parasites in common Kashmiri domestic fowl (*G. g. domesticus*), have been completed by: Khan and Chishti (1982), Fotedar and Khateeb (1987), Khan and Chishti (1989) and Pandit *et al.* (1991), indicating wide prevalence of trematode infections in domestic fowl.

A study conducted by Fotedar and Khateeb (1987) revealed the prevalence of infection by *Echinostoma azerbaidjanicus, E. revoltum, E. turkestanicus* and

Echinochasmus coaxatum in the local fowl, collected from different places. These authors further reported that percentage of infection and parasite burden increased at optimum temperature, and during rainy season. The highest incidence of infection was recorded in the month of September, and lowest in the month of December and January. Pandit *et al.* (1991) during the course of their study on the helminth parasites in indigenous fowl of Kashmir, have found the occurrence of only one species of trematode (*Echinostoma revolutum*), accounted for about 1 per cent of infection due to this parasites.

2.1.5 Mammals (Bat, Domestic Animals and Wild Mammals) Trematodes

Two species of trematodes have so far been reported from Kashmir, infecting the intestine of bat, *Pipistrellus mimus* by Fotedar and Khan (1974).These parasite are: *Prosthodendrium (Paralecithodendrium) indicum (=Paralecithodendrian indicum)* Khan and Fotedar, belonging to sub-family Lecithodendriinae of family Lecithodendriidae (Prosostomata: Digenea) and *Plagiorchis (Plagiorchis) pipistrellicola* Fotedar and Khan, belonging to the sub-family Plagiorchiinae, family Plagiorchiidae (Prosostomata: Digenea).

An intestinal fluke, belonging to family Echinostomatidae, *viz. Artyfechinostomum malayanum (= Artyfechinostomum surfrartyfex)*, has been reported from Jammu and Kashmir, infecting Indian mongoose, (*Herpests auropunctatus*) (Borkataki *et al.*, 2017).

About 21 species of rumen and liver trematode parasites of domestic ruminants (sheep, goat and cattle) and Indian Yak (domestic), are known to occur in various localities and regions of Kashmir and Ladakh. These species are belonging to 10 genera spread over 3 families and 6 sub-families, under sub-order Prosostomata and order Digenea. The total number of trematode parasites species belonging to various sub-families under different families are as: Paramphistomidae, sub-family Paramphistominae (8 spp.), Gastrothylacinae (3 spp.), Cladorchiinae (2 spp.), Orthocoeliinae (4 spp); Dicrocoeliidae, sub-family Dicrococliinae (1 sp.) and Fasciolide, sub-family Fascioliinae (3 spp.) (**Table 2.8**).

The number of species of trematode, infecting various domestic mammals is cattle (cow/buffalo), sheep, yak and goat is 11, 10, 7 and 7 respectively. Trematode parasites of cattle in this region are, *viz. C. calicophorum, C. papillosum, Gastrothylax compressus, G. crumenifer, Olveria indica, Orthocoelium dicranocoelium, O. dinniki, O. gigantopharynx, O. indonesiens, Paramphistomum epiclitum,* and *P. gracilus* (Table 2.6). The trematode parasite species of sheep include *C. calicophorum, Ceylonocotyle scoloicoelium, Cotylophoron cotylophorum, Dicrocoelium* sp., *Fasciola* sp., *Gastrothglax compressus, G. crumenifer, Olveria thapari, Paramphistomum cervi and P. gracilis.*

Goats of this region, have been found to be infected by 7 species of trematodes and these include. *C. caliciphorum, C. cotylophorum, dicrocoelium,* sp. *Fasiola* sp., *G. compressus, P. epiclitum* and *P. gracilis.* Like common liver fluke, *Fasciola hepatica, Fasciola gigantica,* liver fluke of larger species is found in cattle, sheep and goat in all parts of India including Jammu and Kashmir State, causing much damage to these valuable domestic animals (**Table 2.8**).

The Indian Yak (*Bos grunniens*) in Ladakh region has been shown to be infected by 7 spp. of trematode parasites by Ranga Rao *et al.* (1994). These are : 3 spp. of flukes, infecting liver of Yak, *viz.* as *F. hepatica, F. gigantica* and *Paramphistanum explanatum* (=*Gigantocotyle explanatum*) and 4 Species of stomach fluke (amphistome), as *C. cotylophorum, G. crumenifer, Fischoederius cobboldi* and *P. cervi* (Table 2.8.)

Table 2.8: Trematode Parasites of Mammals of Vetrinary Importance in J&K, Ladakh UTs: Systematic Checklist and Host Diversity

Class: Trematoda

Subclass Digenea

Order: Echinostomida

Superfamily: Echinostomatoidea

Family: Fasciolidae

Subfamily: Fasciolinae

1. *Fasciola gigantica* [Sh, Ya; Ll; K, L; Ranga-Rao *et al.* (1994), Mir *et al.*, 2008]
2. *Fasciola hepatica* [Go, Sh,Ya; Ll, GB; J,K,L; Ranga- Rao (1994), Mir *et al.*, 2008, 2012]
3. *Fasciola* sp. [Go, Sh, Ca; Gl; J, K, L; Makhdoomi *et al.* (1995), Kuchai *et al.* (2011), Khajuria *et al.* (2012)

Superfamily: Gorgoderoidea

Family: Dicrocoeliidae

Subfamily: Dicrocoeliinae

4. *Dicrocoelium dendriticum* [Go, Sh; J, K; Khajuria *et al.* (2012), Mir *et al.* (2012), Gul and Tak (2016 b) and Sheikh *et al.* (2016)]
5. *Dicrocoelium* sp. [Bu, Ca, Go, Sh; J, L; Makhdoomi *et al.* (1995), Mir *et al.* (2013), Allaie *et al.* (2014)

Superfamily: Paramphistomoidea

Family: Gastrothylacidae

6. *Fischoederius cobboldi* [Ya; S; L; Ranga- Rao *et al.* (1994)]

Subfamily: Gastrothylacinae

7. *Gastrothylax compressus* (= *Gastrothylax indicus, G. orientalis, G. glandiformis*) [Dru,Go, Sh, Ya; R, S; Sey (1983), Bambroo (1970), Raina *et al.* (1987)
8. *Gastrothylax crumenifer* [Dru, Ya, R, S, K, L; Bali (1976), Ranga-Rao (1994)]

Family : Paramphistomidae

Subfamily: Cladorchiinae

9. *Olveria indica* [Bu; R, K,; Raina *et al.* (1987)]
10. *Olveria thapari* [Sh; R, K; Bali and Fotedar (1973)]

Subfamily: Orthocoeliinae

11. *Orthocoelium dicranocoelium* [Dru, R, K; Raina *et al.* (1987)]
12. *Orthocoelium dinniki* [Dru, R,K; Raina *et al.* (1987)
13. *Orthocoelium gigantopharynx* [Dru, R, K, Raina *et al.* (1987)]
14. *Orthocoelium indonesiense* [Dru, R, K; Raina *et al.* (1987)

Subfamily : Paramphistominae

15. *Calicophoron calicophorum* [Ca, Go, Sh; R, K; Bali (1971), Bali and Fotedar (1972)] (cf. Raina *et al.* (1987)
16. *Calicophoron papillosum* [Co, Bu, R, K; Raina *et al.* (1987)]
17. *Ceylonocotyle scoliocoelium* [Sh; R, J, K; Bali and Fotedar (1972 a)]
18. *Cotylophoron cotylophorum* [Sh, Go, Ya; R, S; K, L; Bali (1976), Ranga-Rao *et al.* (1994)
19. *Paramphistomum cervi* [Go, Sh, Ya; R; J, L; Bali (1976), Rana-Rao *et al.* (1994), Mir *et al.* (2012)
20. *Paramphistomum epiclitum* [Ca, Go, Sh; R; K; Raina *et al.* (1987)
21. *Paraphistomum explanatum* (= *Gigantocotyle explanatum*) [Ya, S, L; Ranga-Rao (1994)]
22. *Paraamphistomum gracilis* [Ca, Dru, Go, Sh; R; K; Raina *et al.* (1987)
23. *Paramphistomes* [Bo, Bu, Ca, Dru, Go, Sh, Eq; R; J, K; Raina and Khan (1983), Allaie *et al.* (2014, 2016), Bushra *et al.* (2012); Lone *et al.* (2012), Mir *et al.* (2013); Pandit *et al.* (2013); Tramboo (2015)

Key to abbreviations: Bo: Bovines; Bu: Buffalo; Ca: Cattle; Co: Cow; Dru: Domestic Ruminants; Eq: Equines; GB: gall bladder; GI: gastrointestine; GO: goats; J: Jammu; JK: Jammu and Kashmir; K: Kashmir; L: Ladakh; LI: Liver; R: rumen; S: stomach; SH: Sheep; YA: Yak (Indian)

2.1.6 Highly Prevalent Trematode Parasites of Veterinary Importance

(a) *Dicrocoelium dendriticum* (Dicrocoeliasis)

D. dendriticum, commonly known as Lancet fluke or small liver fluke infects grazing animals like cows, sheep, goats and cattle, causing disease known as dicrocoeliasis. Ruminants are the main definitive host of this fluke, however, other herbivorous animals, carnivores and humans can be accidental hosts. In humans, this parasite can infect bile duct. Most of the infections, especially in cows, are asymptomic. Some animals may show symptoms like anaemia, emaciation and liver cirrhosis. Terrestrial snails act as first intermediate host of *D. dentriticum* and ants act as second intermediate host of this fluke.

In Jammu and Kashmir region, the prevalence and distribution, with high infection in age group and due to dicrocoeliasis in sheep, goats, cattle and buffalo, have been studied in detailed by number of workers,covering their published papers (2006, 2011-2017) and, the data is given in **Table 4.1.4**, p. 105. Apart from this, the earlier prevalence of dicocoeliasis in sheep slaughtered in Kashmir region, was found to be 6.31 per cent, with lambs being highly infected and highest prevalence of infection in autumn season (Abdullah *et al.*, 2000). In Jammu region, *D. dendriticum* was found 18.9 per cent of the goats examined. The maximum prevalence was recorded in winter (28.9 per cent), followed by post rainy season (22.2 per cent) and minimum in rainy season (9.4 per cent) (Godara *et al.*, 2014)

(b) *Fasciola hepatica, F. gigantica* (Fascioliasis, distomatosis or liver rot disease)

Fasciolia spp. live primarily in bile duct of sheep, goats and cattle. *F. gigantica* causes tropical fascioliasis and is very harmful to their hosts, particulary sheep and buffaloes. The young or immature flukes migrating through liver tissue and crossing the walls of bile duct, cause major damage, resulting in destruction of tissues and bleeding, cell death and fibrosis. Earlier, the prevalence of 23.54 per cent of fascioliasis in slaughtered sheep, has been found (Abdullah *et al.* 2000). Later, the prevalence of 23.92 per cent of tropical fascioliasis in sheep of Kashmir has been reported (Mir *et al.*, 2008). The detailed investigations on fascioliasis disease, due to *Fasciola gigantica* in slaughtered animals and sheep of Kashmir region, have shown prevalence of this parasite as 20.86 per cent and 23.92 per cent respectively (Mir *et al.*, 2008; Gul *et al.*, 2016).

Fasciola hepatica or common fluke, usually termed as temperate fluke. The symptoms of infection with this fluke include inflammation of liver or hepatitis and bile duct (cholangitis). In chronic fascioliasis, there is anaemia, edema and digestive disturbances. In acute fascioliasis,there is complete organ failure, leading to death of animals. The fascioliasis disease in in cattle, sheep and goats of Kashmir region, due to *F. hepatica*, studied by various authors (2008, 2016, 2017) (**Table 4.1.4**, p. 106), has shown an average prevalence of 18.46 per cent of this parasite. Whereas, in Jammu (low -altitude subtropical area), the prevalence of *F. hepatica* is shown as 5.28 per cent (Mir *et al.*, 2012).

The prevalence percentage data has been made available by various authors in years; 2006, 2011, 2012, 2013, 2015, 2017, see table pertaining to fascioliasis due to *Fasciola* spp. in cattle, sheep, goats and buffalo of Jammu, Kashmir and Ladakh region (**Table 4.1.4**, p. 106). On the basis of prevalence data given by different authors in different years of these regions, the average prevalence of fascioliasis in the domestic animals, belonging to Ladakh, Kashmir and Jammu, is calculated as 33.67 per cent, 21.17 per cent and 4.30 per cent respectively.

(c) *Notocotylus*

Notocotylus sp. is a digenetic trematode, lives in the caecum of aquatic birds and mammals. It causes emaciation, diarrhoea and ataxia. In Kashmir region, infection due to *Notocotylus*,with prevalence rate 2.67 per cent has been dectected in domestic geese (Hamdani *et al.*, 2017).

(d) Paramphistomes (= Amphistomes)

Paramphistomes are the members of the family Paramphistomatidae, including genera: *Paramphistomum, Calicophoron, Cotylophoron* and *Gigantocotyle.* Paramphistomes or stomach flukes are parasites of ruminants, which particularly affect cattle and sheep. The disease caused by the infestation of stomach fluke is known as Paramphistomiasis, causing weight loss, loss of production and anaemia. Immature stages of fluke cause diarrhea. In case of paramphistomes, snails are intermediate hosts.

The earlier detailed study on the prevalence of paramphistomes in the domestic ruminants of Kashmir was done by Raina and Khan (1982). For this purpose, these authors conducted survey from October, 1980 to September, 1981. The collection of amphistomes of sheep and goat was made from slaughter houses present in Srinagar (Chota Bazar and Rainawari).In addition to this,the authors also made on the spot collection of amphistomes of the ruminants from far-flung slaughter houses, present in regions like Baramulla, Sopore, and Handwara. The collection of amphistome parasites from cattle was made at Sheer, Heewan, Drangabal and Tral. The percentage of infection in sheep, goat and cattle was reported to be showing variation from month to month. The peak month of infection was found to be in May in cattle and sheep, as 100 per cent and 97.7 per cent respectively. The peak infection in goat was in the month of June, showing 91.4 per cent of infection.

Raina *et al.* (1987) completed thorough investigation on the paramphistomes, parasitizing the domestic ruminants of Kashmir. They reported and described eleven species of paramphistomes from domestic ruminants (sheep, goat and bovines) collected from their rumina. The collection of trematode parasites was made from slaughter houses situated in various regions of Kashmir, *viz.* Srinagar, Pulwama, Anantnag, Verinag, Tral and Pattern. The various genera and the number of species reported are : *Calicophoron* (2 spp.), *Gastrothylax* (2 spp.) *Olveria* (1 sp.), *Orthocoelium* (4 spp.) and *Paramphistomum* (2 spp.). The termatode parasite, *Gastrothylax compressus* was reported to be as a commonly prevalent species of sheep and cattle in various areas of Kashmir, whereas *Oliveria indica* showed less common prevalence in this region.

Earlier, from Jammu region, the incidence of Paramphistomes (= Amphistomes), with prevalence 3.74 per cent, has been shown in horses of Katra (Jammu), based on faecal examination (Khajuria *et al.*, 2005). Paramphistomes/*Paramphistomum* sp. in ruminants, cattle and equines of Jammu and Kashmir region, has been studied by number of workers during past ten years. The data on prevalence of this disease provided by these authors, according to publications related to years: 2006, 2008-2009, 2011-2017, is encapsulated in **Table 4.1.4**, p. 107. The highest prevalence (46 per cent) of paramphistomiasis due to Paramphistomum sp., has been found in small ruminants in Ganderbal (Lone *et al.*, 2012) (**Table 4.1.4**, p. 107). The highest prevalence (32.26 per cent) paramphistomiasis due to Paramphistomes was detected in buffaloes in Jammu region (R. S. Pora) (Singh *et al.*, 2009) (**Table 4.1.4**, p. 107). Apart from this, earlier the stomach fluke disease caused by *Paramphistomum cervi*, showed prevalence of 42 per cent during the period Jan., 2004 to Dec., 2004, sheep of Kashmir valley, high during rainy season (Ahmad *et al.*, 2008). Later investigations on the prevalence of *P.cervi* in Jammu and Kashmir indicated 6.69 per cent prevalence, highest in summer, found in goats of low-altitude sub-tropical areas of Jammu region (Mir *et al.*, 2012) and 2.71 per cent in sheep of Kashmir (Mir *et al.*, 2008). In north Kashmir, *P. cervi*, has shown prevalence of 17.98 per cent in adult slaughtered cattle (Magray *et al.*, 2017).

2.2 Cestode Parasites

Cestodes or tapeworm parasites belong to class Cestodea (=Cestoda) of Phylum Platyhelminthes and possess an elongated tape–like body, dorsoventrally flattened, usually white in colour however, some species show yellow or gray colouration. Cestodes are endoparasites of various Vertebrates (fishes, amphibians, reptiles, birds and mammals, including humans). The body of the typical cestode parasite is divided into three regions: i) Scolex, the organ of attachment; (ii) the neck region, situated immediately posterior to scolex and (iii) strobila, which constitutes the main portion of the body and is made up of a chain of proglottids. The most anteriorly present proglottids, are generally immature and proglottids posterior to immature ones are sexually mature. During the life-cycle of cestodes, one or two, sometimes even more intermediate hosts are required. The life-cycle pattern and larval habitat show wide variation in different groups of cestodes. The larvae are known to parasitize any organ or part of vertebrate as well as invertebrate host.

In Jammu and Kashmir, Ladakh UTs, a total of 59 species of cestodes, belonging to 28 genera, are known to be parasitic on various vertebrate groups, *viz.* fishes, amphibian (toad), reptile (lizard), birds (wild and domestic), mammals (wild and domesatic) and humans (**Table 2.12**). Birds have been found to be infected with highest number of cestode species *i.e.* 23, followed by mammals, with 18 spp., fishes with 11 species and humans with 5 spp. Besides these, 1 species each of cestode found to infect toad and lizard (**Table 2.12**). These species come under two Sub-classes: Cestodaria (Order, Caryophyllidea); Eucestoda (Order, Pseudophyllidea, Proteocephalidea and Cyclophyllidea). The four orders of class Cestoda, incorporated a total of twelve families (**Table 2.13**).

In Cestodaria group, the cestodes are not divided into segment and contain only one set of reproductive organs; the scolex is lacking; presence of 10-hooked larva (decanth). In Eucestoda, the body is divided into segments or proglottids, each containing a set of male and female reproductive organs; scolex is usually present, with 6-hooks. The Caryophyllid is represented by one family occurring in (Caryophyllaeidae) including one genus, *viz. Adenoscolex*, parasite of fishes in various localities and areas of Kashmir. The Caryophyllaeid cestodes show the presence of single set of reproductive organs and utilize aquatic oligochaetes as intermediate host. The adults are parasitic in fishes. These are usually small unsegmented tapeworms.

The Pseudophyllid cestodes are represented by two families, *viz.* Bothriocephalidae and Diphyllobothriidae, including genus *Bothriocephalus* and *Ligula* respectively. these cestodes are comparatively small, and chiefly parasites of fishes in Kashmir region. Pseudophyllid parasites are characterized by the presence of true sucking grooves (bothria) on the scolices.

The Proteocephalid cestodes include two genera, *viz. Proteocephalus* and *Gangesia,* belonging to the family Proteocephalidae, parasitizing fishes of Jammu and Kashmir region. These cestodes are distinguished from one another by the arrangement of reproductive organs inside each proglottid. The eggs of Proteocephalids are embryonated when laid and hatch on ingestion by the suitable intermediate host, *viz.* copepod.

2.2.1 Fish, Amphibians, Reptile and Bird Cestodes

(a) Fish Cestodes

The earliest systematic survey for the cestode parasites of fishes of earstwhile J&K State, have been conducted by Datta (1936), who for the first time reported the existence of cestode, *viz. Ligula intestinalis* (Diphyllobothriidae) in the intestine of fish, (*Schizothorax*) of Dale lake. Later, Fotedar (1958) gave a description of new genus/

species of Caryophyllaeid, *Adenoscolex oreini,* infecting intestine of *Schizothorax richardsonii* (=*Oreinus sinuatus*) from Kashmir region. In the later years, more cestode parasites such as: *Schyzocotyle acheilognathi* (=*Bothriocephalus acheilognathi*) (Bothriocephalidae); *Gangesia fotedari, G. lucknowi, Gangesia* sp., *G. jammuensis, G. kashmirensis, Proteocephalus kashmirensis* and *P. torulosus* (Proteocephalidae: Proteocephalinae), were recorded from the intestine of number of fish host species,existing in different water bodies of various localities/areas of Jammu and Kashmir (Dhar and Fotedar, 1980; Dhar and Raina, 1983; Dhar and Majidah, 1983 and Ara and Ahmad, 2000; Wali *et al.*, 2016 a; Ahmad *et al.*, 2018).

The Asian fish tapeworm (*Schzocotyle acheilognathi*) poses threat to fresh water fish, mainly cyprinids, across the world (de Leon *et al.*, 2018). A number of detailed studies have been accomplished on fish parasitization by different cestode parasites, occurring in the lakes of Kashmir, covering diversity, occurrence, seasonal variation, with influence of season on the intensity of infection due to parasite species, has been accomplished by Dhar and Majadah (1987), Dhar and Peerzada (1992), Khan and Majadah, 1993 (1994), Majadah and Khan (1998) and Chishti *et al.* (2002).

Till-to-date, the available published data on the occurrence of Cestode species infecting intestine of different host fishes, covering as many as 14 spp., belonging to 7 genera of Jammu and Kashmir, have revealed that a total of about 12 species of cestode parasites, distributed over 5 different genera, belonging to four families: Caryophyllidae, Bothriocephalidae, Diphyllobothriidae and Proteocephalidae, are known to exist in various localities of Jammu and Kashmir region. The cestode parasite- complex of various fish host species, occurring in lakes (Dal, Anchar, Wular, Manasbal), Jehlum river and other water-bodies, in the light of recent revised nomenclature changes, is given in **Table 2.9.**

Table 2.9: Cestode–Complex of Fish Hosts in Lakes, River and other Water-Bodies of Kashmir Region

Sl.No.	*Fish Species*	*Cestode Parasite Species*
1	*Carassius carassius*	*Adenoscolex oreini*
2	*Cyprinus carpio communis*	*Schyzocotyle acheilognathi* (=*Bothriocephalus acheilognath*)*i*
3	*Cyprinus carpio specularis*	*A. oreini, Gangesia* sp.
4	*Glyptothorax*	*Gangesia fotedari, Gangesia* sp. *G. kashmirensis, G. lucknowi*
5	*Salmo trutta*	*A.oreini*
6	*Schizothorax esocinus (=. Schizothorax hugelii*	*A. oreini, Gangesia* sp.
7	*Schizothorax plagiostomum*	*A.oreini, S.acheilognathi*
8	*Schizothorax progastrus*	*S. acheilognathi*
9	*Schizothorax richardsonii (=. Oreinus sinuatus) (=O. plagiostomus)*	*A. oreini, Gangesia* sp.
10	*Schizothorax curvifrons*	*A. oreini, Gangesia* sp.
11	*Schizothorax niger*	*A. oreini, Gangesia* sp.*S. acheilognathi*

Sl.No.	Fish Species	Cestode Parasite Species
12	*Triplophyse kashmirensis (= Nemacheilus-kashmirensis)*	*A. oreini, Gangesia* sp., *Protocephalus kashmirensis, P. torulosus*
13	*Schizothorax sp.*	*Adenoscolex* sp., *Bothriocephalus* sp. *Ligula intestinalis*
14	*Wallago attu*	*Gangesia jammuensis*

From **Table 2.9**, it is clear that the fish parasitization, with decreasing order of number of parasite species in various fish host species is as: *Glyptothorax* sp., found to be infected by highest number of parasites as four, followed by *Schizothorax niger, Schizothorax* sp. and *Triplophys kashmirensis,* by three parasite species each. *C. c. specularis, S. esocinus, S. richardsonii, S. curvifrons,* infected with 2 species each and rest of the fishes like *Carassins carassius, C. c. communis* and *Wallago attu,* infected by only one species each of cestode parasite.

Detailed investigations on the seasonal population fluctuation of *Adenoscolex oreini,* infecting various fishes, *viz. S. niger, S. esocinus* (= *S. hugelii*), *S. curvifrons* and *Schizothorax richardsonii* (=*Oreinus sinuatus*), *O. plagiostomus* in Wular lake, have been carried out for the first time by Dhar and Peerzada (1992). The prevalence and mean intensity of *A. oreini* in various fishes of Wular lake of Kashmir Valley during winter season (1988-89), shown by these authors is given in **Table 2.9**.

Table 2.10: Seasonal Population Fluctuation of *Adenoscolex oreini*, Infecting Fishes of Wular Lake (Kashmir)

Name of the Fish Examined	No. of the Fish	No. of Fish Infected	Percentage	Intensity and No. of Worms
Schizsthorax niger	387	129	33.3	5.8 (753)
S. hugelii	182	65	35.7	5.8 (379)
S. esocinus	30	07	23.3	5.5 (39)
S. curvifrons	51	15	29.4	2.4 (37)
Oreinus plagiostomus	55	10	10.1	5.5 (55)
Oreinus sinuatus	78	23	29.4	6.3 (145)

Source: Dhar and Peerzada (1992).

The initial data presented by Dhar and Peerzada (1992) (*see* **Table 2.10**) on the prevalence and percentage of fish infection covered a total of six species of fishes, infected by *A. oreini,* which is here being modified, keeping in view the recent nomenclatural changes pertaining to fish hosts, reducing the actual number of fish species to 4 instead of 6, which are, *viz. S. niger, S. esocinus, S. curvifrons* and *S. richardsonii.* The highest percentage of infection, therefore, was prevalent in *S. esocinus* (33.9 per cent), followed by *S. niger* (33.3 per cent), *S. curvifrons* (29.41 per cent) and *S. richardsonii* (24.81 per cent), as presented in the **Table 2.10a**.

Table 2.10a: Prevalence Percentage of Infection in Fishes of Wular Lake

Name of Fish	*No. of Fish*	*No. of Fish Infected*	*Percentage*
Schizsthorax niger	387	129	33.3
S. esocinus (= *S. hugelii*)	30+(182)	7+(65)	33.96
Total (*S. esocinus*)	212	72	
S. curvifrons	51	15	29.41
S. richardsonii (= *Oreinus plagiostomus*) (= *O. sinuatus*)	(55+78)	(10+23)	24.81
Total (*S. richardsonii*)	133	33	24.81

Modified after Dhar and Peerzada (1992).

Majidah and Khan (1998) during the course of their investigation on seasonal population dynamics of two cestode parasite, *A. oreini* and *Gangesia* sp. in fishes such as *S. niger, S. esocinus* (= *S. hugelii*), *S. curvifrons, S. richardsonii* (= *O. plagiostomus*), *Triplophysa kashmirenis* (= *Nemacheilus kashmirensis*) and *Cyprinus carpio specularis* of Wular lake found that the infra population concentration was relatively more during summer. Further, the maximum infection was noticed in *S. esocinus, S. curvifrons* and *S. niger.*

The existence of cestode parasite *Adenoscolex oreini,* infecting the intestine of fishes, *viz. Carassius carassius* and *S. curvifrons* from Dal lake (urban eutrophic) and Manasbal lake (rural mesotrophic) cestode species (*A. oreini*) was observed to be having two infection periods as winter and spring (Chishti *et al.*, 2002). In water bodies of Gurez region of Kashmir, the cestode parasite prevalence during summer season due to *Adenoscolex oreini* was as 16.19 per cent and 19.84 per cent in *Salmo trutta* and *Schizothorax plagiostomus* respectively (Ahmad *et al.*, 2018). The prevalence of 30.63 per cent and 32.43 per cent cestode parasites- *Schizocotyle acheilognathus* (= (=*Bothriocephalus acheilognathi*) and *Adenoscolex oreini* in male fishes (*Scizothorax niger, S. esocinus* and *S. curvifrons*), inhabiting Dal Lake and River Jehlum, has been observed (Wali, 2001a).

(b) Amphibian and Reptilian Cestodes

It was Fotedar (1986), who for the first time reported the existence of nematotaeniid cestode, *Nematotaenia kashmirensis,* infecting small intestine of toad (*Bufo viridis*) in various regions of Kashmir Himalaya. The infection due to *N. kashmirensis* was reported to be very common among toads at many localities, *viz.* Srinagar, Baramulla, Sopore, Anantnag, Verinag besides, high-altitude areas (c 7,000–9,000 ft.) like Pahalgam and Gulmarg. Heavy infection of *N. kashmirensis* was also prevalent in toads of these regions and in some cases of infected amphibian, the distal region found bulged out due to the presence of large number of parasites, coiled together in lumen, showing hardly any space for food to pass. The number of cestodes, in such cases was as high as twelve, of different length, and the incidence was found during August to September month. Raina *et al.* (1973, 77) reported the occurrence of an Anocephalid Cestode parasite, *viz. Oochoristica tuberculata,* infecting the intestine of lizard (Rock Agama), *Laudakia tuberculata* (= *Agama tuberculata*) in

Kashmir region besides, a description of teratological abbrration in *O. tuberculata*, collected from this Kashmir Rock Agama, has also been provided by these authors.

(c) Birds (Wild and Poultry) Cestodes

(i) Wild Birds

In Kashmir region, so far four species of cestode, infecting the intestine of *Corvus monedula* (Eastern Jackdaw) and *C. macrorhynchos*, (The Himalayan jungle crow) have been reported. The parasites reported include *Anomotaenia galbulae, Choanotaenia infundibulum, C. microcantha* and *Mayhewia kavinia* (Fotedar and Chishti, 1974, 76; Chishti and Khan, 1984; Chishti *et al.*, 1987). Birds, *viz. Acridotheres tristis* of Kashmir, has been found to be infected with Cestode parasite (*Choanotaenia fotedari*). Cestodes: *Anomotaenia acrocephali* and *Choenotaenia orioli*, are known to infect intestine of birds, *viz.*Great Red Warbler (*Acrocephalus stenotorecus brunnescens*) and *Oriolus oriolus*, of Kashmir respectively (Fotedar and Chishti, 1974 a, 76).

In Kashmir region, Indian Little Grebe or Dabchick, *Podiceps ruficollis capensis* has been observed to be infected with intestinal cestode parasites, *viz. Pseudoschistotaenia indica* and *P. pindchii* (Fotedar and Chishti, 1980; Chishti,1982). In addition to this bird, Fotedar *et al.* (1970) and Chishti (1980), have observed the intestine of birds, Mallard or Wild Duck (*Anas platyrhynchos*) and Common Blue Rock Pigeon (*Columba livia neglecta*), in Kashmir region, infected with parasite, *Dilepsis fotedari* and *Raillietina torquata* var. *kashmirensis* respectively. From Anchar Lake (Kashmir), Fotedar and Chisthi (1973) reported the existence of cestode-*Anomotaenia kashmirensis* in the intestine of bird, *viz. Sturnus vulgaris.* Cestodiasis due to cestodes, *Choanotaenia infundibulum* and *Raillietina cesticillus*, has been observed in Scaly -bellied woodpecker (*Picus squamatus*) in Kashmir (Nashiruddulah *et al.*, 2016). Till to date, a total of 12 species of cestodes,distributed over 6 genera are known to infect about 13 species of wild birds (**Table 2.13**).

(ii) Domestic/Poultry Birds

Fotedar and Chishti (1974 a) reported the existence of cestode parasite, *Choanotaenia infundibulani*, infecting the intestine of domestic fowl, *Gallus domesticus* of Kashmir region. The detailed studies on the occurrence and seasonal variation of Cestode parasite infection of domestic fowl (indigenous), *G. g. domesticus* in this valley, have been done by Fotedar and Khateeb (1987). Their survey has revealed a total of three Cestode parasites, *viz. Amoebotaenia sphenoides, Davainea proglottina* and *Raillietina tetragona*. The highest incidence of these parasites was shown by these authors in the month of September and lowest in the month of December and January besides, the prevalence and mean worm burden showed a decrease with decreasing temperature and rainfall.

The study on the prevalence of intestinal (small) cestode parasites in indigenous fowl of Kashmir, has shown the prevalence percentage as: *Raillietina tetragona*, 20 per cent; *R. echinobothridia*, 10 per cent; *Choanotaenia infundibulum*, 3 per cent; *Hymenolepis carioea*, 2 per cent and *Cotugnia diagonopora*, 1 per cent. (Pandit *et al.*, 1991). The above authors also observed that the cestodiasis infection in the fowl as compared to other helminth parasites was 45 per cent. The infections were single, however, mixed infection with two to three species was not uncommon in Kashmir. Like other helminth parasites, cestode infection is responsible for causing much damage to poultry besides, they become susceptible to other diseases also. The occurrence of helminth diseases has been shown to be influenced by varying climatic and ecological factors. Till-to-date date about 7 species of cestodes, distributed over 6 genera have been found in Kashmir, infecting domestic fowl (indigenous) (*Gallus domesticus*) (**Table 2.13**).

The indigenous domestic fowl of Kashmir, has been found to be infected with four cestode species and showed their prevalence the months- April, June and July as : *Raillietina tetragona* (47.05 per cent); *R. cesticilus* (38.23 per cent); *R. echinobothrida* (17.64 per cent) and *Amboebotaenia cuneata* (= *A. sphenoides*) (Dar *et al.*, 2008). The free-range backyard chicken (*Gallus gallus domesticus*) of Kashmir, has shown the prevalence of different cestodes, with percentage as : *Raillietina tetragona* (65 per cent); *R. cesticillus* (22.50 per cent); *R. echinobothrida* (33.33 per cent); *R. spiralis* (26.66 per cent); *Choanotaenia infundibulum* (18.33 per cent); *Choanotaenia gondwana* (10.83 per cent); *Amoebotaenia cuneata* (20.00 per cent); *A. domesticus* (15 per cent) and *Davainea proglottina* (18.33 per cent) (Dar *et al.*, 2013).The parasitic affection due to cestode, *Raillietina* sp. in domestic pigeon in Jammu, has shown 25 per cent of prevalence (Mehmood *et al.*, 2009).

2.2.2 Wild and Domestic Animal Cestodes

Raina *et al.* (1978) for the first time recorded the existence of cestode parasite, *viz. Mathevotaenia symmetria* from the intestine of common rat, *Rattus rattus* in Kashmir. In Jammu region, the prevalence of cestode, *viz. Dipylidium caninum* has been detected in cat (Borkataki *et al.*, 2013). In Kashmir region, cestode parasite, *viz. Mesocestoides* sp., has been found to be parasitizing wild animal fox (*Vulpus vulpus*) (Pandit *et al.*, 2007).

Eleven species of cestode parasites, infecting intestine and different organs/ parts of ovines, sheep and goat, and bovids and yak of Kashmir, and Ladakh region, has so far been recorded, these species are: *Avitellina centripunctata, A. woodlandi, A. chalmersi, Stilesia globapunctata, Thysaniezia himalayai, Thysaniezia* sp. Thysanosomatina: Anoplocephilidae); *Moniezia expansa, Moniezia* sp. (Anoplocephalinae: Anoplocephalidae); *Coenurus* sp. *Taenia* sp. and *Echinococcus* sp. (Taeniidae) (Fotedar and Bambroo, 1978; Makdoomi *et al.*, 1995; Ranga Rao, 1994; Raina, 1973,75; Sudhan and Shahardar, 1999 and Tak *et al.*, 2003).

In Ladakh, Yak (*Bos grunniens*) is known to be infected with tapeworm, *Moniezia* sp., affecting intestine. In addition to this Cestode species, *Coenurus* sp. in the peritoneum in the form of cyst and *Echinococcus* sp. (cyst) in liver and lungs of yak, have also been reported from Ladakh by Ranga Rao *et al.* (1994).The occurrence of coenurus cyst in palpebral sub-conjunctiva in sheep has been reported from Kashmir (Dar *et al.*, 2016). Tak *et al.* (2003), have shown the wide prevalence of hydatid disease among bovids of Kashmir. The existence of Coenurus cyst in sub-cutaneous tissues of ear of goat, belonging to *Taenia* sp. (metacestode) for the first time, has been found in Kashmir (Sudhan and Shahardar,1999).

In Kashmir Valley, the incidence of cestode, *viz. Avitellina centripunctata, A. woodlandi, A. chalmarsi, Moniezia expansa, Stilesia globipunctata* and *Thysaniezia,* infecting sheep and goat in Kashmir, has thoroughly been studied by Raina (1973). He showed the incidence of *Avitellina* spp., *Moniezia* sp. and *Stilesia* sp., reaching to its maximum during March to August. The overall percentage infection of sheep and goat was 40.83 per cent. These parasites were shown to have little variation in their percentage occurrence, almost equally common. However, *Avitellina* has been shown to possess maximum percentage of incidence, followed by *Stilesia* and lowest percentage being that of *Moniezia* sp.

The above author also investigated the weight-length relationship of three cestodes and correlated with worm burden. The average weight per cm. length was highest in *Moniezia* (14mg/cm) and lowest in *Stilesia* (3.5 mg/cm). The *Avitellina* was intermediate between the two cestode species as 5.1 mg/cm. The high weight per cm length in case of *Moniezia,* indicated a heavy drain on the host and hence lesser number of these worms per host, for a greater number could be lethal.

An abattoir survey, conducted by Tariq *et al.* (2008 a) on cestode parasites of sheep, brought from outside Kashmir valley, slaughtered in Hazratbal locality of Srinagar, has revealed three cestodes, these included *Monezia* (43 per cent prevalence), *Avitellina* (38 per cent) and *Thysaneizia* (31 per cent). The maximum prevalence of infection due to these cestodes was observed from August to November (2006).

2.2.3 Highly Prevalent Cestode Parasites of Veterinary Importance

(a) Anoplocephalids (*Avitellina, Moniezia, Paranocephala, Stilesia, Thysaniezia*)

(i) Avitellina spp. (A. centripunctata, Avitellina sp.)

The infection due to *Avitellina* spp.is called as avitelliniasis, affecting small intestine of sheep and goats. *Avitellina* spp. occasionally infect cattle and other wild and domestic ruminants, are non-pathogenic tapeworm. The intermediate hosts in case of *Avitellina* tapeworm are thought to be psocids (bark-lice or book lice). The detailed studies on the incidence of infection in sheep and goats due to *Avitellina* spp., *Avitellina (A.) centripunctata* and *A. (A.) woodlandi,* in Kashmir region has been accomplished by Raina (1973), besides taxonomical and histomorphological investigations done on the cesatode species (Raina, 1975). In goats of Kashmir, 2.4 per cent of prevalence due to *A. centripunctata,* has been encountered (Ahmed *et al.,* 2009). Avitelliniasis in goats (slaughtered), is reported to be having 42 per cent prevalence, caused by *Avitellina* sp. in Ganderbal area of Kashmir valley (Lone *et al.,* 2012) (**Table 4.1.5**, p. 108).

(ii) Moniezia spp. (M. expansa, Moniezia sp.)

Moniezia is a large tapeworm, inhabiting the small intestine of sheep, goats and cattle. The intermediate host of this parasite is an oribatid mite. The infection caused by *Moniezia* is known as monieziasis, responsible for stock breeding loss through diarrhea and fleshloss. The earliest studies on the incidence of infection in sheep and goats of Kashmir region, with *Moniezia,* has been done by Raina (1973). The prevalence percentages of Monieziasis in small ruminants, cattle, buffalo, hangul (wild mammal) in Jammu and Kashmir, have been made available by various authors during last nine years (2006, 2009, 2011-2017), given in **Table 4.1.5**, p. 108. The highest prevalence *i.e.*, 48 per cent of monieziasis was found in sheep and goats in Ganderbal area of Kashmir (Lone *et al.*, 2012).

(iii) Paranocephala

This cestode parasite in Kashmir region has been found to be infecting equines, with 8.14 per cent prevalence (Pandit, 2008). This parasite resides in small intestine, mainly jejunum and ileum. The infection due to this tapeworm is seldom pathogenic and has indirect life-cycle. The equids are final hosts and oribatids or moss mites or beetle mites act as intermediate host of this parasite (**Table 4.1.5**, p. 109).

(iv) Stilesia spp. (S. globipunctata and Stilesia sp.) (Stilesiasis)

Stilesia is found in the small intestine of small ruminants, cattle and buffalo, remains attached at the junction of duodenum and jejunum membrane and forms the nodule in the intestine. Oribatid mites are suspected to be the main intermediate host of *Stilesia.* Heavey infection due to *Stilesia* causes death of animals. The earliest report on the incidence of infection due to *Stilesia* in sheep and goats of Kashmir, has been provided by Raina (1973). In Kashmir region, the prevalence of stilesiasis in goats, cattle and buffalo, showed prevalence of 3.62 per cent.1.45 per cent and 2.89 per cent respectively (Mir *et al.*2012, 2013) (**Table 4.1.5**, p. 109).

(v) Thysaniezia spp. (T. giardia, Thysaniezia sp.) (Thysanieziasis)

Thysaniezia parsitizes cattle, buffalo, sheep and goats, resides in the small intestine of these domestic animals. Oribatid mites act as intermediate host of this parasite. This parasite is less pathogenic, causing infection, known as thysanieziasis. In heavey infection, the clinical signs are diarrhea, ill-thrift and loss of body weight in young lambs. In Kashmir region, in goats 2.1 per cent and 28 per cent prevalence of thysanieziasis, have been investigated by Ahmad *et al.* (2009) and Lone *et al.* (2012) respectively. In addition to above Anoplocephalids, Khajuria *et al.* (2012), reported infection due to Anoplocephalids in small ruminant of Jammu region, having prevalence as 2.6 per cent (**Table 4.1.5**, p. 108).

(b) *Dipylidium caninum* (Double Pored Dog Tapeworm)

D. caninum mainly infects dogs and cats, resides in the intestine of these animals. This parasite anchor to the wall of the small intestine by hook like mouth parts, causes little apparent damage other than anal irritation. Fleas are intermediate hosts of *D. caninum*. Human cases of zoonotic dog tapeworm, have been reported from different parts of the world, including India (Narasinaham *et al.*, 20,13). In Jammu region, 2 per cent prevalence of *D.caninum* has been detected in cat (Borkataki *et al.*, 2013) (**Table 4.1.5**, p. 108).

(c) Taeniid Parasites (Taeniidae)

(i) *Echinococcus granulosus (Echinococcosis or hydatidosis)*

The hydatid disease *viz.* hydatidosis (unicellular hydatid cyst), caused by cestode, *Echinococcus granualosus* (Taeniidae, Cyclophyllidae), is reported to be commonly occurring in domestic animals in many regions and localities of Kashmir. The hydatid diseases (larval tape worm type), *viz.* hydatidosis (unicellular hydatid cyst) and alveococcosis or echinococcosis (multilocular or alveolar hydatid cyst), caused by *Echinococcus granulosus* and *E. multilocularis* respectively, belong to the family Taeniidae, order Cyclophyllidae, class Cestoda of phylum Platyhelminthes. These hydatid diseases are not only serious parasitic infection but also affecting human and domestic animal populations since the time immemorial. In Kashmir Valley, hydatid diseases are showing wide prevalence in human population, besides hydatidosis is also reported to be commonly occurring in domestic animals in many localities and regions of valley.

E. granulosus infects carnivores like dogs and other Canidae as definitive hosts and intermediate hosts like sheep, goats, horses, camel, cattle and pigs as intermediate hosts. The carnivore hosts do not show clinical symptoms of hydatidosis, however, herbivores can show a range of symptoms, depending upon the severity of infection. In livestocks and herbivors, the parasite causes nausea, vomiting, diarrhea and weight loss. The parasite can also cause fluid or masses in the abdomen, enlargement of liver and abdomen, and difficulty in breathing. *E. granulosus* is zoonotic disease of humans, infected by accidently ingesting the eggs of this parasite.

The prevalence of hydatidosis in slaughtered domestic animals, *viz.* cattle, sheep, goat and buffaloes, at different abattoirs in Kashmir Valley, during August, 2001 to July, 2002, was investigated by Tak *et al.* (2003). The overall percentage infection rate of the disease was shown as 38.59 per cent, 26.00 per cent, 14.47 per cent and 8.00 in cattle, sheep, goats and buffaloes respectively. The organ-wise distribution of hydatid cysts as shown by authors was that simultaneous involvement of both lung and liver, was more frequent in sheep (71.15 per cent) and goats (54.54 per cent). In cattle, lungs showed (63.63 per cent), comparatively higher infection than liver. These studies also highlighted that right basal lungs were more infected than left ones. Spleen, kidney and heart of cow were only exceptional organs for the existence of hydatid cyst. In buffaloes, 100 per cent infection was observed in lungs. Also, the authors showed that the fertility of hydatid cyst was highest on sheep (98.36 per cent), followed by buffaloes 50 per cent, goat (27.27 per cent) and the least in cattle (9.09 per cent).

Ranga Rao *et al.* (1994) during the course of their extensive investigations on parasitic infection of Indian Yak- *Bos* (*Poephagus*) *grunniens* in Ladakh region, have detected the occurrence of hydatid cyst in this domestic animal. They recovered majority of these cysts from liver (fertile) as well as lung (infertile) of these animals. In Jammu and Kashmir, the cystic echinococcosis has been found in sheep and goats.The prevalence percentage in sheep in Kashmir was 14.3 per cent (Rashid *et al.*, 2017). In Jammu region, the average prevalence (17 per cent) of hydatid disease in small ruminants, calculated on the basis of prevalence percentage data provided by various authors during the year 2013-2014 (**Table 4.1.5**, p. 108).

(ii) *Taenia hydatigena (= Cysticercus tenuicollis)*, *Taenia solium/Cysticercus cellulosae* (procine cysticercosis)

T. hydatigena or thin necked bladderworm or dog tapeworm, parasitizes dogs, cats and other domestic animals. This tape worm was formerly known as *Cysticercus tenuicollis*, is the larval stage of *T. hydatigena*. The larval stages or cysticerci stages of *T. hydatigena* parasitize wild or domestic animals, whereas the adult tape worm, are parasites of dogs and cats. There are no clinical signs observed with *T. hydatigena* infection. However, pathological changes due to *T. hydatigena* cysticerci, causing cysticercosis, take place in animals and severe infection can be dangerous. In this cysticercosis, cysts occur, varying in number 1-75, in the omenta and mesenteries. In heavey infection, the viscera may become knotted together and functioning of organs become impaired. In case of large numbers of larvae migrating through live parenchyma, tissue will damage and acute or fatal hepatitis may result.

The prevalence percentage of *T. hydatigena* infection in dogs reported from J&K and Ladakh are : in Jammu region as 10 per cent (Shah *et al.*, 2017); Ladakh as 22.7 per cent (Rebecca *et al.*, 2014)., T. *hydatigena* infection has also been found in sheep of Jammu and percentage of infection in different location of slaughtered animal was : 5.1 per cent in liver; 21.3 per cent (mesentry) and 16.7 per cent (omentum) (Sharma *et al.*, 2008). In Kashmir region, the overall prevalence of *T. hydatigena* (= *Cysticercus tenuicolis*) in slaughtered sheep was 24.6 per cent and high in adults, and the percentage of infection in different sites observed to be as: 66.2 per cent (mesentry), 18.6 per cent (omentum); 13.3 per cent (liver); 0.7 per cent (rumen); 0.5 per cent (lungs); 0.5 per cent (peritoneum) (Rashid *et al.*, 2014). Cysticercus cellusae or the pork bladder worm worm, is the larval stage (cysticercoid, metacestode) of *Taenia solium*, human tape worm, having swine and wild boars as intermediate hosts. *Cysticercus cellulosae* are found in many parts of pig's body, mostly in muscles and subcutaneous besides, brain, heart, liver, *etc. Cysticercus cellulosae*, causing procine cysticercosis in pigs, with 1.16 per cent prevalence, has been reported from Jammu region (Sharma *et al.*, 2017) (**Table 4.1.5**, p. 109).

(iii) *Taenia pisiformis*

*T. pisiform*is is commonly called as rabbit tape worm, parasitizing rabbits, rodents and carnivores. The adults of this parasite reside in the small intestine of definitive hosts. *T. pisiformes* cysticercus (larval stage) causes sever damage to rabbits, serving as intermediate host. The infection due to *T. pisiformis* is asymptomic and severe infection is caused due to larval damage by migrating and resulting inflammation. Mir *et al.* (2006), reported the prevalence of *T. pisiformis* infection in rabbits in Kashmir, having 16 per cent of prevalence in this animal.

(iv) *Raillietina* spp. (*R. cesticillus, R. tetragona*)

Railliettina cesticillus and *R. tetragona*, are cestode parasites of poultry and other wild birds. These parasites reside in the small intestine of these birds. The *Raillietina* spp. have indirect life cycle and the intermediate hosts are ants, flies or beetles.The clinical symptoms of infection caused by *R. cesticillus* include emaciation, weakness, stunted growth and weight loss. The intermediate hosts of his parasites are beetles. This parasite has affected indigenous chickens (*Gallus gallus domesticus*) in Kashmir. Earlier, the widely prevalent cestodes, showing high seasonal occurrence in the indigenous domestic fowl of Kashmir, included *Raillietina tetragona* (47.05 per cent) and *Raillietina cesticilus* (38.23 per cent) (Dar *et al.*, 2008) (**Table 4.1.5**, p. 109).

The infection percentage due to this parasite was found to be 23.22 per cent, either singly or in association with other parasites like *Amoebotaenia cuneata* (= *Amoebotaenia sphenoides*), *Raillietina tetragona* and *Choanotaenia infundibulim* (Salam *et al.*, 2010) *Raillietina tetragona* is not highly harmful for poultry. Low infection remains without clinical signs. Heavy and chronic infection with *R. tetragona* may cause diarrhea, anaemia, wight loss, in testinal inflammation and haemorrahge. In Gurez (Kashmir), 27 per cent prevalence of *R. tetragona* in domestic ducks, has been found (Sheikh *et al.*, 2016 a) (**Table 4.1.5**, p. 109).

2.2.4 Highly Prevalent Cestode Parasites of Medical Importance

(a) *Echinococcus* spp. (Hydatid Cyst Diseases) : *E. granulosus* (Hydatidosis) and *E. multilocularis* (Multilocular or Alveolar Cyst Disease)

(i) Unicelluar Hydatid Cyst Disease

In human population, E.*granulosus* is responsible for the formation of unicellular hydatid cyst, causing disease known as hydatidosis, measures from 2-8 mm in length and having single gravid proglottid, longer than wide and lodges typical taenia like eggs. The definite/natural or wild hosts of this cestode are dogs, wolves, foxes, *etc.* The parasite occurs in hundreds/thousands in these hosts and may be practically harmless. The infection of this disease reach to humans and domestic animals as their intermediate host through close association with dogs or through eggs from feces of dogs, wolves, or foxes, gaining entry into the intermediate hosts like humans, further it may be either through contaminated water or forage.

The development of hydatid cyst is slow. The young larva changes into hollow bladder, around which the host adds an enveloping, fibrous cyst wall. The fluid of the cyst is almost colourless, however, in older cysts in human tissues, is characterized by multiple daughter bladders or 'blood capsules', with multiple invaginated protoscoleces, budding from their walls (cystic cavity or inner layer of germinal epithelium) and the material consisting of dead and generated scoleces in milky fluid of parent cyst is referred to as hydatid sand.

In Jammu and Kashmir region, the hydatid cyst shows wide diversity in different sites and organs of human body, include abdominal wall, brain, kidney, lungs, liver, neck, pancreas, peritoneum, ovary, sigmoid mesocolon, spleen, urinary bladder and uterus (**Table 2.11**). Besides these, multiple cysts in liver and lungs, liver and spleen, liver and peritoneum, have also been reported from Kashmir Valley (**Table 2.11**). The hydatid cysts cause much harm only when they grow to outrageous size and press on other organ/s or when they liberate their fluid by leaking or rupturing, precipitating allergic symptoms. The formation of hydatid cysts in human and domestic animal represents the end of the life-line for the parasite.

Table 2.11: Diversity of Hydatid Cyst Infection Site in Human Population of Jammu and Kashmir

Sl.No.	*Infection Site in Human Body/Locality*	*Author(s) Reporting*
I.	**Unilocular hydatid cyst disease (Hydiotosis)**	
1	Lungs	Tak (2000) Fomda *et al.* (2002); Chishti *et al.* (2001); Showkat *et al.* (2012)
2	Sigmoid mesocolon	Ahmad *et al.* (2001)
3	Brain	Wani *et al.* (1999); Fomda *et al.* (2001)
4	Neck-Cervicofacial region under sternomastoid muscle	Pampori *et al.* (2003)
5	Uterus (wall and vault)	Fazili *et al.* (2002)
6	Pancreas	Yattoo, *et al.* (1999); Misger *et al.* (2000); Fomda *et al.* (2002)
7	Kidney	Gupta *et al.* (2002); Fomda *et al.* (2002)
8	Liver	Khuroo *et al.* (1991,99); Malik *et al.* (1997); Wani *et al.* (2005); Bari *et al.?* (2005)
9	Spleen	Chishti *et al.* (2001); Sarmast *et al.* (2012); Wani *et al.* (2005); Wani (1993)
10	Urinary bladder	Fomda *et al.* (2002)
11	Abdominal wall	Fomda *et al.* (2002)
12	Adrenal	Wani *et al.* (2005)
13	Mesocolon	Wani *et al.* (2005)
14	Peritoneum/Pelvis	Wani *et al.* (2005)
15	Gal bladder	Wani *et al.* (2005)
16	Retroperitoneal	Sherwani *et al.* (2015)
17	Lumbar paraspinal muscle	Parry *et al.* (2010)
18	Ovary	Fomda *et al.* (2002)
19	Liver and lungs	Chishti *et al.* (2001); Fomda *et al.* (2002)
20.	Liver and spleen	Chishti *et al.* (2001)
21.	Liver and peritoneum	Chishti *et al.* (2001)
22	Interperitoneal pelvic space	Parry *et al.* (2011)
II.	**Multiple/Alveolar cyst Disease (Alveococcosis)**	
1.	Alveolar hydatid disease of liver, with Budd-Chiari syndrome, Uri (Kashmir)	Khuroo *et al.* (1980)

The prevalence and epidemiological studies on human hydatid disease in populations in different areas of all the districts of Kashmir region from 1998-99, have been made by Chishti *et al.* (2001). During this period, these authors found a total of 170 patients suffering from hydatid disease. The observations have been made in this connection, on patients suffering from this disease, and admitted in various surgical wards of Sher-i-Kashmir Institute of Medical Sciences (SKIMS). The district-wise percentage of infection of this disease was shown as: 26.47 per cent,

21.76 per cent, 18.82 per cent, 11.88 per cent and 5.38 per cent, in district Anantnag, Pulwama, Baramulla, Srinagar, Budgam and Kupwara respectively.

The aforementioned authors have also given the percentage of cyst of organ site as: 52.94 per cent, 35.29 per cent, 7.05 per cent, 2.3 per cent, 1.17 per cent, 0.53 per cent and 0.53 per cent in organ liver (isolated), lungs (isolated), liver and lungs, liver and spleen, liver and peritoneum, spleen, ovary respectively. Apart from this, their studies have shown the percentage of age-distribution and sex incidence of disease. The percentage is as: from 0-10 years 8.82 per cent; 11-20 years, 31.76 per cent; 21-30 years, 24.11 per cent; 31-41 years, 18.82 per cent; 41-50 years, 8.87 per cent; 51-60 years, 5.29 per cent and 60-70 years, 2.35 per cent. The high incidence of disease was found in males (60 per cent) and 40 per cent of females were found to be positive for the disease. The susceptibility of males to infection than females, is mainly due to their (males) association with domestic animals during rearing.

The prevalence of human hydatid disease in Kashmir valley, in the recent years, have been updated by Fomda *et al.* (2002), based on clinically detected cases of disease in 5808 patients (4422 male and 1386 female), with age group ranging from 12 to 68 years, attending SKIMS, Soura, Kashmir, between April 1984 – April 2001. Further, the authors have shown the commonest site of infection as liver (60 per cent), followed by lungs (33.23 per cent), kidney (3.71 per cent), ovary (1.18 per cent), brain (1.01 per cent), pancreas (0.34 per cent), multiple cysts in liver and lungs (0.20 per cent), spleen (0.12 per cent), urinary bladder (0.08 per cent) and abdominal wall (0.01 per cent). The results of this study indicated a high prevalence of hydatid disease in the Valley and demands prompt control measures.

In Kashmir region, 3.5 per cent (Dar *et al.*, 2002) and 4.10 per cent (Wani,1993), cases of hydatid disease of spleen were reported. Echinococcosis prevalence pertaining to affected organs/parts of body of the population in Kashmir, based on medical record of SKIMS, 1998-2003, has shown highest prevalence of cases having hydatid cyst in liver (91.3 per cent), followed by spleen (2.2 per cent), pertoneum and pelvis (1.6 per cent), pancreas (1.1 per cent), gall bladder (0.6 per cent), mesocolon (0.6 per cent) and adrenal (0.6 per cent) (Wani *et al.*, 2005). In the seropositivity study of echinococcosis infection in asymptomic population of Kashmir, 5.3 per cent prevalence of infection through ELISA was found, with 9.23 per cent prevalence in Anantnag district, 8.5 per cent in Baramulla and 8 per cent in Bandipore (Fomda *et al.*, 2015). Based on investigation on human cystic echinococcosis Andrabi *et al.* (2020), out of 458 samples, 20 (4.36 per cent) samples, have been found positive for seroprevalence in South Kashmir.

(ii) Alveolar or Multilocular Hydatid Disease (Echinococcosis) in Human

The adult worms pertaining to multilocular or alveolar hydatid cysts of *E. multilocularis*, range from 3-5 mm long, are commonly found in liver of man. *E. multilocularis* causes disease in human known as polycystic echinococcosis. The definitive hosts of this parasite are: fox, cats, dogs and wolves, *etc.* The human cases of multilocular hydatid disease result when fruit and vegetable contaminated by faeces, containing eggs from infected canines are consumed by human or water contaminated by egg-containing faeces may be another source of infection. Apart from this, the intimation of man with dogs and cats can also result into human infection.

The growth and proliferation of *E.multilocularis* are extreamely slow. The larva of this parasite is sponge like, constantly growing mass of small separate vesicles embedded in a fibrous tissues. The vesicle contains a gelatinous substance instead of fluid. Roots grow out into neighbouring tissues. The central portion degenerates and dies, while growth continues on the outside. This result in a cluster of comparative smaller cysts that metastize through the surrounding tissue. In this type of cyst, there is neither protoscoleces nor hydatid sand and this type of hydatid cyst is known as multiloular or alveolar cysts, developing principally in liver. If the development of cyst reaches such a stage that they cannot be removed surgically, in that case death commonly results.

In Kashmir valley, alveolar hydatid disease has been reported from Uri region by Khuroo *et al.* (1980). The patient having multicellular hydatid cyst was also found to be suffering from Budd-Chiari syndrome. This syndrome occurs due to thrombosis of hepatic veins and often the adjacent vena cava. Also, in this syndrome, the liver becomes tender, swollen and ascites, with portal hypertension and esophageal varices. This is usually a serious disease and obstruction can be removed surgically, to prevent death. During the course of investigation on hydatid disease between 6th Jan, 2017 to 22nd May, 2019 in Kashmir region, 6 patients were found to be suffering from alveolar hydatid disease (Jehangir *et al.*, 2020). This finding suggested the prevalence of the disease on the basis of imaging and histopathological examination in two males and four female patients. The diversity of hydatid cyst infection sites, in human population of Jammu and Kashmir region, reported by different authors, is given in **Table 2.1.1** and **4.1.2**, pp. 97-98.

(b) *Hymenoplepis nana* (Hymenolepiasis)

Hymenolepiasis disease in human is caused by Cestode, *Hymenolepis nana,* belongs to the family Hymenolepididae and is also called as dwarf tapeworm of man, because of the fact that it is the smallest adult intestinal tapeworm found in man. The disease caused by this Cestode parasite is, characterized by malnutrition, diarrhoea and abdominal pain. In severe cases, there may be convulsions, epilepsy and insomnia. In human hymenolepiasis there is no intermediate host and the infection is caused directly from patient to patient. *H. nana* is also parasitic on rodents. The young childrens are sometimes susceptible to infection from murine sources. In the hymenolepiasis, the infection is caused by ingestion of *H. nana* eggs from soiled linen, toilet seat or directly from anus to mouth (oro-faecal) and personal hygiene.

H. nana shows a world-wide distribution, including India and UT J&K. According to Faust *et al.* (1974), this parasite is more common in children than in adults and usually found in family, and Institutional group than in the population; more prevalent in warm than cold climate region.Arora *et al.* (1976), during the course of their study on the incidence of intestinal parasites in a rural community in Lamberi village of district Rajouri (Jammu), found *H. nana* infection in 1.8 per cent of population examined and significantly higher prevalence rate was in the age group below 15 years (29 per cent) than in the age group of 15 years and above (0.5 per cent). The prevalence (4.13 per cent) of *H. nana* in human population of Jammu region, based on hospital record of SMGSH (Jad *et al.*, 2015) and 2.05 per cent in Kashmir (Budgam) (Lone *et al.*, 2019) (**Table 4.12**, p. 98).

(c) *Taenia saginata* (Human taeniasis) and Neurocystocercosis

(i) *Human Taeniasis*

Taeniasis disease in human population in Kashmir region is reported to be caused by cestode, *viz. T. saginata* or beef tapeworm of man, belongs to Cyclophylldean group under family Taeniidae.This cestode parasite is the common and large parasite, showing cosmopolitan distribution, infecting small intestine by adult worms, with their heads embedded in the mucosa of the intestine. The human infection due to *T. saginata* is acquired by eating raw/uncooked and 'measly' or insufficiently cooked beef/or meat of cattle, containing viable cysticercus larvae. Cattle become infected from grazing on ground, polluted through faecal matter of human, containing eggs of parasite. Cattle and allied animals act as intermediate host, and man acts as definitive host of *T. saginata.*

The common symptoms of this disease are abdominal pain, excessive appetite, weakness and loss of weight. Due to large size of *T. saginata,* it is responsible for causing considerable disturbance in the normal functioning of digestive tract. At times, the worms may produce acute intestinal stoppage or sometimes the proglottids enter in the appendiceal lumen and cause appendicitis. In taeniasis, usually the absorbed by-products cause systemic intoxication in the patients.

In Kashmir, Ahmad (1978) during the course of his investigation on the helminth parasites of rural population found 1.5 per cent cases of *T. saginata.* The highest incidence was located in school-going age group (5 – 15 years). Koul *et al.* (2000) reported the prevalence of *T. saginata* in 185 patients (122 males and 63 females), having age from 3 – 60 years from this region. During the period of past six years (2007, 2008, 2010, 2012, 2016, 2017), detailed investigations have been carried out by number of reseachers on the prevalence of *Taenia saginata* infection in children populations(rural, urban), between the age group 4-15 years in Kashmir. The prevalence percentages of *T. saginata,* provided by the previous authors, in human populations, are presented in **Tables 4.1.2**, pp. 98-99.

The prevalence of *T. saginata* infection recorded by various authors in childrens (4-15 years) (urban and rural) in various districts and areas of Kashmir region, showed highest prevalence in Kupwara district as 7.69 per cent, followed by Shopian and Gurez 5.3 per cent each,Srinagar city 3.7 per cent and Sonawari Tehsil (Baramulla) 0.61 per cent (Table 4.1.2). In order to study the safe drug for the treatment of *T. saginata* in Kashmir, Lateef (2008) conducted epidemiological survey between the period 2004-2005 in Sonawari Tehsil of Baramula district and found 55 persons out of 8894 persons, having infection due to taeniasis. The safe drug, nitazonide, a new thiazoide antiparasitic agent was suggested as a safe, effective and inexpensive drug for the trearment of niclosomide and praziquantel resistant *T. saginata* infection in Kashmir region,

(ii) *Neurocystocercosis (Taenia solium)*

The larval cysts (enclosed sac, containing immature stage) of *Taenia solium* (pork tapeworm) infect any part of the human body, causing a condition known as cysticercosis. The larval cyst in the brain, cause a form of cysticercosis, known as neurocysticercosis (NCC). In NCC condition, cysts affecting central nervous system (CNS), include parenchyma, subarachnid space or intraventribulum space. The ocular and spine disease also occur, but is less common. The clinical manifestations of NCC range from asymptomatic to life threating condition, causing headaches, seizurs and active epilepsy

In NCC, humans are the definitive hosts for the parasite and swines are the intermediate hosts. The adult tapeworms develop in the human host after they ingest live cysticercus in undercooked pork. The prevalence of NCC, affecting CNS, causing active epilepsy has been reported from Jammu region only. The Kashmir region is free from NCC infection, because of the fact that the consumption of pork is strictly prohibited due to religious reasons. The study conducted on Neurocysticercosis (NCC) in population of Bhore Pind and Bhore villages in Chattah zone of district Jammu, showed among 25 patients, with symptomatic epilepsy, 10 (40 per cent) had CT/MR evidence of past or recent NCC infection, having active epilepsy in these patients due to NCC (Raina *et al.*, 2012).

Table 2.12: Cestode Parasites of Vertebrates (Fishes, Amphibians, Reptiles, Birds, Mammals) and Humans in Jammu, Kashmir and Ladakh Region: Checklist, Host Diverstiy and Distribution

Host type/Cestode Species	*Host Species*	*Locality(ies)*
FISHES		
Adenoscolex oreini	Fishes: *Schizothorax richardsonii(=Oreinus sinuatus), S.plagiostomus,Salmo trutta, S. esocinus (= S. hugelli), S. curvifrons, S. niger, Triplophysa kashmifrensis (= Nemacheilus kashmirensis, Carassius crassius crassius and Cyprinus carpio specularis.*	Dal Lake, Manasbal Lake, and water bodies in Gurez valley of Kashmir region.
Adenoscolex sp.	Fish : *Schizothorax* sp.	Anchar, Manasbal and Wular lake
Schyzocotyle acheilognathi (=Bothriocephalus acheilog-nathi)	Fishes: *Cyprinus carpio communis, S.niger, S. progastrus, S. plagiostomus, Schizothorax sp.*	River Jehlim-(Chattabal, Srinagar), Dal Lake
Gangesia (G.) fotedari	Fish : *Glyptothorax*	Wular Lake
Gangesia (G.) lucknowi	Fish : *Glyptothorax*	Baramulla
Gangesia sp.	Fish : *Glyptothorax, S. esocinus, S. curvifrons, S. richardsonii, Triplophysa kashmirenis (N. kashmirensis), C. c. specularis*	Kashmir
Gangesia (Vermia) jammue-nsis	Fish : *Wallago attu*	Jammu
Gangesia (V.) kashmirensis	Fish : *Glyptothorax*	Baramulla (Kashmir)
Ligula intestinalis	Fish : Schizothorax	Dal Lakr

Host type/Cestode Species	*Host Species*	*Locality(ies)*
Protocephalus-kashmirensis	Fish : *Triplophysa kashmirensis* (= *Nemacheilus kashmirensis*)	Kashmir
Protocephalus torulosus	Fish : *Triplophysa kashmirensis*	Harvan (Kashmir)
AMPHIBIAN (TOAD)		
Nematotaenia kashmirensis	Amphibian (Toad) : *Bufo viridis*	Srinagar, Baramulla, Sopore, Anantnag, Verinag, Pahalgam and Gulmarg
REPTILE (LIZARD)		
Oochoristica tuberculate	Reptile (Lizard) : *Laudakia tuberculata* (= *Agama tuberculata*)	Kashmir
BIRDS (WILD and DOMESTIC/POULTRY)		
Amoebotaenia cuneate (=*Amoebotaenia-sphenoides*)	Domestic fowl: *Gallus gallus domesticus*	Kashmir
Amoebotaenia domesticus	Domestic fowl	Kashmir
Anomotaenia acrocephali	Birds : *Acrocephalus stentoreus brunnescens* (Great Red Warbler)	Kashmir
Anomotaenia galbulae	Bird : *Corvus moneula, C. macrorhynchos*	Kashmir
Anomotaenia kashmirensis	Bird : *Sturnus vulgaris*	Anchar, Srinagar
Choanotaenia fotedari	Bird : *Acridotheres tristis*	Kashmir
Choanotaenia gondwana	Bird (domestic) : *G. gallus domesticus*	Kashmir
Choanotaenia infundibulam	Bird (wild): *Corvus splwndens*; *Picus squamatus* (Scaly bellied wood pecker); Bird (domestic): *Gallus domesticus*	Kashmir
Choanotaenia microcantha	Bird :*Corvus monedula*	Kashmir
Choanotaenia orioli	Bird : *Orioles oriolus*	Kashmir
Cotugnia diagonophora	Domestic fowl (indigneous)	Kashmir
Davainea proglottina	Domestic fowl (indigenous)	Kashmir
Dilepis undula	Birds (wild): *Corvus monedula, C. splendens, C. macrorhynchus*	Kashmir
Hymenolepis carioea	Domestic fowl (indigneous)	Kashmir
Mayhewia kavini	*Corvus monedula*	Kashmir
Pseudoschistotaenia indica	Bird (wild)Podiceps ruficollis capensis	Kashmir
Pseudoschistotaenia pindchii	Bird (wild):*Podiceps ruficollis capensis*	Kashmir
Raillietina echinobothridia	Domestic fowl	Kashmir
Raillietina cesticillus	Domestic duck, Wild bird, *Picus squamatus*	Kashmir
Raillietina tetragona	Domestic fowl	Kashmir
Raillietina spiralis	*Gallus gallus domesticus*	Kashmir
Raillietina torquata kashmirensis	Bird : Blue Rock pigeon	Kashmir
Raillietina sp.	Domestic pigeon, *Columbia livia*	Kashmir

Host type/Cestode Species	*Host Species*	*Locality(ies)*
MAMMALS : WILD and DOMESTICATED		
Methevotaenia- symmetrica	Rat : *Rattu rattus*	Kashmir
Mesocestoides sp.	Fox : *Vulpus vulpus*	Kashmir
Anopholocephalides	Small ruminants	Jammu
Avitellina centripunctata	Sheep and goats	Jammu and Kashmir
Avitellina (A.) woodlandi	Sheep and goats	Jammu and Kashmir
Avitellina chalmersi.	Sheep and goats	Jammu and Kashmir
Coenurus (cyst)	Yak	Ladakh
Echinococcus	Ovines, bovids and yak	Jammu, Kashmir, Ladakh
Moniezia expansa	Sheep and goats	Jammu and Kashmir
Moniezia sp.	Ovines and yak	Kashmir, Ladakh
Paranoplocephala	Equines	Kashmir
Stilesia globipunctata	Sheep, goats	Kashmir
Stilesia sp.	Cattle, buffalo	Jammu
Taenia hydatigena	Sheep, dog	Jammu and Kashmir
Taenia pisiformis	Rabbit	Kashmir
Taenia sp.	Goat	Jammu and Kashmir
Thysaniezia giardia	Goat	Kashmir
Thysaniezia himalayai	Sheep (local)	Kashmir
HUMANS		
Echinococcus granulosus	Humans	Jammu and Kashmir
Echinococcus multilocularis	Humans	Uri (Kashmir)
Echinococcus sp.	Humans	Kashmir
Hymenolepis nana	Humans	Jammu
Taenia saginata	Humans	Kashmir

Source: **(i) Fish Cestodes**: Ahmad and Chishti (1998); Ahmad *et al.* (2018); Ara *et al.* (2000); Chishti and Ahmad (2002); Datta (1936); Dhar and Fotedar (1980); Dhar and Majadh (1983, 84, 87); Dhar and Peerzada (1992); Dhar and Raina (1983); Fotedar (1958); Gupta (1967); Majadh and Khan(1996); Wali *et al.* (2016 a). **(ii) Amphibian (toad) Cestodes**: Fotedar (1966). **(iii) Reptilian (lizard) Cestodes**: Raine *et al.* (1973, 77). **(iv) Bird Cestodes: (a) Domestic/Poultry birds:** Chishti (1980); Dar *et al.* (2008); Dar and Tanveer (2013); Fotedar and Chishti (1974); Fotedar and Khateeb (1987); Kamil *et al.* (2011); Mehmood *et al.* (2019) and Pandit *et al.* (1991); Salam *et al.* (2009); Tanveer and Chishti (1999). **(b) Wild birds:** Chishti (1973, 82); Chishti *et al.* (1987); Chishti and Khan (1984); Dar *et al.* (2013); Fotedar and Chishti (1973, 74, 76 and 80); Fotedar *et al.* (1970). Nashir-uddullah *et al.* (2016); Salam *et al.* (2008, 2010). **(v) Mammals: (a) Wild mammals (rat, fox, dog):** Raina *et al.* (1978); Pandit *et al.* (2007) and Borkataki *et al.* (2013); Shah *et al.* (2017); Rebecca *et al.* (2014). **(b) Domestic mammals (sheep, goat, bovids and yak):** Fotedar and Bambroo(1978), Makdoomi *et al.* (1995). Ranga Rao (1994); Raina (1973,75); Sudhan andShahardar (1999); Tak *et al.* (2003) and Dar *et al.* (2016). **(c) Human:** Ahmad (1978); Ahmad *et al.* (2001); Arora *et al.* (1976); Chishti *et al.* (2001); Fazili *et al.* (2003); Fomda *et al.* (2002); Pampori *et al.* (2003); Koul *et al.* (2000); Khuroo *et al.* (1980, 91, 93); Tak (2000).

Table 2.13: Systematic Position of Cestode Parasite Genera (Class: Cestoidea) of Vertebrates (Fishes, Amphibians, Reptile, Birds, Mammals, including Human) in Jammu, Kashmir and Ladakh Region

Order	Family	Sub-family	Cestode Genus	Host/s
I. Class: Cestodea; Sub-class Cestodaria				
Caryophyllidea	Caryophyllaeidae	Capingentinae	*Adenoscolex*	Fishes
II. Class: Cestodea; Sub-class, Eucestoda				
Pseudophyllidea	Bothriocephalidae	–	*Bothriocephalus*	Fishes
Pseudophyllidea		Diphyllobothriidae	–	Ligula Fishes
Protocephalidea	Proteocephalidae	Proteocephalinae	*Proteocephalus*	Fishes
Protocephalidea	Proteocephalidae	Gangesiinae	*Gangesia*	Fishes
Cyclophyllidea	Mesocestoididae	Mesocestoidinae	*Mesocestoides*	Fox
Cyclophyllidea	Anoplocephalidae	Linstowiinae	*Methevotaenia*	Rat
Cyclophyllidea	Anoplocephalidae	Thysanosomatinae	*Avitellina*	Sheep and Goat
Cyclophyllidea	Anoplocephalidae	Thysanosomatinae	*Stilesia*	Sheep and goats
Cyclophyllidea	Anoplocephalidae	Thysanosomatinae	*Thysaniezia*	Sheep and goats
Cyclophyllidea	Anoplocephalidae	Anoplocephalinae	*Moniezia*	Sheep and goats
Cyclophyllidea	Anoplocephalidae	Anoplocephalinae	*Anoplocephalides*	Small ruminants
Cyclophyllidea	Anoplocephalidae	Anoplocephalinae	*Paranoplocephla*	Equines
Cyclophyllidea	Anoplocephalidae	Linstoniinae	*Oochoristica*	Lizard
Cyclophyllidea	Davaineidae	Davaineinae	*Cotugnia*	Fowl (domestic)
Cyclophyllidea	Davaineidae	Davaineinae	*Davainea*	Fowl(domestic)
Cyclophyllidea	Davaineidae	Davaineinae	*Raillietina*	Wild and domestic birds
Cyclophyllidea	Dilepididae	–	*Dilepis*	Wild and domestic birds
Cyclophyllidea	Dilepididae	–	*Amoebotaenia*	Domestic fowl
Cyclophyllidea	Dilepididae	–	*Anomotaenia*	Wild bird
Cyclophyllidea	Dilepididae	–	*Choanotaenia*	Wild and domestic birds
Cyclophyllidea	Hymenolepididae	–	*Hymenolepis*	Domestic fowl, Human
Cyclophyllidea	Hymenplepididae	–	*Mayhewia*	Wild bird
Cyclophyllidea	Nematotaeniidae	–	*Nematotaenia*	Amphibian (toad)
Cyclophyllidea	Taeniidae	Taeniinae	*Taenia*	Human
Cyclophyllidea	Taenidae	Echinococcinae	*Echinococcus*	Domestic animal, Human
Cyclophyllidea	Amabiliidae	–	*Pseudoschis-totaenia*	Wild birds
Cyclophyllidea	Dipylidiidae	–	*Dipylidium*	Cat

Note: The cestode genera, listed in the above table, have been classified in the light of system of classification given by Yamaguti (1959/85), Maviaux *et al.*, 2017 and Parasitipedia. Net (website for parasite.

Chapter 3

Parasitic Nematodes (Animals and Humans)

3.1 Prevalence, Species Richness and Host Diversity

Nematodes or roundworm belong to phylum Nematoda, are elongated and cylindrical, tappering at both ends. Nematodes, *viz. Capillaria* spp. are uniform, cylindrical and extremely thin. Very often male nematodes are smaller than females. The nematodes are covered outside by very tough and impermeable transparent or semitransparent cuticle. This cuticle is usually marked from outside by markings, bristles, spines, ridges or expansions of various kinds.The nematode parasite species in J&K and Ladakh Union Territories (UTs), are known to infect not only vetebrates (fishes, amphibians, reptiles, birds and mammals, including humans) but also invertebrates like insects, infecting various sites and systems of these hosts in diverse areas and localities in UTs. The nematode affected sites in these hosts include eye, stomach, intestinal tract, respiratory tract and body cavity, *etc.* The total number of nematode species has been found to be as 4 spp., 9 spp., 5 spp., 4 spp., 19 spp., 14 spp., 51 spp.14 spp. and 5 spp., infecting various host groups like insects, fishes, amphibians, reptiles, birds, mammals, domestic mammals, wildlife and humans respectively, for details **Table 3.3**. These nematode species, parasitic on diverse hosts, belong to as many as 66 genera, under 35 families (**Table 3.1**) and for systematic position of genera *see* **Table 3.2**.

All nematodes pass through four moults during their development and fourth stage larva has been assigned to specific forms of nematodes. Vertebrates acquire their nematode parasites in different ways: (i) the penetration of infective third stage larvae into skin of hosts (*Rhabdias* spp.); (ii) through the ingestion of eggs containing infective larva (Oxyuroidea); (iii) through corpophagy (*e.g.* pinworms); (iv) through the ingestion of larvae in the tissues of intermediate and paratenic hosts (Camallanoids); (v) through autoinfection (Cosmocercoids); (vi) transplacental and transmammary transmission (Ascaridoids).

3.1.1 Insect Nematodes

Earlier records of nematode parasites of insects Kashmir, have been given by Basir (1941). He reported two new nematode species, *viz. Galebiella galebiella* and *G. islamabadi* from the intestine of aquatic beetle, inhabiting hill streams (c, 6,000 ft above sea level) in Anantnag region of Kashmir Valley. Gupta (1967), also included these two species of insect nematodes in his paper on helminthic-fauna of Kashmir.

Description of a new species of nematode; *viz. Pseudonymus multiannulata*; infecting aquatic beetle (*Hydrophilus* sp.) in Srinagar, has been provided by Fotedar (1964). So far only three species of nematodes, distributed over two genera are reported from Kashmir region (**Table 3.3**). In addition to *P. multiannulata*, the other two species of genus Pseudonymus, infesting aquatic insects of Kashmir include *Pseudonymus hydrophili* (= *Galebiella galebiella*) and *Pseudonymus islamabadi* (=*Galebiella islamabadi*) (Gantait and Venkataraman, 2013). Oxyurid nematode, *viz. Leidynema appendiculate*, parasitizing cockroaches (*Blatta orientalis*), has been found in Kashmir (Rehana *et al.*, 2016).So far, from Kashmir region, 4 species of parasitic nematodes, belonging to two genera are reported (**Table 3.3**).

3.1.2 Fish Nematodes

The previous systematic and taxonomic studies on fish-nematodes of J&K, Ladakh, UT, have been done by: Chakravorti (1942); Thapar (1950); Gupta (1967); Fotedar and Dhar (1970); Raina and Dhar (1972); Fotedar and Dhar (1975); Dhar and Fotedar (1979); Dhar and Anwari (1982); Chishti and Bakhshi (1990); Khan *et al.* (2004); Peerzada (2009); Ahmad *et al.* (2018). A total of nine species of nematodes, belonging to four genera of 3 different families (Camallanidae, Rhabdochonidae and Cystidicolidae), under order Spirurida,class Secernetea, has so far been recorded from this region of India. These nematodes have been found to infect intestine of eleven species of fishes, endemic as well as exotic of the UTs.

The studies conducted outside India on development and transmission of nematode parasite infection of fishes, have established that fish acquire cystidicolid infection by ingesting arthopods (Ephemeroptera or crustaceans), containing third stage larva. The Camallanids usually involve copepods (crustaceans) as intermediate host (Anderson, 2000). The various nematode species of different fish hosts include *Camallanus cotti* (=*Camallanus fotedari*), *C. oxycephalus*, *C. solmonae*, *Procamallanus kashmirensis*, *Rhabdochona guptai*, *R. kashmirensis*, *R. himalayai*, *Rhabdochona* sp. and *Spinitectus* sp. The host-fish spectrum of nematode parasites, in the light of recent revised fish species nomenclature, is given as under:

Sl.No.	Fish Species	Nematode Parasite Species
1	*Cyprinus carpio specularis*	*Camallanus. cotti* (= *C. fotedari*), *Rhabdochona guptai R. kashmirensis, R. himaalyai*
2	*Triplophysa kashmirensis* (= *Nemacheilus kashmirensis*)	*C. cotti R. kashmirensis* and *R. himalayai*
3	*Nemacheilus sp.*	*R. himalayai*
4	*Schizothorax richardsonii* (= *Oreinus plagiostomus*) (= *Oreinus richardsonii* (= *Oreinus sinuatus*)	*C. cotti, R. kashmirensis R. himalayai, Spinitectus* sp., *R. guptai and Camallanus oxycephalus*

Sl.No.	Fish Species	Nematode Parasite Species
5	*Schizothorax curvifrons*	*C. cotti, R. kashmirensis* and *R. himalayai*
6	*Schizothorax esocinus* (= *Schizothorax hugelii*	*C. cotti. R. himalayai, R. kashmirensis*
7	*Schizothorax niger*	*C. cotti, R. himalayai R. kashmirensis*
8	*Schizothora plagiostomus*	*C. cotti, Rhabdochona guptai*
9	*Schizothorax sp.*	*C. oxypcephalus, R. himalayaiRhabdochona* sp.
10	*Wallago attu*	*Procamallanus kashmirensis, R. kashmirensis*
11	*Salmo trutto fario*	*Rhabdochona* sp., *Rhabdochona guptai.*
12	*Salmo sp.*	*C. salmoe, P. kashmirensis*

Note: The synonyms of fish species in above cited table, given in the parentheses, are in accordance with the nomenclature given by: Talwar and Jhingran (1991) and Kullander *et al.* (1999).

From the above table, it is evident that the highest number, *i.e.*, 6, of different species of nematode, have been reported in the fish, *viz. Schizothorax richardsonii,* followed by 4 spp. in *Cyprinus carpio specularis* and, 3 spp. each in *Triplophysa kashmirensis, Schizothorax curvifrons, S. esocinus, S. niger* and *Schizothorax* sp., 2 spp. each have been found in fishes such as *Schizothorax plagiostomus, Salmo trutto fario, Wallago attu* and *Salmo* sp. The fish host species, *viz. Nemacheilus* sp., is known to be infected with only one species of nematode parasite.

Detailed investigation, for the first time on fish parasitization by different nematode species, *viz. C. cotti* (= *C. fotedari*) from the intestine of *Schizothorax niger* and *R. himalayai* also from the intestine of *Schizothorax* sp. and *Nemacheilus* sp., has been made by Dhar and Majadah (1987). The studies on the seasonal population of nematode parasites of fishes at two Sites of Wular Lake in Kashmir, *viz.* Ningli and Watlab, have been made by Majadah and Khan (1998), showing nematode parasite infection due to the species like *C. cotti, R. himalayai* and *R. kashmirensis* in 6 spp. of Wular Lake fishes: *S. niger, S. curvifrons* (=*S.hugelii*), *S. esocinus. S. richardsonii* (= *Oreinus plagiostomus*), *cyprinus carpio specularis* and *Triplophysa kashmirensis* (= *Nemacheilus kashmirensis*).

Chishti *et al.* (2002) conducted a detailed investigations on the comparative fish parasite diversity, including nematode diversity in two fresh water lakes of Kashmir as urban eutrophic Dal Lake and rural mesotrophic Manasbal Lake, in relation to ecological factors in different seasons. The fish hosts studied during the course of three years were *S. curvifrons* and *Schizothorax richardsonii* (= *Oreinus sinuatus*). The nematode species recorded were *Camallanus oxycephalus* and *R. guptai.* These authors have found that the nematode infection in *Schizothorax* showed existence in both the waterbodies, showing peak availability towards winter and spring season. The infection due to nematode parasite, *C. oxycephalus* occurred during July and August, with resulting peak in population density during late summer and autumn. Further, the parasite richness revealed that fishes of Dal Lake were heavily parasitized due to perhaps eutrophication.

The population study of nematode, *Camallanus cotti* (= *Camallanus fotedari*) in fish, *Triplophysa kashmirenisis* (= *Nemacheilus kashmiensis*) in Lidder Stream showed 39.5 per cent prevalence of parasite, with maximum infection, 61.3 per cent in October and high infection in full grown fishes. In water bodies of Gurez valley, fish (*Salmo trutta*) was found to be having 19.52 per cent of prevalence due to nematode, *Rhabdochona gupti* in summer season (Ahmad *et al.*, 2018). Apart from this, these authors also showed the water bodies of this region, inhabiting fish, *Schizothorax plagiostomus*, having 22.22 per cent and 18.25 per cent prevalence of infection during summer due to *Rhabdochona gupti* and *Camallanus cotti* respectively.

3.1.3 Amphibians and Reptilian Nematodes

In Kashmir region, amphibian *viz.*, *Bufo viridis* (toad) has been found to be parasitized by four species of nematodes in intestine part of alimentary canal. These species are: *Cosmocerca crenshawi*, *C. commutata* (= *C. kashmirensis*) (cosmocercidae). *Oswaldcruzia kashmirensis* (Molineidae); *Oxysomatium srinagarensis* (Oxyuridae). Besides these species, *Rhabdias bufonis* (Rhabdiasidae), is reported to be as a common lung parasite of toad in Kashmir Valley (**Table 3.3**).

Two snake nematode species: *Kalicephalus indicus* and *K. kashmirensis*, have been found in the stomach and oesophagus of *Topidontus platyceps* and *Ptyas mucosus* respectively in Kashmir Valley (Fotedar *et al.*, 1966). These nematode parasites are belonging to the order Strongylida (the bursate nematodes) and family Diaphanocephalidae. In addition to these, two more nematode species of reptile, *viz. Thelandros baylisi* and *Thelandros jummensis*, belonging to the order Oxyurida family Pharyngodonidae, have been found in J&K UT. *T. baylisi* has been reported to parasitize lizards: *Laudakia tuberculata* (= *Agama tuberculata*) and *Calotes versicolor* in Jammu and Kashmir, whereas, *T. jammuensis* is found to be parasitic on lizard, *Hemidactylus flaviviridis* in Jammu region (Fotedar, 1968) (**Table 3.3**).

3.1.4 Birds (Wild and Poultry) Nematodes

A total of 19 species of nematodes, belonging 13 genera, infecting birds (wild and poultry), is known to be prevalent in Jammu and Kashmir region (**Table 3.3**). These nematode species in wild bird species are *viz.,. Acuaria skrjabini,* (Acuariidae), *Capillaria anatis* (Capillaridae), *Contracaecum milvi* (Anisakidae), *Diplotriaena buckelyi* (Diplotriaenidae), *Rusguniella turdi* (Acuariidae), *Syngamus trachea* (Syngamidae), *Amidostomum fulicae, Epomidiostomum uncinatum, E. anatinum* (*Trichostrongylidae*), have been reported from Kashmir, infecting wild birds such as *Passer domestica, Milvus migrans lineatus, Acridotheres tristis, Turdus unicolor, Corvus monedula, Corvus splendens, Fulica atra atra* and *Anas platyrhynchos* respectively. The site of infection due to these nematode species included gizzards caecum, body cavity, stomach and trachea, *etc.* (**Table 3.3**).

There are about 9 species of nematode parasites, known to infect poultry birds like domestic fowl (indigenous), *Gallus domesticus* and *Anser domesticus* in Kashmir Valley. The locations of infection by these nematode species have been found as: small intestine, crop, oesophagus, gizzard, caecum and trachea. The nematodes of poultry importance of Kashmir include *Ascardia galli* (Ascaridiidae), *Capillaria annulata, C. caudinflata* (Trichuridae), *Echinuria uncinata* (Acuariidae), *Heterakis gallinarum, H. indica* (Heterakidae), *Sterptocara indica* (Acuariidae) and *Syngamus trachea* (Syngamidae),*Amidostomum anseris* (Trichostrongylidae) and *Tetrameres fissipina* (Tetrameridae) (**Table 3.3**).

Detailed investigations on prevalence and seasonal variations of nematode parasites of domestic fowl (*Gallus*) (indigenous), have been done by Fotedar and Khateeb (1987) and Pandit *et al.* (1991). The former authors studies about 140 local fowls, collected from different localities and areas of Kashmir Valley and thoroughly examined their gizzards, small intestine, colon and caecum, and reported the existence of two common intestinal (small) and caecal nematode parasites as *Ascaridia galli* and *Heterakis gallinarum* (= *H. gallinae*) respectively, of poultry importance. The above authors have shown the highest incidence of infection in the month of September and lowest in the month of December and January. It was also reported by them that the incidence and mean worm burden indicated a decrease with decreasing temperature and rainfall.

Pandit *et al.* (1991) studied the prevalence of helminth parasites, including nematodes of indigenous fowls of Kashmir Valley after examining about 80 birds and it was found that about seven species of nematode parasites cause much damage to poultry in this region and they also become susceptible to other diseases. The seven species of nematodes found by them, causing infections in different body sites/systems of birds (domestic fowl), with their rate of infection were: *A. galli* (small intestine), 28 per cent; *H. gallinarum* (caecae), 22 per cent; *Syngamus trachea* (trachea), 9 per cent; *Heterakis indica* (caecae), 8 per cent; *Capillaria annulata* (crop and oesophagus), 4 per cent; *C. caudinflata* (small intestine), 2 per cent and *Echinuria uncinata* (gizzard), 1 per cent. The most infections reported was single, however, mixed infections with two or three species was not uncommon. The occurrence of these parasites was described by these authors as greatly influenced by climatic and ecological factors of the region.

The outbreak of fatal duck enteritis due to nematode, *Tetrameres fissipina* in ducks reared on the banks of Wular lake in Bandipore region, has been witnessed and the disease was characterized by 100 per cent morbidity and massive mortality of 45 per cent (Kamil *et al.*, 2011)

3.1.5 Rat, Shrew, Bat, Cat and Dog Nematodes

In Kashmir Valley, two species of parasitic nematode, *viz., Heterakis spumosa* (=*Ganguleterakis spumosa*) and *Trichocephalus muris*, infecting intestine and stomach respectively, of common rat (*Rattus rattus*), have so far been reported. The domestic cat (*Felis domestica*) has been found to be infected in this Valley by a cosmopolitan nematode parasite of felines, *viz. Toxocara cati* (= *T. mystax*), *Ancylostoma caninum* and *Ancylostma* sp (**Table 3.3**). In Jammu region, *Toxocara* nematodes- *T. canis* and *Toxocara* sp., have been found to infect dogs and cat respectively (Shah *et al.*, 2017; Borakataki *et al.*, 2013). *Toxascaris leonina* is found to be parasitizing carnivores (dog) in Ladakh region (Rebecca *et al.*, 2014). In Kashmir region, *T. canis* infection was detected in stray dogs by examining faecal deposits and out of 968 faecal samples,

188 samples of stray dogs were positive for *T. canis* (Dar *et al.*, 2014). In this region, *T. canis* is posing public health risk.

A detailed investigation, based on molecular diagnostic techniques on canine microfilariae in dogs in Ladakh, has revealed a novel species of *Acanthocheilonema* and *Microfilaria auquieri* (Rani-Puteri *et al.*, 2010). These microfilarial nematodes are belonging to the family Onchocercidae. The adult worms live in tissues and circulatory system of vertebrate hosts. They release microfilariae into blood system. In dogs, *Acanthocheilonema* develop into worms in sub-cutaneous tissues, resulting in skin nodules. The microfilariae are transmitted by blood-feeding arthropod vectors. dog fly, *viz. Hippobosca longipennis*, has been reported to be as a potential intermediate host of species of *Acanthocheilonema* in north India (Rani-Puteri *et al.*, 2011). The filarial worms in dogs are considered as zoonotic in nature and have potential to infect humans, hence are important from the point of view of public health.

The capillariinid nematodes: *Capillaria minuta* and *Capillaria* sp., have been found to infect oesophagus and stomach of shrew (*Crocidura* sp.) of Kashmir region. In addition to these, *Capillaria kashmirensis* is reported to infect the stomach of bat, *Myotis muricola* (Vespertilionidae) from Anantnag district of Kashmir (**Table 3.3**).

3.1.6 Ovine and Caprine Nematodes

In Jammu and Kashmir and Ladakh UTs, a total of about 50 species of nematodes, belonging to 31 genera, has been found to infect domestic mammals, *viz.* bovids, equines, ruminants, yak and cattle, *etc.* (**Table 3.3**). These nematodes commonly infect gastro-intestinal tract, abomasum and lungs of sheep, goat and lambs, have so far been recorded from various regions and localities of Jammu, Kashmir and Ladakh region. These nematode species are belonging to 18 genera, distributed over 8 different families. The ovine and caprine nematode species, covering many families, of theses regions are: Ancylostomatids – *Bunostomum triganocephalum, Gaigera pachycelis, Monodontus alatus;* Ascarid- *Ascaris lumbricoides;* Chabertiids - *Chabertia himalayana, Chabertia ovina, Oesophagostomum columbianum;* oxyurid- *Skrjabinema ovis;* Protostrongylids -*Muellerius* sp, *Protostrongylus rufescens Varestrongylus pneumonicus Varestrongylus* sp.; Strongyloidid- *Strongyloides* sp; Trichostrongylids– *Dictyocaulus filaria, Haemonchus contortus, H. C. kashmirensis, Haemonchus* sp., *Marshallagia marshalli,Marshallagia* sp., *Trichostrongylus axei, T. colubriformis, T. trichostrongylus* sp., *Ostertagia* sp., Molineids- *Nematodirus fotedari, N spaththiger;* Trichurid-*Trichuris ovis* and *Trichuris* sp.

The earliest detailed studies on gastro-intestinal nematode parasites in sheep of Kashmir, have been done by Dhar *et al.* (1982). They have revealed a total of 8 species of nematodes belonging to genera, *viz. Bunostonum, Chabertia, Haemonchus, Nematodirus, Skrjabinema, Trichostrongylus* and *Trichuris,* infecting abomasum and intestinal tract of both sexes of sheep. The infection with *Trichostronglus axei* and *T. colubriformis were* 57.7 and 40.2 per cent respectively. Of the total nematode species recovered., the most common infection by *Trichostrongylus* spp. together accounted for 97.9 per cent of the total worms recovered at necropsy. It was further, shown by the above authors that lambs in Kashmir region start building up heavy infection from May onwards reaching maximum levels during September.

Studies on population kinetics, development of juvenile stages and histopathology of bursate nematodes of Kashmir, *viz. Oesophagostomum columbianum* and *Chabertia himalayanum*, common intestinal parasites of sheep and lambs in Kashmir, have been accomplished by: Bali and Fotedar (1974); Sharma and Dhar (1978, 79) and Fotedar and Wali (1984). *O. columbianum*, commonly known as nodule worm, reported to cause disease known as oesophagostomomiasis. The larvae of nematode parasite of *O. columbianum*, cause lesions in the intestinal mucosa and nodules like dense capsules are formed out of these lesions. This disease can lead to chronic diarrhoea, the animal become weak and may even die. The adult worms of *Chabertia* attack intestine of sheep.

Baqui and Shah (2001), reported the incidence and intensity of nematode Species such as: *Haemonchus controtus, Marshallegia marshalli, Dictyocaulus filariae, Trichuris ovis* and *Bunostomum triganocephalum*, in ovines of Kashmir, during January to December 2002. They also studied the alteration of haematological parameter in ovine nematodiasis, such as Hb per cent. EST, PCU, MCH, Eosinophil and RBC counts, in relation to worm burden and faecal egg counts. The nematode infection exerted a great impact on the haematological picture of ovines nemtodiasis.

An abattoir survey conducted by Tariq *et al.* (2008) on gastro-intestinal helminth parasites of sheep (brought from out side Kashmir), slaughtered in Hazaratbal locality of Srinagar, has revealed the infestation of 12 nematode species and these were, with their prevalence percentages as: *Haemonchus* (79 per cent); *Bunostomum* (76 per cent); *Trichuris* (68 per cent); *Nematodirus* (66 per cent); *Trichostrongylus* (58 per cent); *Strongyloides* (69 per cent); *Gaigeria* (52 per cent); *Oesophagostomum* (42 per cent); *Chabertia* (56 per cent); *Ostertagia* (48 per cent); *Marshallagia* (67 per cent). *Gongylonema* (37 per cent) and *Skrjabinema* (36 per cent). The maximum prevalence was observed from the month of August to November (2006).

3.1.7 Bovid Nematodes

In Jammu region, the incidence of nematode parasites, *viz. Parafilaria bovicola*, infecting bullock and *P. sahii*, parasites of buffalo, belonging to the family Filariidae, sub-family Filariinae, were reported by Khajuria (1966). These nematodes were responsible for causing haemorrhagic filariasis in bovids. The disease caused formation of sub-cutaneous nodules, mostly on the neck or back regions and would break open and bleed profusely for months together. The rate of infection was found high during summer and rainy season. Bhadwal and Raina (1994), reported the existence of nematode parasite – *Thelazia* (Thelazioidae: Thelaziidae: Thelaziinae), causing disease in buffalo in J&K, known as thelaziasis. In this disease, the parasite affect eye orbit under the lids, conjunctiva, nictitating membrane, lachrymal glands and ducts of domestic mammal.

The bovine parasitic brochitis disease in bovines of this region is caused by a single species of lungworm parasite, *viz. Dictycaulus viviparous*. The parasite mostly affected the young animals and disease showed high prevalence and intensity during autumn and winter months (Dhar, 1985). Nematode, *viz. Setaria digitata* in cow in Jammu region has been found (Kaur *et al.*, 2015). Toxocariasis due to *Toxocara vitulorum* in a suckling calf in Pulwama district of Kashmir has been reported by Ahmad *et al.* (2016).

3.1.8 Equine and Yak (*Bos grunniens*) Nematodes

Ranga Rao (1994), studied the nematode parasites of Indian Yak (*Bos. grunniens*) in Ladakh region (UT).They found 10 spp. under 10 genera of 4 families, as gastro-intestinal nematode parasites and 1 sp. from peritoneum of Yak. These species were: *Ascaris* sp. (Ascarididae); *Bunostomum* sp., *Chabertia* sp. (Chabertiidaee); *Cooperia, Haemonchus controtus, Mecistocirrus digitatus, Nematodirus* sp. *Ostertagia* sp. (Trichostrongylidae); *Setaria cervi* (Setariidae); *Toxocara vitulorum* (=*Neoascaris vitulorum*) (Ascarididae) and *Trichostrongylus* sp. (Trichostrongylidae). Excepting *S. cervi,* found in peritoneum, all other nematode species were found to be infecting the gastro-intestinal tract of *B. grunniens* (**Table 3.3**).

The equines in Jammu and Kashmir, have been found to infected with a total of 11 species pf nematode parasites, distributed over nine genera, *viz. Cyanthostomum, Dictycaulus, Oxyuris, Parascaris, Setaria equina, Setaria digitata, Strongyles, Strongyloides, Strongylus* and *Triodontophorus*. 3 species of filarid nematodes, *viz. Setaria digitata, Setaria equina* and *Thelazia* sp., harbouring peritoneal cavity of horses and invading central nervous system and also eyes, causing ocular setariasis (= setariosis) and thelaziasis or ocular filariasis in Jammu and Kashmir has been observed (Bhadwal and Raina, 1994;Yadav *et al.*, 2006; Parrah *et al.*, 2004) (**Table 3.3**).

3.1.9 Hangul (Kashmir Stag), Black Bear and Fox Nematodes

The unique and endangered species of mammalian wildlife, *viz. Cervus elaphus hanglu,* has been found to be infected by ten common nematode parasites of gastro-intestinal tract. Earlier, Shahardar *et al.* (1995), for the first time reported the prevalence parasites in Hangul wildlife of Kashmir and have identified these parasitic species of the nematodes upto only genus level. These included *Haemonchus, Marshallagia, Nematodirus,Oesophagostomum,Ostertagia, Strongyloids* and *Trichostrongylus.* The nematode parasites of Hangul (*Cervus elaphus hangul*) in Kashmir so far reported are: *Chabertia ovina, Dictyocaulus ovina, Haemonchus* sp., *Marshallagia* sp., *Nematodirus* sp. *Oesophagostomum cicumcincta, Oesophagostomum* sp. *Stongyloidus* sp., *Trichostrongylus* sp. and *Trichuris* sp. These nematodes are responsible for causing ill health in the wild animal besides, animal becomes susceptible to other diseases (**Table 3.3**).

Dictyocaulosis (lung worm disease) due to *Dictyocaulus viviparus* showed prevalence of 28.40 per cent in young and male of Hangul in Dachigam National Park of Kashmr (Lone *et al.*, 2014. Haemonchosis infection due to nematode, *Haemonchus* sp., has been found in Hangul, with 55.39 per cent of prevalence (Lone *et al.*, 2014). Prevalence of 4.2 per cent of chabertiasis, 13.7 per cent of oesophagotomiasis and 39.75 of trichuriasis infection, due to nematode, *Chabertia ovina, Oesophagostomum cicumcincta* and *Trichuris* respectively, was observed in Hangul (Lone, 2014). In the Himalayan Black Bear (*Ursus thibetanus laniger*) in Kashmir region, Sheikh *et al.*, 2017) found 37.50 per cent prevalence of ancylostomiasis infection due to parasite, *Ancylostoma* (hookworm). In addition to hookworm, these authors have also reported the presence of nematode, *viz*. *Balysascaris transfuga*, in the Himalyan black bear with prevalence rate as 47.32 per cent. Pandit *et al*. (2007), reorded the prevalence of nematode, *viz*. *Physaloptera* in the stomach of fox (*Vulpus vulpus*) (**Table 3.3**).

3.2 Widely and Highly Prevalent Nematode Parasite Diseases of Veterinary Importance

3.2.1 Ascariasis (Sheep)

The high incidence of ascarid infection in sheep in Handwara district of Kashmir due to nematode parasite, *Ascaris lumbricoides* was noticed by Dhar *et al*. (1983). This finding was further supported from the information collected by the authors from medical hospital, which confirmed high incidence of human ascariasis at Handwara region and in the adjoining area. Sheep actually acted as an important agent for transmitting the human Ascarid infection in the area.

3.2.2 Hookworm Diseases (Ancylostomiasis, Bunostomumiasis and Gaigeriasis)

The canine hookworm disease, Ancylostomiasis in this region is caused by nematode, *Ancylostoma caninum*. This parasite infects small intestine of dogs and cats occasionally.

A.caninum larvae my migrate to human intestine. This blood sucking parasite, in the fourth stage can cause anaemia and inflammation in the dog's small intestine. The prevalence of 40 per cent (Shah *et al*., 2017) and 80 per cent (Borkataki, 2013), of infection due to *A. caninum* in dogs and cats respectivey, has been found in Jammu region. In dog population of Ladakh region, 4.7 per cent prevalence of *Ancylostoma* was encountered (Rebecca,2014).

The hookworm disease, Bunostomiasis, caused by *Bunostomum trigonocephalum* and *Bunostomum* sp., has shown wide prevalence in small ruminants of Jammu and Kashmir. In these animals, the parasite causes anaemia and anasarca due to blood loss. An average 13.96 per cent of prevalence of bunostomiasis in small ruminants of Jammu and Kashmir is found, based on the prevalence percentage data provided various authors between the year 2010-2013 (**Table 4.1.3**, p. 99). Gaigeriasis is also a hookworm disease, caused by *Gaigeria pachyscelis* and *Gaigeria* sp., found in sheep of Kashmir region. *Gaigeria* spp. occur in the intestine, duodenum of sheep and goats, and are voracious blood suckers. In Kashmir region, prevalence of 3.33 per cent and 1.81 per cent of gaigeriasis found in sheep, according to Wani *et al*. (2011, 2012).

3.2.3 Lungworm Diseases (*Dictyocaulus* spp.) (Dictyocaulosis); *Mullerius*, *Protostrongylus* and *Varestrongylus*

(a) Lungworms of Ovines

The various species of lungworms, infecting domestic ovines, occurring in different localities and regions of Kashmir and Ladakh are: *Dictyocaulus filaria* (Trichostronglyloidea: Trichostromgylidae); *Muellerius* sp., *Protostrongylus rufuscens*, *Varestrongylus pneumonicus* and *Varestrongylus* sp., belonging to Super family Metastrongyloidea and family Protostrongylidae.

Dictyocaulus harbours the bronchial tubes of sheep, are thread like worm and having length of several inches. These worms cause coughing and bronchitis in hosts, and may block off so much of the lungs as to be fatal. *Muellerius* is a cosmopolitan nematode parasite of sheep and goat, and attacks the parenchyma of lung of these domestic animals. *Protostrongylus rufuscens*, usually occurs in the trachea, bronchi and bronchioles of sheep and goat. This lungworm is also called as red-lungworm, because of its colour due to ingested blood. *P. rufuscens* is pathogenic and its infection can lead to fatal pneumonia. The species of the genus *Varestrongylus,* are parasites of lungs of domesticated sheep and goats, infecting bronchi and bronchioles of these animals of veterinary importance.

In various districts and tehsils of Kashmir and Ladakh, detailed investigations on the prevalence, distribution, pathology and control of lungworm infections in sheep and goat, have been conducted by: Dhar *et al.* (1972, 79); Dhar and Sharma (1978, 78a, 79); Dhar (1980, 82,85); Bansal *et al.* (1982) and Sharma (1994). In Kashmir Valley, Dhar *et al.* (1972), were the first authors to study in detail the prevalence of lungworm diseases in sheep. For this purpose, they selected the Animal Farms of Government as well as those belonging to private breeders. The areas of study included various localities of north-east parts of Kashmir, *viz.* Dachigam, Shuhama, Goabal (Kangan), Chervan and Gund. It was reported that lungworm, *Dictyocaulus filaria* infected 21.3 per cent of lungs and *Varestrongylus* sp. 29.5 per cent of the lungs examined. Mixed infections in 6.6 per cent of the lungs, having *D. filaria* and *Varestrongylus* sp., was also encountered. Apart from this, infection with *Varestrongylus* was found to be more common as compared to *D. filaria.* Further studies on the prevalence of infection due to *D. filaria* in Kashmir were conducted by Dhar and Sharma (1978, 78*a*). They reported the prevalence of infection to be more in goat than in sheep. Later, Dhar *et al.* (1979) showed the occurrence of another lungworm disease caused by nematode, *Muellerius* sp. in sheep of Kashmir Valley.

The epidemiological studies on lungworms in Kargil region of Ladakh, have been carried out by Dhar and Sharma (1979). Their study has revealed higher prevalence of lungworm infection in goats than in sheep. The ovines (domesticated) of the Pandras region, showed highest prevalence of infection than those of the Government Farm at Khumbathang. The common lungworm of sheep and goat in Kargil, reported by Sharma (1994), included *D. filaria, Protostrongylus rufescens, Varestrongylus pneumonicus* and occasionally *Muellerius* sp. The prevalence of lungworm disease in ovines due to *Dictyocaulus viviparous* nematode parasite, for the first time was reported from Lolab area of Kashmir valley by Sharma and Dhar (1987).

In a comprehensive review paper on bovine and ovine parasitic bronchitis, with particular reference to the research work done in India, including earstwhile J&K State, Dhar (1985) has concluded that lungworm disease (*Verminous pneumonia*) is very economically important disease of cattle and sheep in this region. Further, according to this author, sheep and goat are known to be infected by more than one species of lungworm parasite, more sheep of Private breeder as compared to Governemrnt Farms, have been found to be infected with lung worms and, autumn and winter months showed high incidence and intensity of lung worm infection in the animals, mostly young ones.

In Jammu, Kashmir and Ladakh region, lungworm infection due to *Dictocaulus* nematodes, has been found in sheep, goats, equines and wild mammals. Lungworm disease or dictyocaulosis is caused by nematodes, *viz. Dictyocaulus filaria* and *D. viviparus. D. filaria* is a large lungworm of sheep and goats, residing mainly in small bronchi. The clinical symptoms due this nematode include persistent cough and moderate dyspnea. *D. viviparus*, maily affects cattle and bovine. This parasite causes parasitic pneumonia and bronchiolitis. The adult worms reside in trachea and bronchi.

In the recent years, the prevalence of dictyocaulosis due to *D. filaria* in small ruminants of Jammu, Kashmir and Ladakh region has been investigated by various authors (**Table 4.1.3**, p. 100). The prevalence percentage found as: 16.36 per cent in sheep of Kashmir, 32.24 per cent in pashmina goats in Kargil; 3.52 per cent in goats of Jammu and 37.5 per cent in goats of Ladakh (**Table 4.1.3**, p. 100). Apart from this, *Dictyocaulus* sp. affecting Equines in Kashmir, with 14.10 per cent prevalence, has been reported by Pandit *et al.* (2008).

3.2.4 Nodular Worm Disease (Oesophagostomiasis) (*Oesophagostomum* spp.)

Nodular worm disease is caused by nematodes, *Oesophagostomum* spp. (*O. cicumcinata* and *O. columbianum*) in Jammu and Kashmir region, are parasites of sheep and goats, besides wild mammal (hangul). *Oesophagostomum* spp. cause necrotic nodules in the walls of intestines of its infected hosts, resulting in poor condition, passage of soft droppings, with copius mucus, which can lead to serious problems, such as dysentery. An average 70.56 per cent of oesophagostomiasis due to *O.columbianum* has been calculated in sheep and goats of Kashmir region, on the basis of data on prevalence studies, provided by varius authors during the year 2010-2011 (**Table 4.1.3**, p. 102).

In Jammu region, the oesophagostomiasis due to *Oesophagostomum* sp.in small ruminants showed an average 7.92 per cent of prevalence as per prevalence data provided by authors in the year 2013 and 2016. In Kashmir region, *Oesophagostomum* sp. showed 28 per cent (average) prevalence in sheep and goas, based on prevalence data made available by vavious authors during the year 2015, 2011, 2012 (**Table 4.1.3**, p. 102).

3.2.5 Brown Stomach Worms: *Ostertagia* (Oestertagiasis) and *Teladorsagia circumcincta*

Ostertagiosis infection due to nematode, *Ostertagia* has been reported by many workers in sheep and goats of Jammu and Kashmir region. *Ostertagia* is also called as brown stomach worm, it mainly infects gatro-intestinal region of cattle and is potentially fatal to these animals. In type I ostertagiosis, there is diarrhea and a reduced weight loss. In type II ostertagiosis, there is emergence of large number of inhabited larvae from the bomasal mucosa of chronic diarrhea and there may be progressive weight loss. The ostertagiosis prevalence data of small ruminants in Jammu and Kashmir provided by various authors through their research publications, pertaining to 5-year duration, is given **Table 4.1.3**, p. 102. On the basis of this data, the overall average 21.46 per cent of prevalence of ostertagiosis infection in ruminants, predonantly females, in Jammu and Kashmir regin is being established.

Teladorsagia cicumcinata (= *Ostertagia cicumcinata*) is also known as brown stomach worm. This parasitic nematode of sheep and goats infects the gastric glands of stomach, causing weight loss and there is loss of wool production and death. This economically important nematode of veterinary importance has been reported from Kashmir, having 29.8 per cent of prevalence in goats (Tariq *et al.*, 2010).

3.2.6 Haemonchosis (*Haemonchus contortus, Haemonchus* sp.)

Haemonchus contortus and *Haemonchus* sp. or Red stomach worm or wire worm or Barber's Pole worm, infect ruminants, cattle, buffalo and wild mammal (Hangul) in J&K, and Ladakh UTs. These nematode parasites infect abomasum, remaun attached to abomasal mucosa and feed on the blood of animals. The disease caused by these parasites is known as haemonchosis, showing clinical signs as anaemia, oedema and death of infected animals. The prevalence of haemonchosis in domestic cattle and other animals, has been investigated by different authors from time to time, in Jammu, Kashmir and Kargil (Ladakh) regions, showing wide and high prevalence of haemonchosis. The details of prevalence percentage and high infection rate in age group and in gender due to haemonchosis in animals of veterinary importance in UTs, are presented in **Table 4.1.3**, p. 101. On the basis of data recorded by different authors, through publications covering period of 7 years (2010-2016) of haemonchosis infection in domestic animals, belonging to Jammu, Kashmir and Ladakh region (**Table 4.1.3**, p. 101), the average prevalence of this disease, has been calculated as 51.25 per cent, 42.47 per cent and 41.67 per cent of haemonchosis infection in domestic animals of Kashmir, Jammu and Ladakh respectively.

3.2.7 Chabertiasis (*Chabertia ovina, Chabertia* sp.)

Chabertia (Large-mouthed bowl worm) shows high prevalence in Jammu, Kashmir and Ladakh region, causing disease, known as Chabertiasis in sheep, goats, cattle, buffalo and wild mammal (Hangul). The symptoms of this disease include weight loss, passage of soft faeces, containing much mucus. The wide prevalence, with distribution, high infection rate in age group and gender of chabertiasis, affecting domestic animals in UTs, has been reported through number of research publications of various authors, pertaining to as many as 8 years (2010-2017) (**Table 4.1.3**, p. 100). On the basis of published prevalence percentage data, covering 7 years (2010-2013, 2015-2017), of chabertiasis infection in aminals of veterinary importance, the average prevalence is found to be as 16.49 per cent (**Table 4.1.3**, p. 100).

3.2.8 Nematodiriasis (*Nematodirus spathiger* and *Nematodirus* sp.)

Neatodiriasis infection is due to *Nematodirus* parasitic nematodes, live in small intestine of sheep and goats. These small worms cause damage and inflammation of the gut, resulting in persistent diarrhea and wasting. Most of damage is caused by the immature worms in small intestine. In Kashmir region, prevalence of 25.2 per cent of nematodirasis due to *N. spathiger* in goats, is found (Tariq *et al.*, 2010). The high prevalence of *Nematodirus* sp., causing nematodiriasis in small ruminants, particularly sheep in Kashmir region, reported by various authors during the year 2007, 2011. 2012. 2015 and 2018 (**Table 4.1.3**, p. 102). On the basis of published prevalence percentage data of these authors, an average 56.24 per cent prevalence of Nematodirus infection in small ruminants is calcutated (**Table 4.1.3**, p. 102).

3.2 9 Strongylosis (*Strongyles/Strongyls, Strongylus, Triodontophorus*) and Cyathostomiasis (*Cyathostomum*)

Infection with parasitic nematodes, *viz. Strongyles* or *Strongyls, Strongylus* and *Triodontophorus,* is known as Strongylosis. *Stongyles* are parasites, residing in gatro-intestinal tracts of domesticated mammals like sheep, goats, cattle, live-stock, pigs and horses. This parasite is commonly found in horses. The damage caused by this parasite to the animals is due to loss of nutrients or blood resulting in serious health problems.

The prevalence studies on Strongyles infection in domestic mammals of Jammu, Kashmir and Ladakh region, made by various workers from time to time, encapsulated in **Table 4.1.3**, p. 103, 105. On the basis of prevalence percentage data given in this table, the average prevalence of strongylosis due to Strongyles in domestic animals of Kashmir and Jammu, is found to be 40.50 per cent and 46.78 per cent respectively. Besides this, the 8.57 per cent prevalence of Strongyles infection is found in live-stock of Kargil (Ladakh) (Wani *et al.*, 2013).

Strongylus sp. (Red worm or blood worm) is large strongyles of horses, causing disease known as equine strongylosis. The symptoms of this disease include diarrhea, weakness, anorexia, anemia and in severe cases, there may be rupture of intestine and death. Equine strongylosis from Kashmir region, with prevalence 81.19 per cent has been found by Pandit *et al.* (2008).

Triodontophorus, is a large non-migratory strongyles, causing strongyloisis, found in colon. The clinical signs of this disease are: significant damage to mucosal surface, leading to anaemia. The prevalence of 41.39 per cent of this nematode parasite has been found in equines of Kashmir region (Pandit *et al.*, 2008).

Cyathostomum (= *Trichonema*), also known as *Cyathomins* or small red worm, is a parasitic nematode of horses. This nematode predominantly reside in colon and shows clinical signs like lethargy, sudden weight loss, debilitation and diarrhea. The larval stages of *Cyathostomum* can cause even more serious problems, leading to condition, known as cyathostomiasis, which is responsible fo causing damage to gut wall, resulting in high mortality. High prevalence *i.e.*, 96.78 per cent of *Cyathostomum* infection has been found in equines of Kashmir (Pandit *et al.*, 2008)

3.2.10 Strongyloidiasis (*Strongyloides* sp.)

Strongyloides causes infection in domestic and wild mammals and poultry (domestic geese), known as strongyloidiasis. The infection is acquired through skin contact with contaminated soil containing infective strongyloides larvae. After entering circulatory system via skin, larvae migrate to lungs and then ascend to enter gastro-intestinal system. This disease shows clinical signs as dermatitis and diarrhea. The clinical symptoms are most encountered in young animals. Strongyloides infection in cats is asymptomatic and self-limiting.

The prevalence studies on strongyloidiasis in animals of veterinary importance of Jammu and Kashmir region, such as goats, cattle, buffalo, equines and pigs, have been made by various researchers during the period of 8 years (2006, 2008-2010, 2012- 2013, 2015, 2017) (**Table 4.1.3**, p. 103). Besides, data on strongyloidiasis distribution and its high prevalence in age group and gender of hosts, has also been provide (**Table 4.1.3**, p. 103). On the basis of data provided by previous authors, the average prevalence of strongyloides in the animals of economic importance is found to be 8.64 per cent. Strongyloidiasis infection has also been detected in cat in Jammu region, with 28 per cent of prevalence (Borkataki *et al.*, 2013) and in domestic geese in Kashmir, having 74.67 per cent of prevalence (Hamdani *et al.*, 2017).

3.2.11 Trichuriasis (*Trichuris ovis* and *Trichuris* sp.)

The disease caused by parasitic nematodes, *Trichuris* spp. in domestic animals live stock and wild mammals, is known as trichuriasis. These nematodes infect caecum and colon, and heavy infestation may also be seen in very young lambs. The clinical signs are mainly diarrhea, sometimes with mucous and blood. In animals with high worm burden, there is severe anaemia, dehydration and jaundice, this condition may lead to the death of the animal.

The prevalence percentages of *Trichuris ovis*, with high infection in age group and gender, in sheep and goats of Jammu, Kashmir and Ladakh, have been studied by various authors in different years (2010-2012), are presented in **Table 4.1.3**, p. 104. On the basis of prevalence percentage data of trichuriasis due to *T. ovis* in small ruminants of Jammu and Kashmir region, an average 34.18 per cent prevalence of *T. ovis* infection is found. In addition to the wide prevalence of *T.ovis* infection in these regions, *Trichuris* sp. has been reported to unfect sheep, goats,cattle, buffalo and pig in Jammu and Kashmir region. The prevalence percentage, along with seasonal occurrence, high infection rate in different age group and gender of host animals, has been provided by various researchers during the 8 years (2010-2017) (**Table 4.1.3**, p. 104). As per prevalence data in this table, an average 18.43 per cent prevalence of *Trichuris* sp. infection in domestic animals, has been calculated.

3.2.12 Trichostrongyliasis (*Trichostrongylus*) and *Marshallegia* spp.

Trichostrongylus or black scour worm, causes a disease in sheep and goats, known as trichostrongyliasis. This nematode resides in intestine and abomasum of ruminants, damages the lining of the gut, resulting in nutritional disturbances, poor growth, wasting and persistent diarrhea. During the years: 2011-2013, 2015-2017, various authors have provided data on the prevalence percentage, seasonal distribution of *Trichostrongylus* infection in sheep and gloats of Kashmir, besides high infection in age group and gender of host (**Table 4.1.3**, p. 104). On the basis of prevalence data of 6 years incorporated in the table, trichostrongylosis infection, indicating an average of 20.57 per cent in Jammu and Kashmir region.

Marshallegia marshalli and *Marshallegia* sp. are, abomasal parasites of sheep and goats in Kashmir. *Marshallegia* is not responsible for having any significant pathogenicity. In Kashmir region, *M. marshalli* has shown prevalence of 16.6 per cent in goats (Tariq *et al.*, 2010) and 3 per cent in sheep (Wani *et al.*, 2011).

3.2.13 Toxocariasis (*Toxocara canis, T. vitulorum*)

Toxocara canis, also known as dog round worm, causing disease known as toxocariasis. The majority of infection due to this disease are asymptomatic. However, the heavy infection in puppies and kitten may be responsible for abdominal distension, with signs of colic, diarrhea and poor growth. Toxocariasis is classified according to organs affected, having two pathlogical conditions or syndromes : visceral larval migrans (VLM), or associated with major organs and ocular larval migrans (OLM), affecting eyes. In addition to this, there may be neurological toxocariasis, associated with peripheral or central nervous system.

Dogs are definitive hosts of *T. canis*. Humans are accidental hosts who become infected by ingesting infective eggs present in soil.The transmission is oro-faecal via infected dogs or cats. For details of human toxocariasis (*see* page 82). After ingestion, the eggs of *T.canis* hatch and larvae penetrate the intestinal wall and are carried by circulation to wide variety of tissues/organs like liver,heart, lungs, brain, muscle and eyes. In Jammu region, 15 per cent and 32 per cent of toxocariasis, due to *T. canis* and *Toxocara* sp. in dogs and cat respectively, have been encountered (Shah *et al.*, 2017; Borakataki *et al.*, 2013).

Toxocara vitulorum is a large-sized worm, found in small intestine of cattle, buffaloes and ruminants. The larvae of *T. vitulorum*, actually cause great damage to major organs, especially liver and intestine. In these animals, the toxocariasis is mainly transmitted through trans-placental and transmammary tansmission (Ahmad *et al.*, 2016), causing severe anaemia, diarrohea, weight loss and anorexia, especially in buffalo calves. The prevalence of 1.4 per cent of *T.vitulorum*, in young female cattle during winter season in Kashmir region has been found (Bushra *et al.*, 2013).

3.2.14 Ascardiasis (*Aascardia galli*)

A. galli is a parasite of domesticated and wild birds, causing Ascardiasis. This disease is responsible for enteritis, with diarrhea and unthriftness. In severe infection, there is intestinal blockage, loss of blood, retarded growth and these conditions cause reduction in egg production and growth rate, leading to economic loss to the poultry industry. *A. galli* is most prevalent and pathogenic parasite in Kashmir region and prevalence of 32.11 per cent in domestic geese in Gurez region during summer season has been found (Sheikh *et al.*, 2016). 5 per cent prevalence of ascardiasis in domestic pigeons (*Columba livia*) of Jammu region, has been reported (Mehmood *et al.*, 2019).

3.2.15 Baylisascariasis (*Baylisascaris*)

Baylisascariasis disease caused by *Baylisascaris transfuga* is a intestinal nematode found in the Himalayan Black Bear in Kashmir region, showing prevalence of 47.32 per cent (Sheikh *et al.*, 2017). In baylisascariasis infection, the larvae *B. transfuga* have potential to cause visceral ocular and/or neural larval migrans in mammals, birds and human.

3.2.16 Gongylonemiasis (*Gongylonema*)

Gongylonemiasis is caused by *Gongylonema*, thread- like worm commonly known as' gullet or' stitch' worm, found within the tissues of oesophagus or upper alimentary tract of sheep and goats. These parasites burrow through cornified layer of mucosa forming serpiginous tracks. This disease in animals is transmitted by insects (beetles, cockroaches) (intermediate hosts). Animals acquire infection by feeding on intermediate hosts. *Gongylonema pulchrum* can infect humans and has ability to localize in the submucosa of oral cavity or oropharynx. Humans acquire infection of gongylonemiasis accidently by consuming intermediate hosts. There are reports of 50 cases of human gongylonemiasis world wide (Ayala and Yencha, 2012). Tariq *et al.* (2008 a), have found the prevalence of *Gongylonema* in sheep, brought from out side Kashmir valley, slaughtered at local abattoir at Hazratbal (Srinagar), with prevalence of 37 per cent and maximum prevalence from August to November (2006).

3.2.17 Oxyuriasis (*Oxyuris* sp.)

The equine pin worm or *Oxyuris* sp. affects equines, resides in rectum of these animals, causing severe irritation of perianal region, causing rubbing, biting at the tail. Some horses become nervous and stop eating. In Kashmir region, 9.40 per cent prevalence of oxyuriasis infection in equines has been reported (Pandit *et al.*, 2008).

3.2.18 Parascariasis (*Parascaris* sp.)

Parascariasis due to nematode parasite, *Parascaris*, that affects young horses,usually under two year of age. In heavy infestation, with migratory larvae, cause coughing and with heavy burden of adult worm may cause diarrohea, which may lead to even to death of animals. Pandit *et al.* (2008), showed 4.1. per cent of prevalence of *Parascaris* in equines of Kashmir region.

3.2.19 Capillariasis (*Capillaria* sp.)

Capillaria is a parasitic nematode which causes infection in intestine of pultry, dogs, cats and also live-stock, called as capillariasis. The clinical symptoms of this disease mainly include diarrhea, wasting, poor growth and dejection. The morbidity and mortality with the infection of capillariasis are usually low. In Kashmir, prevalence of 5.33 of capillariasis in domestic geese has been observed (Hamdani *et al.*, 2017). In Jammu region, prevalence of 6.67 per cent of capillariasis in domestic pigeon, has been observed (Mehmood *et al.*, 2019)

3.2.20 Heterakiasis (*Heterakis gallinarum*)

H. gallinarum is a parasitic nematode living in the caecum of galliform birds. It causes infection called as heterakiasis, which is mildly pathogenic. However, in heavy infection, there is inflammation and thickening of the caecum and appearance of nodules. This leads to reduction in egg production. *H. gallinarum* is often carries/vector of protozoan parasite, *Histomonas meleagridis*, causing disease known as Histomoniasis or enterohepatitis or black head disease. In Gurez area of Kashmir, Sheikh *et al.* (2016), has shown 32.11 per cent of prevalence of heterakiasis in domestic geese.

3.2.21 Setariasis (*Setaria digitata, S. equina*) and Thelaziasis (*Thelazia*)

Setariasis in animals is caused by filarid nematodes of genus *Setaria.* These nematodes infect body cavity, brain, spinal cord, and their larvae circulate in the blood. Steriasis disease is common in horses and cattle. The causative agent of setariasis is infected animals and the carriers are mosquitoes, and other haematophagus flies. The ocular setariasis or equine ocular setariasis, is a common vision threatening condition in horses, leading to blindness, caused by *Setaria equina, S. digitata* and *Thelazia lacrymalis* (eye worm).

In Jammu region, the horses aged 1-4 years, have been found with partial to complete corneal opacity, due to *S. digitata* and *S. equina* (Yadav *et al.*, 2006). Earlier, from Kashmir region, Parrah *et al.* (2004), reported ocular filariasis in horses, caused by *Setaria* spp. Ocular setariosis in eye of horse in this region, has also bee reported by Shakeebah *et al.* (2018). Thelaziasis disease in Buffaloe in Jammu region has been reported by Bhadwal and Raina (1994). These authors mechanically removed 6 worms from left eye and 11 worms from right eye of 6- year old buffaloe. The animal was suffering from wound infection of eyes. Occurrence of *S. digitata* in cow in Jammu region, suffering from intussusception, has been observed (Kaur *et al.*, 2015).

3.3 Highly Prevalent Nematode Parasites of Medical Importance

3.3.1 *Ancylostoma duodenale* (Human Ancylostomiasis)

The hookworm disease in human is caused by *Ancylostoma duodenale*, affecting lungs, skin and small intestine. The hookworm infection, is acquired by coming in contact with soil/dirt containing larvae of this worm. The larvae enter the skin of a person, travel through blood stream, and enter lungs. They are carried to small intestine after coughing them out of the lung and swallowing them. The mature hookworms attach to the small intestine and derive nutrients through human blood. This causes anaemia. The eggs of hookworms enter the faeces and after passing out the eggs mature into larvae in soil. The symptoms of human ancylostomiasis include skin rash, weight loss, loss of appetite Besides these, breathing complication, cough, fever, stomach pain, diarrhea, anaemia, malnutrition take place. This may lead to heart failure.

Arora *et al.* (1976) were the first investigators to report the prevalence of hookworm infection in rural communities of Lambri village of Rajouri district of Jammu and found 4.6 per cent prevalence in the human population. From Jammu region, the prevalence of 0.82 per cent of ancylostomiasis, basesd on patients attending SMGS Hospital, has been revealed, improved dramatically as compared to findings conducted about 35 years ago (Jad *et al.*, 2015)

3.3.2 *Ascaris lumbricoides* (Ascariasis)

Ascariasis is infection of the small intestine, caused by roundworm, *Ascaris lumbricoides*, acquired by ingesting eggs, passed out by infected people, deposited in soil or dirt. Ascariasis often shows no symptoms, excepting abdomainl discomfort, loss of appetite, weight loss, nausea, diarrhea, fever discomfort in chest and bloody mucus. The other symptoms include cough due to migration of worms through body. In heavy infection, there is intestinal blockage, causing severe pain and vomiting in childrens. Intestinal ascariasis is highly endemic in all parts of Kashmir besides, surgical, hepato-biliary, gall-bladder and pancreatic ascariasis, due to *A. lumbricoides*, have also been widely reported from different regions of Kashmir (Khuroo, 1996). In Kashmir, several studies have been done on: biliary ascariasis (Khuroo and Zargar1985; Hangloo *et al.*, 1989, 92 and Wani and Chrungoo (1992); Hepato-biliary ascariasis (Khuroo,1999); pancreatic ascariasis (Khuroo *et al.*, 1992 *a* and Khuroo,1999); gall-bladder ascariasis Khuroo *et al.* (1992); surgical ascariasis (Mir *et al.*, 2002). The bowl obstruction due to intestinal ascariasis in children in Kashmir, has been found to be very common and major source of morbidity and mortality (Baba *et al.*, 2009; Sheikh *et al.*, 2010). These studies were based on patients (children) who visited tertiary care Centre in Kashmir.

In Jammu region, Arora *et al.* (1976) reported the prevalence of 6 per cent of *A. lumbricoides* infection in rural population in Lambri village of district Rajouri and 3.3 per cent prevalence of ascariasis, based on record of patients who visited SMGS hospital of Jammu (Jad *et al.*, 2015). In Kashmir region, Wani (1970) for the first time conducted a thorough research study on incidence and symptomology of ascariaris due to nematode parasite, *Ascaris lumbricoides* in human population of Kashmir. He found 58 per cent of positive cases of ascariasis in patients who attended S. M. H. S. Hospital (Government College), Srinagar. Further, the high percentage about 80.7 per cent was observed in age group 6-10 years and lesser incidence in higher age group. In this region, Ahmad (1978) made investigations on the incidence of *A. lumbricoides* in the rural population of Kashmir. He revealed that of the 3 intestinal parasites (*A. lumbricoides, Gardia lambia* and *Taenia saginta*), *A. lumbricoides* was very much predominant and accounted for 61.8 per cent of total cases examined. The incidence of ascariasis disease among males and females was noticed as 64 per cent and 59 per cent respectively. The declining trend was further seen with rise in adult age after 15 years. The higher incidence of 89.2 per cent and 73.7 per cent of parasite infestation was found in Kashmir rural population, covering school going age groups and pre-school age groups respectively.

The 82.60 per cent (38/46) prevalence of ascariasis in Srinagar district was found in patients in age group (21-40 years), who visited SKIMS hospital (Ahmed *et al.*,

2002). In Kashmir region, covering various areas such as Valley, Shopian district, Gurez valley, Srinagar and Kupwara district, detailed studies on prevalence of ascariasis in children up to age of 15 years, have been carried out by various authors during the period of 6 years (2007, 2008, 2010, 2012, 2016, 2017) (**Table 4.1.1**, pp. 95-96). On the basis of prevalence data provided by different authors in different years in the said Table, an average (overall) prevalence of ascariasis in Kashmir region is calculated as 59.39 per cent

The hepato-biliary and pancreatic ascariasis (HPA), is a disease of great public importance in endemic areas of Kashmir. In HPA, the ascarides make repeated visits into and out of the bile and pancreatic duct. During its repeated but limited exploration of papillary orifice and advance into ducts, the person comes down with severe hepato-biliary or pancreatic disease. The dead ascarides in the duct lead to the formation of brown pigment 'stones', leading to the syndrome of Recurrent Pyogenic Cholangitis, a disease more often in Oriental region known as Cholangiohepatis (Khuroo, 1999).

The higher incidence of nematode parasite infection among the human population of Kashmir region, is believed to be because of faeco-oral contact of poor personal hygiene and also due to extensive contamination of soil and water in different localities and areas of rural regions of Kashmir. Yaseen (1979) studied the intestinal functions of ascariasis in Kashmir valley and has concluded that *Ascaris* infestation does not lead to functional or structural impairment of small intestine, resulting in mal-absorption syndrome.

3.3.3 *Enterobius vermicularis* (Enterobiasis) (Human Pin Worm)

Enterobius vermicularis infection is also known as enterobiasis, affects intestine of humans, especially childrens. In many persons the infection can be asymptomatic, while others develop mild symptoms. The common symptom is itchiness, leading to disrupt sleep, irritability, bacterial infection from scratching. In rare cases the nematode can spread to viginal areas, causing urinary tract infection. The nematode can also travel to appendex, causing appendicitis like symptoms- such as vomiting, abdomainal pain and reduced appetite, without inflammation of appendix. Enterobiasis infection is aquired by passing between perons through contamination, finger sucking, nail biting and ingestion of infectious eggs of the nematode parasite.

Earlier, in Jammu region, Arora *et al.* (1976) reported the prevalence of enterobiasis in natural population in Lamberi village of Rajouri (Jammu). The detailed studies on the prevalence of human pin worm infection in the children population (up to 15 years) in the rural and urban areas of Kashmir like Shopian district, Gurez vally and district Kupwara, have been made by number of authors. The prevalence percentages of *E.vermicularis* in these areas, pertaining to the publications appeared in the years (2007, 2008, 2010, 2012, 2016, 2017) are presented in the **Table 4.1.1**, p 96. On the basis od prevalence data of *E.vermicularis* in Kashmir region, an average (overall) 11.43 per cent of prevalence of infection has been calculated.

3.3.4 Human Toxocariasis (*Toxocara canis/T. cati*)

Human toxocariasis is a zoonotic nematode infection, caused by *Toxocara canis* and *T. cati*, due to migration of *Toxocara* sp. larvae through human body. The adults of these nematodes live in upper digestive tracts of definitive hosts(canids, felids). The eggs are produced in large numbers, passed in the faeces are not infective and required incubation period in soil to embryonate. The infection in humans is caused due to accidental ingestion of embryonate eggs, containing second stage larvae, from soil, dirty hands or vegetables *etc.* leading to serious clinical manifestations or syndromes. Larvae have been found in liver, lungs, heart, eye (ocular toxocariasis) and brain. They are often associated with migratory tracks, characterized by haemorrhage, necrosis and inflammation and in the eyes a single larva is observed, can cause partial or total retinal detachment with loss of vision (Magnaval *et al.*, 2001).

The prevalence of human toxocariasis in Srinagar distict of Kashmir valley for the first time was detected through ELISA in the highly endemic area of ascariasis (Ahmad *et al.*, 2002). The seroprevalence of toxocariasis in children in Kashmir region has been found to be 32.86 per cent (94/286) (Dar *et al.*, 2008). The seroprevalence of *Toxocara canis* infection in children population of District Srinagar and Kupwara was observed to be as 25 per cent and 34.78 per cent respectively (Ahmad *et al.*, 2008) (**Table 4.1.1**, p. 96). A case of ocular toxocariasis in a child from Kashmir, has been found by Fomda *et al.* (2007). Seroprevalence of toxocariasis in children in earstwhile J&K State, has been reported by Sofi *et al.* (2008). In Kashmir region, out of 514 individuals, 298 (57.97 per cent) tested seropositive for human toxocariasis (Dar *et al.*, 2013).

3.3.5 *Trichuris trichiura* (Human whipworm) Trichuriasis

The infection caused by *Trichura trichiura* is known as trichuriasis, affecting human large intestine. The infection is aquired from soil contaminated with human feces, entering mouth (oro-feacal) or by eating vegetables, fruits not washed or peeled or cooked not properly. In low infection of trichuriasis, there are no symptoms, however, in heavy infection, there is painful passage of stool, containing mucus, water and blood. In severe infection growth retardation or rectal prolapse can also occur.

The prevalence of *Trichuris trichiura* infection in rural population of in Lambri village of Rajouri district of Jammu, for the first time was studied by Arora *et al.* (1976). In Jammu region, the prevalence trichuriasis infection, has been found to be very low as 0.82 per cent, based on the hospital record of patients, visited tertiary hospital (MGSH), during past years (Jad *et al.*, 2015). The high prevalence of *Trichuris trichiura* infection in children population (urban, rural) in the age group of 5-15 years, in various areas of Kashmir province, has been studied by various authors, pertaining to seven years (2007, 2008, 2010, 2012, 2015, 2015, 2016, 2017) (**Table 4.1.1**, pp. 96-97). On the basis of prevalence data provided by these authors in different years, an average (overall) prevalence 21.73 per cent has been calculated.

The species richness, systematic position and host diversity of nematode parasites, infecting insects, fishes, amphibians, wildbirds, poultry, small mammals, domestic animals, wildlife and human are encapsulated in the **Tables 3.1–3.3**.

Table 3.1: Species Richness and Host Diversity of Nematode Parasites, Infecting Insects, Fishes, Amphibians, Reptiles, Birds, Small Mammals, Domestic Animals, Wild Animals (Wildlife) and Humans in Jammu, Kashmir and Ladakh

Sl. No.	*Nematode Genus*	*No. of Species*	*Number of Species of Nematodes in Various Host Group*									
			IN	*FI*	*AM*	*RE*	*WB*	*PO*	*SM*	*DA*	*WL*	*HU*
1	*Acanthocheil-onema*	1	-	-	-	-	-	-	-	1	-	-
2	*Amidostomum*	2	-	-	-	-	1	1	-	-	-	-
3	*Ancylostoma*	2	-	-	-	-	-	-	-	1	1	1
4.	*Ascardia*	1	-	-	-	-	-	1	-	-	-	-
5.	*Ascaris*	2	-	-	-	-	-	-	-	2	-	1
6.	*Acuaria*	1	-	-	-	-	1	-	-	-	-	-
7	*Baylisascaris*	1	-	-	-	-	-	-	-	1	-	-
8.	*Bunostomum*	2	-	-	-	-	-	-	-	2	-	-
9.	*Camallanus*	3	-	3	-	-	-	-	-	-	-	-
10.	*Capillaria*	5	-	-	-	-	-	2	3	1	-	-
11.	*Chabertia*	3	-	-	-	-	-	-	-	3	1	-
12	*Cheilospirura*	1	-	-	-	-	-	1	-	-	-	-
13.	*Cooperia*	1	-	-	-	-	-	-	-	1	-	-
14.	*Controcaecum*	1	-	-	-	-	1	-	-	-	-	-
15.	*Cosmocerca*	2	-	-	2	-	-	-	-	-	-	-
16	*Cyathostomum*	1	-	-	-	-	-	-	-	1	-	-
17.	*Dictyocaulus*	2	-	-	-	-	-	-	-	2	-	-
18.	*Diplotriaena*	1	-	-	-	-	1	-	-	-	-	-
19.	*Echinuria*	1	-	-	-	-	-	1	-	-	-	-
20.	*Enterobius*	1	-	-	-	-	-	-	-	-	-	1
21.	*Epomidiostomum*	2	-	-	-	-	2	-	-	-	-	-
22	*Gaigeria*	1	-	-	-	-	-	-	-	1	-	-
23	*Gongylonema*	1	-	-	-	-	-	-	-	1	-	-
24.	*Haemonchus*	2	-	-	-	-	-	-	-	2	-	-
25.	*Heterakis*	2	-	-	-	-	-	2	1	-	-	-
26.	*Kalicephalus*	2	-	-	-	2	-	-	-	-	-	-
27	*Leidynema*	1	1	-	-	-	-	-	-	-	-	-
28.	*Marshallegia*	2	-	-	-	-	-	-	-	2	1	-
29.	*Mecistocirrus*	1	-	-	-	-	-	-	-	1	-	-
30	*Microfilaria*	-	-	-	-	-	-	-	-	1	-	-
31.	*Monodontus*	1	-	-	-	-	-	-	-	1	-	-
32	*Muellerius*	1	-	-	-	-	-	-	-	1	-	-
33	*Nematodirus*	4	-	-	-	-	-	-	-	4	1	-

Sl. No.	Nematode Genus	No. of Species	Number of Species of Nematodes in Various Host Group									
			IN	FI	AM	RE	WB	PO	SM	DA	WL	HU
34	*Oesophagostomum*	2	-	-	-	-	-	-	-	2	1	-
35	*Ostertagia*	2	-	-	-	-	-	-	-	2	1	-
36	*Oswaldcruzia*	1	-	-	1	-	-	-	-	-	-	-
37	*Oxysomatium*	1	-	-	1	-	-	-	-	-	-	-
38	*Oxyuris*	1	-	-	-	-	-	-	-	1	-	-
39	*Parascaris*	1	-	-	-	-	-	-	-	1	-	-
40	*Parafilaria*	2	-	-	-	-	-	-	-	2	-	-
41	*Physaloptera*	1	-	-	-	-	-	-	-	-	1	-
42	*Procamallanus*	1	-	1	-	-	-	-	-	-	-	-
43	*Prostrongylus*	1	-	-	-	-	-	-	-	1	-	-
44	*Pseudonymus*	1	1	-	-	-	-	-	-	-	-	-
45	*Rhabdias*	1	-	-	1	-	-	-	-	-	-	-
46	*Rhabdochona*	4	-	4	-	-	-	-	-	-	-	-
47	*Rusguniella*	1	-	-	-	-	1	-	-	-	-	-
48	*Setaria*	3	-	-	-	-	-	-	-	1	-	-
49	*Skrjabinema*	1	-	-	-	-	-	-	-	1	-	-
50	*Spinitectus*	1	-	1	-	-	-	-	-	-	-	-
51	*Streptocara*	1	-	-	-	-	-	1	-	-	-	-
52	*Strongyloides*	3	-	-	-	-	-	1	-	2	1	-
53	*Syngamus*	1	-	-	-	1	-	-	-	-	-	-
54	*Strongyles*	1	-	-	-	-	-	-	-	1	-	-
55	*Strongylus*	-	-	-	-	-	-	-	-	1	-	-
56	*Thelandros*	2	-	-	-	2	-	-	-	-	-	-
57	*Thelazia*	1	-	-	-	-	-	-	-	1	-	-
58	*Toxocara*	3	-	-	-	-	-	-	-	3	-	1
59	*Trichocephalus*	1	-	-	-	-	-	-	1	-	-	-
60	*Trichostrongylus*	4	-	-	-	-	-	-	-	4	1	-
61	*Trichuris*	2	-	-	-	-	-	-	-	1	-	1
62	*Teladorsagia*	1	-	-	-	-	-	-	-	1	-	-
63	*Triodontophorus*	1	-	-	-	-	-	-	-	1	-	-
64	*Tetrameres*	1	-	-	-	-	1	-	-	-	-	-
65	*Toxascaris*	1	-	-	-	-	-	-	1	-	-	-
66	*Varestrongylus*	2	-	-	-	-	-	-	2	-	-	-

Abbreviations: IN: Insects; FI: Fishes; AM: Amphibians; RE: Reptiles; WB: Wildbirds; PO: Poultry; SM: Small mammals; DA: Domestic animals (cat, dog, sheep, goat, cattle, buffalo, bovines, equines, live-stock); WL: Wildlife (hangul, the himalyan black bear, fox) and HU: Humans.

Table 3.2: Systematic Position of Animal and Human Parasitic Nematode (Genera) (Nematoda: Secernetea) of J&K, and Ladakh UTs

Order/Sub-order 1 : Rhabdita

    Superfamily 1: Rhabditiodea

        Family 1: Rhabditidae

            Genus: *Rhabdias*

        Family 2: Strongyloidae

            Genus: *Strongyloides*

    Superfamily 2: Diaphanocephaloidea

        Family 1: Diaphancephalidae

            Genus: *Kalicephalus*

    Superfamily 3: Ancylostomatoidea

        Family: Ancylostomatidae

            Subfamily 1: Ancylostomalinae

                Genus: *Ancylostoma*

            Subfamily 2: Bunostominae

                Genus: *Gaigera, Bunostomum* and *Monodontus*

    Superfamily 4: Strongyloidea

        Family 1: Chabertiidae

            Subfamily 1 : Chabertiinae

                Genus: *Chabertia*

            Subfamily 2. Oesophagostominae

                Genus: *Oesophagostomum*

        Family 2: Syngamidae

            Subfamily 1. Syngaminae

                Genus: *Syngamus*

        Family 3: Stongylidae

            Genus: *Triodontophorus, Strongyles, Strongylus*

            Subfamily 2. Cyanthostominae

                Genus: *Cyapthostomum*

    Superfamily 5: Trichostongyloidea

        Family 1: Cooperiide

            Subfamily : Cooperiinae

                Genus: *Cooperia*

        Family 2: Trichostongylidae

            Subfamiy 1: Ostertagiinae

                Genus: *Ostertagia, Marshallagia, Teladorsagia*

Subfamily 2: Trichostrongylinae

Genus: *Trichostongylus*

Subfamily 3: Haemomchinae

Genus: *Haemonchus*

Subfamily 4: Mecistocirrinae

Genus: Mecistocirrus

Family 3: Dictyocaulidae

Subfamily : Dictyocaulinae

Genus: *Dictyocaulus*

Family 4: Amidostomidae

Subfamily: Amidostomatinae

Genus: *Amidostomum, Epomidiostomum*

Family 5: Molineidae

Subfamily 1: Molineinae

Genus: *Oswaldcruzia*

Subfamily 2: Nematodirinae

Genus: *Nematodirus*

Superfamily 7: Metastrongyloidea

Family : Protostongylidae

Subfamily1 :Elaphostongylinae

Genus: *Muellerius*

Subfamily 2: Protostongylinae

Genera: *Protostongylus, Varestrongylus*

Order/Suborder 2: Oxyurida

Superfamily 1 : Oxyuridea

Family 1 : Oxyuridae

Genus: *Enterobius, Skrjabinema, Oxyuris*

Family 2: Phatyngodonidae

Genus: *Thelandros*

Superfamily 2. : Thelastomatoidea

Family: Thelastomidae

Genus: *Leidynema*

Subfamily : Pseudonymidae or Pseudonyminae ?

Genus: *Pseudonymus*

Order/Sub-order 3: Ascaridida

Superfamily1 : Cosmocercoidea

Family : Cosmercidae

Genera: *Oxysomatium, Cosmocerca*

Superfamily 2: Heterakoidea

Family 1 : Heterakidae

Subfamily: Heterakinae

Genus: *Heterakis*

Family 2: Ascaridiidae

Genus: *Ascaridia*

Superfamily 3: Ascaridoidea

Family : Anisakidae

Subfamily : Anisakinae

Genus: *Controcaecum*

Family : Ascarididae

Subfamily: Ascaridinae

Genera: *Ascaris, Balysascaris, Parascaris*

Family: Toxocaridae

Subfamily : Toxocarinae

Genus: *Toxocara*

Order/Sub-order 4: Camallaninada

Superfamily : Camallanoidea

Family : Camallanidae

Genera: *Camallanus, Procamallanus*

Order/Sub-order 5 : Spirurida/Spirurina

Superfamily 1 : Filarioidea

Family : Filariidae

Subfamily : Filariinae

Genus: *Parafilaria*

Family : Onchocercidae

Subfamily : Onchocercinae

Genus: *Acanthocheilonema, Microfilaria*

Subfamily : Setariinae

Genus: *Setaria*

Superfamily 2 : Spiruroidea

Family : Gongylonematidae

Subfamily : Gongylonematinae

Genus: *Gongylonema*

Superfamily 3 : Thelazioidea

Family: Thelaziidae

Subfamily : Thelaziinae

Genus: *Thelazia*

Family: Rhabdochonidae

Subfamily : Rhabdochoninae

Genus: *Rhabdochona*

Superfamily 4 : Habronematoidea
Family: Cystidicolidae
Genus: *Spinitectus*
Family: Tetrameridae
Subfamily: Tetramerinae
Genus: *Tetrameres*
Superfamily 5 : Acuarioidea
Family: Acuariidae
Genus: *Acuaria, Cheilospirura, Rusguniella, Streptocara* and *Echinuria*
Superfamily 6 : Diplotriaenoidea
Family: Diplotriaenidae
Subfamily: Diplotriaeninae
Genus: Diplotriaena
Superfamily 7 : Physaloperoidea
Family: Physalopteridae
Subfamily: Physalopterinae
Genus: *Physaloptera*
Order/Sub-order: Trichinellida
Superfamily: Trichinelloidea
Family: Trichuridae
Subfamily: Trichurinae
Genus: *Trichuris, Trichocephalus*
Family: Capillaridae
Subfamily: Capillariinae
Genus: *Capillaria*

Table 3.3: Nematode Parasites: Checklist, Host Species, Host Infection Site and Distribution in Jammu, Kashmir and Ladakh

Nematode Species	*Host Species*	*Location of Infection*	*Distribution/ Locality (ies)*
(A) Insect Nematodes			
(1) *Leidynema appendiculata*	*Blatta orientalis* (Cockroach)	Intestine	Kashmir
(2) *Pseudonymus hydrophili* (=*Galebiella galebiella*)	Aquatic beetle	Intestine	Kashmir(Hill stream)
(3) *Pseudonymus multiannulata*	*Hydrophilus* sp.	Intestine	Kashmir
(4) *Pseudonymus islambadi* (=*Galebiella islambadi*	*Hydrophilus* sp.	Intestine	Kashmir
(B) Fish Nematodes			
(1) *Camallanus cotti* (=*Camallanus fotedari*)	*Cyprinus carpio*	Intestine	Kashmir Valley Lakes (Wular, *etc.*)

Nematode Species	Host Species	Location of Infection	Distribution/ Locality (ies)
	Triplophysa kashmirensis (= *Nemacheilus kashmirensis*),	–do–	–do–
	Schizothorax richardsonii (= *Oreinus plagiostomus*),	–do–	–do–
	Schizothorax curvifrons,	–do–	–do–
	S. esocinus, (=*S.hugelii*),	–do–	–do–
	Schizothorax niger,	–do–	–do–
	Schizothorax plagiostomus	–do–	–do–
(2) *Camallanus oxycephalus*	*Schizothorax richardsonii, Schizothorax* sp.	Intestine	Kashmir (Dal, Manasbal Lake)
(3) *Camallanus salmoe*	*Salmo* sp.	Intestine	Kashmir.
(4) *Procamallanus kashmirensis*	*Wallago attu, Salmo* sp.	Intestine	Jammu, Kashmir
(5) *Rhabdochona guptai*	*Cyrinus c. spicularis*	Intestine	Hokarsar, Dal lake, Manasbal Lake
	Schizothorax richardsonii (=*Oreinus sinuatus*),	–do–	–do–
	S. plagiostomus, Salmo trutto fario	–do–	
(6) *Rhabdochona kashmirensis*	*C. c. spicularis,*	Intestine	Kashmir valley
	Triplophysa kashmirenis (=*Nemacheilus kashmirensis*),	–do–	–do–
	Schizothorax richardsonii (=*Orienus plagiostomus*),	–do–	–do–
	S. curvifrons and *Schizothorax esocinus* (=*Schizothorax huglii*),	–do–	–do–
	Wallago attu	–do–	–do–
(7) *Rhabdochona himalayai*	*C.c. spicularis,*	–do–	Lakes of Kashmir
	Triplophysa kashmirensis (= *Nemacheilus* sp., *O. plagiostomus*),	–do–	Jammu & Kashmir
	S. esocinus (=*S. hugelii*),	–do–	–do–
	S. curvifrons,	–do–	–do–
	Schizothorax sp.	–do–	–do–
(8) *Rhabdochona* sp.	*Salmo trutto fario, Schizothorax*	Intestine	Kashmir
(9) *Spinitectus* sp.	*Schizothorax richardsonii* (=*Oreinus richardsonii*)	Intestine	Jammu

Nematode Species	Host Species	Location of Infection	Distribution/ Locality (ies)
(C) Amphibians (Toads)			
Nematodes			
(1) *Cosmocerca crenshawi*	*Bufo viridis* (Toad)	Intestine	Kashmir
(2) *Cosmocerca commutata*	–do–	Intestine	–do–
(=*Cosmocerca kashmirensis*)			
(3) *Oswaldcruzia kashmirensis*	–do–	Intestine	–do–
(4) *Oxysomatium srinagarensis*	–do–	Rectum	–do–
(5) *Rhabdias bufonis*	–do–	Lungs	-do
(D) Reptiles (Snakes, lizards)			
Nematodes			
(1) *Kalicephalus indicus*	*Topidontus platyceps* (Snake)	Stomach	Pahalgam (Kashmir)
(2) *Kalicephalus kashmirensis*	*Ptyas mucosus* (Snake)	Oesophagus Stomach	Srinagar (Kashmir)
(3) *Thelandros baylisi*	*Calotes versicolor* (Golden lizard), *Laudakia tuberculata* (=*Agama tuberculata*)(Agama)	Stomach	Jammu and Kashmir
(4) *Thelandros jummensis*	*Hemidactylus flaviviridis* (Yellow belly gecko or northern House gecko)	Not known	Jammu
(E) Birds (Wild and domestic)			
Nematodes			
(1) *Acuaria skrjabini*	*Passer domestica*	Gizzard	Srinagar, Kashmir
(2) *Amidostomum fulicae*	*Fulica atra atra*	Gizzard	Dal Lake
(3) *Amidostomum anseris*	*Anser domesticus*	Intestine	Gund
(4) *Ascardia galli*	Domestic fowl (indigenous)	Intestine	Srinagar
(5) *Capillaria annulata*	Domestic fowl	Crop	Kashmir
(6) *Capillaria anatis*	*Corvus monedula*, *C. splendens*	Oesophagus	Kashmir
(7) *Caprillaria caudinflata*	Domestic fowl	Intestine	Kashmir
(8) *Cheilospirura harmulosa*	Chickens (indigenous)	Intestine	Kashmir
(9) *Controcaecum milvi*	*Milvus migrans linatus*	Stomach	Jammu
(10) *Diplotriaena buckelyi*	*Acridotheres tristis*	Gizzard	Hokersar, Kashmir
(11) *Echinuria uncinata*	Domestic fowl (indigenous)	Gizzard	Hokersar
(12) *Epomidiostomum anatinum*	*Anas platyrhynchos*	Gizzard	Hokersar
(13) *Epomidiostomum uncinatum*	Domestic fowl (indigenous)	Gizzard	Kashmir
(14) *Heterakis gallinarum*	The mallard	Caecum	Kashmir

Nematode Species	Host Species	Location of Infection	Distribution/ Locality (ies)
(15) *Heterakis indica*	Domestic fowl (indigenous)	Caecum	Kashmir
(16) *Rusguniella turdi*	*Turdus unicolor*	Intestine	Kashmir
(17) *Streptocara indica*	*Gallus domesticus*	Intestine	Kashmir
(18) *Syngamus trachea*	*Corvus splendens*	Tracheae	Kashmir
(19) *Tetrameres fissipina*	Domestic duck	Gastro-intestine	Kashmir (Bandipore)
(F) Mammals (Rat, shrew, bat, cat, dog)			
Nematodes:			
(1) *Ancylostoma caninum*	Dog, Cat	Intestine	Jammu
(2) *Ancylostoma* sp.	Cat	Intestine	Kashmir
(3) *Capillaria kashmirensis*	*Myotis muricola* (Bat)	Stomach	Anantnag
(4) *Capillaria minuta*	*Crocidura* sp. (Shrew)	Oesophagus	Kashmir
(5) *Capillaria* sp.	*Crocidura* sp.	Oesophagus	Kashmir
(6) *Heterakis spumosa* (= *Ganguleterakis spumosa*)	*Rattus rattus* (Rat)	Intestine	Kashmir
(7) *Toxocara cati* (= *Toxocara mystax*)	*Felis domestica* (Cat)	Not known	Kashmir
(8) *Toxocara canis*	Dog	Not known	Kashmir
(9) *Toxocara leonina*	Dog	Not known	Ladakh
(10) *Toxocara* sp.	Cat, Dog	Not known	Jammu
(11) *Trichocephalus muris*	*R. rattus*	Stomach	Kashmir
(12) *Strongyloides* sp.	Cat	Not available	Jammu
(13) *Acanthocheilonema*	Dog	Not available	Ladakh
(14) *Microfilaria auquieri*	Dog	Not known	Ladakh
(G) Domestic Mammals (Equines, Sheep, goats, Yak, Buffalo bovids, cattles, *etc.*)			
Nematodes			
(1) *Ascaris lumbricoides*	Sheep	Gastro-intestine	Kashmir (Handwara)
(2) *Ascaris sp.*	*Bos grunniens* (Yak)	Gastro-intestine	Ladakh
(3) *Bunostomum trigano-cephalum*	Sheep and goats	Gastro-intestine	Jammu & Kashmir
(4) *Bunostomum* sp.	Indian Yak	Intestine	Ladakh
(5) *Chabertia himalayana*	Sheep	Intestine	Handwara
(6) *Chabertia ovina*	Sheep, goats, cattle, buffalo	Gastro-intestine	J & K, Ladakh
(7) *Chabertia* sp.	Indian Yak	Gastro-intestine	Ladakh, Kargil
(8) *Cooperia* sp.	Indian Yak	Intestine	Ladakh, Lolab
(9) *Cyathostomum*	Equines	Not known	Kargil (Ladakh) Kashmir
(10) *Dictyocaulus filaria*	Goat, sheep	Lungs	Kargil

Nematode Species	*Host Species*	*Location of Infection*	*Distribution/ Locality (ies)*
(11) *Dictyocaulus viviparous*	Domestic animal (bovid)	Lungs	Kashmir
(12) *Dictyocaulus* sp.	Equines	Gastro-intestine	Handwara
(13) *Gaigeria pachycelis*	Sheep	Gastro-intestine	Kashmir
(14) *Gongylonema* sp.	Sheep (from outside)	Pharynx/upper alimentary tract	Srinagar, Kashmir
(15) *Haemonchns controtus*	Bovids, Indian Yak	Gastro-intestine	Kashmir
(16) *Haemonchus contortus kashmirensis*	Sheep	Gastro-intestine	Kashmir
(17) *Haemonchus* sp.	Sheep, goats, cattle	Not available	Ganderbal, Kashmir
(18) *Marshallagia marshalli*	Sheep, goat	Lung	Ladakh
(19) *Marshallagia* sp.	Sheep	Intestine	Kashmir
(20) *Mecistocirrus digitatus*	Indian Yak	Gastro-intestine	Kargil (Ladakh)
(21) *Monodontus (Pseudo-gaiheria) alatus*	Sheep	Not known	Kashmir
(22) *Muellerius* sp.	Sheep, goats	Lungs	Ladakh
(23) *Nematodirus fotedari*	Sheep	Intestine	Kashmir
(24) *Nematodirus spathiger*	Sheep, goats	Gastro- intestine	Handwara
(25) *Nematodirus* sp.	Sheep, Indian yak	Gastero-intestine	Ladakh,Kashmir
(26) *Oesophagostomum columbianum*	Sheep, goats, lamb	Intestine	Kashmir
(27) *Oesophagostomum* sp.	Sheep, goats	Intestine	Jammu, Kashmir
(28) *Ostertagia* sp.	Indian Yak,sheep, goat	Intestine	J & K, Ladakh
(29) *Oxyuris* sp.	Equines	Not known	Kashmir
(30) *Parafilaria bovicola*	Bullock	Sub-cutaneous nodule	Jammu
(31) *Parafilaria sahaii*	Buffalo	Sub-cutaneous nodules	Jammu
(32) *Parascaris* sp.	Equines	Not available (data)	Jammu & Kashmir
(33) *Prostostrongylus rufescens*	Goats	Lungs	Ladakh
(34) *Setaria equina*	Equines (Horses)	Peritoneum, eyes	Jammu, Kashmir
(35) *Setaria cervi*	Indian Yak	Peritonem	Ladakh region
(36) *Setaria digitata*	Cow, equines (horses)	Peritoneum, eyes	Jammu, Kashmir
(37) *Skrjabinema ovis*	Sheep	Gastro-intestine	Handwara
(38) *Strongyles* sp.	Live stock, Buffalo, equine, pig, sheep,	Not known	Kashmir
(39) *Strongylus* sp.	Equines	Not known	Kashmir
(40) *Strongyloides* sp	Cattle, pig, buffalo, equines, live stock	Gastro-intestine	Jammu &Kashmir

Nematode Species	*Host Species*	*Location of Infection*	*Distribution/ Locality (ies)*
(41) *Teladorsagia ciruminicta*	Goats	Not known	Kashmir
(42) *Toxocara vitulorum (=Neoascaris vitulorum)*	Cattle	Not available	Jammu & Kashmir
(43) *Thelazia sp.*	Buffalo	Eyes	Jammu & Kashmir
(44) *Trichostrongylus axei*	Sheep	Abomasum, Intestinal tract	Handwara
(45) *Trichostrongylus colubriformis*	Sheep,	Abomasum, Intestinal tract	Handwara
(45) *Trichostrongylus* sp.	Indian Yak, sheep, goat	Gastro-intestine	J & K, Ladakh
(46) *Trichostrongylus* sp.	Sheep, goats	Abomasum, intestinal tract	Jammu &Kashmir
(47) *Trichuris ovis*		Abomasum, intestinal tract	Jammu & Kashmir
(48) *Trichuris* sp.	Cattle, buffalo, pig - equines, dog	Not known	Kashmir
(49) *Triodontophorus* sp	Equines	Abomasum, intestine	J & K, Ladakh
(50) *Varestrongylus pneumonicus*	Sheep, goats	Lungs (bronchi/ bronchioles)	Kashmir & Ladakh
(51) *Varestrongylus* sp.	Sheep, goats	Lungs (bronchi/ bronchioles)	Jammu & Kashmir
(H) Wildlife (Hangul, Bear, Fox)			
Nematodes			
(1) *Ancylostoma* sp.	The Himalayan Black Bear (*Selenarctos thibetans laniger*)	Not available	Kashmir
(2) *Baylisascaris transfuga*	The Himalayan Black Bear	Not available	Kashmir (Dachigam)
(3) *Chabertia ovina*	Hangul (*Cervus e.h.*)	Gastro-intestine	Dachigam, Kashmir
(4) *Dictyocaulus viviparus*	Hangul	Not available	Dachigam, Kashmir
(5) *Haemonchus* sp.	Hangul	Gastro-intestine	Dachigam, Kashmir
(6) *Marshallagia* sp.	Hangul	Gastro-intestine	Dachigam, Kashmir
(7) *Nematodirus* sp.	Hangul	Gastro-intestine	Kashmir
(8) *Oesophagostomum circumcicata*	Hangul	Not available	Dachigam, Kashmir
(9) *Oesophagostomum* sp.	Hangul	Gastro-intestine	Dachigam, Kashmir
(10) *Ostertagia* sp.	Hangul	Gastro-intestine	Dachigam, Kashmir
(11) *Physaloptera* sp.	*Vulpes Vulpes* (Fox)	Stomach	Kashmir

Chapter 4

Catalogue on Prevalence of Helminths of Medical and Veterinary Importance

4.1 Distribution Prevalence and, High Infection in Age Group and Gender Due to Helminth Parasites (Nematodes, Cestodes, Trematodes)

4.1.1 Nematodes of Medical Importance

Nematodes of medical importance are given in Table 4.1.1.

Table 4.1 1: Distribution, Prevalence and, High Infection in Age Group and Gender Due to Helminth Parasites (Nematodes and Cestodes) of Medical Importance in Jammu and Kashmir, and Ladakh UTs (NA= data not available)

Distribution/ Study area	*Prevalence (Overall per cent)*	*High Infection*		
		Age Group	*Gender*	*References*
1. *Ancylostoma duodenale* (Dubini) (Old world Hookworm) (Human ancylostomiasis)				
Jammu (hospital SMGS based study)	0.82	NA	NA	Jad *et al.* (2015)
2. *Ascaris lumbricoides* (Large roundworm of Human) (Ascariasis)				
Kashmir Valley	68.30	Children (rural/ urban)	NA	Wani *et al.* (2017)
Shopian District	71.8	Children (4-15 yrs.)	NA	Wani and Amin (2016)
Jammu(hospital SMGS based study)	3.3	NA	NA	Jad *et al.* (2015)

Distribution/ Study area	*Prevalence (Overall per cent)*	*High Infection*		
		Age Group	*Gender*	*References*
Kashmir valley	69.23	Children (rural) (4-15 yrs.)	Male	Lone *et al.* (2012)
Kashmir	28	Children (rural, urban) (5-14yrs.)	NA	Singh *et al.* (2010)
Gurez valley	71.87	Children (rural, urban) (11-15 yrs.)	Insignificant Males/ Females	Wani *et al.* (2010)
Kashmir valley	68.30	Children (rural, urban)	NA	Wani *et al.* (2008)
Pulwama District	69.84	Children (0- 15 yrs.)	NA	Wani and Ahmed (2009)
Budgam District	54.9	Children (school)	NA	Lone *et al.* (2019)
Kupwara District	69.23	Children (rural) (4-15 yrs.)	NA	Wani *et al.* (2007 a)
Srinagar District	28.4	Children (city)	NA	Wani *et al.* (2007 b)
Srinagar (hospital -based study	82.60	Children (21-40 yrs.)	NA	Ahmad *et al.* (2002)
3. *Enterobius vermicularis* (Linnaeus) (Human Pinworm) (Enterobiasis)				
Kashmir Valley	12.67	Children (rural/ urban)	NA	Wani *et al.* (2017)
Shopian District	13.90	Children (4-15 yrs.)	NA	Wani and Amin (2016)
Kashmir Valley	7.76	Children (4-15 yrs.)	Male	Lone *et al.* (2012)
Gurez Valley	13.92	Children (14-15 yrs.)	Insignificant (Males/ Females)	Wani *et al.* (2010)
Kashmir Valley	12.67	Children (rural, urban)	NA	Wani *et al.* (2008)
Kupwara District	7.69	Children (4-15 yrs.)	Male	Wani *et al.* (2007 a)
Pulwama District	16.80	Children (under 15 yrs.)	NA	Wani and Ahmed (2009)
Budgama District	2.6	School children	NA	Lone *et al.* (2019)
4. *Toxocara canis* (Human toxocariasis)				
Kashmir	32.86 (sero-prevalence)	NA	NA	Dar *et al.* (2008)
Srinagar	25 (sero-prevalence)	Children (5 – 16 years)	NA	Ahmad *et al.* (2008)
Kupwara	34.78 (sero-prevalence)	Children (5- 16 years)	NA	Ahmad *et al.* (2008)
5. *Trichuris trichiura* (Human whipworm) (Trichuriasis)				
Kashmir Valley	27.92	Children (rural, urban)	NA	Wani *et al.* (2017)

Distribution/ Study area	*Prevalence (Overall per cent)*	*High Infection*		
		Age Group	*Gender*	*References*
Shopian	26.4	Children (4-15 yrs.)	NA	Wani and Amin (2016)
Kashmir Valley	30.76	Children (4-15 yrs.)	Females	Lone *et al.* (2012)
Kashmir	05	Children (4-15 yrs.)	NA	Singh *et al.* (2010)
Gurez Valley	26.42	Children (rural, urban)	Insignificant (Males/ Females)	Wani *et al.* (2010)
Kashmir Valley	27.92	Children (rural, urban)	NA	Wani *et al.* (2008)
Srinagar	4.9	Children (Srinagar city)	NA	Wani *et al.* (2007b)
Kupwara District	30.76	Children (4-15 yrs.)	NA	Wani *et al.* (2007 a)
Jammu, based on (SMGS Hospital)	0.82	NA	NA	Jad *et al.* (2015)
Pulwama District	31.65	Children (under 15 yrs.)	NA	Wani and Ahmed (2009)
Budgam District	32.5	Children (school)	NA	Lone *et al.* (2019)

4.1.2 Cestodes of Medical Importance

Cestodes of medical importance are presented in Table 4.1.2.

Table 4.1 2: Distribution, Prevalence and, High Infection in Age Group and Gender Due to Cestode Parasites

Distribution/ Study area	*Prevalence (Overall per cent)*	*High Infection*		
		Age Group	*Gender*	*References*
1. *Echinococcus granulosus* (Hydatid cyst)				
District (9) of Kashmir	5.3	Younger (under 15 yrs.)	Males	Fomda *et al.* (2015)
Anantnag	9.23	Younger (under 15 yrs.)	Males	Fomda *et al.* (2015)
Baramulla	8.5	Younger (under 15 yrs.)	Males	Fomda *et al.* (2015)
Bandipora	8	Younger (under 15 yrs.)	Males	Fomda *et al.* (2015)
Kashmir (Based on Medical records of SKIMS from Jan. 1998 to Dec., 2003)	91.3 (Liver)	Adult	Males/ Females	Wani *et al.* (2005)

Nematode Species	*Host Species*	*Location of Infection*	*Distribution/ Locality (ies)*
(12) *Strongyloides* sp.	Hangul	Gastro-intestine	Kashmir
(13) *Trichostrongylus* sp.	Hangul	Gastro-intestine	Kashmir
(14) *Trichuris* sp.	Hangul	Gastro-intestine	Kashmir
(I) Humans Nematodes			
(1) *Ascaris lumbricoides*	Humans	Gastro-intestinal tract	Jammu & Kashmir
(2) *Entrobius vermicularis*	Humans	Intestine	Jammu & Kashmir
(3) Hookworm	Humans	Intestine	Rajouri (Jammu)
(4) *Toxocara canis*	Humans	Intestine	Kashmir
(5) *Trichuris trichiura*	Humans	Intestine	Jammu

Source: **(i) Insect nematodes**: Basir (1941); Fotedar (1964); Gupta (1967); Rehana *et al.* (2016). **(ii) Fish nematodes**: Ahmad *et al.* (2018); Chakravorti (1942); Chishti and Bakshi (1990); Chishti *et al.* (2002); Dhar and Anwari (1982); Dhar and Fotedar (1979); Dhar and Majidah (1987); Fotedar and Dhar (1970, 75); Gupta (1967); Khan *et al.* (2004); Majidah and Khan (1996); Khan and Majidah (1974); Peerzada (2009); Raina and Dhar (1972); Thapar (1950). **(iii) Amphbian nematodes:** Fotedar (1960, 1965, 1971, 1973 and 1980). **(iv) Reptile (snakes, lizard) nematodes**: Fotedar *et al.* (1966, 68); Fotedar (1968). **(v) Poultry bird nematodes**: Fotedar and Khateeb (1987); Pandit *et al.* (1991); Salam *et al.* (2009);Tanveer and Chishti (1999); Kamil *et al.* (2011). **(vi) Wild bird Nematodes**: Chishti (1978, 81) Chishti and Abdullah (1986); Fotedar and Chishti (1974, 80); Fotedar *et al.* (1972); Fotedar and Kaw (1965); Gupta (1967); Tanveer and Chishti (1999); Dar *et al.* (2013). **(vii) Rat and Shrew Nematodes**: Raina (1970); Raina and Kaul (1974) and Raina and Raina (1985). **(viii) Cat Nematodes**: Fotedar and Chishti (1974); Borkataki *et al.* (2013). **(ix) Dog nematodes**: Shah *et al.* (2017); Rebeeca *et al.* (2014); Rani-Putri *et al.* (2010). **(x) Bat nematode**: Raina and Kaul (1982). **(xi) Wild mammals (Hangul, Black Bear, Fox) nematodes** : Lone *et al.* (2014); Shahadar *et al.* (1995); Pandit *et al.* (2007); Sheikh *et al.* (2017). **(xii) The Indian Yak (Bos grunuiens) nematodes** : Ranga-Rao *et al.* (1994). **(xiii) Domestic animals (mammals) nematodes** : Ahmad *et al.* (2016); Bambroo (1971); Bali (1972); Bhadwal and Raina (1994); Baqui and Shah (2002); Dar (1980, 1982); Dhar and Sharma (1978, 1978 a, 1979); Dhar *et al.* (1972, 1979, 1982, 1982 a, 1983); Fotedar and Bambroo (1965, 1968); Fotedar and Wali (1984); Gupta (1967); Kaur *et al.* (2015); Khajuri (1966); Makhdomi *et al.* (1995); Parrah *et al.* (2004, 2006); Rebecca *et al.* (2014); Sharma (1994); Sharma and Dhar (1978, 1979, 1987); Tariq *et al.* (2008 a) Yadav *et al.* (2006). **(xiii) Human nematodes:** Arora *et al.* (1976); Hangloo *et al.* (1989); Khuroo (1999); Mir *et al.* (2002); Wani and Chrungoo (1992) Wattal *et al.* (1986); Ahmad *et al.* (2002).

Distribution/ Study area	*Prevalence (Overall per cent)*	*High Infection*		
		Age Group	*Gender*	*References*
	2.2 (Spleen)	Adult	Males/ Females	Wani *et al.* (2005)
	1.1 (Pancreas)	Adult	Males/ Females	Wani *et al.* (2005)
	1.6 (Peritoneum and Pelvic)	Adult	Males/ Females	Wani *et al.* (2005)
	0.6 (Gall Bladder)	Adult(51yrs.)	Females	Wani *et al.* (2005)
	0.6 (Meso-colon)	Adult (30yrs.)	Females	Wani *et al.* (2005)
	0.6 (Adrenal)	Adult (49 yr.)	Females	Wani *et al.* (2005)
Kashmir (Tertiary Care Hospital Based study from Jan 1994 to Dec 2001)	12.1 (Surgically confirmed cases of cystic echino-coccosis)	Adult/Young	Males/ Females	Formada *et al.* (2002)
Kashmir	3.5 (Spleen)	NA	NA	Dar *et al.* (2002)
Srinagar	4.10 (spleen)	NA	NA	Wani (1993)
2. *Hymenolepis nana* (Bilharz) (Dwarf Tape worm)				
Jammu (Hospital study of SMGS Hospital)	4.13	NA	NA	Jad *et al.* (2015)
Budgam District	2.05	Children (school)	NA	Lone *et al.* (2019)
3. *Taenia saginata* Goeze (Beef tapeworm) (Taeniasis)				
Kashmir Valley	4.60	Children (rural, urban)	NA	Wani *et al.* (2017)
Shopian	5.3	Children (4-15 yrs.)	NA	Wani and Amin (2016)
Kashmir Valley	7.69	Children (4-15 yrs.)	Males/ Females	Lone *et al.* (2012 b)
Kashmir	4	Children (5-14 yrs.)	NA	Singh *et al.* (2010)
Gurez Valley	5.39	Children (11-15 yrs.)	Insignificant (Males/ Females)	Wani *et al.* (2010)
Baramulla	.61	NA	NA	Lateef *et al.* (2008)
Kashmir Valley	4.60	Children (rural, urban)	NA	Wani *et al.* (2008)
Kupwara District	7.69	Children (rural) (4-15 yrs.)	NA	Wani *et al.* (2007 a)
Srinagar	3.7	Children (Srinagar City)	NA	Wani *et al.* (2010 b)

Distribution/ Study area	Prevalence (Overall per cent)	High Infection		
		Age Group	Gender	References
Kashmir	2.25	All age group	NA	Khan *et al.* (2008)
Pulwama District	3.01	Under 15 yrs.	NA	Wani and Ahmed (2009)
Budgam District	9.1	Children (school)	NA	Lone *et al.* (2019)

4.1.3 Nematodes of Veterinary Importance

Nematodes of veterinary importance are incorporated in Table 4.1.3.

Table 4.1.3: Distibution, Prevalence and, High Infection in Age Group and Gender Due to Nematodes
(Key to the abbreviations given at the end of the table)

Host	Distribution/ Study Area	Prevalence (per cent)	High/Maximum Infection			Reference
			Period	Age Group	Gender	
1. ***Ancylostoma caninum*** **(Ereolani) (Hookworm) (Ancylostomiasis)**						
DO	Jammu	40	NA	NA	NA	Shah *et al.* (2017
CT	Jammu	80	NA	NA	NA	Borkataki *et al.* (2013)
2. ***Ancylostoma*** **sp.**						
BB	Kashmir	37.50	NA	NA	NA	Sheikh *et al.* (2017)
DO	Ladakh	4.7	NA	NA	NA	Rebecca *et al.* (2014)
3. ***Ascardia galli*** **Schrank (Ascaridiasis)**						
DG	Gurez	32.11	Summer	Y	NA	Sheikh *et al.* (2016)
4. Ascarids						
DG	Kashmir	02.67	NA	NA	NA	Hamdani *et al.* (2017)
PI	Jammu	30.96	NA	Y	NA	Khajuria *et al.* (2010)
BU (calf)	R. S. Pora	7.05	Rainy Season	NA	NA	Singh *et al.* (2009)
5. ***Baylisascaris transfuga*** **(Baylisascariasis)**						
BB	Kashmir	47.32	NA	NA	NA	Sheikh *et al.* (2017)
6. ***Bunostomum trigocephalum*** **(Hookworm disease) (Bunostomiasis)**						
SH	Kashmir	7.27	NA	NA	NA	Wani *et al.* (2012)
GO	Kashmir valley	30.01	Summer	Y	NA	Tariq *et al.* (2010)
7. ***Bunostomum*** **sp.**						
SH, GO	Jammu (Middle Agro Climatic Area)	3.83	Monsoon Season	Y	F	Khajuria *et al.* (2013)
SH	Gandarbal	14.66	Summer	A	F	Wani *et al.* (2011)

Host	Distribution/ Study Area	Prevalence (per cent)	High/Maximum Infection			Reference
			Period	Age Group	Gender	
8. *Capillaria* sp. (Capillariasis)						
DG	Kashmir	5.33	NA	NA	NA	Hamdani *et al.* (2017)
9. *Chabertia ovina* Fabricius (Large-mouthed bowl- worm) (Chabertiasis)						
SH, GO	Gurez	19.5 (SH) 17.7 (GO)	Autumn	YY	NA	Sheikh *et al.* (2016) -do-
SH	Kashmir Valley	20	NA	NA	NA	Wani *et al.* (2012)
GO	Gurez	17.70	NA	NA	NA	Sheikh *et al.* (2016)
PG	Ladakh	18.78	Rainy Season	Y	F	Kuchai *et al.* (2012)
GO	Jammu	4.57	Summer	NA	NA	Mir *et al.* (2012)
GO	Ladakh	22.35	Rainy Season	Y	F	Kuchai *et al.* (2011a)
GO	Kashmir	29.08	Summer	Y	NA	Tariq *et al.* (2010)
HA	Dachigam	04.02	Summer	Y	F	Lone *et al.* (2014)
10. *Chabertia* sp.						
GO	Jammu and Kashmir (TA)	08.47	Summer	NA	NA	Bihaqi *et al.* (2017)
CA	Jammu	2.91	NA	NA	NA	Mir *et al.* (2013)
BU	Jammu	6.35	NA	NA	NA	Mir *et al.* (2013)
SH, GO (SL)	Gandarbal	52	Summer	NA	NA	Lone *et al.* (2012)
GO	Baramulla	01	NA	NA	NA	Lone *et al.* (2012)
SH	Gandarbal	20.83	Summer	A	F	Wani *et al.* (2011)
SH	Kashmir Valley	6.67	Summer	A	F	Tamboo *et al.* (2015)
11. *Cyathostomum or Cyclicostephanus* (= *Trichonema*)						
EQ	Kashmir Valley	96.78	NA	NA	NA	Pandit *et al.* (2008)
12. *Dictyocaulus filaria* (Lungworm) (Dictyocaulosis)						
SH	Kashmir	16.36	NA	NA	NA	Wani *et al.* (2012)
PG	Kargil	32.24	Rainy Season	Y	F	Kuchai *et al.* (2012)
GO	Jammu	3.52	Summer	NA	NA	Mir *et al.* (2012)
GO	Ladakh	37.05	Rainy Season	Y	F	Kuchai *et al.* (2011a)
13. *Dictyocaulus viviparus* (Bovine Lungworm)						
HA	Dachigam	28.40	Summer	Y	M	Lone *et al.* (2014)
14. *Dictyocaulus* sp.						
EQ	Kashmir	14.10	NA	NA	NA	Pandit *et al.* (2008)

Host	*Distribution/ Study Area*	*Prevalence (per cent)*	*High/Maximum Infection*			*Reference*
			Period	*Age Group*	*Gender*	
15. *Gaigeria pachycelis* Railliet and Henrey (Hookworm) (Gaigeriasis)						
SH	Kashmir	1.81	NA	NA	NA	Wani *et al.* (2012)
16. *Gaigeria* sp.						
SH	Ganderbal	**3.33**	Summer	A	F	Wani *et al.* (2011)
17. *Haemonchus controtus* (Cobb) (Barber's Poleworm) (Red stomach worm) (Haemonchosis)						
SH, GO	Gurez	27.64 (SH) 26.04 (GO	Autumn Autumn	Y	NA	Sheikh *et al.* (2016b)
SH	R.S. Pora	76	NA	NA	NA	Vohra *et al.* (2016)
SH (SL)	Anantnag Pulwama Srinagar	61.61	Summer	NA	M	Tak *et al.* (2013)
SH	Kashmir	50.90	NA	NA	NA	Wani *et al.* (2012)
GO	Gurez	26.4	Autumn	Y	NA	Sheikh *et al.* (2016)
GO	Baramulla	60	NA	NA	NA	Lone *et al.* (2011)
GO	Jammu	28.87	Summer	NA	NA	Mir *et al.* (2012)
GO	Ladakh	41.02	Rainy Season	Y	F	Kuchai *et al.* (2011)
GO	Kashmir	48.03	Summer	Y	NA	Tariq *et al.* (2010)
PG	Kargil	42.15	Rainy Season	Y	F	Kuchai *et al.* (2012)
HA	Dachigam	55.39	Summer	Y	M	Lone *et al.* (2014)
SH, GO	Jammu (MAC)	61.18	Monsoon	Y	F	Khajuria *et al.* (2013b)
18. *Haemonchus* spp.						
SH (S)	Srinagar	53.3	Summer	NA	NA	Gul and Tak (2016)
SH	Kashmir	55	Summer	A	F	Tramboo *et al.* (2015)
SH	Jammu	58.20	NA	NA	F	Singh *et al.* (2014)
SH, GO	Gandarbal	82	Summer	NA	NA	Lone *et al.* (2012)
SH	Ganderbal	41.50	Summer	Y	F	Wani *et al.* (2011)
GO	Jammu	61.65	NA	NA	F	Singh *et al.* (2014)
GO	Ganderbal	82	Summer	NA	NA	Lone *et al.* (2012)
GO	Jammu	28.87	Summer	NA	NA	Mir *et al.* (2012)
CA	Jammu	14.59	NA	NA	NA	Mir *et al.* (2012)
BU	Jammu	9.82	NA	NA	NA	Mir *et al.* (2012)
19. *Heterakis gallinarum* Schrank (Heterakiasis)						
DG	Gurez	32.11	Summer	Y	NA	Sheikh *et al.* (2016)
20. *Marshallegia marshalli* (Ranson)						
GO	Kashmir	16.6	Summer	Y	NA	Tariq *et al.* (2010)

Host	*Distribution/ Study Area*	*Prevalence (per cent)*	*High/Maximum Infection*			*Reference*
			Period	*Age Group*	*Gender*	
21. *Marshallegia* sp.						
SH	Ganderbal	03?	Summer	A	F	Wani *et al.* (2011)
22. *Nematodirus spathiger* (Railliet) (Thin-necked intestinal worm) (Nematodiriasis)						
GO	Kashmir	25.2	Summer	Y	NA	Tariq *et al.* (2010)
23. *Nematodirus* spp.						
SR	Kashmir	70.63 (NK) 70.8 (SK) 70.84 (CK)	NA NA NA	NA NA NA	NA NA NA	Allaie *et al.* (2018)
SH, GO (SL)	Ganderbal	60	Summer	NA	NA	Lone *et al.* (2012)
SH	Kashmir	57.75	Summer	A	F	Tramboo *et al.* (2015
SH	Kashmir	89.07	NA	NA	NA	Bhat *et al.* (2007)
SH	Kashmir	9	Monsoon	Y	NA	Bhat *et al.* (2012)
SH	Gandarbal	13.16	Summer	A	F	Wani *et al.* (2011)
24. *Oesophagostomum circumcincta* (Nodular worm) (Oesophagostomiasis)						
HA	Dachigam	13.7	Summer	NA	NA	Lone *et al.* (2014)
25. *Oesophagostomum columbianum* (Nodular worm) (Oesophagostomiasis)						
SH	Kashmir	12.72	NA	NA	NA	Wani *et al.* (2012)
GO	Kashmir	23.5	Summer	Y	NA	Tariq *et al.* (2010)
GO	Srinagar	34.34	NA	NA	NA	Lone *et al.* (2011
26. *Oesophagostomum* sp.						
SH	R. S.Pora	12	NA	NA	NA	Vohra *et al.* (2016)
SH	Kashmir	9.17	Summer	A	F	Tramboo *et al.* (2015)
SH	Gandarbal	11.83	Summer	A	F	Wani *et al.* (2011)
SH, GO	Jammu (MCZ)	3.84	Monsoon	Y	F	Khajuria *et al.* (2013b)
SH, GO (S)	Ganderbal	46	Summer	NA	NA	Lone *et al.* (2012)
GO	Baramulla	45	NA	NA	NA	Lone *et al.* (2011)
27. *Ostertagia* sp. (Brown stomach worm) (Oestergiasis)						
SH, GO	Jammu (LST)	12.17	Monsoon	Y	F	Khajuria *et al.* (2013 b)
SH (SL)	Srinagar	41.2	Summer	NA	NA	Gul and Tak (2016 a)
SH, GO	Gurez	25.20 (SH)22.91 (GO)	Autumn	Y	NA	Sheikh *et al.* (2016 b)
SH	Kashmir	11.67	Summer	A	F	Tramboo *et al.* (2015)
SH	Ganderbal	28	Summer	A	F	Wani *et al.* (2011)
GO	Jammu and Kashmir (TA)	07.62	Summer	NA	F	Bihaqi *et al.* (2017)
GO	Gurez	22.9	Autumn	Y	NA	Sheikh *et al.* (2016)

Host	Distribution/ Study Area	Prevalence (per cent)	High/Maximum Infection			Reference
			Period	Age Group	Gender	
28. *Oxyuris* sp. (Equine pin worm) (Oxyuriasis)						
EQ	Kashmir	9.40	NA	NA	NA	Pandit *et al.* (2008)
29. *Gongylonema* (Gongylonemiasis)						
SH, OK, SL	Srinagar	37	Aug. to Nov.	NA	NA	Tariq *et al.* (2008 a)
30. *Parascaris* (Parascariasis)						
EQ	Kashmir	4.1	NA	NA	NA	Pandit *et al.* (2008)
31. *Strongyles/Strongyls* (Strongylosis)						
CA	Kashmir	73.85	Winter	?	F	Bushra *et al.* (2013)
SS	Kashmir	57.75	Summer	?	F	Tramboo *et al.* (2015)
BO	Gurez	11.1	NA	NA	NA	Allaie *et al.* (2014)
BU	R.S.Pora	49.19	Rainy Season	NA	NA	Singh *et al.* (2009)
EQ	Gurez	28.06	NA	NA	NA	Allaie *et al.* (2014
LS	Kargil	8.57	NA	NA	NA	Wani *et al.* (2013)
SH, GO	Jammu	57.86	Monsoon	NA	NA	Khajuria *et al.* (2012)
SH	Kashmir	24.61	Monsoon	Y	NA	Bhat *et al.* (2012)
PI	Jammu	35.48	NA	Y	NA	Khajuria *et al.* (2010)
SH, GO	Jammu	44.62	Rainy	NA	NA	Jadav *et al.* (2006) ?
32. *Strongyloides* sp. (Strongyloidiasis)						
GO	JK (TA)	5.9	Summer	A	F	Bihaqi *et al.* (2017)
CT	Jammu	28	NA	NA	NA	Borkataki *et al.* (2013)
DG	Kashmir	74.67	NA	NA	NA	Hamdani *et al.* (2017)
CA	Kashmir	3.58	Winter	Y	F	Bushra *et al.* (2013)
EQ	Kashmir	6.19	NA	NA	NA	Pandit *et al.* (2008)
PI	Jammu	10.39	NA	Y	NA	Khajuria *et al.* (2010)
BU	Jammu	0.1	Rainy	NA	NA	Singh *et al.* (2009)
SH, GO	Jammu (MCZ)	4.14	Monsoon	Y	F	Khajuria *et al.* (2013
SH, GO	Jammu (LST)	4.16	Monsson	NA	NA	Khajuria *et al.* (2012)
SH, GO	Jammu	1.05	Rainy	NA	Y	Yadav *et al.* (2006)
SH, GO (SL	Ganderbal	42	Summer	NA	NA	Lone *et al.* (2012)
SH	Kashmir	1.67	Summer	A	F	Tramboo *et al.* (2015)
SH	Kashmir	15.5	Mansoon	Y	NA	Bhat *et al.* (2012)
33 *Strongylus* sp. (Strongylosis)						
EQ	Kashmir	81.19	NA	NA	NA	Pandit *et al.* (2008)
34. *Teladorsagia circumcincta* (Stadelman) (= *Ostertagia circumcincta*) (Brown Stomach worm)						
GO	Kashmir	29.08	Summer	Y	NA	Tariq *et al.* (2010)

Host	*Distribution/ Study Area*	*Prevalence (per cent)*	*High/Maximum Infection*			*Reference*
			Period	*Age Group*	*Gender*	
35. *Trichostrongylus* (Black Scour worm) (Trichostrongyliasis)						
SHGO	Jammu (MCZ)	13.67	Monsoon	Y	F	Khajuria *et al.* (2013b)
SHGO (SL)	Ganderbal	58	Summer	NA	NA	Lone *et al.* (2012)
SH	R.S.Pora	09	NA	NA	NA	Vohra *et al.* (2016)
SH	Kashmir	17.5	Summer	A	F	Tramboo *et al.* (2015)
SH	Kashmir	03.63	NA	NA	NA	Wani *et al.* (2012)
SH	Gandabal	24.83	Summer	NA	F	Wani *et al.* (2011)
GO	JK (TA)	12.86	Summer	NA	F	Bihaqi *et al.* (2017)
GO	Kashmir	25.01	Summer	Y	NA	Tariq *et al.* (2010)
36. *Trichuris ovis* (Whip worm) (Trichuriasis)						
SH	Kashmir	30.9	NA	NA	NA	Wani *et al.* (2012)
PG	Ladakh	37.46	Rainy	Y	F	Kuchai *et al.* (2012)
GO	Ladakh	48.3	Rainy	Y	F	Kuchai *et al.* (2011)
GO	Jammu (STA)	14.43	Summer	NA	NA	Mir *et al.* (2012)
GO	Baramulla	55	NA	NA	NA	Lone *et al.* (2011)
GO	Kashmir	19	Summer	Y	NA	Tariq *et al.* (2010)
HA	Dachigam	39.75	Summer	Y	M	Lone *et al.* (2014)
37. *Trichuris* sp.						
SH, GO	Srinagar	59.37	Winter	Y	F	Gul and Tak (2016 a)
SH, GO	Jammu	14.53	Mansoon	NA	NA	Khajuria *et al.* (2012)
SHGO (SL)	Ganderbal	74	NA	NA	NA	Lone *et al.* (2012)
SH	Kashmir	1.5	Summer	A	F	Tramboo *et al.* (2015)
SH	Ganderbal	7	Summer	A	F	Wani *et al.* (2011)
GO	J&K (TA)	4.29	Summer	A	F	Bihaqi *et al.* (2017)
GO	Gurez	15.3	NA	NA	NA	Allaie *et al.* (2014)
SH, GO (S)	Ganderbal	74	NA	NA	NA	Lone *et al.* (2012)
CA	Jammu	8.02	NA	NA	NA	Mir *et al.* (2013)
CA	Kashmir (CK)	5.41	Wet Season	Y	F	Bushara *et al.* (2013)
BU	Jammu	2.89	NA	NA	NA	Mir *et al.* (2013)
PI	Jammu	10.32	NA	Y	NA	Khajuria *et al.* (2010)
SH, GO	Jammu	3.8	Rainy	NA	NA	Yadav *et al.* (2006)?
38. *Toxocara canis* (Dog round worm) (Toxocariasis)						
DO	Jammu	15	NA	NA	NA	Shah *et al.* (2017)

Host	Distribution/ Study Area	Prevalence (per cent)	High/Maximum Infection			Reference
			Period	Age Group	Gender	
39. *Toxocara vitulorum* (Toxocariasis)						
CA	Kashmir	1.04	Winter	Y	F	Bushra *et al.* (2013)
40. *Toxocara* sp.						
CT	Jammu	32	NA	NA	NA	Borkataki *et al.* (2013)
41. *Triodontophorus* sp.						
EQ	Kashmir	41.39	NA	NA	NA	Pandit *et al.* (2008)

4.1.4 Trematodes of Veterinary Importance

Trematodes of veterinary importance are given in Table 4.1.4.

Table 4.1.4: Distribution, Prevalence and, High Infection in Age Group and Gender Due to Trematode Parasites

Host	Distribution/ Study Area	Prevalence (per cent)	High/Maximum Infection			Reference
			Period	Age Group	Gender	
1. ***Dicrocoelium dendriticum***						
SH	Kashmir	4.45	Summer	A	NA	Mir *et al.* (2008)
SH	Gurez	23.57 (SH)	Autumn	Y	NA	Sheikh *et al.* (2016 b)
GO	Gurez	22.91 (GO)	Autumn	Y	NA	Sheikh *et al.* (2016 b)
GO	Jammu (MCZ)	3.87	Summer	NA	NA	Mir *et al.* (2012)
2. ***Dicrocoelium* sp.**						
GO	JK (TA)	3.44	Summer	A	F	Bihaqi *et al.* (2017)
SH	Kashmir	11.58	Summer	A	NA	Tramboo *et al.* (2015)
SH	Ganderbal	9.33	Winter	A	F	Shahnawaz *et al.* (2011)
SH, GO	Jammu (MCZ)	5.4	Monsoon	Y	F	Khahuria *et al.* (2013)
SH, GO	Jammu	3.90	Monsoon	NA	NA	Khajuria *et al.* (2012)
SH, GO	Jammu	1.92	Rainy	NA	NA	Yadav *et al.* (2006)
SH, GO	Ganderbal	52	Summer	NA	NA	Lone *et al.* (2012)
CA	Jammu	4.37	NA	NA	NA	Mir *et al.* (2013)
BU	Jammu	1.15	NA	NA	NA	Mir *et al.* (2013)
GO	Gurez	3.5	NA	NA	NA	Allaie *et al.* (2014)
CA	Kashmir	3.8	Winter	Y	F	Bushra *et al.* (2013)
3. ***Fasciola gigantica* (Cobbold)**						
CA (SL)	Kashmir	20.86	Autumn	NA	NA	Gul *et al.* (2016)
SH	Kashmir	23.92	NA	NA	NA	Mir *et al.* (2008)

Host	*Distribution/ Study Area*	*Prevalence (per cent)*	*High/Maximum Infection*			*Reference*
			Period	*Age Group*	*Gender*	
4. *Fasciola hepatica* (Linnaeus)						
SH	Gurez	21.13	Autumn	Y	NA	Sheikh *et al.* (2016b)
GO	Gurez	21.87	Autumn	Y	NA	Sheikh *et al.* (2016 b)
CA (SL)	Kashmir	3.36	Autumn	NA	NA	Gul *et al.* (2016)
CA (SL)	Kashmir (north)	20.89	NA	A	NA	Magray *et al.* (2017)
SH	Kashmir	9.96	NA	NA	NA	Mir *et al.* (2008)
GO	Jammu (LST)	5.28	Summer	NA	NA	Mir *et al.* (2012)
GO	Gurez	21.87	Autumn	Y	NA	Sheikh *et al.* (2016)
HA	Dachigam	17.3	Summer	Y	M	Lone *et al.* (2014)
5. *Fasciola* spp.						
CA	Ladakh	51.42 (SL)	Wet Season	Y	F	Kuchai *et al.* (2011)
CA	Ladakh	27.69 (FA)	Wet Season	Y	F	Kuchai *et al.* (2011)
CA	Ladakh	21.91 (HH)	Wet Season	Y	F	Kuchai *et al.* (2011)
CA	Kashmir (CK)	16.45	Wet Season	A	F	Bushra *et al.* (2013)
SH, GO	Jammu (LST)	05.62	Monsoon	NA	NA	Khajuria *et al.* (2012)
SH, GO - (SL)	Ganderbal	60	Summer	NA	NA	Lone *et al.* (2012 b)
SH, GO	Jammu (MCZ)	8.2	Monsoon	Y	F	Khajuria *et al.* (2013)
SH	Ganderbal	04.66	Winter Season	A	NA	Shahnawaz *et al.* (2011)
SH	Kashmir	03.58	Summer	A	F	Tramboo *et al.* (2015)
GO	JK	(TA) 02.75	Summer	A	F	Bihaqi *et al.* (2017)
CA	Jammu	02.18	NA	NA	NA	Mir *et al.* (2013)
BU	Jammu	01.73	NA	NA	NA	Mir *et al.* (2013)
SH GO	Jammu	03.8	Rainy Season	NA	NA	Yadav *et al.* (2006)
6. *Notocotylus* sp.						
DG	Kashmir	2.67	NA	NA	NA	Hamdani *et al.* (2017)
7. *Paramphistomes* (= *Amphistomes*)						
BU	R.S. Pora	32.26	Rainy	NA	NA	Singh *et al.* (2009)
SH, GO	Jammu (MCZ)	8.2	Monsoon	Y	F	Khajuria *et al.* (2013)
SH	Ganderbal	6.5	Winter	A	F	Shanawaz *et al.* (2011)

Host	Distribution/ Study Area	Prevalence (per cent)	High/Maximum Infection			Reference
			Period	Age Group	Gender	
SH, GO	Jammu (LST)	21.14	Monsoon	NA	NA	Khajuria *et al.* (2012)
SH, GO	Jammu	8.7	Rainy	NA	NA	Yadav *et al.* (2006)
SH	Kashmir	4.83	Summer	A	F	Tramboo *et al.* (2015)
GO	JK (TA)	3.05	Summer	A	F	Bihaqi *et al.* (2017)
CA	Kashmir	13.12	Winter	Y	F	Bushara *et al.* (2013)
CA	Jammu	29.92	NA	NA	NA	Mir *et al.* (2013)
BU	Jammu	12.71	NA	NA	NA	Mir *et al.* (2013)
EQ	Kashmir	0.91	NA	NA	NA	Pandit *et al.* (2008)
BO	Gurez	5.5	NA	NA	NA	Allaie *et al.* (2014)
SH, GO	Ganderbal	46	Summer	NA	NA	Lone *et al.* (2012)
8. *Paramphistomum cervi* (Zeder)						
GO	Jammu (LST)	6.69	Summer	NA	NA	Mir *et al.* (2012)
SH	Kashmir	2.71	Summer	A	NA	Mir *et al.* (2008)
CA (SL)	Kashmir (north)	17.98	NA	A	NA	Magray *et al.* (2017)
9. *Paramphistomum* sp.						
SHGO	Ganderbal	46	Summer	NA	NA	Lone *et al.* (2012)
SH	Kashmir	4.83	Summer	A	F	Tramboo *et al.* (2015)
GO	J&K (TA)	3.05	Summer	A	F	Bihaqi (2017)
BO	Gurez	5.5	NA	NA	NA	Allaie *et al.* (2016)

4.1.5 Cestodes of Veterinary Importance

Cestodes of veterinary importance are included in Table 4.1.5.

Table 4.1.5: Distribution, Prevalence and, High Infection in Age Group and Gender Due to Cestode Parasites

Host	Distribution/ Study Area	Prevalence (per cent)	High/Maximum Infection			Reference
			Period	Age Group	Gender	
1. *Anophlocephalides*						
SH, GO	Jammu (LST)	20.6	Monsoon	NA	NA	Khajuria *et al.* (2012)
2. *Avitellina centripunctata* (Rivolta)						
GO	Kashmir valley	2.4	Autumn	Y	NA	Ahmad *et al.* (2009)
3. *Avatellina* sp.						
SH, OK, SL	Srinagar	38	Aug. to Nov.	NA	NA	Tariq *et al.* (2008 a)
GO (SL)	Ganderbal	42	NA	NA	NA	Lone *et al.* (2012)

Host	*Distribution/ Study Area*	*Prevalence (per cent)*	*High/Maximum Infection*			*Reference*
			Period	*Age Group*	*Gender*	
4. *Dipylidium caninum*						
CT	Jammu	2	NA	NA	NA	Borkataki *et al.* (2013)
DO	Jammu	20	NA	NA	NA	Shah *et al.* (2017)
5. *Echinococcus granulosus*						
SH (Local)	Kashmir	14.3	NA	NA	NA	Rashid *et al.* (2017)
SH	Jammu	20.45	NA	A	F	Khajuria *et al.* (2013a)
GO	Jammu	17.77	NA	A	F	Khajuria *et al.* (2013a)
GO (SL)	Jammu	14.1	NA	A	NA	Godara *et al.* (2014)
6. *Moniezia expansa* Rudolphi (Double-pored ruminant tapeworm)						
SH	Gurez	1.3	Autumn	Y	NA	Sheikh *et al.* (2016 b)
GO	Gurez	10.41	Autumn	Y	NA	Sheikh *et al.* (2016 b)
GO	JK (LST)	2.11	Summer	NA	NA	Mir *et al.* (2012)
HA	Dachigam	6.5	Summer	Y	F	Lone *et al.* (2014)
GO	Kashmir	4.1	Autumn	Y	NA	Ahmad *et al.* (2009)
7. *Moniezia* sp.						
SH, OK, SL	Srinagar	43	Aug. to Nov.	NA	NA	Tariq *et al.* (2008 a)
SH	Kashmir valley	7.92	Summer	A	F	Tramboo *et al.* (2015)
SH	Kashmir valley	3.3	Monsoon	Y	NA	Bhat *et al.* (2012)
SH	Ganderbal	11.83	Winter	A	F	Shahnawaz *et al.* (2011 a)
GO	JK (TA)	9.44	Summer	A	F	Bihaqi *et al.*, 2017)
SH, GO	Ganderbal	48	Summer	NA	NA	Lone *et al.* (2012)
SH, GO	R. S. Pora	0.96	Rainy Season	NA	NA	Yadav *et al.* (2006)
CA	Kashmir (CK)	4.10	Winter Season	Y	NA	Bushra *et al.* (2013)
CA	Jammu	3.64	NA	NA	NA	Mir *et al.* (2013)
BU	Jammu	1.15	NA	NA	NA	Mir *et al.* (2013)
GO	Gurez	14.1	NA	NA	NA	Allaie *et al.* (2014)
8. *Paranoplocephala*						
EQ	Kashmir	8.14	NA	NA	NA	Pandit *et al.* (2008)
9. *Raillietina cesticillus* Molin (Broad-headed tapeworm)						
DD	Kashmir	33.62	Summer	NA	NA	Salam *et al.* (2010)
DD	Kashmir	38.23	April	NA	NA	Dar *et al.* (2008)
10. *Raillietina tetragona* (Molin)						
DD	Gurez	27	Summer	Y	NA	Sheikh *et al.* (2016 a)
DD	Kashmir	47.05	July	NA	NA	Dar *et al.* (2008)

Host	Distribution/ Study Area	Prevalence (per cent)	High/Maximum Infection			Reference
			Period	Age Group	Gender	
11. *Stilesia globipunctate*						
GO	Jammu (ST)	3.62	Summer	NA	NA	Mir *et al.* (2012)
GO	Kashmir Valley	2	Autumn	Y	NA	Ahmad *et al.* (2009)
12. *Stilesia* sp.						
CA	Jammu	1.45	NA	NA	NA	Mir *et al.* (2013)
BU	Jammu	2.89	NA	NA	NA	Mir *et al.* (2013)
13. *Taenia hydatigena*/Cysticercus tenuicollis (Dog tape worm or thin necked bladder worm)						
SH (SL)	Kashmir	24.6	NA	NA	NA	Rashid *et al.* (2014)
DO	Jammu	10	NA	NA	NA	Shah *et al.* (2017)
DO	Ladakh	22.7	NA	NA	NA	Rebecca *et al.* (2014)
SH	Jammu	25.6	NA	A	M	Sharma *et al.* (2008)
14. *Taenia pisiformis* Bloch						
RB	Kashmir	16	NA	NA	NA	Mir *et al.* (2006)
15. *Taenia solium*/Cysticercus cellulosae (Pork tapeworm)						
PI	Jammu	1.16	NA	NA	NA	Sharma *et al.* (2017)
16. Taeniid						
CT	Jammu	40	NA	NA	NA	Borkataki *et al.* (2013)
17. *Thysanzeia giardi* (Moniez)						
GO	Kashmir	2.1	Autumn	Y	NA	Ahmad *et al.* (2009)
18. *Thysanzeia* sp.						
SH, OK, SL	Srinagar	31	Aug. to Nov.	NA	NA	Tariq *et al.* (2008 a)
GO	Ganderbal	28	NA	NA	NA	Lone *et al.* (2012 a)

Key to the abbreviations: A: Adult; **BB**: The Himalayan BlacK Bear; **BO**: Bovine; **BU**: Buffalo; **CA**: Cattle; **CK**: Central Kashmir; **CT**: Cat; **DD**: Domestic Duck; **DG**: Domestic Geese; **DO**: Dog; **EQ**: Equine; **F**: Female; **FA**: Farm animal; **GO**: Goat; **HA**: Hangul (*Cervus elaphus hanglu*); **PI: Pig**; **HH**: House hold animal; **J**: Jammu Province; **JK**: Jammu and Kashmir State; **K**: Kashmir Province; **L**: Ladakh Province; **LST**: Low-altitude subtropical area; **LS**: Live stock; **M**: Male; **MCZ**: Middle climatic zone; **NA**: Not available data; **NK**: North Kashmir;**OK**: Out side Kashmir (brought); **PI**: Pig; **PG**: Pashmina goat (*Capra siberica*); **RB**: Rabbit; **SH**: Sheep; **SL**: Slaughtered animal; **SK**: South Kashmir; **STA**: Subtropical area; **TA**: Temperate area; **Y**: Young.

Chapter 5

Acanthocephalan Parasites and Plant Parasitic Nematodes

5.1 Acanthocephalan Parasites

5.1.1 Systematic Survey, Diversity and Species Richness

Acanthocephalan parasites are commonly known as spiny-headed worms. The adults are endoparasites of various vertebrates, living in the alimentary tract, characterized by the presence of protrusible armed proboscis at the anterior terminal. The sexes are separate. The females are longer in size than males. van Cleave (1948), Hymen (1951) and Cheng (1973) gave Acanthocephalan a status of phylum, Cheng (1973) included three orders under this phylum as: Archiacanthocelphala (parasites of terrestrial hosts). Palaeacanthocephala (parasites of fishes, reptiles birds, some terrestrial reptiles and mammals and invertebrate crustacean). Eoucanthocephala (parasites of fishes, reptiles and crustacean) Petrotschenko (1956) regarded Acanthocephala as a class and divided it into three sub-classes, *viz.* Neoechinorhynchinae, Echinorhynchinae and Giganthorhynchinae. Yamaguti (1963) also regarded Acanthocephala as a class, however, included under it four orders as: Apororhynchidea, Neoechinorhynchidea, Echinorhynchidea and Giganthorhynchidea. For the classification and nomenclature of acanthocephalan parasite species, reported from various vertebrate hosts (fishes, amphibians and birds) so far from Jammu, Kashmir and Ladakh region, Yamaguti (1963) and Amin (2013) hare been followed, and these species belong to orders *viz.* Neoechinorhynchidea and Echinorhynchidea.

The acanthocephalans are injurious to the host by damaging the tissues of the intestinal wall by embedding the hooks present on the proboscis. There is inflammation at the site of attachment when the worms move and reattach, the old sores become infected by bacteria. In heavy infection, there is loss of appetite and interference with digestion, leading to the unthriftiness. Fotedar and Dhar (1977) reported heavy infection with about 200 *Pomphorhynchus* worms in the intestine of single fish, *viz. Schizothorax richardsonii* (=*Oreinus plagistomus*) from River Tawi of Jammu. Hyman (1951), reported massive infection from other region of the world. Occasionally,worms cause perforation of the gut wall and leading to a fatal peritonitis (Chandler and Read, 1961). A single worm species can exist in more than one species of host, as has been reported from Kashmir fishes. For example, Acanthocephalan, *viz. Pomphorhynchus kashmirensis,* has been found to infect as many as eight (8) different species of fishes in Kashmir. The same acanthocephalan species has been found in different families of the fishes.

Acanthocephalan worms parasitize almost all groups of vertebrate hosts as definitive hosts and reside in their alimentary tract after ingestion. The life-cycle of Acanthocephalan requires an intermediate host, which is usually an arthropod-(crustacean), for parasite species of aquatic vertebrate hosts and grubs, roaches *etc.*, for those of land animals. Some vertebrate hosts also act as intermediate hosts in which larvae remain encapsulated inside body cavity or in the muscle, for example in case of *Polymorphus.* The eggs of acanthocephalans after passing out from vertebrate hosts are ingested by invertebrate hosts (arthropod). In this host, egg develops into acanthor stage (spindle shape embryo stage). In host's gut, it penetrates the gut wall and move into haemocoel and thereafter transformed into a stage known as acanthella. After further development, the acanthella changes to juvenile known as cystocanth. This development in intermediate host takes from 6-12 weeks. When the arthropod host, harbouring with cystocanth is ingested by the appropriate vertebrate host, the cystocanth in the vertebrate intestine loses its sheath and develops into adult, with sexual maturity. In some cases, a second intermediate host is required.

It was Datta (1936, 37), who for the first time studied the Acanthocephalans of Kashmir fishes. Various new species of Acanthocephalans described by him include *Acanthocephala kashmirensis, Neoechinorhynchus devdevi* (=*Eosentis derdevi*), *N. yalei* (=*E. yalei*). Besides these, he also recorded *N. rigidus* and *N. rutili* from Kashmir Himalayan regions. Detailed investigations of Acanthocephala of fishes and Amphibians of Kashmir have been done by Kaw (1941, 51). He described various new species of these worms from fish hosts, which are, *viz. Echinorhynchus orientalis, Neoechinorhynchus manasbalensis* and *Pomphorhynchus kashmirensis.* Apart from these, Kaw (1951) also reported *Pomphorhynchus sp.* (encysted young form) from fish (*Botia birdi*). The first checklist of acanthocephalan parasites of vertebrates (fishes and amphibians) of Kashmir was provided by Gupta (1967). He listed about nine species of acanthocephala, infecting eight different species of fishes besides, two species parasitising frog (*Rana cyanophylictis*). This checklist was based on the investigations carried out by earlier workers on these parasites from different regions and areas of Jammu, Kashmir and Ladakh.

Various authors who described more new acanthocephalans from different fish species and amphibian (toad) are: Fotedar (1968) on *Neoechinorhynchus oreini* from *Schizothorax richardsonii* (=*Oreinus sinuatus*); Fotedar *et al.* (1970) on *Pomphorhynchus bufonis* from *Bufo viridis* (toad); Fotedar and Dhar (1977) on

Pomphorhynchus megacanthus from *Cyprinus carpio spicularis,* on *Pomphorhynchus oreini* from *S. richardsonii* (=*O. sinuatus*), on *P. orientalis* from *Schizothorax niger* and on *P. tori* from *tor tor;* Dhar and Kharoo (1985) on *Neoechinorhynchus glyptosternum* from *Glyptosternum* sp.; Chishti and Bakshi (1992) on *Pomphorhynchus fotedari* from *Cyprinus carpio communis;* Amin *et al.* (2017) on *Acanthogyrus* (*Acanthosentis*) *kashmirensis,* infesting cyprinids, *viz. Schizothorax plagiostomus* and *S. labiatus,* from tributary of River Jehlum (Sandras).

Earlier, the descriptions of two new acanthocephalans parasitizing the fishes of Jammu have been provided by Fotedar and Dhar (1977) on *Pomphorhynchus jammuensis* from *Schiothorax richardsonii* (=*Oreinus plagiostomus*); Koul *et al.* (1992) on *Pallisentis jagni* from *Channa channa. Pomporhynchus bufonis,* described as a new species from the intestine of toad, *Bufo viridis* in Kashmir region (Fotedar *et al.*, 1970), has been declared as a invalid species (Amin *et al.*, 2003)

For the first time, Fotedar and Dhar (1973) recovered a large number of Juvenile form of Acanthocephla, *viz. Polymorphus* sp. from frog (*R. cyanophlyctis*). These authors also found large number of Juvenile form of *Polymorphus* at different stages of development in the body cavity of Amphipod (*Gammarus* sp.) occurring in Murran area of Kashmir. They also successfully completed the life-cycle studies of this worm by artificially infecting young ducks. Late juveniles and adults were identified as *Polymorphus magnus*. These authors presume that present infection might be accidental, since the infected frog also inhabited the same water body in which the amphipods in large number were found to be infected.

The acanthocephalan species diversity in fish hosts of Jammu, Kashmir and Ladakh, in rivers and other water-bodies of these regions, showed higher number *i.e.*, 5 spp. in *S. richardsoni,* followed by 4 each in *S. niger* and *Triplophysia kashmirensis,* 3 in *S. esocinus,* 2 each infecting *S. curvifrons, S. plagiostomus* and *Schinothorax* sp. Rest of the 10 species of fishes showed the incidence of single acanthocephalan species infection, in each fish host species The acanthocephalan species diversity in fish hosts of Jammu, Kashmir and Ladakh,. In addition to these, Frog (*Rana cyanophlyctis*) showed the incidence of 3 parasite species and toad was reported to be parasitized by single species. Also, invertebrate (Amphipod) and Duck were found to be infected with *Polymorphus* sp. and *Polymorphus magnus* respectively, in Murran region of Kashmir Valley (**Table 5.3**).

5.1.2 Acanthocephalans of Fishes of Wular Lake

Datta (1936) was the first researcher to investigate the acanthocephalans parasitizing the fishes of Wular Lake in Kashmir. Later, a detailed survey of acanthocephans of fishes of Wular Lake, for the first time, was conducted by Dhar and Majidah (1987). They reported the wide existence of two acanthocephalans, *viz. Neoechinorphnchus manasbalensis* from the intestine of fishes, *viz. S. esocinus, S. niger, Triplophysa kashmirensis* (=*Nemacheilus kashmirensis*) and *Nemacheilus* sp. (**Table 5.1**).

The thorough investigations on seasonal population dynamics of *N. manasbalensis, P. kashmirensis* and *Pomphorhynchus* sp., parasitizing the fishes of Wular Lake have been conducted by Dhar and Peerzada (1994), Chishti and Peerzada (1998) and Majidah and Khan (1998). Detailed findings on host spectrum and seasonal dynamics of *P. kashmirensis* and *N. manasblensis* by Chishti and Peerzada (1998) have revealed that of the six different fish species of Wular Lake, with both the parasites, were only prevalent in fishes, *viz. S. curvifrons, S. niger* and *S. richardsoni. Cyprinus carpio* and *S. esocinus* were only parasitized by *P. kashmirensis*. However, *Crossochiclus latias* did not show the prevalence of either of the acanthocephan (Table 5.1). These authors have also shown that the maximum prevalence of infection due to *P. kashmirensis,* was spring season in all fish hosts. Further, it was reported that the recruitment of infection was dominated by the presence of juveniles and post juveniles, and also males. The distribution of parasites was over dispersed in all the hosts so in length group of *Schizothorax niger*. An increase in the mean number of parasites per host with an increase in host length was also indicated by Chishti and Peerzada (1998).

Table 5.1: Host Spectrum of *Pomphorhynchus kashmirensis* and *Neoechinorhynchus manasbalensis* in Wular Lake (Kashmir)

Sl.No.	*Host Fish Species*	*P. kashmirensis*	*N. manasbalensis*
1.	*Cyprinus carpio*	+	-
2.	*Crossochilus latius*	-	-
3.	*Schizothorax curvifrons (= Schizothorax hugelii)*	+	+
4.	*Schizothorax esocinus*	+	-
5.	*Schizothorax niger*	+	+
6.	*Schizothorax richardsonii (=Oreinus plagiostomus) (= Oreinus sinuatus)*	+	+
7.	Hybirds *(Schizothorax and Oreinus)*	-	-

+: Present in the fish host intestine, -: Absent in the fish host intestine

Source: Modified after Chishti and Peerzada (1998).

The high incidence of infection due to three species of acanthocephalan parasites, *viz. Pomphorhynchus kashmirensis, Pomphorhynchus* sp.and *Neoechinorhynchus manasbalensis,* in four fish species, *viz. Schizothorax labiatus, S. esocinus, S.niger* and *S.plagiostomus,* inhabiting River Jehlum, has been reported by Qayoom *et al.* (2015). In these fish hosts, the *Pomphorhynchus* parasites showed highest prevalence *i.e.,* 27 per cent, followed by *Neoechinorhynchus,* with 24 per cent prevlalence. The gender-wise prevalence was found to be highest in males of *Pomphorhynchus* (29.59 per cent), followed by *Neoechinorhynchus manasbalensis* (22.44 per cent).

An annual occurrence of *P. kashmirensis* in different *Schizothorax* spp. (*S. niger, S. curvifrons* and *S. esocinus*) in Dal Lake and River Jehlum was studied in detailed during the period Dec., 2009 to Nov., 2016 (Ali *et al.,* 2012). These authors showed the overall prevalence of *P. kashmirensis* as 25.89 per cent (94/363). Of these, in Dal Lake, prevalence was 20.68 per cent (42/203) and in Jehlum river, 32.5 per cent. The seasonal prevalence showed highest in summer (39.62 per cent) (*S. niger,* 46.34 per cent; *S.curvifrons,* 31.11 per cent; *S.esocinus* 30 per cent). The males of these fishes showed prevalence as 30.30 per cent and juveniles as 27.27 per cent.

The distribution,with seasonal prevalence of *Pomphorhynchus kashmirensis* infesting three fish species,viz. *Schizothorax niger, S. esocinus* and *S. curvifrons,* occurring in Dal Lake and River Jehlum, have been studied by Wali *et al.* (2016). These authors found overall prevalence of *P.kashmirensis* in these fishes as 27.47 per cent and in Dal Lake, the prevalence reported as 20.98 per cent and in River Jehlum 34.07 per cent. The acanthocephan infection in the fishes was highest during summer season. Apart from this, *P. kashmirensis* in *Schizothorax* spp. in Dal Lake revealed an overall prevalence of this parasite in male fishes as 23.1 per cent and in females 17.7 per cent. In River Jehlum, the overall prevalence in males was 35.8 per cent,whereas females showed 32.03 per cent and males dominated in infection.

Table 5.2: Fish Host Spectrum of Acanthocephalans of Wular Lake of Kashmir Valley

Sl.No.	*Fish Species*	*Acanthocephalan Species*
1.	*Cyprinus carpio*	PK
2.	*Triplophysa marmorata (=Nemacheilus vittatus)*	NR
3.	*Salmo trutta fario*	PK
4.	*Schizothorax curvifrons*	NM, PK
5.	*Schizothorax esocinus*	NM, PK
6.	*Schizothorax niger*	NM, PK
7.	*Schizothorax plagiostomus*	NM, PK
8.	*Schizothorax richardsonii*	NM, PK
9.	*Schizothorax* spp.	PK
10.	*Triplophysa kashmirensis*	NM, NR, PK

NM: *Neoechinorhynchus manasbalensis*; NR: *Neoechinorhynchus rutil*; PK; *Pomphorhynchus kashmirensis.*

Through the scientific literature published so far on the acanthocephalans, parasitizing the fish species, existing in Wular lake of Kashmir, it is evident that three acanthocephans are known *viz. Neoechinorhynchus manasbalansis, N. rutili* and *Pomphorhynchus kashmirensis.* Of these worms, *P. kashmirensis* has shown maximum number of fish host species *i.e.,* 9, followed by *N. manasbalensis* as 6 (six) and *N. rutili,* with 2 (two) fish host species only (**Table 5.2**).

As per annotated checklist of acanthocephala-fauna, incorporated in **Table 5.3**, it is evident that 25 species of acanthocephalan parasites, infecting different vertebrates (fishes, amphibians and duck) and invertebrate hosts (crustacean), is known to occur in different areas of Jammu and Kashmir (UT). These species, belong to 5 families, with total number of species as: Quadrigidae (2 spp.), Echinorhynidae (2 spp.), Neoechinorhynchidae (9 spp.), Pomphorhynchidae (10 spp.) and Polymorphidae (2 spp.). 21 species of acanthocephalan parasites, have been found to be infecting 24 species of fishes, belongingto different families, inhabiting waterbodies (rivers, lakes and streams, *etc.*). of Jammu, Kashmir and Ladakh region. 3 species, *viz. Neoechinorhynchus cyanophlyctis, Pomphorhynchus dubius* and *Polymorphus,* are known to infect frog (*Rana cyanophlyctis*). In addition to these, *Polymorphus minutus* and

Polymorphus sp., parasitize duck (*Anas querquedula*) and crustacean (*Gammarus* sp.) respectively in Murran region of Kashmir valley (Table 5.3).

Table 5.3: Acanthocephalan Parasites of Vertebrates (Fishes, Amphibians and Birds) of Jammu, Kashmir and Ladakh

Sl. No.	*Acanthocephalan Species and Host Species*	*Locality(ies)*	*Reference(es)*
1.	*Acanthocephalus kashmirensis* Datta Host Fishes: *Schizopygopsis stoliczke* Steindachner, *Schizothorax niger, S. esocinus, S.curvifrons, S. labiatus*	Kashmir, Ladakh Shallabug welland, Sind stream	Datta (1936), Khurshid and Ahmad (2014)
2.	*Acanthogyrus* (*Acanthosentis*) *kashmirensis* (= *Neoechinorhynchus kashmirensis* Fotedar and Dhar) Host Fishes: *Tor tor, Bangura diplostomus* (= *Labeo diplostomum*), *Labeo rohita, Ptychobarbus, Schizothorax plagiostomus, S. labiatus*	Jehlum River and its tributries	Amin *et al.* (2017)
3.	*Echinorphynchus orientalis* Kaw Host Fish: *Schizothorax* sp.	Dal Lake, Kashmir	Kaw (1951)
4.	*Neoechinorhynchus cyanophlyctis* Kaw Host Frog: *Rana cyanoplychtis*	Srinagar, Kashmir	Kaw (1951)
5.	*Neoechinorhynchus devdevi* (Datta) (= *Neoechinorhynchus yalei* (*Datta*) (= *Eosentis devdevi* Datta) Host Fish: *Schizothorax niger* (Heckel) (= *Schizothorax planifrons* Heckel), *S. plagiostomus, Diptychus maculatus*	Dal Lake, Srinagar Manasbal, Suru River (Ladakh)	Datta (1936), Tripathi (1956), Dar *et al.*, (2012, 2013)
6.	*Neoechinorphynchus glyptosternumi* Dhar and Kharoo Host fish: *Glyptosternum* sp.	Kashmir	Dhar and Kharroo (1985)
7.	*Neoechinorhynchus hutchinsoni* Datta Host fishes: *Diptychus maculatus Steindachner, Triplophysa kashmirensis* (= *Nemacheilus kashmirensis*)	Leh, Ladakh	Tripathi (1956) Raina and Koul (1983)
8.	*Neoechinorhynchus manasbalensis* Kaw Host fishes: *Schizothorax curvifrons* (= *Schizothorax higelli*), *S.esocinus, S.niger, S.plagiostomus, S.richardsonii* (= *Oreinus pagiostomus, O. sinuatus*), *Triplophysa kashmirensis* (= *Nemacheilus kashmirenis*)	Wular Lake	Cheshti and Peerzada (1998), Dhar and Majidah (1987), Kaw (1951), Dhar and Majidah (1987)
9.	*Neoechinorhynchus oreini* Fotedar Host fish: *Schinothorax richardsonii* (= *Oreinus sinuatus* Heckel)	Arapat stream Anantna-g	Fotedar (1987)

Sl. No.	*Acanthocephalan Species and Host Species*	*Locality(ies)*	*Reference(es)*
10.	*Neoechinorhynchus rigidus* (van Cleave) (=*Eosentis rigidus* van Cleave) Host fish: *Schizothorax zarudnyi*	Saistan, Kashmir	Datta (1937) Tripathi (1956)
11.	*Neoechinorhynchus dattai* Golvan (=*Neoechinorhynchus rutili* (Muller) Host fishes: *Triplophysa kashmirensis* (= *Nemacheilus stolliczkae* Steindaehna *Nemacheilus vittatus*	Wular Lake	Datta (1936)
12.	*Neoechinorhynchus* spp. Host fishes: *Schizothorax niger, S. plagiostomus, S. curvifrons, S. esocinus, S.labiatus*	Water bodies of Kashmir valley	Yousuf *et al.* (2011)
13.	*Pallisentis jagni* Koul, Raina and Bambroo Host fish: *Channa channa*	Jammu	Koul *et al.* (1992)
14.	*Polymorphus minutus* (Goez) (=*Polymorphus magnus* Skrjabin) Host duck: *Anas querquedula*	Murran, Kashmir	Fotedar and Dhar (1973)
15.	*Polymorphus* sp. Hosts: Frog-*Rana cyanophylctis;* Crustacean-*Gammarus* (Amphipod)	Murran, Kashmir	Fotedar and Dhar (1973)
16.	*Pomphorhynchus dubius* Kaw (A juvenile form) Host Frog: *Rana cyanophlyctis*	Srinagar	Kaw (1941) Yamaguti (1963)
17.	*Pomphorhynchus fotedari* Chishti and Bakshi Host fish: *Cyprinus carpio communis* Linnaeus	Hokersar, Kashmir	Chishti and Bakshi (1992)
18.	*Pomphorhynchus laevis* (Zoega) (=*Pomphorhy-nchus jammuensis* Fotedar and Dhar) Host fishes: *Schizothorax richardsonii* (= *Oreinus plagiostomus*), *Schizothraxlabiatus, S eocinus*	Tawi River, Jammu; River Jehlum	Fotedar and Dhar (1977), Qayoom *et al.* (2015)
19.	*Pomphorhynchus kashmirensis* Kaw Host fish: *Cyrinus carpio* Linnaeus	Wular	Chishti and Peerzada (1998)
	Host fish: *Salmo trutta fario* Linnaeus	Telbal, Harvan, Kashmir	Fotedar and Dhar (1977)
	Host fish: *Schizothorax curvifrons* Heckel (=*Schizothorax hugelli* Heckel)	Wular, Dal Lake, River Jehlum	Chisthi and Peerzada (1998), Ali *et al.* (2012), Wali *et al.* (2016)
	Host fish: *Schizothorax esocinus* Heckel	Wular	Chishti and Peerzada(1998)
	Host fish: *Schizothorax niger* (Heckel)	Wular, Dal Lake, River Jehlum	Chishti and Peerzada(1998), Ali *et al.* (2012) Wali *et al.* (2016)
	Host fish: *Schizothorax pleagiostomus* Heckel (= *Oreinus sinuatus* Heckel)	Wular, Anchar, Manasbal, River Jehlum	Dhar and Majidah (1987), Amin *et al.* (2012)

Sl. No.	*Acanthocephalan Species and Host Species*	*Locality(ies)*	*Reference(es)*
	Host fish: *Schizothorax richardsonii* (Gray) (= *Oreinus plagiostomus* Heckel)	Wular	Chishti and Peerzada (1998)
	Host fish: *Triplophysa kashmirensis* Hora (= *Nemacheilus kashmirensis* Hora)	Wular	Kaw (1941), Dhar and Majidah (1987)
20.	*Pomphorhynchus kawi* Fotedar and Dhar Host: Fish	Water bodies Kashmir	Fotedar and Dhar (1977)
21.	*Pomphorhynchus megacanthus* Fotedar and Dhar Host fish: *Cyprinus carpio spicularis* Lacepede	Telbal, Srinagar	Fotedar and Dhar (1977)
22.	*Pomphorhynchus oreini* Fotedar and Dhar Host fish: *Schizothorax richardsonii* (Gray) (= *Oreinus sinuatus* Heckel)	Ganderbal	Fotedar and Dhar (1972)
23.	*Pomphorhynchus orientalis* Fotedar and Dhar Host fish: *Schizothorax niger* Heckel	Ganderbal	Fotedar and Dhar (1972)
24.	*Pomphorhynchus tori* Fotedar and Dhar Host fish: *Tort tor*	River Jehlum	Fotedar and Dhar (1972)
25.	*Pomphorhynchus* spp. Host fishes: *Schizothorax niger, S. plagiostomus, S. curvifrons, S. eocinus, S. labiatus, Botia birdi*	Dal Lake, and water bodies of Kashmir	Kaw (1951), Yousuf *et al.* (2011), Qayoom *et al.* (2015)

Systematic Position of Acanthocephala: Serial No/s in the above table	*Order*	*Family*	*Sub-family*
2, 13	Gyracanthocephala	Quadrigyridae	Pallisentinae
4, 5,6,7, 8, 9,10, 11,12	Echinorhynchida	Neoechinorhynchidae	Neocehinorhynchinae
1, 3	Echinorhynchida	Echinorhynchida	Echinorhynchinae
14, 15	Polymorphida	Polymorphidae	—
16, 17, 18, 19, 20, 21, 22, 23, 24, 25	Echinorhynchida	Pomphorhynchidae	—

Note: 1. Synonyms in case of hosts and acanthocephalan species are given in parentheses, below listed species in Table 5.2. For nomenclature of Acanthocephala, Yamaguti (1963), Bhattacharya (2007), Amin *et al.* (1985, 2003, 2012), Amin (2013) have been consulted and, for nomenclature of fishes, Kullandar *et al.* (1999) and Talwar and Jhingran (1991), have been followed.

5.2 Plant-Parasitic Nematodes

5.2.1 General Account

Plant- parasitic nematodes belong to the group Nemathelminthes, class Nematodes. The nematode size ranges from 300-1000 μ. They have eel-shaped unsegmented bodies, without legs and other appendages. The form of some species, however, become swollen pear shaped or spheroid. Majority of plant-parasitic nematode feed on roots of different plant species by feeding apparatus known as stylets, responsible for causing damage to higher plants, crops and other economically important plants, and also transmit virus diseases. The plant-parasitic nematode diversity in subtropical and tropical countries of the world, have been studied by Luc (1990). In the recent years, the diversity of these economically important nematodes, affecting agricultural crops - food, vegetables, fruits, fibres and other economically important plants in Indian sub-continent, have been provided by Swarup and Das Gupta (1986), Chandniwala (1996) and Trivedi (1998 and 98 a). The Jammu, Kashmir and Ladakh regions show great diversity of plant-parasitic nematodes. Many species have been reported to be as serious pests of temperate and crops. However, detailed studies on a actual yield loss estimates have not been conducted thoroughly. In these regions, like other nematodes, plant-parasitic nematodes have successfully colonized in great variety of habitats.

Table 5.4:The Major Plant-Parasitic Nematode Groups of Agricultural Crops of J&K (UT)

Ectoparasitic Nematodes		*Endoparasitic Nematodes*	
I	**Ring Nematodes**		**Semi-Endoparasitic**
(i)	*Criconemoides(=Criconemella)*	**I**	**Reniform Nematode**
(ii)	*Hemicriconemodies*	(i)	*Rotylenchulus*
(iii)	*Merocriconema (=Neolobocriconema)*		**Migratory Endoparasitic**
II	**Spiral Nematodes**	**I**	**Root-lesion Nematode**
(i)	*Helicotylenchus*	(i)	*Pratylenchus*
(ii)	*Rotylenchus*	**II**	**Ear-cockle or Wheat-gall Nematode**
III	**Lance Nematode**	(i)	*Anguina tritici*
(i)	*Hoplolaimus*	**III**	**Burrowing Nematode**
IV	**Needle Nematode**	(i)	*Radopholus*
(i)	*Longidorus*	**IV**	**Leaf or White-tip Nematode**
V	**Pin Nematode**	(i)	*Aphelenchoides*
(i)	*Paratylenchus*	**V**	**Paddy-root Nematode**
VI	**Stunt Nematode**	(i)	*Hischmanniella*
(i)	*Tylenchorhynchus*		**Sedentary -Endoparasitic**
VII	**Dagger Nematode**	**I**	**Cyst-forming Nematodes**
(i)	*Xiphinema*	(i)	*Heterodera*
VIII	**Root – rot Nematode**	(ii)	*Globodera*
(i)	*Scutellonema*	**II**	**Root-Knot or gall Nematode**
		(i)	*Meloidogyne*

The phytonematodes attack the roots of plants, through roots some of these parasites attack and damage aerial and foliage part of plant, such as stem, leaf, seed, causing gall - formation, lesions and rots, twisting or distortion of leaves and stem besides, stunting and abnormal development of floral parts. The nematodes infestation in plants and various crops leads to reduced plant vigour, yellowing of foliage, wilting, delayed flowering and low yield of poor quality. On the basis of feeding habits, the various species of plant-parasitic nematodes, attacking valuable crops in J&K, Ladakh UTs, are grouped as ectoparasitc nematodes and endoparasitic nematodes. The ectoparasitic nematodes are external feeders and do not enter plant tissues, remain in soil and intermittently feed on roots, such as: Ring-nematodes (*e.g.*, Criconemella, *Hemicriconemoides*, *Neolobocriconema*); Spiral-nematode (*Helicotylenchus, Rotylenchus*); Lance- nematodes (*Hoploliamus*); Needle-nematode (*Longidorus*); Pin-nematode (*Paratylenchus*); Stunt-nematode (*Tylenchorhynchus*); Dagger-nematode (*Xiphinema*) and Root-rot nematode (*Scutellonema*) (**Table 5.4**).

The endoparasitic nematode may be either semi–endoparasitic or completely endoparasitic. The semi-endoporasitic nematode species partially penetrate the root, head and neck, remain fixed in cortex and posterior region remain free in soil. These types of nematodes damage cortex, endosperm, pericycle and pholem parenchyma of its hosts. The semi-endoparasitic nematodes include Reniform-nematodes (*Rotylenchulus reniform*). The endoparasitic nematodes enter host's roots during parasitization and these group of nematodes may be migratory endoparasitic nematodes and sedentary endoparasitic nematodes. The migratory endo- parasitic nematodes cover: Root–lesion nematodes, such as *Pratylenchus*; Burrowing nematode (*Radopholus*),Leaf or white-tip nematode (*Aphelenchoides*). Paddy – root nematode (*Hirschmanniella*) and Ear-cockle/seed or wheat-gall nematode (*Anguina tritici*). The sedentary endoparasitic nematodes of this region include Cyst–forming nematodes (*Heterodera* and *Globodera*) and Root-knot or gall-nematode (*Meloidogyne*).

Number of serious plant parasitic- nematodes reported from J&K (UT) regions, are known to cause considerable damage to various parts of number of valuable crops. These mainly include root damaging nematodes. and aerial or foliage part damaging nematodes. The root-infesting nematodes like *Meloidogyne, Xiphinema* and are responsible for producing characteristic galls on the roots of host plants. The nematodes, *viz. Pratylenchus, Radopholus* and *Merocriconema* (=*Neolobocriconema*), damage the cortex of roots by penetration and movement, and sometimes small lesions are also produced. *Rotylenchulus reniform* (Reniform-nematode) causes damage to cortex endosperm, pericycle and phloem parenchyma of roots. Sometimes nematode species, *viz. Meloidogyne hapla* and *Pratylenchus* spp. kill the distal part of root by girdling and remaining portion of root that is alive develops branch rootlets. This deformity is sometimes called as 'witches' broom 'or 'bearding'. *Xiphinema,* may at occasion causes root-surface necrosis in host-plant by killing the surface cells of large areas. *Meloidogyne* and *Xiphinema*, cause curly- tips in some cases, due to devitalize roots.

Root-rot nematodes–*Scutellonema* and *Radopholus* are responsible for extensive tissue damage, leading to rots by other organisms *i.e.* secondary invaders. The other root-attacking nematodes include *Heterodera* spp., *Hoplolaimus* spp., *Tylenchorhynchus* spp., *Rotylenchus* spp. and *Helicotylenchus* spp., widely occurring in many parts of Jammu and Kashmir region. The above-ground infesting nematodes or aerial or foliage-parts damaging nematodes, are responsible for showing symptoms of different types. The stunt nematode (*Tylenchorhynchus*). causes stunted growth in plants and reduce the ability to withstand adverse condition. The nematode, *viz.*

Globodera and *Aphelenchoides*, cause nutritional deficiency in plants. The larval stages of gall- nematode (*Anguina*), live in growing points of wheat and develop distortion in stem and leaves. Also, *A. tritici* survives for long period in wheat cockles or galls.

5.2.2 Systematic Survey and Diversity

The earlier authors who conducted surveys and studies on plant-parasitic nematodes (PPNs) of Kashmir, and also gave descriptions of new nematode include Jairajpuri and Siddiqi (1963) Khan and Siddiqi (1963) Jairajpuri (1965) Jairajpuri and Baqri (1966), Raina (1966) and Jairajpuri and Loof (1967). Mahajan (1971) for the first time gave a checklist of plant-parasitic nematodes of Jammu and Kashmir State based on surveys made by him in different – major crop and fruit – growing areas and localities of the State. However, this list does not include data on specific genus and species of nematode, affecting the particular host plant species. Later, several workers made thorough taxonomically studies, newly recorded many species and also providing descriptions of new species of phytonematodes of this region, affecting number of plants and trees of agricultural, horticultural, forestry impotrance, besides other economically important plant species. In this connection, the valuable contribution have been made by : Mahajan (1972, 73) Fotedar and Mahajan (1971, 72,. 72 b, 73 and 74); Fotedar and Handoo (1974,77); Sheikh (1979) Gupta and Uma (1980 and 81), Fotedar and Kaul (1984,85); Kaul and Wali-ullah (1995) and Akhtar *et al.* (1996).

A list of Plant-parasitic nematodes, reported previously by various authors from Kashmir Valley, covering about 22 species of nematodes, affecting 12 spp. of crops and economically important plants, has been given by Fotedar and Kaul (1982). Further, their work also incorporated a checklist of 16 different genera of plant nematodes of Kashmir, with their distribution, population density, from 21 different localities, damaging agricultural crops (food, fruit, vegetable, oil seed) and, aromatic and medicinal plants. This list was mainly based on surveys conducted by them from 1979-81. In additional to these, these authors have also reported the wide distribution of nematodes, *viz.* Spiral-nematodes, belonging to the genus *Helicotylenchus,* Lesion-nematodes of the genus *Pratylenchus* and Stunt- nematode species, belonging to genus, *Tylenchorhynchus.*

Kaul (1998) studied the diversity and host-range of plant- parasitic nematodes, affecting vegetables, fruits and other crops. He indicated the existence of a total of 49 species of nematodes belonging to 17 different genera in Jammu and Kashmir State. 30 nematode species have been shown to be attacking a wide variety of vegetable crops, 14 species affecting fruits crops, 10 spp. damaging food (cereal) crop, 2 spp. attacking pulse crop (*Phaseolus vulgaris*) and 1 sp. each affecting *Crocus sativus* (saffron), *Cynodon doctylon* and a grass species in earstwhile J&K State.

In the recent years, many authors have done indepth investigations on PPNs of economically important crops/plants and trees of Jammu and Kashmir. The important research papers are on: cereal plant (paddy) (Hassan *et al.*, 2008); fruits (Zaki and Mantoo, 2003, Singh *et al.*, 2012 a, Zalpuri *et al.*, 2013 b and Askary *et al.*, 2014); vegetables (Wali-ullah *et al.*, 2008, Mantoo *et al.*, 2008, Hassan *et al.*, 2016); medicinal and aromatic plants (Zaki and Mantoo, 2008, Singh and Gupta, 2011,

Sheikh *et al.*, 2014); ornamental plants (Abbas and Wali-ullah, 2010, Rashid *et al.*, 2014); forest range plantations (Mushtaq *et al.*, 2007, Askary and Haider, 2010.

The data compiled during the course of present work, showed the prevalence of 131 species, belonging to 42 genera of plant parasitic nematodes (PPNs) under class Enoplea (order Dorylaimida, Triplochida) and Class Chromadorea (order Rhabdita), incorporating 9 and 12 families respectively. These PPNs have been found in vast areas and localities of J&K, and Ladakh UTs, associated with agricultural, horticultural, forestry plants/trees and other economically important host plant species. The data on PPNs has been updated in the light of nomenclatural/systematic changes. For this purpose, latest reference books/monographs and websites on world database of plant nematodes, such as : Luc *et al.* (1990); Siddiqi (2000); De Ley and Blaxter (2004), Geraert (2008); Websites (world databases of nematodes) - Nemaplex (Ferris, 2020) and Nemys (Bezerra *et al.*, 2020), have been consulted (**Table 5.11**).

The species richness of PPNs, affecting different host trees and plant species are : cereal plants (maize, paddy and wheat), associated with 21 spp. (15 genn.); 22 species of fruit trees (temperate, tropical), with 54 spp. (25); 22 kinds of vegetable plants/crops (brassicas, solanaceous, cucurbits and other vegetables), with 68 spp. (29); medicinal/aromatic plants (saffron, coriander, mint, *Rosa* spp., ginger) with 14 spp. (10); forest trees (deodar, pine, mulberry, salix), with 5 spp. (5); 7 spp.of) ornamental plants, with 16 spp. (13) and various grasses, with 5 spp. (5) (**Tables 5.8 and 5.10**) The diversity, host range, distribution and population dynamics of PPNs, affecting cereals, fruit and forest range plantations, brassicas, cucurbits and other important vegetables, medicinal, aromatic and ornamental plants in Jammu, Kashmir and Ladakh region, are given as under :

5.2.3 Affecting Cereals Plants, Forest Range Plantations and Grasses

(a) Paddy (*Oryza sativa*) (Poaceae)

So far 11 species of nematodes belonging to 7 genera, have been reported from Jammu and Kashmir, damaging paddy or rice. These include 6 spp. of genus *Hirschmanniella* and 1 sp. each under genera, *viz. Aphelenchus, Hemicriconemoides, Meloidogyne, Tylenchorhynchus, Tylenchus* and *Xiphinema* (**Tables 5.8 and 5.10**). Mahajan (1971) was first worker to record nematodes, *viz. Hirschmanniella mucronota* and *H. oryzae*, damaging paddy in Kashmir region. Wali-ullah *et al.* (1991), during the course of his survey in various localities in districts (Srinagar, Badgam and Baramulla) of Kashmir, found the incidence of four species of nematode, belonging to geuus *Hirschmanniella*. The prevalence of *H. oryzae* was found 71 per cent in Badgam district followed by *H. gracilis*. In Srinagar district (Shalimar and Harvan locality), these authors also found the incidence of nematodes, *viz. H mucronota, H. gracilis* and *H. shamimi*, with population frequency 65 per cent, 20 per cent and 5 per cent respectively. In Baramulla district, all the three species except *H. shamimi* were found and these nematode species, showed varying frequencies and densities. The existence of *Tylenchorhynchus oryzae*, attacking rhizosphere of paddy roots in Kashmir, has been found by Kaul and Wali-ullah (1995).

From Jammu region, Singh *et al.* (2007), for the first time recorded the root-knot nematode, *Meloidogyne graminicola* infesting paddy. From Kashmir region, reports on the prevalence of *Hirschmanniella* sp. and *Tylenchus* on paddy, with variations in its dimention for an altered habitat, and control strategy, have been provided (Hassan *et al.*, 2008). These authors made observations on the effect of nematicides/pesticides – Nemagon, Nemafos and Vapam (soil fumigant) for controlling these nematode parasites and the order of efficacy was Nemofos > Vapam.

(b) Maize (*Zea mays*) (Poaceae)

Fourteen species of nematodes under 9 genera, affecting *Z. mays,* have till date, been reported by various authors from Jammu and Kashmir regions. The highest number of species, *i.e.* 3, belongs to the genus *Helicotylenchus,* 2 spp. each to genera – *Longidorus, Pratylenchus* and rest of the genera (7), having one species each (**Tables 5.8 and 5.10**). The earliest recorded species of nematode of maize from erstwhile J&K State were: *Longidorus reneyii, Hoplolaimus* sp., *Helicotylenchus steineri, H. pteracercusoides,* by Raina (1996), Mahajan (1970), Fotedar and Mahajan (1974) and Fotedar and Kaul (1985) respectively. Later, Fotedar and Kaul (1982) recorded four more genera of nematodes attacking maize plant in Kashmir. The incidence of species of nematodes, affecting maize crop in Kashmir Valley has been shown by : Fotedar and Kaul (1982) (*Longidorus* sp. and *Xiphinema;* Kaul (1987) (*Paratylenchus minimus*) and Kaul and Fotedar (1987) (*Helicotylenchus erythrinae*). The three species of nematodes, *viz. Pratylenchus sativus, Aphlenchoides srinagarensis* and *Tylenchorhynchus zeae,* affecting maize, have been found in Kashmir region (Hassan *et al.,* 2009).

(c) Wheat (*Triticum vulgare*) (Poaceae)

So far, from Jammu and Kashmir, a total of 2 species, belonging to 2 genera (*Anguina, Basiria*) of parasitic nematodes has been recorded, (Tables 5.8 and 5.10). Fotedar and Mahajan (1973) were first to record the nematode, *viz. Basiria tritici* of wheat in Kashmir. Later, the occurrence of *Anguina tritici* (seed or wheat-gall nematode), causing 'Ear cockles' disease of wheat, was noticed by Sheikh (1979) in some wheat fields in the vicinity of Srinagar. The disease affected plant showed more or less stunted growth and leaves exhibited characteristics wrinkled, rolled or curled condition. In the recent years, the distribution of *A. tritici* in different wheat fields of various areas of Kashmir, have also been reported by: Fotedar and Kaul (1982 and Zaki (1999). *A. tritici,* reported from a single village in Jammu, having infestation 5- 80 per cent. In the wheat fields in Leh and Kargil Districts of Ladakh (UT), severity of disease in wheat due to *A. tritici* was found to be 5- 8 per cent (Zaki and Mir, 1994). In Jammu region, Singh *et al.* (2011) has shown the prevalence of *A. tritici.*

(d) Forest Range Plantations and Grasses

Parasitic nematode, *Tylencholaimellus* is known to infest forest trees like Scot pine (*Pinus sylvestris*) and the Himalayan cedar/deodar (*Cedrus deodara*), whereas genus *Tylenchorhynchus,* is found to be associated with Salix (*Salix babylonica*) and mulberry (*Morus* spp.) in Kashmir. In addition to these genera, *Hemicriconemoides* and *Promumtazium* affect the deodar plantations in Kashmir (**Tables 5.8 and 5.10**). Grasses in Kashmir have been observed to be infested with as many as 6 species, under 5 genera of parasitic nematodes. The genera included *Boleodorus, Helicotylenchus, Heterodera, Proleptonchus* and *Tylencholaimellus* (**Tables 5.8 and 5.10**).

5.2.4 Affecting Brassicas, Cucurbits, Solanaceous and other Vegetable Crop Plants

(a) General

As many as 68 spp. under 29 genera of plant parasitic nematodes (PPNs), have been found to affect 7 different brassicas vegetables, 4 solanaceous vegetable crops, 3 cucurbits and 8 other kinds of vegetables, in diverse habitats of vast areas and localities of J&K, Ladakh UTs (Tables 5.5 and 5.6). A large number of research publications devoted to PPNs, associated with vegetable plants/crops of these region include Brassicas (cabbage, cauliflower, kale/'hak', knol khol, mustard, radish, turnip), Cucubits (bottle gourd, pumpkin, winter squash), solanaceous vetables (brinjal, chilli, potato, tomato) and other kinds of vegetables- carrot, French bean, garlic, okara, pea, soya bean and spinach. The highest number of genera *i.e.*, 21 of PPNs, is associated with brinjal vegetable plant in J&K, followed by cauliflower, with 15 genn., tomato with 14 genn. and radish, with 12 genn. each, and for rest of the vegetable plants (brassicas, cucurbits and solanaceous), affected by total number PPN genera. The data is given in **Tables 5.5, 5.6 and 5.10**.

In Kashmir region, Wali-ullah (1992) for the first time conducted a detailed survey for PPNs, affecting fourteen vegetables, *viz.* potato, tomato, brinjal, chilli, okara, kale/hak, cabbage, cauliflower, radish, spinach, turnip, onion, pea and pumpkin, revealed about 33 species, under 17 genera of PPNs. The PPNs were found to be having high prevalence in diverse fields in vast areas and localities of different districtsofKashmir.The nematode genera: *Tylenchus, Basiria, Cephalenchus, Tylenchorhynchus, Nothotylenchus, Helicotylenchus, Hirschmanniella, Hoplolaimus, Meloidogyne, Globodera, Rotylenchulus* and *Pratylenchus*, were observed in 98 per cent of samples, and *Aphelenchus* and *Aphelenchoides* were in 20 per cent of samples. Occurrence and distribution of PPNs around rhizosphere of vegetable crops (brinjal, tomato, knol khol, chilli, potato, cabbage, beans, radish, turnip and bottle gourd), in two districts of Kashmir, *viz.*, Anantnag and Srinagar, showed 6 species of PPN (Mantoo *et al.*, 2008) (Tables 5.5 and 5.6).

(b) Brinjal (*Solanum melongena*) (Solanaceae)

As many as 48 species of nematodes, distributed over 21 different genera have so far been reported affecting Brinjal in Jammu and Kashmir. Various genera, with total number of species are : *Aphelenchoides* (2 spp.), *Aphelenchus* (2), *Aporcelaimellus* (1), *Basira* (2), *Criconemoides* (1), *Dorylaimus* (1), *Helicotylenchus* (7), *Heterodera* (1), *Hirschimanniella* (1), *Hoplolaimus* (2), *Longidorus* (1) *Meloidogyne,* (2) *Nothotylenchus,*(2) *Paravulus* (1), *Pratylenchus* (4), *Psilenchus* (1), *Rotylenchus* (1), *Sakia* (1), *Tylenchorhynchus* (6), *Tylenchus* (1) and *Xiphinema* (1) (Tables 5.9 and 5.10).

The earliest records of PPNs, damaging Brinjal crop in Kashmir have been given by Fotedar and Kaul (1982 1984, 1985), Zaki *et al.* (1991) and Kaul (1998). In Jammu district, the diversity of PPNs on brinjal has shown the prevalence as many as 25 species, under 18 genera, distributed in diverse localities of this district (Wali-ullah *et al.*, 2008). Of the various phytonematodes species, *Meloidogyne, Tylenchorhynchus, Helicotylenchus, Rotylenchus* and *Hoplolaimus*, have been found to be invariably in all the samples collected.

Table 5.5: Generic Diversity and Richness of Parasitic Nematodes Associated with Brassicas and Cucurbit Vegetable Plants in Jammu and Kashmir, UT

Nematode Genus	*Bg*	*Ca*	*Cl*	*Ka*	*Kn*	*Mu*	*Pu*	*Ra*	*Tu*	*Ws*
Aglenchus	–	–	+	–	–	–	–	–	–	–
Aphelenchoides	+	+	–	–	+	–	+	+	+	–
Aphelenchus	–	–	+	–	–	+	+	–	–	–
Basiria	–	–	+	+	+	–	–	+	–	–
Boleodorus	–	–	+	–	–	–	–	+	–	–
Filenchus	–	–	+	–	–	–	–	–	–	–
Globodera	–	–	–	+	–	–	–	+	–	–
Helicotylenchus	+	+	+	+	+	–	+	+	+	+
Hirschmanniella	–	–	+	+	+	–	+	+	–	–
Hoplolaimus	–	+	+	+	+	–	+	+	+	–
Longidorus	–	+	–	+	–	–	–	–	–	–
Meloidogyne	–	+	–	+	–	–	+	+	–	+
Merlinius	–	–	+	–	–	–	–	–	–	–
Nothotylenchus	–	–	+	+	–	+	–	+	+	–
Paratylenchus	–	–	–	–	–	+	–	–	+	–
Pratylenchus	+	+	+	+	+	+	–	+	+	–
Psilenchus	–	–	+	–	–	+	–	–	–	+
Rotylenchus	–	–	–	–	–	–	–	–	–	–
Rotylenchulus	–	–	–	–	–	–	–	–	+	+
Trichodorus	–	+	–	–	–	–	+	–	–	–
Tylenchorhynchus	+	+	+	+	+	+	+	+	+	+
Tylenchus	+	+	+	+	+	–	–	–	+	–
Xiphinema	+	–	+	–	+	–	–	+	+	–
Total number of Genera	**6**	**9**	**15**	**11**	**9**	**6**	**8**	**12**	**10**	**5**

+: Nematode genus present; –: Nematode genus absent.

Bg: Bottle gourd; Ca: Cabbage; Cl: Cauliflower, Ka: Kale/'hak'; Kn: Knol khol; Mu: Mustard (field); Pu: Pumpkin; Ra: Radish; Tu: Turnip; Ws: Winter squash.

(c) Cabbage (*Brassica oleracea* var. *capitate*) (Brassicaceae)

In all 9 genera, covering 18 spp. of PPNs, prevalent in Jammu and Kashmir. have so far been reported, affecting cabbage. These are: *viz. Aphelenchoides* (2),

Helicotylcnchus (3), *Hoplolaimus* (2), *Meloidogyne* (1), *Pratylenchus* (2), *Trichodorus* (1), *Longidorus* (1) *Tylenchorhynchus* (4), *Tylenchus* (2) (Table 5.10).

Table 5.6: Generic Richness and Diversity of Plant Parasitic Nematodes, Affecting Vegetable Plants other than Brassicas and Cucurbits in Jammu and Kashmir, UT

Nematode Genus	*Br*	*Ch*	*Cr*	*Fb*	*Ga*	*Ok*	*On*	*Pe*	*Po*	*Sb*	*Sp*	*To*
Aglenchus	–	–	–	+	–	–	–	–	+	–	+	–
Aphelenchoides	+	+	–	+	–	+	+	+	+	–	+	+
Aphelenchus	+	–	–	–	–	+	+	–	+	–	+	+
Aporcelaimellus	+	–	–	–	–	–	–	–	–	–	–	–
Basiria	+	–	–	–	–	+	+	–	+	–	–	–
Boleodorus	–	+	–	–	–	–	–	–	–	–	+	–
Criconemoides	+	–	–	+	–	–	–	+	–	–	–	–
Dorylaimus	+	–	–	–	–	–	–	–	–	–	–	–
Globodera	–	–	–	–	–	–	+	–	+	–	–	+
Filenchus	–	–	–	–	–	–	–	–	–	–	–	–
Helicotylenchus	+	+	–	+	+	+	+	+	+	–	+	+
Heterodera	+	–	–	–	–	–	–	–	–	–	–	–
Hirschmanniella	+	+	–	+	–	–	–	–	–	–	–	+
Hoplolaimus	+	+	–	–	–	+	+	+	+	–	+	+
Longidorus	+	–	–	–	–	–	–	+	–	–	+	–
Lordellonema	–	–	–	–	–	–	–	–	–	+	–	–
Meloidogyne	+	+	+	+	–	+	–	+	+	–	+	+
Merlinus	–	–	–	–	–	–	–	–	–	–	+	–
Nothotylenchus	+	–	–	–	–	+	+	–	–	–	–	+
Pratylenchus	+	+	–	+	–	+	–	+	+	–	–	+
Paravulus	+	–	–	–	–	–	–	–	–	–	–	–
Psilenchus	+	+	–	+	–	–	–	–	–	–	+	–
Rotylenchus	+	–	–	–	–	–	–	–	–	–	–	–
Rotylenchulus	–	–	–	–	–	+	–	–	–	–	–	–
Sakia	+	–	–	–	–	–	–	–	–	–	–	–
Scutellonema	–	–	–	–	–	–	–	–	–	–	–	+
Trichodorus	–	–	–	–	–	–	+	–	–	–	–	+
Tylenchorhynchus	+	+	–	+	–	+	–	+	+	–	+	+
Tylenchus	+	+	–	+	–	+	–	+	+	–	+	+
Xiphinema	+	+	–	+	–	–	–	+	+	–	–	+
Total number of Genera	**21**	**11**	**1**	**11**	**1**	**11**	**8**	**10**	**12**	**1**	**12**	**14**

+: Nematode genus present; –: Nematode genus absent.

Br: Brinjal; Ch: Chilli; Cr: Carrot; Fb: French bean; Ga: Garlic; Ok: Okara; On: Onion; Pe: Pea; Po: Potato; Sb: Soya bean; Sp: Spinach; To: Tomato.

(d) Cauliflower (*Brassice oleracea* var. *botrytis*) (Bassicaceae)

A total of 20 spp. under 15 genera of PPNs, affecting cauliflower vegetable in Jammu and Kashmir has so far been reported. These include *Aphelenchus* (1 sp.), *Aglenchus* (1), *Basiria* (1), *Boleodorus* - (2), *Filenchus* (1), *Helicotylenchus* (2), *Hoplolaimus* (1), *Hirschmanniella* (1), *Merilinus* (1), *Nothotylenchus* (1), *Psilenchus* (1), *Pratylenchus* (2), *Tylenchorhynchus* (3), *Tylenchus* (1), *Xiphinema* (1) (**Tables 5.10 and 5.11**).

In Wagam locality of District Budgam of Kashmir valley, cauliflower crop has been observed to be infested with six species of PPNs. These parasites, with their absolute density includeded *Boleodorus cylindricus* (18 per cent), *Psilenchus* (31 per cent), *Helicotylenchus* (34.5 per cent), *Merlinius* (35 per cent), *Aglenchus* (35 per cent), and *Filenchus* (31.5 per cent) (Hassan *et al.*, 2018 b).

(e) Chilli (*Capsicum annuum*) (Sotanaceae)

About 21 spp. of PPNs, under 11 genera, damaging chilli crop are known to be prevalent in J&K. The genera, with number species associated with this vegetable crop are : *Aphelenchoides* (2), *Boleodorus* (1), *Helicotylenchus*(2), *Hirchmanniella* (2), *Hoplolaimus* (1), *Meloidogyne* (2), *Pratylenchus* (2), *Tylenchorhynchus* (4), *Tylenchus* (4) and *Xiphinema* (1) (**Tables 5.6, 5.10 and 5.11**). Zaki *et al.* (1991) have reported the incidence of *Meloidogyne hapla*, damaging severely chillies in the fields of Division of Olericulture, (SKUAST), Shalimar (Kashmir).

The population analysis of the genera build- up on chilli crop in Kashmir valley has been studied, showing the absolute densities percentage of genera as *Tylenchus* (86 per cent), *Helicotylenchus* (96.6 per cent), *Psilenchus* (100 per cent), *Pratylenchus* (100 per cent) and *Hirschmanniella* (96.6 per cent) (Hassan *et al.*, 2016). Apart from this, in Kashmir region, the absolute prevalence of PPNs, *viz. Tylenchus arcuates* affecting chilli has been found (Hassan, 2018).

(f) French Bean (*Phaseolus vulgaris*) and Soya Bean (*Glycine max*) (Fabaceae)

The parasitic nematodes, belonging to 11 genera, including 13 species, associated with French bean in Jammu and Kashmir region, have been found as: *Aphelenchoides* (1 sp.), *Aglenchus* (1), *Circonemoides* (1), *Helicotylenchu* (3 spp.),*Hirschmanniella* (1) *Meloidogyne* (1), *Psilenchus* (1), Pratylenchus (1), *Tylenchorhynchus* (1), *Tylenchus* (1), *Xiphinema* (1) (Tables 5.10 and 5.11). The absolute prevalence densities of various nematode genera build-up analyzed on French bean crop in Kashmir was found to be as : *Aglenchus* (27 per cent), *Tylenchus* (19.5 per cent), *Hirschmanniella* (33 per cent), *Helicotylenchus* (33 per cent) and *Psilenchus* (35.4 per cent) (Hassan *et al.*, 2016). Nematode, *viz. Lordellonema bashirensis*, has been found in Safapore (Kashmir), inhabiting soil around the roots of soya bean (Akhtar *et al.*, 1994).

(g) Kale (*Brassica oleracca* var. *acephala*) (Brassicaceae)

In Kashmir region, a total of fifteen species of PPN, under 11 genera, has been found to damage the root of this local Cultivar, Kale/Hak.The various genera, with total number of species are : *Aphelenchoides* (1), *Basiria* (1), *Globodera* (1), *Hirschmanniella* (1), *Hoplolaimus*(1), *Longidorus* (1), *Meloidogyne* (1), *Pratylenchus* (1), *Tylenchorhynchus* (2), *Helicotylenchus* (3) (**Tables 5.10 and 5.11**).

(h) Okara (Lady Finger) (*Ablemoschus esculentus*) (Malvaceae)

As many as sixteen species, belonging to 11 genera of PPNs have so far been recorded from J&K, affecting okra vegetable in different areas and localities. The various genera,with number of species are: *Aphelenchus* (1 sp.), *Aphelenchoides* (1), *Basiria* (1), *Helicotylenchus* (2 spp.), *Hoplolaimus* (2), *Meloidogyne* (1), *Nothotylenchus* (1), *Pratylenchus* (2), *Rotylenculus* (1), *Tylenchorhynchus* (3) and *Tylenchus* (1) (**Tables 5.6, 5.10 and 5.11**). During the course of survey in Jammu and Samba district, *Meloidogyne incognita* infesting okara, has been more frequently found in maximum localities of these Districts (Singh *et al.*, 2012). These authors, have also studied the pathogenic effect and management of *M. incognita*.

(i) Onion (*Allium cepa*) and Garlic (*Allium sativa*) (Liliaceae)

In Jammu and Kashmir, onion crop, has been found to be infested with a total of 9 species, under 8 genera of PPNs. The genera, with number of species are : *Aphelenchoides* (1 sp.), *Aphelenchus* (1), *Basiria* (1), *Globodera* (1), *Helicotylenchus* (2), *Hoplolaimus* (1), *Nothotylenchus* (1) and *Trichodorus*. In Kashmir, garlic is affected by a nematode parasite, *viz. Helicotylenchus* (**Tables 5.6, 5.10 and 5.11**).

(j) Potato (*Solanum tuberosum*) (Solanaceae)

About 28 species, belonging to 12 genera of PPNs have been found to be damaging potato in Jammu and Kashmir. The various genera pertaining to this crop reported, with total number of species are: *Aglenchus* (1sp.), *Aphelenchus* (1), *Aphelenchoides* (3 spp.), *Basiria* (1 sp.), *Globodera* (1 sp.) *Helicotylenchus* (3 spp.) *Hoplolaimus* (2 spp.) *Meloidogyne* (2 spp.), *Pratylenchus* (7 spp.), *Tylenchrhynchus*(3 spp.), *Tylenchus* (3 spp.) and *Xiphinema* (1 sp.) (Tables 5.10 and 5.11). High root population of number of nematodes attacking potato in wide field areas of Kashmir, have been studied by number of workers. It was Krishna- Prasad (1986) who for the first time recorded the existence of Root-knot nematode (*Meloidogyne hepla*) from Kashmir. Of the various lesion nematodes, *Pratylenchus zeae* has been found to be a dominant species in the field by Wali-ullah and Bhat (1990).

A detailed survey conducted by Zaki *et al.* (1991) in potato crop fields of eight localities of Kashmir in the year 1988, revealed valuable information on population dynamics of nematode, parasitic on potato crop of this region. The maximum population of *Pratytenchus* spp. was noticed by them in early season and declining towards the harvest of potato *i.e.* August. According to these authors, *Tylenchorhynchus* spp. (Stunt-nematodes) showed high population during winter months, through October to April. In case of *Helicotylenchus* spp (spiral-nematodes), maximum population was found by them during June and July months and minimum during December and January.

Potato losses due to PPNs, *viz. Pratylenchus pentrans* and *P. thornei* in Kashmir region, have been estimated (Zaki *et al.*, 1993). The prevalence of nematode generic build-up on *S. tuberosum* has been statistically studied by calculating absolute frequency, absolute density, prominence value, relative density and mean intensity four nematode genera (Hassan *et al.*, 2016 a). In this study, the absolute frequency of these genera was found as : *Helicotylenchus*, 80 per cent; *Tylenchus*,90 per cent; *Psilenchus*, 86.6 per cent and *Aglenchus*, 76.6 per cent.

(k) Pea (*Pisum sativum*) (Fabaceae) and Carrot (*Daucus carota sativus*) (Apiaceae)

In J&K, vegetable pea has been found to be affected by 10 spp., under 10 genera. The various genera, with total number of species are : *Aphelenchoides* (2 spp.), *Criconemoides* (1 sp.), *Helicotylenchus* (1), *Hoplolaimus* (1), Longidorus (1), *Meloidogyne* (1), *Pratylenchus* (1), *Tylenchorhynchus* (1), *Tylenchus* (1) and *Xiphinema* (1) (Tables 5.10 and 5.11). In Jammu region, the occurrence of *Meloidogyne incognita* infesting carrot (Singh, 2009) and pea (Singh and Satyapariya, 2008), have been recorded.

(l) Radish (*Raphanus sativus*) (Brassicaceae)

Radish crop in J&K, has been shown by many authors to be affected by about 23 species, under 12 genera of PPN, *viz. Aphelenchoides* (3 spp.), *Basiria* (1 sp.), *Boleodorus* (1), *Helicotylenchus* (2), *Hirschmanniella* (1), *Hoplolaimus* (1), *Globodera* (1), *Meloidogyne* (1), *Nothotylenchus* (1), *Rotylenchulus*(1), *Pratylenchus* (3), *Tylenchorhynchus*(3), *Tylenchus* (1) and *Xiphinema* (2) (**Tables 5.10 and 5.11**).The populations of three PPNs, *viz. Helicotylenchus*, *Pratylenchus* and *Boleodorus*, affecting radish crop in Kashmir have statistically been studied, covering their varied frequencies, density and prominence value, showing *Boleodorus*, with highest frequency as 93.3 per cent (Hassan *et al.*, 2016 a)

(m) Spinach (*Spinacea oleracea*) (Amaranthaceae)

About thirteen species, under 12 genera of PPNs in J&K, UT, are known to be associated with spinach crop. The prevalent genera, with number species are: *Aglenchus* (1), *Aphelenchus* (1), *Aphelenchoides* (1), *Boleodorus* (1), *Helicotylenchus* (2), *Hoplolaimus* (1), *Longidorus* (1), *Merilinus* (1), *Meloidogyne* (1), *Psilenchus* (1), *Tylenchorhynchus* (1), *Tylenchus* (1) (Tables 5.10 and 5.11).

(n) Tomato (*Lycopersicum esculentum*) (Solanaceae)

In J&K, 26 spp., under 14 genera of PPNs, have been found to be associated with tomato plant. The genera, with number species reported from these region are : *Aphelencoides* (2 spp.), *Aphelenchus* (1), *Globodera* (1), *Helicotylenchus* (5), *Hirschmanniella* (1), *Hoplolaimus* (1), *Meloidogyne* (1), *Nothotylenchus* (1), *Pratylenchus* (2), *Scutellonema* (1), *Tylenchorhynchus* (4), *Tylenchus* (3), *Trichodorus* (1), *Xiphinema* (2) (Tables 5.10 and 5.11).

Earlier in Kashmir, Zaki and Bhatti (1989) observed *Meloidogyne javanica* infesting tomato plantation and these authors also studied the effects of castor leaves in combination with different fertilizers on *M. javanica*. Later, studied tomato crop loss due to four species of PPN as *P. penetrans*, *H. taperua*, *T. kashmirensis and Tylenchus*. Bhat and Kaul (1994), studied the influence of soil temperature on the population behaviour of *Meloidogyne incognita* infesting tomato crop in Kashmir.

(o) Turnip (*Brassica campertris rapa*) and Field Mustard (*Brassica campestris*) (Brassicaceae)

Turnip crop in this region has been reported to be affected by 19 spp., under 10 genera of PPNs. The various genera, with number of species are: *Aphelenchoids* (3 spp.), *Helicotylenchus* (2), *Hoplolaimus* (1), *Nothotylenchus* (1), *Paratylenchus* (1), *Pratylenchus* (4), *Rotylenchulus* (1) and *Tylenchorhynchus* (2), *Tylenchus* (2) and *Xiphinema* (2) (**Tables 5.10 and 5.11**). Parasitic nematodes associated with field mustard in Kashmir valley, have been found to be as *Aphelenchus radicicolus* and *Psilenchus haki* (Hassan *et al.*, 2008).

(p) Winter Squash (*Cucurbita maxima*), Bottle Gourd (*Lagenaria siceraria*) and Pumpkin (*Cucurbita pepo*) (Cucurbitaceae)

The 5 species, including 5 genera of PPNs damaging *C. maxima*, have been encountered in J&K, these are: *Helicotylenchus* (1 sp.), *Meloidogyne* (1), *Psilenchus* (1), *Rotylenchulus* (1) and *Tylinchorhynchus* (1). 6 spp., belonging to 6 genera of PPNs, are known to be associated with bottle gourd and these include *Aphelenchoides* (1 sp.), *Helicotylenchus* (1), *Pratylenchus* (1), *Tylenchorhynchus* (1), *Tylenchus* (1) and *Xiphinema* (1) (Tables 5.10 and 5.11). 12 spp., under 9 genera of PPN in J&K, have been reported to infest pumpkin plants and these genera, with number of species are: *Aphelenchus* (1 sp.), *Aphelenchoides* (1), *Tylenchorhynchus* (1), *Hoplolaimus* (2), *Trichodorus* (1), *Helicotylenchus* (3 spp.), *Hirschmanniella* (1), *Meloidogyne* (1) and *Rotylenchulus* (1).

(q) Knol Khol (*Brassica oleracea* var. *gongylodes*)

In J&K, UT, a total of 16 species belonging to 10 genera, are known to infest knol Khol vegetable. Various genera, incorporating number of species of different genera included *Aphelenchoides* (3 spp.), *Basiria* (1 sp.), *Hoplolaimus* (1), *Helicotylenchus* (2), *Hirschmanniella* (1), *Longidorus* (1), *Pratylenchus* (1), *Tylenchorhynchus* (2), *Tylenchus* (2), *Xiphinema* (2 (Tables 5.10 and 5.11)

5.2.5 Affecting Fruit (Temperate and Tropical) Plantations

(a) General

In Jammu and Kashmir region, a total of 54 spp., distributed over 26 genera of PPNs, is known to occur, affecting fruit plantations (temperate and tropical), *viz.* almond, apple, apricot, banana, cherry, citrus, kiwi, mango, papaya, peach, pear, plum, pomegranate, strawberry and walnut (**Tables 5.7 and 5.10**). The total number of PPN genera, affecting rest of the fruit kinds plantations, is presented in **Table 5.7**. The highest number of PPN genera *i.e* 14, associated with fruit has been found in case of Kiwi, followed by 13 genn. each in fruit plantations *viz.* apple, cherry and walnut. Earlier, in Kashmir region, a detailed study on plant parasitic nematodes (PPNs) associated with temperate fruit trees, *viz.* apple, apricot, cherry, pear, plum, pear, orchards and nurseries has been accomplished (Zaki and Mantoo, 2003). These authors reported about 8 spp. of PPNs in rhizospheres of these fruit trees in three districts of Kashmir, *viz.* Srinagar, Anantnag, and Baramulla, found population of these nematodes highest in July and lowest in December. This study has also shown that there was a positive effect of ambient temperature and a negative effect of rainfall and relative humidity on nematode build-up.

Table 5.7: Generic Richness and Diversity of Plant Parasitic Nematodes, Affecting Fruit Plantations (Temperate and Tropical) in Jammu and Kashmir, UT

Nematode Genus	*Different Kinds of Temperate and Tropical Fruit Trees/Plants*															
	Al	*Ap*	*Ar*	*Ba*	*Ch*	*Ci*	*Gv*	*Ki*	*Ma*	*Pa*	*Pe*	*Pr*	*Pl*	*Po*	*St*	*Wa*
Aglenchus	–	+	–	–	+	–	–	–	–	–	–	–	–	–	–	+
Aphelenchoides	–	+	+	–	+	–	–	+	–	–	–	+	+	–	–	+
Aphelenchus	–	–	–	–	+	–	+	+	–	–	–	–	–	–	–	
Basiria	–	+	–	–	–	–	–	–	–	–	–	–	–	–	–	–
Boleodorus	–	+	–	–	–	–	–	–	–	–	–	–	–	+	–	+
Criconemoides	+	+	+	–	+	–	–	–	–	–	–	+	+	–	–	+
Globodera	–	–	–	–	+	–	–	–	–	–	–	+	–	–	–	–
Helicotylenchus	+	+	+	–	+	–	–	+	–	–	+	+	+	–	+	+
Hoplolaimus	+	+	+	–	+	+	–	+	–	–	–	+	+	–	–	+
Longidorus	+	+	–	–	–	–	–	+	–	–	–	+	–	–	+	+
Merocriconema	+	–	–	–	–	–	–	–	–	–	–	–	–	–	–	–
Meloidogyne	+	–	+	+	+	+	+	+	–	+	–	–	+	–	+	–
Merlinius	–	–	–	–	–	–	–	+	–	–	–	–	–	–	–	–
Nothotylenchus	–	–	–	–	–	–	–	+	–	–	–	–	–	–	–	–
Paratylenchus	+	–	+	–	+	–	–	–	–	–	+	–	+	–	–	+
Pratylenchus	+	+	+	–	+	+	–	+	–	–	+	+	+	–	–	+
Psilenchus	–	–	–	–	–	–	–	+	–	–	–	–	–	–	–	+
Quinisulcius	–	–	–	–	–	–	–	–	–	–	–	–	–	+	–	–
Radopholus	–	–	–	+	–	–	–	–	–	–	–	–	–	–	–	–
Rotylenchus	–	–	–	–	+	–	–	–	–	–	–	–	–	–	–	–
Rotylenchulus	–	–	–	–	–	–	–	–	–	–	–	–	–	–	–	–
Sakia	–	–	–	–	–	–	–	+	–	–	–	–	–	–	–	–
Scutallonema	–	–	–	–	–	–	–	–	+	–	–	–	–	–	–	–
Tylenchorhynchus	+	+	+	–	+	–	–	+	–	–	+	+	+	+	–	+
Tylencholaimellus	–	+	–	–	–	–	–	–	–	–	–	–	–	+	–	–
Tylenchus	–	+	–	–	–	+	–	+	–	–	+	–	–	+	–	+
Xiphinema	–	+	+	–	+	+	–	+	–	–	–	+	+	–	–	+
Total number of Genera	9	13	9	2	13	5	1	14	1	1	5	9	9	5	3	13

+: Nematode genus present; –: Nematode genus absent.

Al: Almond; Ap: Apple; Ar: Apricot; Ba: Banana; Ch: Cherry; Ci: Citrus; Gv: Grapevine Ki: Kiwi; Ma: Mango; Pa: Papaya; Pe: Peach; Pr: Pear; Pl: Plum; Po: Pomegranate; St: Strawberry; Wa: Walnut.

Table 5.8: Generic Richness and Diversity of Parasitic Nematodes, Associated with Cereal and Forest Range Plantations in Jammu and Kashmir

Nematode Genus	*Cereals (Maize, Paddy, Wheat), Forest Trees and Grasses*							
	De	*Gr*	*Ma*	*Mu*	*Pa*	*Pi*	*Sa*	*Wh*
Anguina	–	–	–	–	–	–	–	+_
Aphelechoides	–	–	+	–	–	–	–	–
Aphelenchus	–	–	–	–	+	–	–	–
Basiria	–	–	–	–	–	–	–	+
Boleodorus	–	+	–	–	–	–	–	–
Helicotylenchus	–	+	+	–	–	–	–	–
Hemicriconemoides	+	–	–	–	+	–	–	–
Heterodera	–	+	–	–	–	–	–	–
Hirschmanniella	–	–	–	–	+	–	–	–
Hoplolaimus	–	–	+	–	–	–	–	–
Longidorus	–	–	+	–	–	–	–	–
Meloidogyne	–	–	–	–	+	–	–	–
Paratylenchus	–	–	+	–	–	–	–	–
Pratylenchus	–	–	+	–	–	–	–	–
Proleptonchus	–	+	–	–	–	–	–	–
Promumtazium	+	–	–	–	–	–	–	–
Tylenchorhynchus	–	–	+	+	+	–	+	–
Tylencholaimus	–	–	–	–	–	+	–	–
Tylencholaimellus	+	+	–	–	–	–	–	–
Tylenchus	–	–	+	–	+	–	–	–
Xiphinema	–	–	+	–	+	–	–	–
Total number of Genus/Genera	**3**	**5**	**9**	**1**	**7**	**1**	**1**	**2**

+: Nematode genus present; –: Nematode genus absent.

De: Deodar/The Himalayan cedar (*Cedrus deodara*); Gr: Grasses; Ma: Maize (*Zea mays*); Mu: Mulberry (*Morus* spp.); Pa: Paddy/rice (*Oryza sativa*); Pi: Pine (Scot) (*Pinus sylvestris*); Sa: Salix (*Salix babylonica*); Wh: wheat (*Triticum aestivum*).

In the recent study on distribution of PPNs, associated with stone fruit trees (peach, plum, apricot and cherry) and nut fruit trees (almond and walnut), for the period of 3 years in old nurseries of Kashmir, as many as 10 species of parasites have been encountered (Askaray *et al.*, 2014). During the course of this study, these authors found the absolute density and prevalence of PPNs, *viz. Helicotylenchus indicus* in walnut trees, followed by *Pratylenchus penetrans* in cherry and *Tylenchorhynchus* spp. in almond plantations.

(b) Apple (*Pyrus malus*) (Rosaceae)

About 24 spp. of PPNs have been recorded by different authors, affecting the valuable apple crop in Kashmir, spread over 13 different genera, occurring in different localities and areas of this region. Various genera, incorporating number of species are : *Aphelenchoides* (1 sp.), *Aglenchus* (1), *Criconemoides* (1), *Tylenchorhynchus* (3), *Helicotylenchus* (5), *Pratylenchus* (2), *Basiria* (1), *Boleodorus* (1), *Hoplolaimus* (2), *Longidorus* (1), *Tylencholaimellus* (1), *Tylenchus* (2), *Xiphinema* (3) (Table 5.10) *X. index.* has been observed infesting neglected apple orchards in Baramulla (Kashmir) (Lone *et al.*, 2012) (**Table 5.10**).

(c) Almond (*Prunus amygdalus*) (Rosaceae)

A total 14 species, belonging to 9 different genera of PPNs, have been reported from J&K UT, damaging almond fruit trees. The genera, with total number of species, under each genus are : *Circonemoids* (1 sp.), *Hoplolaimus* (1), *Longidorus* (1), *Meloidogyne* (2 spp.), *Merocriconema* (1), *Pratylenchus* (2), *Paratylenchus* (3 spp.), *Helicotylenchus* (2), *Tylenchorhynchus* (1) (**Table 5.10**).

(d) Apricot (*Prunus armeniaca*) (Rosaceae)

In J&K UT, a total of 15 spp., under 9 genera of PPNs, are known to be associated with apricot fruit trees. The break-up of these species, belonging to each genus is as : *Aphelenchoides* (2 spp.), *Criconemoides* (2), *Tylenchorhynchus* (1), *Helicotylenchus* (3), *Meloidogyne* (1), *Paratylenchus* (1), *Pratylenchus* (3), *Hoplolaimus* (1), *Xiphinema* (1) (**Table 5.10**).

(e) Banana [*Musa* sp. (Musaceae), Papaya (*Carica papya*) (Caricaceae), Mango (*Magnifera indica*) (Anacardiaceae) and Citus (*C. sinensis, C. reticulate, C. limon*) (Rutaceae)]

The tropical fruit trees like banana, papaya, mango, and citrus in Jammu region, has shown the the prevalence 7 genera, including 9 species of PPNs (**Tables 5.7 and 5.10**). Banana plants in Jammu region have been found to be infested with 2 species of PPNs, under 2 genera. These species are : *Meloidogyne incognita* and *Rhodopholus similis*. In this region, papaya and mango plants showed infestations of nematode parasites *Meloidogye javanica* and *Scutellonema brachyurum* respectively. Citrus plants (*Citrus* spp.) in Jammu region, have been found to be infested with PPNs : *Meloidogyne javanica, Hoplolaimus* sp., *Pratylenchus, Tylenchus semipenetrans* and *Xiphinem* (**Table 5.10**).

(f) Cherry (*Prunus avium*) (Rosaceae)

About 24 species, under 13 genera of PPNs, have been reported to damage cherry fruit tree in Kashmir by various authors. These include highest number of species *i.e.*, 5 spp., in case of genus *Helicotylenchus*, followed by 3 spp. each of *Pratylenchus, Tylenchorhynchus* and *Xiphinema*, 2 species under genus *Criconemoides*. Rest of the genera include 1 species as : *Aglenchus, Aphelenchus, Aphelenchoides, Globodera, Hoplolaimus, Rotylenchus, Meloidogyne, Paratylenchus* (**Table 5.10**).

(g) Kiwi (*Actinida deliciosa*) (Actinidiaceae) and Grapevine (*Vitis vinifera*) (Vitaceae)

Kiwi plantations in Kashmir region have found to be associated with a total of 19 species, distributed over 14 genera of plant-parasitic nematodes. Various genera, with number of species in case of each genus are : *Aphelenchus* (1 sp.), *Aphelenchoides* (1), *Tylenchus* (1), *Hoplolaimus* (1), *Merilinus* (1), *Sakia* (1), *Helicotylenchus* (4), *Nothotylenchus* (1), *Longidorus* (1), *Pratylenchus* (1), *Tylenchothynchus* (3), *Psilenchus* (1), *Meloidogyne* (1), *Xiphinema* (1). Grapevine plants in Kashmir has shown the infestation of PPN, *viz. Meloidogyne hapla* (Tables 5.7 and 5.10).

(h) Peach (*Prunus persica*) (Rosaceae)

8 species of PPNs have so far been found to damage peach trees in J&K (UT). These include 4 spp. of genus *Helicotylenchus* and 1 sp. each pertaining to genus *Paratylenchus, Pratylenchus, Tylenchorhynchus* and *Tylenchus* (Tables 5.7 and 5.10)

(i) Pear (*Prunus communis*) (Rosaceae)

15 spp., belonging to 9 genera of PPNs are known to damage pear plantations in this region. Various gerera, with total species in each genus is as: *Aphelenchoides* (1 sp.), *Criconemoides* (2), *Globodera* (1), *Helicotylenchus* (4 spp.), *Hoplolaimus* (1), *Longidorus* (1), *Pratylenchus* (1), *Tylenchorhynchus* (1) and *Xiphinema* (2) (Tables 5.7 and 5.10).

(j) Plum (*Prunus domestica*) (Rosaceae)

In J&K UT, a total of 14 spp., belonging to 9 genera of parasitic nematodes, has been found to be associated with plum fruit plantations. Various genera, and number of species in each genus are : *Aphelenchoides* (2 spp.), *Criconemoides* (2), *Hoplolaimus* (1), *Meloidogyne* (1), *Tylenchorhynchus* (1), *Helicotylenchus* (2), *Pratylenchus* (3), *Paratylenchus* (1) and *Xiphinema* (1) (Tables 5.7 and 5.10).

(k) Pomegranate (*Punica granata*) (Punicaceae)

Pomegranate fruit trees in J&K have been found to be associated with 5 species of PPNs, belonging to five different genera : *Boleodorus, Quinisulcus, Tylencholaimellus, Tylenchorhynchus* and *Tylenchus*. Earliest nematode species, *viz. Tylencholaimellus raski* and *Quinisulcius punici*, have been reported by Jairajpuri and Siddiqi (1963) and Gupta and Uma (1980) respectively. Later, more nematode species like *Boleodorus* sp, *Tylenchorhynchus* sp. and *Tylenchus filiformis* on the *P. granata* have been recorded by Fotedar and Kaul (1982) and Kaul (1998) (Tables 5.7 and 5.10).

(l) Strawberry (*Fragaria vesca*) (Rosaceae)

Various parasitic nematodes associated with strawberry plants in Kashmir Valley include *Helicotylenchus caroliniensis, longidorus, Meloidogyne hapla, M. javanica* and *Xiphinema* (Tables 5.7 and 5.10).

(m). Walnut (*Juglans regia*) (Juglandaceae)

Walnut fruit trees in Kashmir region have been found to be associated with a total of 18 spp., distributed over 13 genera of parasitic nematodes Various genera, with number of species in each genus are : *Aglenchus* (1 sp.), *Aphelenchoides* (1), *Boleodorus* (1), *Criconemoides* (2 spp.), *Helicotylenchus* (1), *Hoplolaimus* (2), *Longidorus* (1), *Pratylenchus* (2), *Paratylenchus* (1), *Psilenchus* (1), *Tylenchorhynchus* (2), *Tylenchus* (2) and *Xiphinema* (1) (Tables 5.7 and 5.10).

5.2.6 Medicinal/Aromatic and Ornamental Plants

(a) Saffron (*Crocus sativus*) (Iridaceae)

Saffron plants in Kashmir have been observed to be affected by a total number of 16 species, under 8 genera of PPNs. These genera, with their number of species are : *Aphelenchus* (2 spp.), *Tylenchus* (4), *Helicotylenchus* (2), *Hirschmanniella* (1), *Pratylenchus* (2), *Tylenchorhynchus* (2), *Psilenchus* (1), *Hemicericonemoides* (1), *Xiphinema* (1) (Tables 5.9 and 5.10). Earlier, the detailed survey on parasitic nematodes infesting saffron plantations in Pampore region of Kashmir region was conducted by Zaki and Mantoo (2008). These authors reported 7 species of PPN in this locality, with their prevalence percentages as : *Helicotylenchus vulgans* (16.6 per cent); *Pratylenchus thornei* (8.8 per cent); *Tylenchus* sp. (13 per cent); *Tylenchorhynchus* (10.7 per cent) *Xiphinema* (14.6 per cent); *Aphelenchus avenae* (5.8 per cent) and *Hemicericonemoides* (3.2 per cent). An absolute frequency of 86.6 per cent of Tylenchus arcuates, infesting saffron plants in Kashmir has been observed (Hassan, 2018).

Table 5.9: Generic Richness and Diversity of Plant Parasitic Nematodes Affecting Medicinal, Aromatic and Ornamental Plants of Jammu and Kashmir

Nematode Genus	*Medicinal, Aromatic and Ornamental Host Plant Species*												
	Am	*Bs*	*Co*	*Dv*	*Ej*	*Gh*	*Gi*	*Hi*	*Mi*	*Ri*	*Rs*	*Sa*	*Te*
Aphelenchoides	–	–	–	+	–	–	–	+	–	–	+	–	–
Aphelenchus	–	–	–	–	–	–	–	–	–	–	–	+	–
Basiria	–	–	–	–	–	+	–	–	–	–	–	–	–
Helicotylenchus	+	–	–	–	–	+	–	–	–	–	–	+	+
Hemicriconemoides	–	–	–	–	–	–	–	–	–	–	–	+_	–
Hirschmanniella	–	–	–	–	–	–	–	–	–	–	–	+	–
Hoplolaimus	+	+	–	+	–	–	–	–	–	–	+	–	+
Longidorus	–	–	–	–	+	–	–	–	–	–	–	–	–
Meloidogyne	–	–	+	–	–	+	+	–	–	+	–	–	+
Paratylenchus	–	–	–	–	–	+	–	–	–	–	–	–	–
Pratylenchus	–	–	–	–	–	+	–	–	+	–	–	+	–
Psilenchus	–	–	–	–	–	–	–	–	+	–	–	+	–
Rotylenchus	–	+	—	+	–	–	–	–	–	+	+	+	–
Rotylenchulus	–	+	–	+	–	–	–	–	–	–	–	–	+
Tylenchorhynchus	–	+	–	–	–	–	–	–	+	–	–	+	+
Tylenchus	–	–	–	+	+	–	–	–	–	–	–	–	+
Xiphinema	+	–	–	–	–	+	–	–	–	–	–	+	–
Total number of Genera	**3**	**4**	**1**	**5**	**2**	**6**	**1**	**1**	**3**	**2**	**3**	**8**	**6**

+: Nematode genus present; –: Nematode genus absent.

Medicinal/aromatic plants: Sa: Saffron (*Crocus sativus*); Mi: Mint (*Mentha sativus*); Co: Coriander (*Coriander sativa*; Ri: *Rosa indica*; Rs: *Rosa sinensis*; Gi: Ginger (*Zingiber officinale*); **Ornamental plants**: Te: *Tegetes erecta*; Dv: *Dahalia variabilis*; Bs: *Bougainvillea specttabilis*; Am: *Antirrhinum majus*; Ej: *Euonymus japonicus*; Hi: *Hibiscus* sp.; Gh: *Gladiolus hortulanus*.

A detailed investigation was carried out in 10 Districts of Kashmir for PPNs affecting saffron and various genera found included *Helicotylenchus* spp., *Pratylenchus* spp., *Tylenchus* spp. and *Hirschmanniella* spp. (Sheikh *et al.*, 2014). The abundance of PPN genera was indicated to be the main parameter to estimate their distribution in soil or rhizoshere system, based on the frequency by these authors. These authors have also given the description of new species, *viz. Helicitylenchus chishti*, having varied frequency, density and prominence value over different areas/ regions of Kashmir valley.

(b) Ginger (*Zingiber officinale*) (Zingiberaceae), Coriender (*Coriender sativa*) (Apiaceae) and Mint (*Mentha arvensis*) (Lamiaceae)

The prevalence of PPNs, *viz. Meloidogyne javanica* and *M. incognita* has been reported, affecting ginger and coriander respectively, in Gharana village, Chatha Farm in Jammu district. Mint (*Mentha*) in Kashmir region is affected by PPNs, *Pratylenchus* sp., *Psilenchus* and *Tylenchorhynchus* (Tables 5.9 and 5.10).

(c) Ornamental Plants

A total of 16 spp., under 13 genn. of PPNs, has so far been reported from Srinagar (Kashmir) and Rajouri (Jammu), infesting ornamental plants. Jammu). The various ornamental plans affected by nematode genera and their number of species are : *Antirrhinum majus* (Plantaginaceae)- *Helicotylenchus, Hoplolaimus, Xiphinema; Bougainvillea spectabilis* (Nyctaginaceae)- *Hoplolaimus, Rotylenchus, Rotylenchulus, Tylenchorhynchus; Dahlia variabilis* (Asteraceae)- *Aphelenchoides, Hoplolaimus, Rotylenchus, Rotylenchulus, Tylenchus; Euonymus japonicus* (Celastraceae) – *Longidorus, Tylenchus; Gladiolus hortulanus* (Iridaceae)- *Helicotylenchus* (2 spp.), *Basiria* (2 spp.), *Meloidogyne, Paratylenchus, Pratylenchus* and *Xiphinema; Hibiscus* (Malvaceae)- *Aphelenchoides; Rosa* spp. (*R. indica, R. sinensis*) (Rosaceae)- *Aphelenchoides, Meloidogyne, Hoplolaimus* and *Rotylenchus; Tagetes erecta* (Asteraceae) – *Helicotylenchus, Hoplolaimus, Meloidogyne, Rotylenchus, Tylenchorhynchus, Tylenchus* (Tables 5.9 and 5.10).

Table 5.10: Species Diversity of Plant-Parasitic Nematodes of Agricultural, Horticultural, Forestry Importance, and Other Economically Important Plants of Jammu and Kashmir, and Ladakh (UTs)

Type of Host Plant/Crop	*Plant-Parasitic Nematode Species*
I. Cereal plants	
Maize (*Zea mays*)	*Aphelencoides srinagarensis, Criconemoides Helicotylenchus dihystera, Helicotylenchuserythrinae, H. pteracercusoides, Hoplolaimus* sp., *Longidorus reneyii, Longidorus brevicaudatus, Longidorus elongatus, Longidorus* sp. *Pratylenchus sativus, Pratylenchus minimus,. Paratylenchus* sp., *Pratylenchus, Pratylenchus brachyurus, Pratylenchus zea, Tylenchus* sp., *Xiphinema, Xiphinema diversicadatum, Tylenchorhynchus, Tylenchorhynchus zeae*
Paddy (*Oryza sativa*)	*Aphelenchus* sp., *Hemicriconemoides* sp., *Hirchmanniella gracilis, Hirschmanniella* sp., *Hischmanniella. mucronata, H. oryzae, H. shamimi, Meloidogyne, Tylenchorhynchus oryzae, Tylenchus* sp., *Xiphinema.*

Type of Host Plant/Crop	Plant-Parasitic Nematode Species
Wheat (*Triticum aestivum*)	*Anguina tritici, Basiria tritici.*
Ulndetermined cereal	*Heterodera avenae*
OIL SEED PLANT	
Field mustard *Brassica campestris/ Brassica rapa*	*Aphelenchus radicicolus, Psilenchus haki, Nothotylenchus basiri, Paratylenchus barkati, Paratylenchus* sp., *Pratylenchus similis, P. thornei, Rotylenchus reniformis, Tylenchorhynchus kashmirensis*
VEGETABLE PLANTS	
Brinjal (*Solanum melongena*)	*Aphelenchoides sanwali, Aphelenchoides, Aphelenchus, Aphelenchus avenae, Aporcelaimellus, Basiria kashmirensis, Basiria* sp., *Criconemoides* sp., *Dorylaimus, Helicotylenchus haki, H. kashmirensis, H. erythtinae, H. thornei, H. mucronatus, H. trapezoidicaudatus, H. indicus, H. jammuensis, H. silvactus, H. zeae, Helicotylenchus* sp, *Heterodera* sp., *Hirschmanniella gracilis, H. oryzae, Hoplolaimus indicus* (= *Basirolaimus indicus*), *Hoplolaimus* sp., *Longodorus* sp., *Meloidogne incognita, M. javanica, Nothotylenchus basiri, Nothotylenchus* sp., *Paravulvus confusus, Pratylenchus penetrans, P. zeae, Psilenchus* sp., *Rotylenchus* sp., *Sakia typica, Tylenchorhynchus haki, Tylenchorhynchus irruglaris T. kashmirensis, T. indicus, T. dubius, T. mashhoodi, Tylenchorhynchus* sp., *Tylenchus andrassyi, T. filiformis, Tylenchus* sp., *Xiphinema basiri* and *Xiphinema* sp.
Bottle gourd (*Lagenaria siceraria*)	*Aphelenchoides* sp., *Helicotylenchus* sp., *Pratylenchus* sp., *Tylenchorhynchus* sp., *Tylenchus* sp., *Xiphinema* sp.
Cabbage (*Bassica oleracea* var. *capitata*)	*Aphelenchoides saprophilus, Aphelenchoides, Helicotylenchus indica, Helicotylenchus* sp., *H. silvactus, Hoplolaimus indicus, Hoplolaimus* sp., *Longidorus* sp., *Meloidogyne incognita, Pratylenchus zeae, Pratylenchus* sp., *Trichodorus* sp., *Tylenchorhynchus haki, T. mashhoodi, Tylenchorhynchus bassicae, Tylenchorhynchus* sp. and *Tylenchus filiformis, Tylenchus* sp.
Carrot (*Daucus carota*)	*Meloidogyne incognita*
Cauliflower (*Brassica-oleracea* var. *botrytis*)	*Aphelenchus avenae, Aglenchus, Basiria kashmirensis, Boleodorus cylindricus, Boleodorus, Filenchus, Helicotylenchus, Hoplolaimus* sp., *Helicotylenchus mucronotus, Hirschmanniella gracilis, Merlinius, Nothotylenchus basiri, Psilenchus* sp., *Pratylenchus brachyurus, P.similis Tylenchorhynchus brassicae, T. kashmirensis* and *T. mashhoodi, Tylenchus andrassyii, Xiphinema basiri.*
Chilli (*Capsicum annuum*)	*Aphelenchoides saprophilus, Aphelenchoides, Boleodorus, Heliocotylenchus* sp., *Helicotylenchus haki, Hirschmanniella gracilis, Hirschamanniella sp. Hoplolaimus indicus, Meloidogyne hapla, M. incognita, Pratylenchus barkati, Tylenchorhynchus brassicae, Pratylenchus* sp., *Tylenchorhynchus* sp., *Tylenchorhynchus kashmirensis, T. meshhoodi, Tylenchus arcuates, Tylenchus andrassyii, T. filiformis, Tylenchus* sp. and *Xiphinema* sp.
French Bean (*Phaseolus vulgaris*)	*Aphelenchoides* sp., *Criconemoides, Helicotylenchus indicus, H. minzi, Helicotylenchus* sp., *Meliodogyne incognita, Aglenchus* sp., *Tylenchus, Hirschmanniella, Pratylenchus* sp., *Psilenchus, Tylenchorhynchus* sp., *Xiphinema* sp.
Soya Bean (*Glycine max*)	*Lordellonema bashirensis*
Spinach (*Spinacia-oleracea*)	*Helicotylenchus silvaticus, Helicotylenchus* sp., *Hoplolaimus* sp., *Aphelenchus avenae, Aphelenchoides sanwali, Aglenchus* sp., *Boleodorus, Longidorus, Merlinius, Meloidogyne inconita, Psilenchus, Tylenchus filiformis, Tylenchothynchus brassicae*

Type of Host Plant/Crop	*Plant-Parasitic Nematode Species*
Kale (*Brassica oleracea* var. *sabellica*)(*acephala*)	*Aphelenchoides, Basiria haki, Globodera rostochiensis, Helicotylenchus dihystera, H. carenessi, H. haki, Hirschmanniella gracilis, Hoplolaimus indicus, Longidorus, Meloidogyne incognita, Nothotylenchus basiri, N. srinagarensis, Pratylenchus zeae, Tylenchorhynchus kashmirensis, Tylenchorhynchus haki, Tylenchus andrassyi* and *Tylenchus filiformis*
Knol khol (*Brassica oleracea gongylodes*)	*Aphelenchoides sanwali, Aphelenchoides composticola, Aphelenchoides, Basiria haki, Hoplolaimus indicus, Helicotylenchus haki, Helicotylenchus* sp., *Hirschmanniella gracilis, Longidorus, Pratylenchus* sp., *Tylenchorhynchus kashmirensis, Tylenchorhynchus* sp., *Tylenchus andrassyii, Tylenchus* sp., *Xiphinema basiri, Xiphinema* sp.
Okra (lady finger) (*Ablemoschus esculentus*)	*Aphelenchus avenae, Aphelenchoides saprophilus, Basiria. kashmiransis, Helicotylenchus mucronotus, Helicotylenchus* sp., *Hoplolaimus indicus, Hoplolaimus* sp., *Meloidogyne incognita, Nothotylenchus basiri, Pratylenchus barkati, Pratylenchus similis, Rotylenchulus reniformis, Tylenchorhynchus brassicae, T. kashmirensis, T. mashhoodi, Tylenchus filiformis*
Onion (*Allium cepa*)	*Aphelenchoides saprophilus, Aphelenchus avenae, Basiria kashmirensis, Globodera rostochiensis, H. dihystera, Helicotylenchus* sp., *Hoplolaimus* sp., *Nothotylenchus basiri, Trichodorus* sp.
Garlic (*Allium sativum*)	*Helicotylenchus* sp.
Pea (*Pisum sativum*)	*Aphelenchoides avenae, A. sanwali, Criconemoides* sp., *Helicotylenchus. dihystera, M. incognita, Pratylenchus brachyurus, Tylenchorhynchus mashhoodi, Hoplolaimus indicus, Longidorus* sp., *Tylenchus andrassyii*
Potato (*Solanum tuberosum*)	*Aglenchus, Aphelenchus avenae, Aphelenchoides sanwali, Aphelenchoides saprophilus, Aphelenchoides* sp., *Basiria kashmirensis, Globodera rostachiensis, Helicotylenchus dihystera, H. indicus, Helicotylenchus* sp., *Hoplolaimus indicus, Hoplolaimus* sp., *Meloidogyne hapla, M. incognita, Pratylenchus barkati, P. brachyurus, P. penetrans, Pratylenchus similis, P. thornei, P. zeae, Pratylenchus* sp., *Tylenchorhynchus kashmirensis, T. mashhoodi, Tylenchorhynchus* sp., *Tylenchus andrassyi, T. filiformis, Tylenchus* sp. and *Xiphinema* sp.
Pumpkin (*Cucurbita pepo*)	*Tylenchorhynchus brassicae, Aphelenchus avenae, Aphelenchoides sanwali, Hoplolaimus indicus* (= *Basiriolaimus indicus*), *Trichodorus* sp., *Helicotylenchus dihystera, H. mucronotus, Helicotylenchus* sp., *Hirschmanniella gracilis, Hoplolaimus, Meloidogyne incognita, Rotylenchulus reniform*
Winter squash (*Cucurbita maxima*)	*Helicotylenchus* sp., *Meloidogyne incognita, Psilenchus* sp., *Rotylenchulus reniformis, , Tylenchorhynchus* sp.
Radish (*Raphanus sativus*)	*Aphelenchoides sanwali, Aphelenchoides saprophilus, Aphelenchoides* sp., *Basiria kashmirensis, Boleodorus, Helicotylenchus. dihystera, Hoplolaimus indicus, Helicotylenchus* sp., *Hirschmanniella gracilis, Globodera rostochiensis, M. incognita, Nothotylenchus basiri, Pratylenchus zeae, Tylenchorhynchus kashmirensis, Tylenchorhynchus mashhoodi, Pratylenchus brachyurus, Pratylenchus, Tylenchorhynchus. brassicae, Tylenchus* sp., *Xiphinema basiri, Xiphinema* sp.
Tomato (*Lycopersicum esculentum*)	*Aphelenchoides composticola, Aphelenchoides, Aphelenchus avenae, Globodera rostochiensis, H. dihystera, H. erythrinae, H. indicus, H. trapezoidicaudatus, Helicotylenchus* sp., *Hirschmanniella gracilis, Hoplolaimus, M. incognita, Nothotylenchus basiri, Pratylenchus penetrans, Pratylenchus* sp., *Scutellonema petersi, Tylenchorhynchus haki, T. kashmirensis, T. mashhoodi, Tylenchorhynchus irregularis, Tylenchus andrassyi, T. filiformis, Tylenchus* sp., *Trichodorus* sp. and *Xiphinema basiri, Xiphinema* sp.

Type of Host Plant/Crop	Plant-Parasitic Nematode Species
Turnip (*Brassica campestris rapa*)	*Aphelenchoides* sp., *Aphelencoides sanwali, A. composticola, Helicotylenchus indicus, Helicotylenchus* sp., *Hoplolaimus indicus, Nothotylenchus basiri, Paratylenchus* sp, . *Pratylenchus barkati, P. penetrans, P. similis, P. thornei, R. reniformis, Tylenchorhynchus* sp., *T. kashmirensis, Tylenchus filiform, Tylenchus* sp, *Xiphinema basiria, Xiphinema* sp.
FOREST TREES	
Scot pine (*Pinus sylvestris*)	*Tylencholaimus pini*
Salix (*Salix babylonica*)	*Tylenchorhynchus rugosus*
Mulberry (*Morus alba*)	*Tylenchorhynchus*
Himalayan cedar/ deodar (*Cedrus deodara*)	*Promumtazium indicum, Tylencholaimellus striatus, Tylencholaimellus sayeedi, Hemicriconemoides magniferae*
MUSHROOM	*Aphelonchoides* sp.
FRUIT TREES	
Apple (*Pyrus malus*)	*Aphelenchoides, Aglenchus costatus, Criconemoides cnoensis*, , *Helicotylenchus retusus*, , *Basiria kashmirensis, Boleodorus* sp., *Helicotylenchus hazratbalensis, H. papillocauda, H. caroliniensis, H. indicus, Hoplolaimus indicus, Hoplolaimus* sp., *Longidorus brevicaudatus, Longidorus elongatus, Longidorus mirus, Longidorus* sp. *Tylenchorhynchus nudus, Pratylenchus zeae, Pratylenchus valvus, Tylenchorhynchus nudus Tylencholaimellus eskei, Tylenchorhynchus* sp., *Tylenchus filiformis, Tylenchus* sp. and *Xiphinema insigne Xiphinema americanus, Xiphinema index, Xiphinema* sp.
Almond (*Prunus -amygdalus*)	*Criconemoides* sp., *Hoplolaimus* sp., *Longidorus* sp., *Meloidogyne javanica, M. hapla, Merocriconema* (=*Neolobocriconema laterale*), *Pratylenchus pseudopratensis, P. vulnus, Paratylenchus penetrans, P. juglansi, Paratylenchus* sp., *Helicotylenchus dihystera, Helicotylenchus indicus*, , *Tylenchorhynchus*
Apricot (*Prunus armeniaca*)	*Aphelencoides, A. costatus Criconemoides cnoensis, Criconemoides, Tylenchorhynchus nudus, Helicotylenchus jammuensis, Meloidogyne hapla, Paratylenchus juglansi, Pratylenchus pseudopratensis, Pratylenchus penetrans, P. zea, Helicotylenchus indicus, Hoplolaimus, H. retusus, Xiphinema insigne*
Cherry (*Prunus avium*)	*Aglenchus costatus, Aphelencoides, Aphelenchus* sp., *Criconemoides cnoensis, Criconemoides, Globodera rostochiensis, Hoplolaimus, Helicotylenchus dihystera, H. erythrinae, H. indicus, H. kashmirensis, H. retusus, Meloidogyne hapla, Paratylenchus juglansi, Pratylenchus zeae, Pratylenchus penetrans, Pratylenchus pseudopratensis, Rotylenchus* sp. *Tylenchorhynchus kashmirensis, T. mashhoodi, T.nudus, Xiphinema basir, X insigne* and *Xiphinema* sp.
Kiwi (*Actinidia deliciosa*)	*Aphelenchoides, Aphelenchus* sp., *Hoplolaimus indicus* (= *Basirolaimus indicus*), *Merlinius, Sakia typica, Heliocotylenchus dihystera, H. kashmirensis, H. indicus, Helicotylenchus* sp., *Nothotylenchus* sp., *Longidorus, Pratylenchus penetrans, Tylenchorhynchus elegans, Tylenchorhynchus oryzae, Tylenchorhynchus kashmirensis, Tylenchus filiformis, Psilenchus, Meloidogyne hapla, Xiphinema basiri*
Mango (*Mangifera indica*)	*Scutallonema brachyurum*

Type of Host Plant/Crop	Plant-Parasitic Nematode Species
Peach (*Prunus persica*)	*Helicotylenchus hazratbalensis, Helicotylenchus indicus, H. kashmirensis, Helicotylenchus* sp., *Paratylenchus juglansi, Pratylenchus penetrans. Tylenchorhynchus pruni, Tylenchus filiformis*
Pear (*Prunus communis*)	*Aphelencoides, A. costatus, Criconemoides cnoensis, Criconemoides, Globodera rostochiensis, Helicotylenchus dihystera, , H. retusus, P. zeae, Hoplolaimus, Helicotylenchus* sp., *Longidorus* sp, *Pratylenchus zeae, Tylenchorhynchus nudus, Xiphinema insigne* and *Xiphinema* sp.
Plum (*Prunus domestica*)	*Aphelencoides, A. costatus, Meloidogyne hapla, Tylenchorhynchus nudus, Helicotylenchus. retusus, Pratylenchus zeae, Paratylenchus juglansi, Pratylenchus penetrans, Pratylenchus* sp., *Helicotylenchus indicus, Criconemoides, C. cnoensis. Hoplolaimus, Xiphinema insigne*
Pomegranate (*Punica granata*)	*Boleodorus* sp., *Quinisulcius punici, Tylencholaimellus raskii, Tylenchorhynchus* sp. and *Tylenchus filiformis*
Strawberry (*Fragaria vesca*)	*Helicotylenchus caroliniensis, Longidorus* sp., *Meloidogyne hapla, Meloidogyne javanica.*
Walnut (*Juglaus regia*)	*Aglenchus costatus, Aphelencoides, Boleodorus* sp., *Criconemoides cnoensis, Criconemoides, Helicotylenchus retusus, Hoplolaimus indicus, Hoplolaimus* sp., *Longidorus* sp., *Pratylenchus penetrans, Paratylenchus juglansi, Psilenchus* sp., *Pratylenchus zeae, Tylenchorhynchus nudus, Tylenchorhynchus* sp., *Tylenchus filfiform, Tylenchus* sp., *X. insigne.*
Citrus fruits (*Citrus sinensis, C. reticulata, C. limon*	*Meloidogyne javanica, Hoplolaimus* sp., *Pratylenchus, Tylenchus semipenetrans, Xiphinema* sp.
Papaya (*Carica papaya*)	*Meloidogyne javanica*
Banana (*Musa* sp.)	*Meloidogyne incognita, Radopholus similis*
Grapevine (*Vitis vinifera*)	*Meloidogyne hapla, Pratylenchus, Tylenchorhynchus, Rotylenchulus, Tylenchus*
MEDICINAL/AROMATIC PLANTS	
Saffron (*Crocus sativus*)	*Aphelenchus avenae, Aphelenchus, Tylenchus ? saffroni* (= *Aerocotylenchus saffroni*), *Helicotylenchus chisthi, Helicotylenchus vulgaris, Hirshemanniella, Pratylenchus, Pratylenchus thornei, Tylenchus, Tylenchus kashmirensis, Tylenchus arcuates, Tylenchorhynchus elegans, Psilenchus* sp., *Tylenchorhynchus, Hemicriconemoides, Xiphinema*
Mint (*Mentha arvensis*)	*Psilenchus, Pratylenchus, Tylenchorhynchus*
Ginger (*Zingiber officinale*)	*Meloidogyne javanica*
Coriender (*Corindrum sativum*)	*Meloidogyne incognita*
ORNAMENTAL PLANTS	
Chrysanthemum (cultivars)	*Meloidogyne incognita*
Rosa indica	*Meloidogyne, Rotylenchus*
Rosa sinensis	*Aphelenchoides, Rotylenchus, Hoplolaimus*
Tagetes erecti	*Meloidogyne, Helicotylenchus, Tylenchorhynchus, Tylenchus, Rotylenchus, Hoplolaimus*
Dahlia variabilis	*Rotylenchulus reniformis, Tylenchus, Aphelenchoides, Rotylenchus, Hoplolaimus*

Type of Host Plant/Crop	*Plant-Parasitic Nematode Species*
Bougainvillea spectabilis	*R. reniformis, Tylenchorhynchus, Rotylenchus, Hoplolaimus*
Antirrhium majus	*Helicotylenchus, Xiphinema, Hoplolaimus*
Euonymus japonicus	*Longidorus, Tylenchus* sp.
Hibiscus	*Aphelenchoides*
Gladiolus hortulanus	*Meloidogyne, Helicotylenchus, Pratylenchus, Heliocotylenchus pseudorobustus, Basiria graminophila, Xiphinema basiri, Basiria* sp., *Paratylenchus.*
GRASSES	
Cyanodon dactylon, meadow grass, undetermined grass species	*Boleodorus cynodoni, Helicotylenchus graminophilus, Tylencholaimellus striatus, Tylencholaimellus sayeedi, Heterodera trifolii, Proleptonchus sclerouterinus*

Table 5.11: An Annotated Checklist of Plant-Parasitic Nematodes and their Host Plants/Crops of Jammu and Kashmir, and Ladakh (UTs)

Sl. No.	*Nematodes Species*	*Host Plant/ Crop Species*	*Distri-bution*	*Author (s)/ Reference (s)*
1	*Aglenchus costatus* (=*Tylenchus costatus*)	Apricot, apple, cherry, plum, pear, walnut	Kmr.	Zaki and Mantoo (2003)
2	*Aglenchus* sp.	Cauliflower, spinach, potato, French beans	Kmr.	Hassan *et al.* (2014, 2016). Sheikh *et al.* (2016), Hassan and Ahangar (2018 b)
3	*Anguina tritici*	Wheat	Srg, Jmu Kmr., Ldk.	Shiekh (1979), Fotedar and Kaul (1982), Kalha *et al.* (1992), Zaki and Mir (1994) and Zaki (1999)
4	*Aphelenchoides composticola*	Tomato, Hak/Kale, turnip	Kmr	Wali-ullah (1992)
5	*Aphelenchoides sanwali*	Potato, brinjal, hak, radish, spinach, turnip, pea, pumpkin	Kmr	Wali-ullah (1992)
6	*Aphelenchoides saprophilus*	Potato, chilli, okara, cabbage, radish, onion	Kmr.	Wali-ullah (1992)
7	*Aphelenchoides* spp.	Apple, apricot, brinjal, chilli, cabbage, bean plum, pear, radish, cherry, tomato, walnut, Potato, Kiwi, Knol khol, *Hibiscus, Rosa sinensis, Dahlia variabilis,* turnip, bottle gourd	Kmr., Jmu.	Mahajan (1971), Zaki *et al.* (1991), Mantoo *et al.* (2008), Wali-ullah (2005), Zaki and Mantoo (2003), Rashid *et al.*- (2014)
8	*Aphelenchoides srinagarensis*	Maize	Kmr	Hassan *et al.* (2009), Sheikh et -al. (2009)
9	*Aphelenchus* sp.	Cherry, kiwi, paddy and brinjal	Srg., Kmr.	Fotedar and Kaul (1982), Wali-ullah (2005), Wali-ullah *et al.*, – (2008)

Sl. No.	Nematodes Species	Host Plant/ Crop Species	Distri-bution	Author (s)/ Reference (s)
10	*Aphelenchus avenae*	Brinjal, cauliflower, saffron, spinach, onion, okara, pea, potato, pumpkin, tomato	Kmr.	Wali-ullah (1992), Zaki and Muntoo (2008)
11	*Aphelenchus radicicolus*	Field mustard	Kmr	Sheikh *et al.* (2008)
12	*Aporcelaimellus* sp.	Brinjal	Jmu.	Wali-ullah *et al.* (2008)
13	*Basiria haki*	Hak/kale	Kmr.	Fotedar and Mahajan (1973)
14	*Basiria kashmirensis*	Okara, onion, cauliflower, apple, radish, brinjal, potato	Srg., Kmr.	Mahajan (1971), Jairajpuri (1965), Wali-ullah (1992), Fotedar and Kaul (1982)
15	*Basiria graminophila*	*Gladiolus hortulans*	Kmr.	Abbas and Wali-ullah (2008)
16	*Basiria* sp. (I)	*Gladiolus hortulans*	Kmr	Walli-ullah (2010)
17	*Basiria tritici*	Wheat	Kmr.	Fotedar and Mahajan (1973)
18	*Basiria* sp. (II)	Brinjal	Jmu.	Wali-ullah *et al.* (2008)
19	*Boleodorus cylindricus*	Cauliflower	Kmr.	Hassan and Ahangar (2018 b)
20	*Boleodorus cynodoni*	*Cynodon dactylon* (Bermuda grass)	Jmu.	Fotedar and Mahajan (1972)
21	*Boleodorus sp.*	Apple, cauliflower, chilli, radish, spinach, pomegranate and walnut	Jmu., Kmr.	Fotedar and Kaul (1982), Hassan *et al.* (2014, 16), Sheikh *et al.* (2016)
22	*Cephalenchus rotundus*	Unidentified Plant species	Kmr.	Siddiqui and Khan (1983)
23	*Criconemoides cnoensis*	Apple, apricot, cherry, plum, pear, walnut	Kmr.	Zaki and Mantoo (2003)
24	*Criconema* sp.	Stone and nut fruit trees.	Kmr	Askary *et al.* (2014)
25	*Criconemoides* sp. (=*Criconemella*)	Almond, apple, bean, cherry, brinjal, apricot, plum, pear, walnut, maize	Kmr., Raj.	Fotedar and Kaul (1982), Zaki and Mantoo (2003), Allie *et al.* (2020)
26	*Ditylenchus* sp.	Maize	Raj.	Allie *et al.* (2020)
27	*Dorylaimus* sp.	Brinjal	Jmu.	Wali-ullah *et al.* (2008)
28	*Filenchus* sp.	Cauliflower	Kmr.	Sheikh *et al.* (2016), Hassan and Ahangar (2018b), Hassan *et al.* (2014)
29	*Filenchus cylindricaudus*	Onion	Kmr.	Hassan *et al.* (2018)
30	*Globodera rostochiensis*	Cherry, onion, kale/hak, pea, radish, tomato, potato, pear	Kmr.	Kaul (1978), Wali-Ullah (1992), Wali-ullah and Kaul (1997)
31	*Helicotylenchus caroliniensis*	Apple, strawberry	Kmr.	Kaul (1984), Kaul (1998)
32	*Helicotylenchus chishti*	Saffron	Kmr.	Sheikh *et al.* (2014)
33	*Helicotylenchus carenessi*	Kale/hak	Kmr.	Zaki *et al.* (1997), Wali-ullah (1992) and Kaul (1998)

Sl. No.	Nematodes Species	Host Plant/ Crop Species	Distri-bution	Author (s)/ Reference (s)
34	*Helicotylenchus dihystera*	Onion, Kale/hak, tomato, almond, Cherry, pea, potato, kiwi, radish, pumpkin, maize	Kmr., Raj.	Wali-ullah (1992, 2005), Wali-ullah and Kaul (1997), Kaul (1998), Munib *et al.* (2016), Allie *et al.* (2020)
35	*Helicotylenchus erythrinae*	Cherry (sweet), tomato, brinjal, maize	Jmu., Kmr.	Kaul and Fotedar (1987) Wali-ullah (1992), Bhat *et al.* (1994), Wali-ullah *et al.* (2008)
36	*Helicotylenchus graminophilus*	Unidentified grass species	Kmr.	Fotedar and Mahajan (1974)
37	*Helicotylenchus haki*	Kale/hak, chilli and Brinjal	Kmr.	Fotedar and Mahajan (1974) Zaki *et al.* (1991), Kaul (1998)
38	*Helicotylenchus. hazratbalensis*	Apple, peach	Kmr.	Fotedar and Handoo (1974), Fotedar and Kaul (1982)
39	*Helicotylenchus indicus*	Apricot brinjal, cabbage, kale, potato, tomato, almond, cherry, plum, walnut, turnip	Kmr., Jmu	Zaki *et al.* (1991), Wali-ullah and Kaul (1997), Askary *et al.* (2014), Munib *et al.* (2016), Wali-ullah (2005), Wali-ullah *et al.* (2008)
40	*Helicotylenchus jammuensis*	Apricot, Brinjal	Jmu	Mahajan (1971), Wali-ullah (2008)
41	*Helicotylenchus kashmirensis*	Cherry, peach, Kiwi	Kmr.	Fotedar and Mahajan (1974), Fotedar and Kaul (1982), Wali-ullah (2005)
42	*Helicotylenchus minzi*	Bean	Kmr.	Zaki *et al.* (1991)
43	*Helicotylenchus mucronatus*	Okara (Lady-finger), brinjal, cauliflower, pumpkin	Kmr.	Kaul (1998), Wali-ullah (1992)
44	*Helicotylenchus papilliocauda*	Apple	Kmr.	Kaul (1987)
45	*Helicotylenchus pseudorobustus*	*Glodius hortulans*	Kmr.	Abbas and Wali-ullah (2010)
46	*Helicotylenchus pteracercusoides*	Maize	Kmr.	Fotedar and Kaul (1982, 85)
47	*Helicotylenchus retusus*	Cherry, apple, apricot, pear, plum, Walnut	Kmr.	Zaki and Mantoo (2003)
48	*Helicotylenchus silvaticus*	Cabbage, brinjal, spinach	Kmr.	Kaul (1998), Wali-ullah (1992)
49	*Helicotylenchus trapezoidicaudatus*	Tomato, brinjal	Kmr.	Fotedar and Kaul (1983), Zaki *et al.* (1991, 93)
50	*H. steineri*	Maize	Kmr.	Fotedar and Mahajan (1974)
51	*H. vulgaris*	Saffron	Kmr.	Zaki and Mantoo (2008)

Sl. No.	Nematodes Species	Host Plant/ Crop Species	Distri-bution	Author (s)/ Reference (s)
52	*Helicotylenchus* spp.	Garlic, winter squash, tomato, French beans, pear, radish, peach, potato, chilli, *Gladiolus hortulsm*, Kiwi, *Tegetes erecta*, *Antirrhium* majus, cauliflower, cabbage, brinjal, Knol Khol, okara, turnip, onion, bottle gourd, spinach, pumpkin	Kmr.	Fotedar and Kaul (1982), Zaki *et al.* (1991, 93), Rashid etal. (2014), Abbas and Wali-ullah (2010), Hassan *et al.* (2014, 2016, 2018), Sheikh *et al.* (2016), Wali-ullah *et al.* (2008), Wali-ullah (1992), Mantoo *et al.* (2008)
53	*Hemicriconemoides magniferae*	*Cedrus deodera*	Kmr.	Kurma and Uma (1988)
54	*Hemicriconemoides* sp.	Rice, saffron	Kmr.	Fotedar and Kaul (1982), Zaki and Mantoo (2008)
55	*Heterodera avenae*	Cereal, and unknown host plant	Jmu, Kmr, Ldk.	Singh *et al.* (1976), Zaki (1999)
56	*Heterodera trifolii*	Unidentified medow grass	Kmr.	Swarup *et al.* (1964)
57	*Heterodera* sp.	Brinjal	Jmu	Wali-ullah *et al.* (2008)
58	*Hirschmanniella gracilis*	Cauliflower, Kale/hak, brinjal, chilli, radish, pumpkin, khol	Kmr.	Wali-Ullah (1992) Wali –Ullah *et al.* (1991)
59	*Hirschmanniella mucronata*	Paddy	Kmr.	Mahajan (1971), Wali-ullah (1991) Luc and Goodey (1963)
60	*Hirschmanniella oryzae*	Paddy, brinjal	Kmr., Jmu	Mathur and Prasad (1971), Mahajan (1971), Wali-ullah *et al.* (1991), Wali-ullah *et al.* (2008)
61	*Hirschmanniella shamimi*	Paddy	Kmr.	Wali-ullah *et al.* (1991)
62	*Hirschamanniella* sp.	Paddy, chilli, French beans	Kmr.	Hassan *et al.* (2008), Sheikh *et al.* (2016), Hassan *et al.* (2016)
63	*Hoplolaimus indicus* (=*Basirolaimus indicus*)	Apple, okra, kale, cabbage, walnut, brinjal, potato, knol khol, radish	Kmr.	Kaul (1984), Kaul (1998), Wali-ullah (1992), Munib *et al.* (2016)
64	*Hoplolaimus* spp.	Apple, walnut, maize, almond, cherry, apricot, potato, tomato, brinjal, okara, cabbage, cauliflower, spinach, onion, pumpkin pear, plum, *Antirrhum magus*, *Bougainvillea spectabilis*, *Rosa sinensis*, *Tegetes eretes*, *Citrus* spp.	Jmu, Kmr.	Mahajan (1970), Fotedar and Kaul (1982), Wali-ullah (1992) Askary *et al.* (2014), Rashid *et al.* (2014), Nehru *et al.* (2005), Zalpuri *et al.* (2013a, b, c)
65	*Longidorus brevicaudatus*	Apple, maize	Kmr., Raj.	Lone *et al.* (2018), Allie *et al.* (2020)
66	*Longidorus elongatus*	Apple, maize	Kmr., Raj	Lone *et al.* (2018) , Allie *et al.* (2020), Lone and Zaki (2014)
67	*Longidorus mirus*	Apple	Kmr.	Lone *et al.* (2018)

Sl. No.	Nematodes Species	Host Plant/ Crop Species	Distri-bution	Author (s)/ Reference (s)
68	*Longidorus reneyii*	Maize	Kmr.	Raina (1966)
69	*Longidorus* spp.	Strawberry, apple, Kiwi, maize, walnut, almond, pear, *Euonymus japonicus,* brinjal, hak, cabbage, spinach, pea	Kmr.	Fotedar and Kaul (1982), Wali-ullah *et al.* (1993), Kaul (1984), Askary *et al.* (2014, Wali-ullah (1992, 2005)
70	*Lordellonema bashirensis*	Soybean	Kmr.	Akhtar *et al.* (1994)
71	*Meloidogyne graminicola*	Paddy	Kmr.	Singh *et al.* (2007)
72	*Meloidogyne hapla*	Apricot, grapevine, strawberry, potato, almond, cherry, plum, Kiwi, Chilli	Kmr.	Krishna and Prasad (1986), Wali-ullah *et al.* (1993,), Wali-ullah(2005), Zaki *et al.* (1991), Askary *et al.* (2014), Askary *et al.* (2018)
73	*Merlinius* sp.	Cauliflower, spinach, Kiwi	Kmr.	Sheikh *et al.* (2016), Wali-ullah (2005), Hassan *et al.* (2014, 2016) Hassan and Ahangar (2018b)
74	*Meloidogyne incognita*	Banana, cabbage, carrot, cauliflower coriander, chilli, *Chrysanthemum* (cultivars) cucurbit, radish, brinjal, okra, pea, bean, spinach, potato, pumpkin tomato, kale	Jmu., Kmr.	Wali-ullah (1992, , Kaul (1998), Bhat and Kaul. (1994), Singh and Satyapriya(2008), Singh *et al.* (2012a, b), Singh (2009b), Singh and Gupta(2011), Singh *et al.* (2012), Wali-ullah *et al.* (2008), Khan *et al.* (2005)
75	*Meloidogyne javanica*	Papaya, brinjal, Citrus spp., almond, strawberry, ginger	Jmu, Kmr.	Wali-ullah (1992), Nehru *et al.* (2005), Singh and Gupta (2011), Zalpuri *et al.* (2013a, b, c), Singh (2009a), Singh *et al.* (2012)
76	*Meloidogyne* sp.	*Rosa indica, Tagetes, Gladiolus hortulans*	Jmu, Kmr	Rashid *et al.* (2014), Abbas and Wali-ullah (2010)
77	*Merocriconema ? laterale*(= *Neolobociconema laterale*)	Almond	Kmr	Khan and Siddiqi (1963), Mehta and Raski (1971), Geraert (2019)
78	*Nothotylenchus basiri* Khan	Okra, onion, kale, cauliflower, mustard, Tomato, radish, brinjal, turnip	Kmr.	Wali-ullah (1992), Kaul (1998)
79	*Nothotylenchus* sp.	Kiwi, brinjal	Kmr.	Wali-ullah (2005), Wali-ullah *et al.* (2008)
80	*Nothotylenchus. srinagarensis*	Kale	Kmr.	Fotedar and Mahajan (1972), Zaki *et al.* (1991)
81	*Paratylenchus juglansi*	Walnut, apricot, cherry, almond, peach, plum	Kmr.	Kaul and Wali-ullah (1989), Askary *et al.* (2014)
82	*Paratylenchus sp.*	Mustard (field), almond, maize, *Gladiolus hortulans,* walnut	Kmr.	Fotedar and Kaul (1982). Abbas and Wali-ullah (2010), Askary *et al.* (2014)

Sl. No.	*Nematodes Species*	*Host Plant/ Crop Species*	*Distri-bution*	*Author (s)/ Reference (s)*
83	*Paravulvus confusus*	Brinjal	Kmr.	Akhtar *et al.* (1994)
84	*Pratylenchus barkati*	Potato, mustard, Chilli, kale, okara, turnip	Kmr.	Wali-ullah (1992), Wali-ullah and Bhat (1990), Kaul (1998)
85	*Pratylenchus brachyurus*	Cauliflower, radish, potato, pea, maize	Kmr., Raj.	Wali-ullah (1990), Wali-ullah (1992), Allie *et al.* (2020)
86	*Pratylenchus minimus*	Maize	Kmr.	Kaul (1987)
87	*Patylenchus penetrans*	Brinjal, Kiwi, potato, tomato, almond, cherry, peach, apricot, plum, walnut	Kmr., Jmu	Wali-ullah (1992), Zaki *et al.* (1993), Kaul (1998), Wali-ullah(2005), Askary. (*et al* 2014), Wali-ullah *et al.* (2008)
88	*Pratylenchus. pseudopratensis*	Cherry, almond	Kmr.	Mahajan (1971), Wali-ullah and Kaul (1997), Kaul (1998)
89	*Pratylenchus sativus*	Maize	Kmr.	Hassan *et al.* (2009)
90	*Pratylenchus similis*	Okara, cauliflower, mustard, , potato, okara, turnip	Kmr.	Wali-ullah and Bhat (1990), Kaul (1998), Wali-ullah (1992)
91	*Pratylenchus thornei*	Mustard, brinjal, potato, saffron.	Kmr.	Zaki *et al.* (1991, 93), Kaul (1990), Wali-ullah (1992)
92	*Pratylenchus vulnus*	Almond, apple, brinjal	Kmr.	Kaul (1984, 98)
93	*Pratylenchus zeae*	Cabbage, brinjal, potato, radish, cherry, apple, walnut, apricot, pear, plum, hak, maize	Kmr., Raj.	Wali-ullah and Bhat (1990), Wali-ullah (1992), Kaul (1998), Zaki and Mantoo(2003), Allie *et al.* (2020)
94	*Pratylenchus* sp.	*E. japanicus*, mint, *Gladiolus hortulanus*, potato, chilli, brinjal, tomato, knol Khol, cabbage, bean, radish, turnip, bottle gourd, pumpkin, *Citrus sinensis*, *C. reticulate*, *C. limon*), grapevine	Kmr., Ldk.	Fotedar and Kaul (1982), Zaki (1999), Sheikh *et al.* (2016), Abbas and Wali-ullah (2010), Nehru *et al.* (2005), Zalpuri *et al.* (2013a, b, c), Munib *et al.* (2016), Wali-ullah (1992), Wali-ullah *et al.* (2008), Hassan *et al.* (2016), Mantoo *et al.* (2008), Hassan and Wali-ullah (2010), Askary *et al.* (2018)
95	*Proleptonchus sclerouterinus*	Grass	Kmr	Mustaq *et al.* (2007)
96	*Promumtazium indicum.*	*Cedrus deodara*	Kmr.	Mustaq *et al.* (2007)
97	*Psilenchus haki*	Mustard (field)	Kmr.	Hassan *et al.* (2008)
98	*Psilenchus* sp.	Cucurbit, walnut, mint, kiwi, potato, spinach, chilli, cauliflower, French bean, brinjal, mint (*Mentha arvensis*)	Kmr., Jmu	Fotedar and Kaul (1982), Wali-ulla (2005), Sheikh (2016), Hassan *et al.* (2014, 2016), Wali-ullah *et al.* (2008), Sheikh *et al.* (2016)
99	*Quinisulcius punici*	Pamegranate	Kmr.	Gupta and Uma (1980)

Sl. No.	Nematodes Species	Host Plant/ Crop Species	Distri-bution	Author (s)/ Reference (s)
100	*Radopholus similis*	Banana	Jmu.	Kaul (1998)
101	*Rotylenchulus reniformis*	Okara, Kale/hak, *Cucurbita* spp., *Dahalia variabilis*, grapevine	Kmr., Jmu	Wali-ullah (1992), Kaul (1998), Rashid *et al.* (2014), Wali-ullah *et al.* (2008), Askary *et al.* (2018)
102	*Rotylenchus* sp.	Brinjal, cherry, stone and nut fruit trees, *Bougainvilla spectabilis, D.variabilis, Rosa sinensis, Rosa indica, T.eretus*	Kmr., Jmu	Mahajan (1971), Wali-ullah and Kaul (1997), Wali-ullah *et al.* (2008), Askary *et al.* (2014), Rashid *et al.* (2014)
103	*Sakia typica*	Brinjal, Kiwi	Kmr.	Wali-ullah (2005), Wali-ullah *et al.* (2008)
104	*Scutellonema brachyurum*	Mango	Jmu.	Mahajan (1971)
105	*Scutellonema petersi*	Tomato	Kmr.	Mahajan (1971)
106	*Trichodorus* sp.	Cabbage, tomato, onion, pumpkin	Kmr.	Wali-ullah (1992)
107	*Tylenchorhynchus brassicae*	Okara, cauliflower, cabbage, chilli, radish, spinach and pumpkin	Kmr.	Siddiqui (1963), Wali-ullah (1992), Kaul (1998)
108	*Tylenchorhynchus dubius*	Brinjal	Jmu	Wali-ullah *et al.* (2008)
109	*Tylenchorhynchus elegans*	Kiwi	Kmr	Wali-ullah (2005)
110	*Tylenchorhynchus haki*	Cabbage, hak tomato, brinjal	Kmr.	Fotedar and Mahajan (1971), Mahajan (1974), Zaki *et al.* (1991), Wali-ullah (1992)
111	*Tylenchorhynchus indicus*	Brinjal	Jmu	Wali-ullah *et al.* (2008)
112	*Tylenchorhynchus irruglaris*	Brinjal, Tomato	Jmu.	Bhat *et al.* (1994), Wali-ullah *et al.* (2008)
113	*Tylenchorhynchus kashmirensis*	Okra, kale, cauliflower, mustard, chilli, brinjal, potato, tomato, cherry, radish, turnip	Kmr.	Mahajan (1974), Zaki *et al.* (1991, 93), Wali-ullah (1997), Munib *et al.* (2016), Wali-ullah (1992)
114	*Tylenchorhynchus mashhoodi*	Okara, cabbage, chilli, cherry, brinjal, cauliflower, pea, tomato, potato, radish	Kmr.	Wali-ullah and Kaul (1997), Wali-ullah (1992), Wali-ullah *et al.* (2008)
115	*Tylenchorhychus oryzae*	Rice, Kiwi	Kmr.	Kaul and Wali-ullah (1995), Wali-ullah (1992, 2005)
116	*Tylenchorhynchus pruni*	Peach	Kmr.	Gupta and Uma (1981)
117	*Tylenchorhynchus nudus*	Apple, apricot, pear, plum, cherry, walnut	Kmr.	Zaki and Mantoo (2003)
118	*Tylenchorhynchus rugosus*	Salix (*Salix babylonica*)	Kmr.	Siddiqui (1963), Fotedar and Kaul (1982)

Sl. No.	*Nematodes Species*	*Host Plant/ Crop Species*	*Distri-bution*	*Author (s)/ Reference (s)*
119	*Tylenchorhynchus zea*	Maize	Kmr	Shiekh *et al.* (2009)
120	*Tylenchorhynchus* sp.	Almond, cucurbit, walnut, mint, mulberry, pomegranate, apple, brinjal saffron, Tegetes *erecta, Bougainvillea. spectabilis*, knol khol tomato, turnip, chilli, potato, cabbage, bean, radish, bottle gourd, grapevine, maize	Kmr., Jmu	Fotedar and Kaul (1982), Zaki and Muntoo (2008), Askary *et al.* (2014), Rashid *et al.* (2014), Wali-ullah *et al.* (2008), Askary *et al.* (2018), Allie *et al.* (2020)
121	*Tylencholaimus pini.*	*Pinus sylvestris*	Kmr.	Muataq *et al.* (2007)
122	*Tylencholaimellus raskii*	Pomegranate	Kmr.	Jairajpuri and Siddiqui (1963)
123	*Tylencholaimellus eskei*	Apple	Kmr.	Mustaq *et al.* (2007)
124	*Tylencholaimellus sayeedi*	*Cedrus deodara*	Kmr.	Mustaq *et al.* (2007)
125	*Tylencholaimellus striatus*	Grasses	Jmu	Mustaq *et al.* (2007)
126	*Tylenchus andrassyi*	Cauliflower, Kale/hak, chilli, brinjal, potato, tomato, pea	Kmr.	Fotedar and Kaul (1984), Wali-ullah (1992)
127	*Tylenchus arcuates*	Saffron, chilli	Kmr.	Hassan and Ahangar (2018 a)
128	*Tylenchus kashmirensis*	Saffron	Kmr.	Mahajan (1971, 73)
129	*Tylenchus filiformis*	Cabbage, tomato, okara, kale/hak, chilli, brinjal, kiwi, potato, spinach, turnip, apple, peach	Kmr.	Kaul (1982), Wali-ullah (1992), Kaul (1998), Wali-ullah *et al.* (2005, 2008), Sheikh *et al.* (2016)
130	*Tylenchus ? safroni* (= *Aerotylenchus safroni*)	Saffron	Kmr.	Fotedar and Handoo (1979), Fotedar and Kaul (1982), Siddiqi (2000), Geraert (2008)
131	*Tylenchus semipenetrans*	*Citrus* spp.	Jmu	Nehru *et al.* (2005), Zalpuri *et al.* (2013 a, b, c)
132	*Tylenchus* sp.	Saffron, E. japonica, walnut, paddy, tomato, potato, brinjal, maize, *D. variablilis, T. erecta,* chilli, French bean, cabbage, radish, turnip, knol- khol, bottle gourd, grapevine	Kmr.	Fotedar and Kaul (1982), Zaki *et al.* (1993), Sheikh *et al.* (2016). Hassan *et al.* (2008, 2016), Rashid *et al.* (2008), Mantoo *et al.* (2008), Askary *et al.* (2018)
133	*Xiphinema americanum*	Apple	Kashmir	Lone *et al.* (2013, 2018)

Sl. No.	Nematodes Species	Host Plant/ Crop Species	Distri-bution	Author (s)/ Reference (s)
134	*Xiphinema basiri*	Apple, brinjal, cherry, Kiwi, *Gladiolus hortulanus,* tomato, hak, cauliflower, radish, turnip	Kmr., Jmu	Wali-ullah and Kaul (1997), Abbas and Wali (2010), Askary *et al.* (2014), Wali-ullah (2005), Wali-ullah (1992), Wali-ullah *et al.* (2008), Lone *et al.* (2018)
135	*Xiphinema diversicaudatum*	Apple, maize	Kmr., Raj,	Lone *et al.* (2018), Allie *et al.* (2020)
136	*Xiphinema index*	Apple	Kmr.	Lone *et al.* (2012)
137	*Xiphinema insigne*	Cherry, walnut, apple, apricot, plum and pear	Kmr.	Zaki and Mantoo (2003), Lone *et al.* (2018), Lone and Zaki (2014)
138	*Xiphinema* sp.	Saffron, *Antirrhum majus,* paddy, tomato, bean, pear, cherry, apple, brinjal, potato, *Citrus* spp., knol khol, chilli, cabbage, radish, turnip, bottle gourd	Kmr.	Fotedar and Kaul (1982), Zaki *et al.* (1991), Kaul (1998), Zaki and Mantoo(2008), Munib *et al.* (2016). Nehru *et al.* (2005), Zalpuri *et al.* (2013 a, b, c), Rashid *et al.* (2014), Mantoo *et al.* (2008)

Kmr.: Kashmir; Srg.: Srinagar; Jmu: Jammu; Ldk: Ladakh; Raj.: Rajouri.

Table 5.12: Systematic Position of Genera (Species), with Total Number of Species of Plant-Parasitic Nematodes (Phylum Nematoda, Class Enoplea and Chromadorea) Prevalent in J&K, and Ladakh UTs

Order	Superfamily	Family	Sub-family	Genus/Genera, with Total Number of Species
*Dorylaimida	Dorylaimoidea	Aporcelaimid-ae	Aporcelaiminae	*Aporcelaimellus* (1 sp.)
–do–	–do–	Dorylaimidae	Dorylaiminae	*Dorylaimus* (sp.)
–do–	–do–	Longidoridae	Longidorinae	*Longidorus* (2 spp.)
–do–	*–do–*	–do–	Xiphineminae	*Xiphinema* (4 spp.)
–do–	*–do–*	Lordellonemat-idea	Lordellone-matina-e	*Lordellonema* (1 sp.)
–do–	*Nygolaimoidea*	Nygolaimidae	Nygolaiminae	*Paravulvus* (1 sp.)
–do–	*Tylencholaimoi-dea*	Leptonchidae	Leptonchinae	*Proleptonchus* (1 sp.)
–do–	*–do–*	Tylencholaimi-dae	Tylencholaiminae	*Tylencholaimus* (1 sp.)
–do–	*NA*	Tylencholaim-ellidae	Tylencholaimell-inae	*Tylencholaimellus* (4 spp.)
–do–	*–do–*	–do–	Mumtaziinae	*Promumtazium* (1 sp.)
Triplochida	*Diptherophoide-a*	Trichodoridae	Trichodorinae	*Trichodorus* (1sp.)
** Rhabdita	Aphelenchoidea	Aphelenchidae	Aphelenchinae	*Aphelenchus* (3 spp.)
–do–	–do–	Aphelenchoididae	Aphlenchoidinae	*Aphelenchoides* (5 spp.)

Order	Superfamily	Family	Sub-family	Genus/Genera, with Total Number of Species
–do–	Criconematoid-ea	Criconematid-ae	Criconematinae	*Criconemoides* (2 spp.), *Criconema* (1 spp.)
–do–	–do–	–do–	Hemicriconenoid-inae	*Hemicriconemoides* (2 spp.)
–do–	–do–	–do–	Macrophosthoni-inae	*Merocriconema* (1 sp.)
–do–	Tylenchuloidea	Paratylenchid-ae	Paratylenchinae	*Paratylenchus* (2 spp.)
–do–	- do -	Anguinidae	Anguininae	*Anguina* (1 sp.), *Ditylenchus* (1 sp.)
–do–	–do–	–do–	NA	*Nothotylenchus* (3 spp.)
–do–	Tylenchoidea	Belonolaimid-ae	NA	*Quinisulcus* (1 sp.)
–do–	–do–	Heteroderidae	Punctoderinae	*Globodera* (1 sp.)
–do–	–do–	–do–	Heteroderinae	*Heterodera* (3 spp.)
–do–	–do–	–do–	Meloidogyninae	*Meloidogyne* (5 spp.)
–do–	–do–	Hoplolaimidae	Hoplolaiminae	*Helicotylenchus* (22 spp.), *Hoplolaimus* (2 spp,), *Scutellonema* (2 spp.), *Rotylenchus* (1 sp.)
–do–	–do–	–do–	Rotylenchulinae	*Rotylenchulus* (spp.)
–do–	Tylenchoidea	Dolicodoridae	Merliniinae	*Merlinius* (1 sp.)
–do–	–do–	Telotylenchid-ae	Telotylenchinae	*Tylenchorhynchus* (14 spp.)
–do–	Tylenchoidea	Pratylenchidae	Pratylenchinae	*Pratylenchus* (12 spp.), *Radopholus* (1 sp.), *Hirschmanniella* (5 spp.)
–do–	–do–	Tylenchidae	Atylenchidae	*Aglenchus* (sp.)
–do–	Tylenchoidea	Tylenchidae	Boleodorinae	*Boleodorus* (1 sp.), *Psilenchus* (1 sp.), *Sakia* (1 sp.)
–do–	–do–	–do–	Tylenchinae	*Basiria* (6 spp.), *Filenchus* (1 sp.), *Tylenchus* (7 spp.)
–do–	–do–	–do–	Tylodorinae	*Cephalenchus* (sp.)

*: Class Enoplea; **: Class Chromadorea; –do–: As ditto; NA: Not available.

Kmr.: Kashmir; Jmu.: Jammu; Ldk.: Ladakh; J&K: Jammu and Kashmir State.

i. Synonyms are given in parentheses, against each species/genera in the table; ii. For classification, Luc *et al.* (1990), Siddiqi (2000), Nemaplex (Ferns, 2020) and Nemys (Bezerra *et al.*, 2020) have been consulted.

Chapter 6

Parasitic Leeches, Crustaceans, Pentastomida Parasites and Parasitic Acari (Ticks and Mites)

6.1 Parasitic Leeches: Systematic Survey and Diversity

The parasitic leeches of Jammu and Kashmir region, are included in Phylum Annelida, Class Clitellata, covering two orders and families, Arhynchobdellida (family Hirudinidae) and Rhynchobdellida (Gloossiphoniidae). These are ectoparasites of warm and cold blooded animals, feeding on their blood. The leeches are slender, having leaf-like body, without setae. They show the presence of large posterior sucker and smaller or pseudosucker at anterior end, surrounding mouth. In the medicinal leech (*Hirudo medicinalis*), the anterior sucker consists of three saw like jaws (tripartite), with teeth or denticles on each jaw. The teeth are used by leech to bite and siphon blood from its hosts (frogs, molluscs, fishes and occasionally human).

The parasitic leech-fauna of Jammu and Kashmir region include a total of 7 spp., belonging to two different orders and families. The Hirudinids (order Arhynchobdellida) are : *Hirudinaria granulosa* and *Hirudo medicinalis,* as parasites of vertebrates, cattle and occasionally humans. Glossphonids (Rhynchobdellida), having five species, belonging to 3 genera, *viz. Glossiphonia* (2 spp.), *Helobdella* (1 sp.) and *Theromyzon* (2 spp.) (Table 6.1). The Glossiphonids are ectoparasites of molluscs, frogs, fishes and water fowl. The species of *Theromyzon* enter throats and nostrils of water fowl/birds like geese and ducks. Kumar *et al.* (2015) have found a tribal man from Rajouri-Poonch belt, with live leech in the larynx, suffered hirudiniasis, causing severe anemia.

The leech-fauna of Jammu and Kashmir region, including parasitic species, have been studied by Moore (1924), Bhatia (1930, 1934, 1939) and Soota (1959).

Chandra (1982), dealt with 52 species of leeches from Indian sub-continent, including earstwhile Jammu and Kashmir State. The leech species found in India, including J&K region are called " endemic leeches", as 32 species of leech occurring in India, are not found anywhere in the world (Mandal, 2004). The widely used medicinal leech (*Hirudo medicinalis*) is a European species, which also finds use in Kashmir region. In addition to this species, Indian leech, *Hirudinaria granulosa,* prevalent in Kashmir region has got medicinal properties. These leeches are being utilized for curing of number of ailments.

Table 6.1: Systematic Checklist of Parasitic Leech-Fauna of Jammu and Kashmir (UT)

Phylum : Annelida

Class : Clitellata

Order 1.: Arhynchobdellida

Family : Hirudinidae

(1) *Hirudinaria granulosa* (Savigny)(Indian or Asian Medicinal leech)
Host : Cattle

(2) *Hirudo medicinalis* (Linnaeus) (European Medicinal leech)
It is widely used medicinal leech in different parts of the world, including J and K

Order 2: Rhynchobdellida

Family : Glossiphonidae

(3) *Glossiphonia complanata* (Linnaeus)
Host : Molluscs

(4) *Glossiphonia weberi* (Blanchard)
Host : Molluscs and aquatic beetle

(5) *Helobdella stagnalis* (Linnaeus)
Hosts : Molluscs, frogs and fishes

(6) *Theromyzon mathai* Bhatia
Hosts : Temporary parasites of vertebrate (waterfowl)

(7) *Theromyzon sexoculata* (Moore)
Host : Waterfowl

Source: Moore (1924); Bhatia (1930, 1934, 1939), Soota (1959).

6.2 Medicinal Leech Therapy (MLT) or Hirudo Therapy

The saliva of medicinal leeches contain number of bio-active substances, having therapeutic values in the treatment of scores of human diseases. The medicinal leech therapy (MLT) is a traditional techinique, however, part of multidisciplinary approach in treatment of diseases. The effects of leech saliva are : analgesic, anti-flammatory, platelet inhibatory, anti-coagulant, thrombin regulatory functions, as well as extra-cellular matrix degradative and antimicrobial effects (Sig *et al.*, 2017). The important substances in the saliva include hirudin (potent anti-coagulant), calin (antiplatelet), anaesthetic compounds and antihistamine vasodilators (increase blood flow to an area). In addition to these, the presence of some important neurotransmitter in saliva, like dopamine, serotonin, acetocholin and enkephalin, induce pain relieving and relaxing effects.

In the recent years, leeches find use in treatment of serious diseases like cadiovascular problems, cancer, in plastic and microsurgery (Abdualkader *et al.*, 2013). In Kashmir Valley, MLT is gaining popularity and many hospitals in this region have initiated this age old practice in treating patients suffering from skin diseases, arthritis, chronic headaches and sinusitis, *etc.* In Kashmir region, a study has established the safety and efficacy of MLT in treatment and management of frostbite ailment (Munshi, 2005).

6.3 Parasitic Crustaceans (Copepods and Branchiurans)

From Jammu and Kashmir region, both parasitic copepods as well as branchiurans are reported, parasitizing the fresh water fish-fauna of these regions. The parasitic copepods are common, generally small in size to microscopic, having both parasitic and free living stages in their life cycle. In case of Ergasilid and Lernaeid copepods, only females are parasitic, males are free living. The branchiurans (Argulids), have comparatively large size and both sexes are parasitic throughout life, however, leave host body to moult or to lay eggs.

The parasitic copepods damage their hosts (fishes) by their attachment mechanism and by feeding activities. The attachment is by means of clawed limbs and penetration of skin by claws causes local lesions. There is destruction of tissues and copepods feed by rasping at the surface of the host using mandibles, removing blood and cellular fluid, leading to weakening, also fish becomes expose to other infections, resulting in decrease fish production. The parasitic Branchiurans (*Argulus* spp.) attach to the host fish with the aid of appendages (first maxillae) and little spines, present on their underside. These parasites have also long slender pre-oral sting to pierce the fish skin and inject toxins (anti-coagulant). They use their mouth and mandibles to consume blood, mucous and tissues at their puncture site. The much damage to fish is due to fungal infection. These parasites also transmit fish viral diseases and other fish parasites.

6.3 1 Systematic Survey and Diversity of Parasitic Crustacean Genera/ Species of Jammu and Kashmir

Phylum Arthropoda: Subphylum- Crustacea: Super Class 1- Multicrustacea, Class- Hexanauplia, Sub-class- Copepoda, Infra class- Neocopeda, Superorder- Podoplea, Order- Cyclopoida, Family- Lernaeidae, genera / species – *Pseudolamproglena simplex, Lernaea kashmirensis, Lernaea gambusiae*; Sub-order- Poecilostomatoida, Family- Ergasilidae, species- *Ergasilus putitori*; Superclass 2- Oligostracea, Class- Ichthyostraca, Subclass- Branchiura, Order- Anguloida, Superfamily- Anguloidea, Family- Angulidae, species- *Argulus foliaceous, Argulus* sp. The classification adopted in this checklist, is based on Walter and Boxshall (2018). However, in this world data base, the parasitic crustaceans, *viz. E. putitori, L. gambusiae* and *L.kashmirensis*, reported and described as new species from earstwhile J&K State, have not been included.

The main crustacean parasites of fresh water fish-fauna of J&K, UT, commonly occurring in lakes, rivers and other water bodies include copepods, belonging to the family Ergasilidae and Lernaeidae, and Branchiurans under family Argulidae. The Ergasilid, described as new taxa, parasitic on fish from this region is as *Ergasilus putitori* Gupta *et al.* (2002). The new species of crustacean parasites of fishes of J&K region, belonging to the family Lernaeidae, recorded and described by various workers are : *Lernaea kashmirensis* Malhotra and Joyti (1972); *Lernaea gambusiae* Koul and Raina (1989); *Pseudolamproglena simplex* Boxhall (1976). In the recent years, the prevalence of Branchiurans (Argulidae), *viz. Argulus foliaceous, Argulus* sp., have been reported, parasitizing three different species of fishes in diverse water bodies of Jammu and Kashmir, causing lot of damage to the fishes (Palaq *et al.*, 2016a, 2016b; Tak *et al.*, 2014) (**Table 6.2**).

In Jammu and Kashmir, a total of 6 spp. under 4 genera of parasitic crustaceans (Copepoda, Branchiura), spread over 3 families as Lernaeidae, Ergasilidae and Argulidae, are known to parasitize 6 spp., under 5 genera of fishes in diverse aquatic habitats. Branchiuran parasite, *Argulus foliaceous* and *Argulus* sp., have found to infest skin and gills of fishes- *Labeo rohita* and *Schizothorax niger*. Copepods like *Lernaea gambusiae, L. kashmirensis* and *Pseudolamproglena simplex*, are known to parasitize fishes as *Gambusia holbrooki* (= *Gambusia affinis*), *Triplophysa kashmirensis* (= *Nemacheilus kashmirensis*) and *Labeo dero* repectively. Apart from these, Copepod- *Ergasilus putitori* and Branchiuran- *Argulus foliaceous*, are reported to cause damage to the gills of fish *Tor putitori* in Jammu region (Table 6.2).

Table 6.2 : Crustaceans parasitizing fishes of Jammu and Kashmir

Host Fish Species	*Crustacean Species (Region)*	*Reference(s)*
Gambusia holbrooki (=*Gambusia affinis*)	*Lernaea gambusiae* (K)	Koul and Raina 1989)
Labeo dero (=*Labeo diplostomum*)	*Pseudolamproglena simplex* (K)	Boxhall (1976)
Labeo rohita	*Argulus foliaceous* (J)	Palaq *et al.* (2016 a, b)
Schizothorax niger	*Argulus* sp. (J&K)	Tak *et al.* (2014)
Triplophysa kashmirensis (=*Nemacheilus kashmirensis*)	*Lernaea kashmirensis* (K)	Malhotra and Joyti (1972)
Tor putitori	*Ergasilus putitori* (J)	Gupta *et al.* (2002)

J: Jammu region; K: Kashmir region; J&K: Jammu and Kashmir (Union Territory).

6.4 Parasitic Pentastomida (*Linguatula serrata*), Lingualulosis, "Pentastomiasis"

Earlier, Pentastomida group was classified between annelids and arthropods, later Pentastomida group was kept under Subphylum Crustacea of Phylum Arthropoda (WoRMS, 2018). On the basis of phylogenetic and systematization of Pentastomida, based on authors : Almeida and Christoffersen (1999) and Christoffersen and Assis (2013), this enigmatic group of parasites, has been given

Phylum rank. Consequent upon this new systematic arrangement, the endoparasite of animals and humans, *viz. Linguatula serrarta* belongs to Phylum Pentastomida, Class- Eupentastomida, Order- Porocephalida, Superfamily- Lingualutoidea, Family – Linguatulidae.

Linguatula serrata is a cosmopolitan zoonotic endoparasite, commonly known as tongue-worm, due to its characteristic transparent tongue like shape. The adult body is dorsoventrally flattened and tapering backwards. The males of *L. serrata* are 18-20 mm in length and females 80-120 mm. The adults feed on blood and fluid of hosts, live in nasopharyngeal region of mammals, cats, dogs, foxes and other carnivores, are normal hosts of this parasites. Almost any mammals is a potential intermediate host of *L. serrata*. Often, the intermediate hosts are herbivores, such as cattle, goats, sheep and other ruminants, that have ingested parasite eggs contained in contaminated plants. In Jammu and Kashmir region, *L. serrata* has shown its prevalence, affecting cattles (bovines) and goats, however, there is no information available about this parasite infecting the humans of this region. This zoonotic parasite has public health importance.

Linguatulosis is a condition associated with *L. serrata*, a form of pentastomiasis, which refer to all diseases caused by Pentastomids, including porocephaliasis (condition associated with closely related genera, like *Porocephalus* and *Armillifer*). The female parasite produces million of eggs. The eggs exit the host in nasal secretion. When swallowed by intermediate host, the larvae (nymphs) hatch in the small intestine, penetrate the intestine wall, lodged in tissue, particularly lungs, liver and lymph nodes. The nymphal stages develop when eaten by definitive host, infective nymphs either attach in the upper digestive tracts to encyst on viscera, causing visceral linguatulosis or travel from stomach reaching nasopharyngeal region, leading to a pathological condition known as as nasopharyngeal linguatulosis.

L. serrata can also infect humans, causing respiratory pentastomiasis and humans become infected by ingestion of eggs contained in the respiratoty secretions, blood, saliva, feces of the definitive hosts (Christoffersen and Assis, 2003). In humans, nasopharyngeal and visceral linguatulosis, also known as Halzoun or Murrara syndrome, due to *L. serrata*, have been reported from different parts of the world (Acha and Szyfres, 2003). This disease results from ingestion of infected raw or under cooked liver or lymph nodes of sheep or goats, and is characterized by block of nasal pathways, oedema and unfeelingness of head region (Mehlhorn, 2009). This parasite can also invade human subconjunctiva and reach the anterior chamber of eye, causing ocular linguatulosis (Lazo *et al.*, 1999; Koehsler *et al.*, 2011).

Mir *et al.* (2004) were first to report the occurrence of *L. serrata* in cow in Kashmir valley. Later, two more cases of bovines suffering from visceral linguatulosis were reported from this region. The parasites were found vigorously wringled into parenchyma of the nodes of bovines, showed severe pathological effects on mesenteric lymph nodes (Mir *et al.*, 2012). The third instar nymphs of *L. serrata*, causing visceral liguatulosis and paratuberculosis (bovine Johne's disease), due to *Mycobacterium avium paratuberculosis*, have been observed in cross alpine goats, *Capra hircus hircus* (= *Capra hircus*) of Kashmir (Mir *et al.*, 2009). From Jammu region, the prevalence of *Linguatula serrata* in goats, has been reported by Godara *et al.* (2013).

Out of the 19 adult goats examined by these research workers, 6 (31.57 per cent), were diagnosed with nymphal parasite stages in the lymph nodes and one goat had parasite nymphs in liver. The infested lymph nodes were enlarged and swollen, and liver of affected animal showed pathological condition.

6.5 Parasitic Ticks

Ticks belong to Ixodida (order), under super-order Anactinotrichida, sub-class Acari, Class Arachnida of Phylum Arthropoda. They are blood-sucking ectoparasites of vertebrates. Ticks are characterized by: leathery integument; presence of movable capitulum or gnathosoma, consists of a base (basis capituli) and the mouth-parts having a hypostome, paired 4-segmented palps and a pair of chelicerae; hypostome with rows of recurved teeth; genital and anal opening present on ventral side and the presence of Haller's organ, a pit like sensory organ (chemoreceptor) on the tarsus of the first pair of legs. The hard-bodied ticks are included in the family Ixodidae and show the presence of a dorsal scutum or shield, which is absent in soft-bodied ticks, belonging to the family Argasidae.

6.5.1 Faunistic Survey

The earliest faunistic study on the tick-fauna of Kashmir region, present in the collection of the Indian Museum, has been accomplished by Sharif (1928). Later, Sen (1938) published a checklist and host-list of some tick species from Jammu and Kashmir region. Dutta *et al.* (1963), recorded 8 species of ticks, including *Ornithodoros tholazani* var. *crossi* and *O. lahorensis,* as vector of human relapsing fever. Besides these, other species reported by them are: *Hyalomma isaaci, Rhipicephalus* (*Boophilus*) *microplus*) (= *Boophilus microsplus*), *Boophilus* sp., *Rhipicephalus haemaphysaloides, R. sanguineus* and *Haemaphysalis montgomeri.* Tick species from Kashmir region *viz. Haemaphysalis cornupuncta* and *H. kashmirensis,* have been recorded and described (Hoogstraal and Verma, 1962; Hoogstraal and McCarthy,1965). Dhanda and Kulkarni (1969) and Dhanda and Bhat (1971) made taxonomical and ecological studies on ticks (*H. corpupuncta* and *H. sundrai*), parasitizing sheep and goats, occurring in various localities of Kashmir.

First comprehensive account of tick-fauna of Kashmir, has been provided by Das and Sharma (1970). These authors have surveyed 22 different species, including some new records, belonging to the family Argasidae as well as Ixodidae, parasitizing birds (poultry, pigeons), rabbit, bat, rodents and even man, occurring in diverse localities/regions of districts, *viz.* Srinagar, Baramulla, Budgam and Anantnag of Kashmir Valley. This paper has listed 7 species of soft-bodied ticks, distributed over 3 genera and 16 species of hard-bodied ticks spread over six genera. Sharma and Das (1971) recorded for the first time tick, *Rhipicephalus decoloratus* (=*Boophilus decoloratus*), parasitizing cows at Lagama (Uri) and also from cows in places like Hazratbal, Safakadal, Saidakadal localities of district Srinagar and in district Baramulla and Anantnag. Dhanda *et al.* (1971) reported tick – *Dermacentor raskemensis* infesting Yak in Leh, in Batalik and goat in Poyen region of Ladakh. The earlier recorded species of ticks of J&K are incorporated in the checklist prepared by Sharma and Das (1972).

A general review, including a checklist of tick-fauna of earstwhile J&K State, has been given by Rao *et al.* (1973), showing the prevalence of 23 species from different districts, *viz.* Srinagar, Anantnag, Baramulla (Kashmir Province), Doda, Rajouri and Udhampur (Jammu Province) besides, Ladakh Province, attacking hosts like birds, mongoose, sheep, goat, shrew, rodents, horse, dogs, buffaloes, small borrowing mammal from these Districts/Provinces. Further, Province-wise break-up of tick species was as: 13 species in Kashmir, 17 species in Jammu and 5 species in Ladakh. A faunal-list of ticks of Jammu region has been contributed by Sharma and Sharma (1976), incorporated about 6 species of Argasids, belonging to 2 genera and 5 species of Ixodids, pertaining to 2 genera, attacking different vertebrate hosts like poultry, birds, bats, sheep, goat, buffalo, cow, rodents, domestic dogs, from Poonch, Jammu, Rajouri and Bhadarwa region.

Dhanda *et al.* (1979) collected nymphs of tick (*Nosomma monstrosum*) from *Mus saxicola* and adults from Buffaloes in Jammu for the first time. Srivastava and Wattal (1981) surveyed ticks in Gulmarg and Tangmarg areas of Kashmir, and recorded four specimens of nymphal stages of *Haemaphysalis bispinosa* from *Rattus rattus*, and of *Boophilus microplus* from cows and buffaloes. A survey of ticks for Crimean-Congo haemorrhagic fever virus activity in earstwhile J&K State, was conducted by Kaul *et al.* (1990), reported 16 species of ticks covering 7 genera as ectoparasites of small and large mammals from Rajouri, Poonch and Udhampur. The newly recorded species by them include *Haemaphysalis intermedia, Hyalomna detritum and H. dromendarii.* Mir *et al.* (1993) showed the prevalence of tick (*Argas persicus*), infesting indigenous fowls of Kashmir. An annotated list of tick-fauna, associated with various hosts in different Districts of Jammu and Kashmir region with descriptions of some species, have been provided by Sharma (1993). Prevalence (40.18 per cent) of Ixodid tick, *Rhipicephalus* (*Boophilus*) *microplus,* infesting cattle in Jammu region, has been observed (Godara *et al.*, 2018).

During the course of survey of tick-fauna associated with dairy animals, belonging to organized and unorganized units of Jammu region, only one species of Ixodid tick species (*Rhipicephalus microplus*), has been recorded. The overall infestation of tick parasite was 42.18 per cent. Highest percentage of infestation was recorded in Mansoon season as compared to other seasons. Animals of age less than 6 months showed highest prevalence and lowest more than 1 year of age, and males showed higher infestation than female (Khajuria *et al.*, 2015).

6.5.2 Species Diversity and Host Range

After scanning number of research papers, including valuable monograph and other works on tick parasites of Jammu, Kashmir and Ladakh region, it was established that a total of about 49 species of ticks, distributed over 11 genera are known to exist in these regions. Of these species, 10 species are Argasids (soft-bodied ticks), distributed over 3 genera and 37 species are Ixodid (hard-bodied ticks), spread over 7 genera. The total number of tick species known to occur in three regions of UTs (J&K, Ladakh) is as 15, 29 and 13 in Jammu, Kashmir and Ladakh respectively. The total number of species under various genera (11), belonging to family Argasidae and Ixodidae of J&K, Ladakh, UTs is as: Argasids – *Argas* (5 spp.), *Ornithodoros* (4),

Otobius (1); Ixodids – *Amblyomma* (1), *Boophilus* (1), *Dermacentor* (3), *Haemaphysalis* (12), *Hyalomma* (8), *Ixodes* (6), *Nosomma* (1) and *Rhipicephalus* (7) (Tables 6.3 and 6.4).

Table 6.3. : Checklist of Parasitic Tick-fauna (Acarina : Ixodida), with their host diversity and distribution Jammu, Kashmir, and Ladakh UT

Tick Species	*Host/s*	*Distribution*
1. *Argas (Carios) vespertilionis*	*Pipistrellus vespertilionis* (Bat)	P, S
2. *Argas (Carios)* sp.	*Pipistrellus minus* (Bat)	P
3. *Argas persicus* (Fowl tick)	Ca, Go, Po, Sh	P, B, K, J
4. *Argas reflexus indicus* (Pigeon tick)	Pi	S
5. *Argas* sp.	Po	P
6. *Ornithodoros crossi*	Ca, Go, Sh	K
7. *Ornithodoros lahorensis*	Ca, Go, Ra, Sh	B, D, J, K, L, S
8. *Ornithodoros savignyi*	Ca, Go, Sh	K
9. *Ornithodoros tholazoni*	Do, *Rattus rattus*, Sh	B, J, S
10. *Otobius megnini* (Spinose ear-tick)	Ca, Go, Sh	S
11. *Amblyomma gervaisi*	*Python molurus*, *Varanus monitor* (= *Aponoma gervaisi*), *Spaerosophis articeps*	U, J
12. *Boophilus* sp.	Ca	K
13. *Dermacentor auratus*	Go, Ra	S
14. *Dermacentor raskamensis*	Ca, Do, Go, Mo-h, Ro, Sh, Ya, Go	Le/L
15. *Dermacentor* sp.	Go, Sh, Ro	S
16. *Haemaphysalis bispinosa*	Ca, Go, Mo, *R, rattus*, Sh, Co, *Herpestes edwardsii*	B, J, P, R, S, U
17. *Haemophysalis cornupunctata*	Ca, Go, Dz. Mo-h, Ro, Mu, Sh, Bu-	B, D, J, P, R, U
18. *Haemophysalis howletti*	Bi, Go, Mo, Sh	R, U
19. *Haemophysalis indica*	Mo, Ro, Shr	U
20. *Haemophysalis intermedia*	Ca	R
21. *Haemophysalis leachi*	*Vulpes bengalensis, H. edwardsii*	J
22. *Haemophysalis microplus*	Ca, Go, *R. rattus*, Sh	D, P
23. *Haemaphysalis montogomeryi*	Ca, Co, Do, Dz, Go, Ho, Ma, Sh Hu, *Apodemus flavicollis*, *Suncus murinus*	A, B, D, J, R, S, P, U
24. *Haemaphysalis paraturtiris*	Bi, Mo, Ro	U
25. *Haemaphysalis kashmirensis*	Ca, Go, Li, Sh, *Laudakia tuberculata* (= *Agama tuberculata*), *Agama melaura*, *Petaurista albiventer*	A, B, J, S, U
26. *Haemaphysalis sulcata*	Go, Sh	B, D, R, U
27. *Haemaphysalis sundrai*	Sh	K
28. *Hyalomma anatolicum*	Ca, Bu, Go, Ho, Sh	B, J, S, U
29. *Hyalomma detriticum*	Ca, Bu	P, R
30. *Hyalomma domendarii*	Cam	J
31. *Hyalomma excavatum*	Ca, Go, Sh	S

Tick Species	*Host/s*	*Distribution*
32. *Hyalomma isaaci*	Ca, Bu, Go	J, R, U
33. *Hyalomma kumari*	Go, Ro	U
34. *Hyalomma marginatum*	Ca, Co, Go,	S
35. *Hyalomma* sp.	Do, Go, Ro	U
36. *Ixodes acutitarsus*	Co, Do, Sh	S
37. *Ixodes himalayensis*	*Rattus* sp., Ro, Shr	A, B
38. *Ixodes ovatus*	Ca, Go, Sh	P, U
39. *Ixodes kashmiricus*	Ca, Go, Hu, Ma, Sh, Ro	A, B, S
40. *Ixodes ricinus*	Ca, Do, Sh	S
41. *Ixodes* sp.	Go, Mo-h, Sh, Sm-m	B, D
42. *Nosomma monostrosum*	Bu, *Mus saxicola*	J, R
43. *Rhipicephalus* (*Boophilus*) *annulatus*	Ca, Sh	S
44. *Rhipicephalus* (*B.*) *decorates*	Ca	A, B, S
45. *Rhipicephalus* (*B.*) *microplus*	Ca, Go, Sh, Co, Bu, *Felis, Suncus*	A, B, D, P, R,U
46. *Rhipicephalus haemaphysaloids*	Ca, Co, Go, Cam, Sh, Sm-m, Mo	B, D, J, P, R, U
47. *Rhipicephalus sanguineus*	Bu, Ca, Do, Gor, Shr, Rab, Sh, Ro, Hu	J & K, D, J, U, B
48. *Rhipicephalus turanicus*	Do (domestic), Go, Sh, Sm-m, Hu	R, S
49. *Rhipicephalus* sp.	Sh, Sm-m, Do	B, J, L, R

Abbreviations: A: Anantnag (district); B: Baramulla, (district); Bi: Bird; Bu: Buffalo; Ca: Cattle; Cam: Camel; Co: Cow; D: Doda (district); Do: Dog; Dz: Dzo; Go: Goat; Gor: Goral; Ho: Horse; Hu: Humans; J: Jammu; J&K: Jammu and Kashmir State; K: Kashmir; L: Ladakh; Le: Leh; Li: Lizard; Ma: Mare; Mo: Mongoose; Mo-h: Mouse hare; Mu: Mule; P: Poonch (district); Po: Poultry; R: Rajuri (district); Ra: Rat; Rab: Rabbit; Ro: Rodent; S: Srinagar (district); Sh: Sheep; Shr: Shrew; Sm.-m: Small mammals; U: Udhampur (district); Ya: Yak.

Source: Das and Sharma (1970a). Dhanda and Bhat (1971); Dhanda and Kulkarni (1969); Dhanda and Rao(1964); Dhanda *et al.* (1979); Dutta *et al.* (1963); Hoogstraal (1962); Hoogstraal and McCarthy (1965); Hoogstraal and Verma (1962); Kaul *et al.* (1978, 90): Mir *et al.* (1993); Rao (1951); Rao *et al.* (1973); Rao and Kalra (1949); Sharma (1993); Sharma and Das (1971); Sharma and Sharma (1976); Srivastava and Wattal (1981).

In Jammu and Kashmir, of the total 5 species of ticks belonging to the genus *Argas*, 3 species have been found to infest birds/poultry and 2 species parasitize bat. In addition to these hosts, *Argas persicus* has sheep, goat and cattle as common hosts, both in Jammu as well as Kashmir region. Genus *Ornithodoros* includes four different species, attacking sheep and goat. Apart from these hosts, 3 species of this genus have been reported from cattle and 1 species each from dog and rodent (*Rattus rattus*). *Otobius megnini* was recorded from Srinagar, attacking sheep, goat and cattle.

Amongst Ixodids, *Boophilus* has four species from Jammu and Kashmir, infesting cattle, of these species 2 species, *viz. Boophilus annulatus* and *Haemophysalis microplus (=B. microplus),* have been found to infest sheep and goat of this region. *Dermacentor* of these areas, has 3 species, all infesting not only the small/burrowing mammals but also sheep and goats. Majority of the species *i.e.*, 11, under the genus *Haemaphysalis* have been reported to attack wide host groups like lizard, birds, small burrowing mammals, cattle, sheep/goats and beasts of burden showing, 1, 1, 9, 6, 8, 2 species respectively, in various Districts of Jammu and Kashmir, *viz*. Poonch, Rajouri, Baramulla, Srinagar, Doda, Udhampur and Anantnag (Tables 6.3 and 6.4).

Table 6.4: Species Diversity of Ticks on Various Groups of Vertebrate, Including Humans Hosts in Jammu, Kashmir and Ladakh

Tick Genus	*Number of Species of different Tick Genera, Parasitizing different Groups of Vertebrates and Humans*								
	(I)	*(II)*	*(III)*	*(IV)*	*(V)*	*(VI)*	*(VII)*	*(VIII)*	*(IX)*
Abylomma	1	–	–	–	–	–	–	–	–
Argas	–	3	2	–	–	1	1	–	1
Ornithodoros	–	–	–	1	1	3	4	–	–
Otobius	–	–	–	–	–	1	–	–	–
Boophilus	–	–	–	–	–	1	1	–	–
Dermacentor	–	–	–	3	–	1	3	1	–
Haemaphysalis	1	1	–	11	1	6	8	3	1
Hyalomma	–	–	–	2	1	6	7	2	–
Ixodes	–	–	–	3	2	4	6	–	1
Nosomma	–	–	–	1	–	1	–	–	–
Rhipicephalus	–	–	–	4	4	4	6	3	1
Total number of species	**2**	**4**	**2**	**25**	**9**	**28**	**36**	**9**	**4**

(I): Lizards and snakes; (II): Domestic fowl, chicken, poultry, Gallus gallus, hens, cocks, pigeon and birds (generals); (III): Vespertilionidi bats; (IV): Rodents, insectivores, mongoose, rabbit, squirrel; (V): Cat, dog, fox; (VI): Cow, buffaloes, yak, dzo, bull; (VII): Sheep, goat and goral; (VIII): Horse, mare, mule and camel; (IX): Human.

Genus *Ixodes,* represented by 6 species, found to parasitize sheep and goats of Jammu and Kashmir region. Besides these hosts, species of genus *Ixode*s have also been reported from cattle, small mammals, dogs and man, with 4, 3, 2 and 1 species respectively, occurring in wide localities of various districts of Jammu and Kashmir. 8 species of ticks, pertaining to genus *Hyalomma* are reported from various Districts of J&K. Of these species, 6 species have been found on sheep and goats, followed by 5 species, affecting cattle, 2 species each parasitizing small mammals and beasts of burden and 1 species parasitic on dogs.

Nosomma monostrosum has been found to attack *Mus saxicola* and buffaloes in Jammu region. *Rhipicephalus* of this region includes 4 species, reported from Jammu, Kashmir and Ladakh region, with wide host range such as small/borrowing mammals, dogs, cattle, sheep, beast of burden and man, having 4, 4, 2, 4, 1, 1 species of genus *Rhipicephalus* respectively (Tables 6.3 and 6.4)

6.5.3 Ticks of Veterinary and Medical Importance

The ticks are responsible for causing injury and wounds from bites to domestic animals, poultry and other mammals. The wounds caused by ticks, if the capitulum is torn off in forcible detachment from the hosts, are likely to become infected and result in inflammed sores or extensive ulcers. When ticks are present in large number, these cause serious or even fatal anaemia in cattle or horses. Ticks have also been found to cause paralysis in domestic animals due to toxins secreted by specific cells in the salivary glands. Apart from causing different diseases in domestic animals, ticks also act as vector of viral, bacterial spirochaetal, rickettsian and protozoan diseases of humans and domestic animals. Different tick-borne diseases, reported from J&K, Ladakh (UTs), are briefly highlighted as under:

6.5.3.1 Ticks of Veterinary Importance

Argas species are considered to be economically important tick species in this region, parasitizing not only the poultry but also domestic animals (sheep, goats and cattle). Tick paralysis in fowl and sheep through *Argas persicus* and *Ornithodoros lahorensis* in Kashmir, has for the fist time been reported by Das and Sharma (1970a). *A. persicus* was earlier recorded by Rao and Kalra (1949), present in most fowl houses examined by these authors. Bhat *et al.* (2014) has reported 3.28 per cent of prevalence of *A. persicus* in back-yard poultry in Jammu region. *A. persicus* acts as a vector of fowl spirochaetosis, caused by *Borrelia anserina*.

Another important Argasid tick, *viz. Otobius megnine* (spinose ear-tick) reported from Srinagar, infesting cattle, sheep and goats, is known to choke the ear of these animals, with tick individuals, wax and other debris, leading to rupture of ear drum and also to secondary infection.

Dermatosis disease, characterized by inflammation, itching, swelling and ulceration of site of bite or skin ulceration and lesion resulting from improper or partial removal of tick mouth, found in domestic animals of Kashmir, is due to ticks, *viz. Ornithodoros lahorensis, Otobius megnini* and *Hyalomma anatolicum* (Das and Sharma, 1970).

The Hard-bodied ticks (Ixodidae) of Jammu and Kashmir are represented by important genera, *viz. Boophilus, Dermacentor, Haemaphysalis, Hyalomma, Ixodes, Nosomma* and *Rhipicephalus.* Of these tick genera, *Haemaphysalis* is having maximum number of species *i.e.,* 11. The species of this tick genus, have wide spectrum of hosts, including small/burrowing mammals, birds, cattle, lizard, dogs and other domestic mammals. The second abundantly found genus of this region is *Hyalomma,* with 8 species. These are large tough and hardy ticks, ectoparasitic on domestic animals, small burrowing mammals and beasts of burden (horses and camel) of Jammu and Kashmir.

The genus *Ixodes,* with 6 species, have been found to be occurring in wide localities of J&K on diverse hosts. These ticks have small size, easily overlooked when searching a host, feeding cattle, sheep, dogs, small mammals and human in this region. *Ixodes ricinus* also known as the European castor-bean tick is an important tick, feeding in its all stages on cattle, sheeps and dogs. The genera, *viz. Boophilus* and *Rhipicephalus,* each represented by 4 species in this region, *Rhipicephalus microplus* (=*Boophilus microplus*), *R. decoloratus* (= *B. decoloratus*) and *R. annulatus* (= *B. annulatus*), parasitize large mammals especially cattle. *Rhipicephalus* species are distributed not only in various localities of Jammu and Kashmir, but also in Ladakh region. *R. sanguineus* (the brown dog tick) most frequently feed on carnivores, mammals and man.

Dermacentor, including three species from J&K, are ornate and bear eyes. These are not host specific. *D. raskamensis* shows wide occurrence and hosts in Ladakh region as: Yak in Leh and Batalik, and goats in Poyen. Besides, this species of tick parasitizes cattle (buffalo, goats and sheep) in Kashmir Valley. Genus *Nosomma* having one species, *viz. N. monostrosum,* has been recorded from Jammu (Rajouri), attacking buffaloes and small mammal (*Mus saxicola*).

(a) Tick-borne Spirochaetal, Rickettsial, Bacterial Diseases in Animals

The tick-borne rickettsial disease, anaplasmosis in cattle has been reported from Kashmir as well from Jammu region. This disease is transmitted by tick species, *viz. Rhipicephalus microplus,* casued by Rickettsian, *Anaplasma marginae* (Rickettisale, Anaplasmataceae). This parasite infects the red blood cells, causing severe anemia and can progress to various clinical complications and death of cattles. Previously, the prevalence of this disease from Kashmir region was reported by Das and Sharma, 1970; Shaw, 1971; Shaw, 1980; Shah, 1989. From Jammu District, the occurrence of o bovine anaplasmosis was reported in cows and buffaloes (116), showed 4.31 per cent of infection (Raina *et al.,* 2005). An investigation on the sero-prevalence of Anaplasma in cattle, sheep and goats (local, exotic and cross breed) in Kashmir region, showed 18 (7.25 per cent) infection due to Anaplasma in cattle, however, sheep showed 1 (0.48 per cent) infection, absent in goats 0 (0 per cent) and highest (14.92 per cent) in exotic cattle and, absent in local and cross breed sheep (Rather *et al.,* 2015).

The avian spirochaetosis or borreliosis is also a tick-borne disease of poultry, showed its prevalence in Kashmir and had rendered considerable economic loss to poultry (Das and Sharma, 1993). This disease is caused by spirochaete, *Borrelia anserina* and transmitted by fowl tick, *Argas persicus*. The symptoms of spirochaetosis include anemia, diarrhea and severe neurological dysfunctions. There is also reduced egg production. The detailed survey of canine tick-borne diseases in the regions of India like Delhi, Mumbai, Skkim and Ladakh, has revealed six species of canine tick-borne pathogens as: *Hepatozoon canis, Ehrlichia canis, Mycoplasma haemocanis, Anaplasma platys, Babesia gibsoni* and *Bebesia vogeli.* Out of these, two tick-borne pathogens, have been encountered in Ladakh region, *viz. H. canis* and *M. haemocanis,* with 24 per cent and 12 per cent infection respectively in dogs, transmitted by ticks (*Rhipicephalus* spp.) (Rani-Puteri *et al.,* 2011). *Hepatozoon canis* (Phylum Apicomplexa), infects leucocytes and parenchymal tissue, transmitted to dogs by the ingestion of ticks. This tick-borne disease of dogs is known as hepatozoonosis. *Mycoplasma haemocanis* is a parasitic bacterium causes infection of red blood cells in dogs and the disease caused is known as Haemotropic mycoplasmosis, transmitted by ticks (*Rhipicephalus* spp.). A detailed investigation on ticks in livestock and dog population of Kashmir, has revealed the prevalence of ticks : *Haemophysalis* spp. (73.68 per cent) and *Ornithodoros* spp. (26.32 per cent), infesting sheep; *Haemophysalis* spp. (85.74 per cent), *Rhipicephalus* spp. (14.26 per cent), parasitic on goats; *Haemophysalis* spp. (100 per cent), with cattle and equines; *Haemophysalis* spp. (88.88 per cent) and *Rhipicephalus* spp. (11.12 per cent), with dogs. (Tramboo *et al.,* 2018).

(b) Tick-Borne Apicomplexan Diseases (Babesiosis and Theileriosis) in Animals

Babesiosis or redwater or cattle tick fever, is tick-borne Apicomplexan disease, caused by *Babesia* - *B. bovis* (= *B. argentina*), *B. bigemina* in cattle and, *B. ovis* and *B. motasi* in sheep and goats. *Babesia* parasite belonging to Apicomplexa, family Babesiidae and transmitted by ticks (*Rhipicephalus* spp.). This parasite infects the red blood cells, causing anaemia, loss of appetite, red-coloured urine and in severe cases, there are signs like nervous incoordination, paralysis and coma, leading to death. The previous reports concerning prevalence of babesiosis in cattle (exotic, pure and cross bred) of Kashmir, and the incidence of infection was attributed to seasonal temperature for the propagation of tick vector, given by Das and Shaw (1968); Shaw (1980,1989). From Jammu region, Raina *et al.* (2005) showed the prevalence of Apicomplexan diseases (babesiosis and theleriosis) in 116 bovines (cows and buffaloes), 16.37 per cent of infection.

The seroprevalence of babesiosis infection in cattle and sheep and goats (local, exotic and crossbred) of Kasmir region, is reported as 13.30 per cent in cattle and, in sheep 2.43 per cent and in goat 0.75 per cent (Rather *et al.*, 2015). Severe type of ovine babesiosis in local as well as exotic and, cross bred sheep and goats, has been encountered in Kashmir region (Haq *et al.*, 2017 c; Tufani *et al.*, 2017). A detailed investigations on the ovine babesiosis of Kashmir region, covering seasonal observation on prevalence of disease in different breeds and age group have been conducted. In a total of 185 cases of ovine babesiosis: the infection rate was 26.77 per cent, highest (91.35 per cent) in cross bred sheep, followed by Corriedale (8.11 per cent), South down (0.54 per cent) and Bakerwal (0.00 per cent); the highest overall prevalence in summer (42.70 per cent) and lowest in winter season (Haq *et al.*, 2017 a). In 185 ovine babesiosis cases diagnosed, showed 26.77 per cent of prevalence rate, adult sheep above the age of 12 months (58.37 per cent), were mostly infected with babesiosis as compared to sheep 0-3 months (24.32 per cent), 3-6 months (8.65 per cent) and 6-12 months (8.6 per cent). (Haq *et al.*, 2017b).

Apart from above mentioned studies on ovine babesiosis, thorough clinico-haematological profiles have been made, with therapeutic management of acute babesiosis cases in bovines and, sheep and goats of Kashmir, and standard treatment protocol showed encouraging results (Haq *et al.*, 2017 c; Tufani *et al.*, 2009, 2015, 2017). An investigation on the cause of anemia in cattle of Jammu region by Singh (2014), have found the major cause of anemia due to helminths (*Trichostrongylus*, *Bunostomum*), followed by ectoparasites (lice, ticks, fleas) and *Babesia bigemina*.

Another tick-borne Apicomplexan disease of cattle, sheep and goats of J&K (UT) is Theileriosis. This disease is also kown as Egyptian fever or Mediterranean coast fever, caused by of parasites, *viz. Theileria parva* and *T. annulata*, belonging to family Theileriidae. In this disease, the parasites reside within lymphocytes and macrophages. The main symptoms of this disease are fever and lymph node enlargement, leading to other clinical complications in cattles and ruminants.The disease is curable, if treated promptly. The main vector of theileriosis are ticks of the genus *Hyalomma*. The previous reports of prevalence of theilerosis in cross-bred Jersy cows and, sheep and goats of Kashmir, transmitted by tick species, *viz.*

Rhipicephalus (*Boophilus*) *annulatus* (= *Boophilus annulatus* and *Hyalomma anatolicum*, have been given by Das and Sharma (1970 a), Shaw and Das (1970), Shaw (1971) and Shaw (1989).

From Jammu region, the prevalence of theileriosis in cows and buffalos, transmitted by ticks (*Hyalomma*) and caused by *Theileria annulata*, has been reported by Gupta *et al.* (2004) and Raina *et al.* (2005). An investigation on the seroprevalence of *Theileria* in cattle, sheep and goats of Kashmir valley, has indicated infection rate of 9.27 per cent in cattle, 1.46 per cent in sheep and in goats the prevalence as 1.51 per cent (Rather *et al.*, 2015).

6.5.3.2 Ticks of Medical Importance

(a) *Tick-Typhus*

The tick-borne typhus is known as Rocky Mountain spotted fever (RMSF), occurring in every continent and zoogeographical region of the world. This disease is caused by *Rickettsia rickettsii*. The rickettsiae are transmitted transtadially and transovarially in the tick vector. The symptoms of RMSF are fever, eschar, lymphadentitis, headach and rash. RMSF can be a deadly disease, if not treated early with proper antibiotics. They cause benign infections of non-human and severe disease in humans which may prove fatal. The principal vectors of this disease are ticks, *viz. Amblyomma* and *Dermacentor*. A case of tick- typhus in Srinagar was reported by Rao (1951). This author found a patient who was in habit of sleeping with dog in his bed and detailed examination of this patient revealed a tick, *viz. Rhipicephalus sanguineus*, sticking inside his ear and another unattached found by him between his toes. The similar type of ticks were also found on dog and collected from sofa of the patient's room.

(b) *Relapsing Fever*

The tick-borne relapsing fever (TBRF) in humans shows the symptoms as bleeding, coma, headach, joint aches, sagging on one side of face, stiff neck, sudden high fever, shaking chills, seizure. TBRF is curable with antibiotics. An outbreak of TBRF that took place amongst troops in Jammu was for the first time reported by Kaul (1949). He reported more than hundred cases of this disease, carried by ticks and caused by spirochaetes, *viz. Borrelia duttoni*. Latter, Rao and Kalra (1949), during their course of investigation on the role of ticks in transmission of this disease established that tick, *viz. Ornithodoros crossi* was a vector of this form of relapsing fever in Kashmir. The infection of this disease was described due to the bite of tick and not through coxal fluid or faeces. Tick itself is an effective and important reservoir of infection.

The period between 1948-49, also showed the high-incidence of disease from Army camps in Kashmir. An appreciable density of tick (*O. tholozani* var. *crossi*) was shown to influence the incidence of relapsing fever, starting from foothills of Jammu province and extended into Ladakh. The breeding season of the vector tick did not influence the incidence of disease. This type of tick-borne relapsing fever, transmitted by *Ornithodores tholozani* showed occurrence from Uzbekistan to Kashmir (*cf.* Harwood and Maurice,1979). The detailed studies on bionomics of vector tick, relapsing fever disease and epidemiology of this disease in J&K and Ladakh UTs,have been done by: Rao and Kalra (1950); Agarwal (1951), Kalra and Rao (1951) and Dutta *et al.* (1963).

(c) Crimean–Congo Haemorrhagic Fever

Crimean-Congo haemorrhagic fever(CCHF), human disease is a tick-borne virus (Nairovirus), belonging to the family Bunyaviridae. It is a deadly haemorrhagic fever and is endemic in Africa, Asia, Eastern Europe and middle East.The signs and symptoms of CCHF include headach, high fever, back, joint and stomach pain, and vomiting. The disease can progress to bleeding into the skin and liver failure. According to World Health Organization, ribavirin is antiviral medication of choice for CCHF (Appannanavar and Mishra, 2011). CCHF has earlier been reported from Western Border districts of Jammu, *viz.* Rajouri and Poonch districts, based on the samples collected from general populations of the area as long term resident of the area (Mohan Rao, 1986). However, in the later study on CCHF activity in these areas by Rodrigues *et al.* (1986), no CCHF virus incidence in the border regions surveyed, was reported either among humans or domestic animals. According to these authors that CCHF virus or closely related virus may be present, however, it was not of major health importance. Kaul *et al.* (1990) during the course of their thorough survey of ticks for CCHF virus activity in Jammu and Kashmir, have reported tick species, *Hyalomma anatolicum,* as an important biological vector of CCHF disease.

The virus of CCHF is enzootic in many regions of the world, including Palaearctic region, chiefly in steppe, semidesert and foothill biotope, where one or two *Hyalomma* species are the predominant ticks, parasitizing domestic and wild animals. CCHF disease involves fever, severe headache and bleeding from every orifice and venipuncture site. The infection occurs in man through the bite of the infected tick or by crushing ticks in contact with the skin or from shearing tick-infected sheep.

6.6 Parasitic Mites

The parasitic mites are small in size and their size ranges from 1–3 m m. Mites parasitize not only animals, poultry/birds but also humans. The body of a mite is typically composed of an anterior gnathosoma or capitulum and posterior region, known as idiosoma. The gnathosoma shows laterally the presence of palps and more medially the chelicerae. The mouth-parts are adapted for piercing and sucking blood of the host. The legs of a typical mite are six-segmented. The mites belong to the Sub-class Acari of Class Arachnida under phylum Arthropoda (Invertebrata). The Acari has been further divided into two Super orders, *viz.* Anactinotrichida (=Parasitiformes) and Actinotrichida (= Acariformes). The Actinotrichida includes three Orders, out of these, two Orders are parasitic as: Astigmata and Prostigmata, covering various species of mites from different regions and host of Jammu, Kashmir and Ladakh. The Anactinotrichida has four Orders, the only two orders, *viz.* Mesostigmata and Ixodida, have representative species from Jammu and Kashmir. The former Order incorporates parasitic mites of this region.

6.6.1 Mites Parasitic on Insects (Grasshopper, Housefly and Bees)

In Jammu, Kashmir and Ladakh regions, so far a total of 2 spp., 19 spp., 1 sp., have been found to parasitize housefly, Bees (honey bee, carpenter bee) and grasshopper respectively. Earlier, the ectoparasitic mites; *Marcrocheles muscae domesticae,* belonging to the family Macrochelidae (Mesotigmata) and *Microtrobium* sp. under family Trombidiidae (Prostigmata), parasitising houseflies (*Musca domestica*), have been reported from Jammu and Kashmir (Sharma and Sharma, 1972). The nymphs and adults of mites prey upon the eggs and first stage larvae of houseflies. The ectoparasitic mite on honeybees (*Apis* spp.), *viz. Varroa jacobsoni,* has been reported by Punjabi and Saraf (1969) from Kashmir and Abrol and Kotwal (1996) from Jammu. This mite is a blood-sucking parasite of *Apis cerana, A. mellifera,* causing varroatosis. The wounds caused by *V. jacobsoni* invite microorganisms, leading to infections. Also, *Varroa* is responsible for transmitting bacteria and viruses to honeybees, which cause septicemia and paralysis in them.

Other mites, attacking honeybees in this region belong to genus *Tropilaelaps,* feed on the haemolymph of larvae and pupae of the honeybees and are considered to be most destructive ectoparasites of bees in J&K. *Tropilaelaps clareae,* infesting *Apis mellifera* in Kashmir, has been studied by Shah and Shah (1991) and *T. koenigerum* reported from subtropical areas of Jammu by Abrol and Putatunda (1995). They found that this species of mite was feeding on four species of honeybees, *viz. Apis cerana, A. dorsata, A. florae* and *A. mellifera.*

Parasitic mite of honeybees – *Acarapis woodi* (Prostigmata: Scutacaridae), is responsible for causing disease, known as Isle of Wright disease in *A. cerana* and *A. mellifera* of Kashmir region, infesting not only workers, but also drones as well as queen bees (Shah, 1996). This mite harbours tracheae of bees, it feeds on the haemolymphs of adult bees by penetrating the tracheal wall, leading to passing of bacteria in haemolymph, secretes toxic substances, also clogs air passages mechanically and kills the host.

The bee mites (ectoparasitic and phoretic), associated with different types of bees : Honey bees (*A. cerana indica, A. dorsata, A. florae, A. mellifera*); *Bombus* sp.; *Xylocopa*; Megachilid; solitary bees, prevalent in earstwhile J&K State, have been recorded (Putatunda and Abrol, 2002). Putatunda and Abrol (2003) gave a list of 17 spp. of mites, parasitic on honeybees of Jammu and Kashmir. The newly recorded species *i.e.,* other than those mentioned above are: *Caryoglyphus lactis, Delfindo* sp., *Euvarroa sinhai, Hypopus* sp., *Kuzinia evae, Kuzinia* sp., *Neocypholaelaps indica, Neocypholaelapsp., Schulzea* sp., *Sennertia* sp., *Tyrophagus longior, T. tropicus, Varroa sinhai* and *Vidia* sp. A detailed studies on seasonal incidence of mites parasitic on honey bees (*Apis mellifera*) in Jammu region, covering mite species as *V. destructor, T. clareae, V. jacobsoni* and *T. koenigeruni,* have been done (Kotwal *et al.,* 2013). These authors have found *V. destructor,* occurring throughout the year and as the most destructive mite of *A. mellifera* in Jammu region.

Ectoparasitic mite, *T. clareae* is reported to be associated with carpenter bee (*Xylocopa iridipennis*) Abrol and Putatunda, 1996). A mite species of insect pest, *viz. Podapolipus diander* in larval stages, has been found for the first time parasitizing insect pest of rice and vegetables, *viz. Acrida exalta* (grasshopper) in Anantnag District of Kashmir valley (Singh *et al.*, 2005).

6.6.2 Mites Parasitic on Small Mammals, Poultry Birds, Sheep and Humans

In J&K, Ladakh (UTs), parasitic mites of vertebrates (bat, small burrowing mammals, birds, sheep and man), are belonging to the orders, *viz.* Astigmata, Prostigmata and Mesostigmata. The order Astigmata includes families of parasitic mites, well-defined, showing weakly sclerotized body, slow moving and causing mange and scrub disease in animals. The Order Prostigmata covers heterogeneous acarines. The family Trombiculidae, under this Order is parasitic in the larval stages on vertebrates and some are as vector of scrub typhus. The order Mesostigmata has important families: Dermanyssidae and Macronyssidae, as parasites of Poultry-birds and Laelapidae, as ectoparasite of small mammals in various localities and regions of Jammu, Kashmir and Ladakh (Table 6.5) An annotated and updated checklist, pertaining to parasitic mites of vertebrates (birds, mammals and humans) of Jammu, Kashmir and Ladakh, has been incorporated in Table 6.5), depicting, a total of 28 spp. of parasitic mites of vertebrates, under 14 genera, belonging to orders and families : Astigmata (Knemidokoptidae, Psoroptidae); Prostigma (Demodicidae, Sarcoptidae, Trombiculidae) and Mesostigma (Dermanyssidae, Macronyssidae, Laelapidae), known to occur in different localities/regions of J&K, Ladakh, UTs. The parasitic mites have been found on vertebrates like rodents, other small burrowing mammals, birds (poultry), sheep and humans. The highest number of mite species *i.e* 14, has been observed parasitizing rodents and other small mammals. It is followed by Birds (poultry), with 5 spp., 3 spp. on humans, 2 spp. on rabbit. Besides these, 1 sp. each parasitizes hosts such as buffalo, dog, fox, leopard, sheep (Table 6.5).

(a) Mites Parasitic on Small Mammals

Rao *et al.* (1973), during the course of his detailed survey on haematophagous arthropods of Western-Himalaya, reported 10 species of Mesostigmatid mites from Jammu, Kashmir and Ladakh Himalayan regions. These included the existence of 5 spp. of mites from Ladakh, 3 spp. occurring in Kashmir and 6 spp. in Jammu, parasitizing rats, shrews, mice, bat and other small mammals of these regions.

Parasitic mites of small burrowing mammals from pastures of Tangmarg and Gulmarg localities of Kashmir, have been studied by Srivastava and Wattal (1981). These authors reported 3 spp. of trombiculid mites, belonging to genus *Leptotrombium*, parasitizing *Alticola roylei montosa* from Gulmarg and 2 spp. each of mesostigmatid mites, under families, *viz.* Dermanyssidae and Laelapidae as haematophagous ectoparasites of small mammals: *A. r. montosa, Bandicota bengalensis, Rattus rattus* and *R. r. gangutrianus*, existing in the pastures of these localities.

(b) Poultry Bird Mites

The earliest study on the prevalence of a mite of veterinary importance, *viz. Ornithonyssus bursa* (tropical-poultry-mite) was conducted by Sharma (1972), parasitizing the poultry birds in Kashmir. *O. bursa* occurs either on birds or in nest. Mir *et al.* (1993) investigated the prevalence of ectoparasitic mites in indigenous fowl in Kashmir. These include mange-causing mites (*Knemidocoptes gallinae* and *K. mutans*) and a red-chicken mite or roost mite (*Dermanyssus gallinae*), (Table 6.5). Bhat *et al.* (2014), reported 1.69 per cent of prevalence of *D. gallinae* in backyard poultry region in Jammu region. The red mite of poultry has got zoonotic importance as it causes skin lesions in humans. Besides, it transmits encephalitis diseases (Saint Louise encephalitis, Eastern and Westren encephalitis) in different parts of the world.

Table 6.5: Parasitic Mites (Class Archnida : Subclass Acari) on Hosts (Bat, Burrowing Mammals, Birds, Sheep and Humans) of Jammu and Kashmir, Ladakh (UTs)

Classification and Mite Species	*Host(s) and Locality(ies)*	*Reference*
Order : Astigmata; Family Knemidokoptidae		
Knemidokoptes gallinae (= *Cnemidokoptes gallinae*)	Domestic Fowl (indigenous) (K)	Mir *et al.* (1993)
Knemidokoptes mutans (= *Cnemidokoptes mutans*)	Domestic Fowl (indigenous) (K)	Mir *et al.* (1993)
Order : Astigma; Family Psoroptidae		
Psoroptes ovis	Marino sheep, sheep (K, J)	Sharma *et al.* (1996), Khajuria *et al.* (2005)
Order : Prostigma; Family Demodicidae		
Demodex brevis	Humans (K)	Rather and Hassan (2014)
Demodex canis	Leopards (J)	Soodan *et al.* (2000)
Demodex cati	Dog (K)	Goswami *et al.* (2015)
Order : Prostigma; Family Sarcoptidae		
Notoedres cuniculi	Rabbit (A)	Darzi *et al.* (200)
Sarcoptes scabie	Rabbit, Buffalo (J)	Nazir *et al.* (2014, 16)
Sarcoptes sp.	Fox (K)	Pandit *et al.* (2007)
Order : Prostigma, Family Trombiculidae		
Leptotrombidium deliense (= *Trombicula delicense*)	Rodents, Humans (K)	Kalra and Rao (1951) Anon (2009, 2016)
Leptotrombidium irregulare	*Alticola roylei montosa* (G)	Srivastava and Watal (1981)
Leptotrombidium nupestre	*A. r. montosa* (G)	Srivastava and Watal (1981)
Leptotrombidium titwaleasis	*A. r. montosa* (G)	Srivastava and Watal (1981)
Order: Mesostigma; Family Dermanyssidae		
Dermanyssus gallinae	Domestic fowl (indigenous) (J, K)	Mir *et al.* (1993), Bhat *et al.* (2014)
Hirstionyssus sunci	*Rattus rattus* (G, T)	Srivastava and Watal (1981)
Hirstionyssus sp.	Unknown (L)	Rao *et al.* (1973)
Liponyssoides (= Allodermanyssus)	House mice (L)	Rao *et al.* (1973)
Order : Mesostigma; Family Macronyssidae		
Ornithonysus bursa	Poultry birds (L)	Sharma (1972)
Order: Mesostigma; Laelapidae		
Androlaelaps marshalli	*R. rattus* (G, T)	Srivastave and Wattal (1981)

Classification and Mite Species	Host(s) and Locality(ies)	Reference
Echinolaelaps echidnius	Rat, shrew (D, U, R)	Rao *et al.* (1973)
Eulaelaps stabularis	Rat, mouse (G, L, T)	Rao *et al.* (1973)
Laelaps agilis	Bat, Bird (B, L)	Rao *et al.* (1973)
Laelaps algericus	Small mammals, *R. r. gangutrianus* (D, L)	Rao *et al.* (1973)
Laelaps jugalis	Rat (B, U)	Rao *et al.* (1973)
Laelaps nuttali	*A. r. montosa* , *Bandicota – bengalensis* (T, G)	Srivastava and Wattal (1981)
Laelaps turkestanicus	Rat (D, B)	Rao *et al.* (1973)
Laelaps sp.	Rat (U, R)	Rao *et al.* (1973)

A: Anantnag; B: Baramulla; D: Doda; G: Gulmarg; J: Jammu; K: Kashmir; L: Ladakh; R: Rajouri; T: Tangmarg; U: Udhampur.

The *Knemidokoptes* spp. are also spelled as *Knemidocoptes* spp. or *Cnemidocoptes* spp. (Harwood and Maurice, 1979). *K. gallinae* is a depluming mite, attacks the skin at the bases of feathers of poultry birds. There is intense itching due to this and birds are impelled to pluck feathers. *K. mutans* is also known as the scaly leg mite, microscopic burrowing mite, attacking the legs of domestic chickens, causing lifting of scales and swollen condition of shanks, with deformity and also encrustations.

The Roost mite (*D. gallinae*), during day time hide in crevices of poultry houses and roosts, however, attacks birds during night. The heavy parasitization of this mite causes mortality in newly hatched chickens and the adult became weak due to loss of blood and susceptible to diseases. Another poultry mite of Kashmir, *viz. Ornithonysscus bursa,* measuring 1mm in length. This parasitic has not only been found on bodies of the birds but also in their nests in different parts of the world. This species is both a day as well as night time feeder. *O. bursa* is also known to attack humans, causing only temporary irritating effect, because the mite cannot survive on host other than birds

(c) Ovine Mites

Sharma *et al.* (1996) reported the infestation of parasitic mite, *viz. Psoroptes ovis* (= *P. communis var. ovis*) in Kashmir Marino sheep, in Ganderbal and Kangan localities of Kashmir. *P. ovis* is of great economic importance, known to cause highly cantagious psorptic mange disease, responsible for the formation of lesions on any part of the body of sheep. The adult mites puncture the epidermis and feed on lymph and tissue fluids, resulting in the formation of scab. This condition leads to loss of fleece and markedly reducing wool production. An outbreak of Psoroptic mange disease in sheep, has been reported by Khajuria *et al.* (2005) from Jammu region

(d) Parasitic Itch Mite of Cattle (Buffalo)

Sarcoptes scabie or itch mite, is a parasite that burrow and cause scabies or mange (sarcoptic mange), extremely pruritic, contagious and debilitating disease of water buffalo (*Bubalus bubalis*). This serious disease leads to decreasing milk, meat production, also morbidity or mortality. The sarcoptic mite infestations in buffaloes pertaining to 72 unorganized Dairy Farma of Jammu, with prevalence rate 3.33 per cent, was observed (Nazir *et al.*, 2014). The control of mange on buffaloes with pour-on eprinomectin and ivermectin, has shown encouraging results (Nazir *et al.*, 2016).

(e) Rabbit Mange Mites

Rabbit mange in Angora and Newzealand white rabbits in rabbitry in Srinagar, due to parasitic mites, *viz. Sarcoptes scabie* var. *cuniculi* and *Notoedres cati* var. *cuniculi*, have been reported by Darzi *et al.* (2007). These mites are highly contagious and burrowers, the affected rabbits showed scales, scab formation, atopecia and pruritis, resulting in considerable loss due to rabbit weight, productivity and wool quality. The treatment of this disease with Doramectin was shown to be effective by the authors

(f) Feline and Canine *Demodex* Mite

In Jammu region, mammal wildlife, leopards in Manda Wildlife Sanctuary were observed to be infested with parasitic mite, *Demodex cati*, showed symptoms of demodicosis. This wild animal had skin lesions almost all over body. The disease was cured with the application of Ivermectin and Deltamethrin (Soodan *et al.*, 2000). *Sarcoptes* sp. infestation has been reported in Jammu and Kashmir, affecting non-hairy portion of skin of fox (*Vulpus vulpus*) (Pandit *et al.*, 2007). In Kashmir valley, concomitant infection of demodicosis or demodectic mange or red mange in a dog (pet), caused by parasitic mite, *Demodex canis* has been observed (Goswami *et al.*, 2015). Demodicosis in dogs does not cause much health problem, however, this can harm the young dogs that have poor immunity or older dogs that are suffering from suppressed immune system.

6.6.3 Mites of Medical Importance

Scrub typhus or bush typhus is a mite-borne human disease, transmitted by the bite of infected trombiculid mites, 'chiggers', *Leptotrombidium deliense* and others. This disease is also known as tsutsugamushi disease, caused by an obligate gram-negative type of proteobacterium,order Rickettsiale family Rickettsiaceae, *Orientia tsutsugamushi* (= *Rickettsia tsutsugamushi*). These mites are both vectors as well as the natural reservois of *O. tsutsugamushi*. Scrub typhus is endemic in many regions of India, including Jammu and Kashmir (Anon 2009, 2016; Rathi and Rathi,2010; Dasan *et al.*, 2014). The earliest record of a mite of medical importance, *Leptotrombidium deliense* (=*Tromsicula deliense*), showing wide distribution in Kashmir and Ladakh region, has been given by Kalra and Rao (1951a). These authors also confirmed the cases of Scrub typhus disease transmitted by *L. deliense* in this region. Scrub typhus is an acute febrile disease, transmitted by mite (trombiculid). When infected mite feed a person, ulcer like eschar develop at the site of feeding besides, the disease shows fever, severe headach, rash and lymphadenopathy, also leading to other severe clinical symptoms and complications. In untreated cases, the mortality may exceed 30 per cent. Disease can be cured with antibiotics. Scrub typhus is zoonotic in nature, where mite, *Leptotrombidium* and wild rats play important role.

An outbreak of scrub typhus disease in Jammu region during the year 2009 (Sept.- Oct.) has been reported, covering a total of 20 adult cases, confirmed serologically for antibodies (IgM) to *O. tsutsugamushi*, using ELISA (Mahajan *et al.*, 2010). The majority of case belonged to district Rajouri and were successfully treated with antibiotics. In Jammu region, 21 childrens tested ELISA positive for IgM against scrub typhus, covering period between 1st Oct., 2008 to 30th Sept., 2009. The children cases belonged to Sunderbani-Naushera- Rajouri belt and other hilly areas such as Jindrah and Chenani areas of Jammu, showed variety of clinical signs and features (Digra *et al.*, 2010)

In Kashmir region, Farhana *et al.* (2016), have reported the prevalence of scrub typhus among the cases of PUO (Pyrexia of Unkown Origin) through serological tests, *viz.* Weil-Felix Agglutination Test (WFT) and Enzyme linked Immunosorbent Assay (ELISA), based on hospital study, during 1st March 31st Oct., 2015. Of the 162 serum samples tested, 22.8 per cent was positive for scrub typhus by WFT and, IgM ELISA and IgG was positive in 8 (4.9 per cent) and 15 (9.3 per cent) respectively. The comments on the results of these serological tests for the prevalence of scrub typhus cases in Kashmir region have been put forth by Kalala and Shivalli (2017).

The human *Dermodex* mite, including two species, *viz. Demodex folliculorum* and *Demodex brevis*, are obligate ectoparasites found on humans. Both these species are contagious to humans. These are commonly known as eye lash mites. *D. folliculorum* live in hair follicles, whereas *D. brevis* lives in sebaceous gland connected to hair follicle. The disease caused by these mites is known as dermodicosis or dermodicidosis. Both these species infest face near nose, eyelashes and can occur elsewhere on the human body. *Demodex* infestation may usually remains asymptomic and may have pathogenic role only when present in high densities and also because of immune imbalance, causing varied clinical manifestations (Rather and Hassan, 2014).

A human case in Kashmir regarding dermatological involvement – acute superficial lymphaginitis, following pigeon mite bite, has been feported by Kaul and Qadri (2010). These authors have reported *Argas refluxus* as a "pigeon mite". However, *A. refluxus* is not a mite, it is a pigeon tick species. The pigeon tick is a pathogenic transmitting disease that typically feed on pigeons, but also parasitize humans, causing local and systematic reaction (Tavassoli *et al.*, 2011). There can also be variety of symptoms corresponding to an Ig E-mediated allergy, with increasing severity resulting in death (Weckesser, 2010).

Chapter 7

Parasitic Insects (Fleas, Lice and Bugs)

7.1 Fleas

The fleas belong to insect order Siphonaptera, are haematophagous ectoparasites of birds and mammals, having small size, 1-9 mm and the females are larger than males of the same species. The body of the adult fleas is strongly sclerotized, laterally flattened, covered with rows of bristles and that point backwards and the wings are absent. The mouth-parts of fleas are modified for piercing host skin and sucking their blood. The legs are strong, the posterior pair is strikingly adapted for jumping. The fleas possess the ability to transferring from one host to another, mostly rats to human. The host association of fleas in J&K, Ladakh (UTs) include wide range of birds and mammalian hosts, such as poultry and wild bird species, small burrowing mammals, mongoose, bats, canines, cattle (bovines), small ruminants, horses, camel and human.

Fleas are vectors of viral, rickettsial, bacterial diseases of humans and other animals, besides serving as intermediate host of pathogenic zoonotic agents between vertebrate hosts, including humans. Fleas can have direct effect by causing allergic dermatitis. The important diseases transmitted by fleas in different parts of world are: murine or endemic typhus, human plague, tularemia, cat scratch disease and Myxomatosis virus. Fleas have ability to transmit diseases through contaminated faeces, soiled mouth-parts, regurgitation of gut contents and infectious saliva.

7.1.1 Systematic Survey, Diversity and Host Range

The earliest records of fleas, parasitizing wide range of mammalian and bird hosts, distributed in different regions of Jammu, Kashmir and Ladakh, are given in outstanding monographs published by: Sharif (1930); Kalra and Rao (1951) and Hopkins and Rothschild (1953, 62, 66, 66 a and 71). Iyengar (1973), published a monograph on the Siphonaptera of Indian regions, describing 76 spp. of fleas, including 10 spp. from Kashmir and Ladakh. The detailed surveys and systematic investigation on flea-fauna of this region, have been conducted by Das and Sharma (1970), Kulkarni and Bhat (1972); Rao *et al.* (1973); Smith (1977); Srivastava and Wattal (1981) and Mir *et al.* (1993).

Das and Sharma (1970), reported for the first time, the occurrence of flea (*Ctenocephalides felis orientalis*), parasitizing goats from various localities of district Srinagar. Kulkarni and Bhat (1972), collected large number of flea, *viz. Neopsylla secura kashmirensis* (= *Neopsylla kashmirensis* Kulkarni and Bhat) from hosts like rodent, shrew like long-tailed mouse (*Apodemes flavicollis*), house mouse (*Mus musculus*). Rao *et al.* (1973), in his detailed review paper on haematophagous arthropods of Western-Himalayan region, have listed about 26 species of fleas, from various districts of earstwhile J&K State, parasitizing rodents, shrew, mongoose, dog, goat, sheep, cattle, bat, man, pika, small burrowing mammals, mouse hare and marmot, *etc.* The new taxa and geographical records of some Siphonaptera of small mammals, rodents (6 spp.) of Kashmir, were given by Smith (1977). These included description of a new parasite subspecies *viz., Amphipsylla phaiomydis, limoria,* parasitizing rodent – *Hyperacrius* and taxonomic down grading of *Neopsylla kashmirensis* Kulkarni and Bhat to *N. secura kashmirensis* Kulkarni and Bhat. Srivastava and Wattal (1981) reported the existence of three species of fleas, *viz. Nosopsyllus punjabensis, N. fasciatus* and *N. simla,* from *Rattus rattus gangutrianus, Mus musculus* and *Alticola roylei montosa,* occurring in the Tangmarg and Gulmarg areas of Kashmir Valley. Recently, the prevalence of flea, *Echidnophaga gallinacea* in indigenous fowl of Kashmir Valley, has been studied by Mir *et al.* (1993).

A total of 34 species of fleas, distributed over 18 different genera, belonging to 7 families, are now known to occur in different districts and localities of Jammu, Kashmir and Ladakh. The various genera, with total number of species are: *Acropsylla* (1 sp.), *Amphipsylla* (3 spp.), *Callopsylla* (2), *Ceratopsyllus* (3), *Citellophilus* (2), *Ctenocephalides* (3) *Echidnophaga* (1), *Frontopsylla* (2), *Ischnopsyllus* (1), *Neopsylla* (4), *Nosopsyllus* (3), *Paradoxopsylla* (1), *Peromyscopsyllus* (1), *Pulex* (1), *Stivalius* (2), *Thaumapsylla* (1), *Xenodaeria* (1) and *Xenopsylla* (2 sp.). These species are belonging to the families, *viz.* Ceratophyllidae, Ctenophthalmidae, Hystrichopsyllidae, Ischnopsyllidae, Leptopsyllidae, Plucidae and Stivaliidae. The total number of fleas species, parasitizing birds and mammalian hosts, including humans prevalent in Jammu, Kashmir and Ladakh, is 16 spp., 19 spp. and 20 respectively (Table 7.2)

The dominant family of fleas in J&K, Ladakh UTs, is found to be Ceratophyllidae, incorporated a total of 10 spp., having wide range of hosts (birds and mammals). This is followed by family Leptopsyllidae including 8 spp., Plucidae (7 spp.), Ctenophthalmidae (4 spp.), Stivaliidae (3 spp.). The families like Hystrichopsyllidae and Ischnopsyllidae, covered 1 sp. each. The total number of fleas species parasitizing various types of birds and mammalian hosts, including human, belonging to different families, occurring in Jammu, Kashmir and Ladakh region, are given the Table 7.2. Of the overall flea-fauna of Jammu, Kashmir and Ladakh, covering 34 species, under 7 families, highest number of flea species *i.e.*, 22, is found to be parasitizing small burrowing mammals (Rodentia) and this accounts for 64.70 per cent of total fleas-fauna. This is followed by dogs showing parasitization by 5 spp. (14.70 per cent), the cattle, goat, sheep, and wild bird, parasitized each with 4 spp. (11.76 per cent) of fleas. Mongoose found to be infested with 3 spp. (8.82 per cent). Bats, human and domestic/poutry birds, each showed parasitization by 2 spp. fleas. (5.88 per cent). Rest of the hosts as, cat and horse were known to have single species (2.94 per cent) each (Table 7.1).

Table 7.1: The Total Number of Fleas Species, with Percentage, Parasitizing Various Kinds of Birds (Poultry and Wild) and Mammalian Hosts, Including, Humans in J&K, and Ladakh (UTs)

Host Type (Total Flea) [Percentage]	*Total Number of Flea Species in J&K, Ladakh*		
	In Jammu	*In Kashmir*	*In Ladakh*
Bat (2) [5.88 per cent]	02	01	-
Cat (1) [2.94]	01	-	-
Cattle (bovines) (4) [11.76)	04	04	01
Dogs (5) [14.70]	05	05	01
Goats (4) [11.76]	04	04	01
Horses (1) [2.94]	01	-	-
Humans (2) [5.88]	01	02	01
Mongoose (3) [8.82]	03	01	-
Poultry birds (2) [5.88]	01	01	-
Sheep (4) [11.76]	04	04	01
Small burrowing mammals (22) [64.70]	10	14	17
Wild birds (4) [11.76]	04	03	04

7.1.2 Fleas of Medical Importance

The main fleas of medical importance found in Jammu, Kashmir and Ladakh region are *Pulex irritans, Xenopsylla astia and X. cheopis,* belonging to the family Pulicidae. *Pulex irritans* also known as the human- flea parasitizes not only humans but also domestic animals. Much irritation is caused to the hosts by the piercing sucking types of mouth-parts of adult *P. irritans. Xenopsylla cheopis* (Oriental rat-flea) is a sole important vector of plague and murine typhus of human in different parts of the world. Murine or endemic typhus in human is transmitted by *X. cheopis* and caused by *Rickettsia typhi.* The infected persons show symptoms of fever, headache, chills, body aches and pains. Besides, there can be rash on chest, back, arms and/ or legs. Murine typhus can treated with appropriate antibiotic therapy.

Human plague or" Black Death" is serious disease in humans and is transmitted between animals (small mammals) and humans by the bite of infected fleas, caused by bacterium, *Yersinia pestis.* The symptoms of this disease include three forms: i) bubonic plague, most common form, showing swollen lymph nodes in the groin, armpit, or neck, called buboes, characterized by fever and headache; ii) septicemic plague, bacteria multiply in the blood, showing symptoms like fever, pain and shock; iii) pneumonic plague, rarest and deadlist form, affecting the lungs, there is shortness of breath, chest pain, cough and sometimes bloody or watery sputum, besides symptoms like fever, headach and weakness. The pneumonic plague spreads from person to person and there is death due to this form of plague, however, early antibiotics treatment in infected plague cases can save lives.

Kalra and Rao (1953) were the earlier researchers who reported the existence of murine typhus epidemic in Kashmir Valley. Due to murine typhus disease in Kashmir, death was reported to be 8-17 per cent besides, *X. cheopis* and lice also played a part in its spread. The high death was not due to the virulence of infection but due to the lack of any medical aid. This disease in Kashmir was confirmed after the trapping of rats, where the cases of murine typhus occurred and rats were found to be parasitized by fleas, *viz. X. cheopis and X. astia.* The individuals of *X. cheopis,* have the habit of easily leaving their host and may transfer to other host of same or of different species. The causative agent of murine typhus disease is a Rickettsian, *viz. Rickettsia mooseri.* It is also known that after the initial infection by rat flea (*X. cheopis*), lice played a role in transmitting disease from person to person. The murine typhus in man is much milder type of disease.

The human plague in Jammu region was reported to be virulent in the winter months of the year 1901. In Poonch region, plague had broken out in the autumn of the year 1902. From 19th November, 1903 to 24th January, 1904, 56 persons died of plague disease in Srinagar. There was total death of 1314 persons belonging to Moffasil districts of Kashmir from 18th December to 11th April, 1904 (Mitra, 1904). In Kashmir region, there were reports of bubonic, septicaemic and pneumonic plague, included areas like Awantipur, Srinagar, Baramulla, Islamabad, Tehsil Khas, Uttar Machipura. During the course of epidemic those who came directly in contact with plague disease, caught infection and nearly all died of after short illness. Those who helped the burial of the corpse next got the disease.

7.1.3 Fleas of Veterinary Importance

The important flea species of veterinary importance of J&K are: *Echidnophaga gallinacea, Nosophyllus nilgirionsis* and *Ceratopsyllus hirudinis,* belonging to the family Ceratophyllidae; *Ctenocephalides canis* and *Ctenocephalides felis* under family Pulicidae. The *E. gallinacea* (the stick- tight flea) of poultry has been found to parasitize the indegineous fowl in Kashmir. Sometimes, it is also called as burrowing fleas, due to impregnated female fleas into the host's skin, causing sever irritation and the skin of the host become swollen. Mir *et al.* (1993) studied the prevalence of *E. gallinacea* in Kashmir fowl, attacking in groups, comb and wattle of the host.

Ctenocephalides canis (dog-flea) and *C. felis* (cat-flea), have been found to attack dogs in Jammu and Kashmir region. These flea species are not host specific and both these species also parasitize sheep, goat and cattle in this region. *C. canis* is also suitable intermediate host for dog tapeworm, *Diphylidium caninum.* The other fleas of veterinary importance are: *Ceratophyllus hirudinis,* showing wide distribution in J&K and has been reported to parasitize birds, known to cause serious injury to poultry (hen); *Nosopsyllus nilgirionsis* is responsible for biting cattle in various localities of Jammu, Kashmir and Ladakh. Parasitic dermatitis disease in canines of Jammu region due to cat flea, *Ctenocephalides felis,* has been reported by Sharma *et al.* (2009). The parasite-induced dermatitis is a skin condition that is caused by allergic reaction to flea bite. It is the most common dermatological disease of dogs.

Table 7.2: Fleas (Siphonaptera: Insecta): Checklist and Vertebrate Host Diversity, Including Human in J&K, Ladakh UTs

Sl. No.	*Fleas Species (Family)*	*Distribution of Host/s of the Fleas-Parasite*		
		Jammu	*Kashmir*	*Ladakh*
1	*Acropsylla girshami* (v)	Rodent, Shrew	–	–
2	*Amphipsylla phaiamydis imonia* (v)	–	*Hyperacrius*	–
3	*Amphipsylla* sp.-I (v)	–	–	Rodent
4	*Amphipsylla* sp-II (v)	–	–	Rodent
5	*Callopsylla* sp. (i)	–	–	Rodent
6	*Callopsylla oreinus* (i) (= *Orneacus oreinus*)	-	-	Bird
7	*Ceratopsyllus hirudinis* (i)	Birds	Birds	Birds
8	*Ceratophyllus orites* (i)	-	-	Bird
9	*Ceratophyllus caliotes* (i)	-	-	Bird
10	*Citellophilus* sp-I (i)	–	–	Rodent
11	*Citellophilus* sp-II (i)	–	–	Marmot
12	*Ctenocephalides canis* (vi)	Dog, Sheep, Goat, Cattle, Human	Dog, Sheep, Goat, Cattle	–
13	*Ctenocephalides felis* (vi)	Cat, Goat, Sheep, Mongoose, Bird, Cattle, *Suncus-murinus* (Shrew)	Cat, Dog, Cattle, Bird, Goat, Mongoose *Acridotheres-tristis* (Bird), *S. murinus*	–
14	*Ctenocephalides felia orientalia* (vi)	Cat, Dog, Cattle, *Gallus domesticus* Sheep and Goat	Cattle, Dog, goat, Sheep	–
15	*Echidnophaga gallinacea* (vi)	–	Kashmir Fowl	–
16	*Frontopsylla ambigua* (v)		Rodent	Rodent
17	*Frontropsylla spadix* (v)	–	–	Rodent
18	*Ischnopsyllus octactenus-* (iv)	*Nyctalus-noctule* (Bat) *Pipistrellus-pipistrellus*-(Bat) *Ploceus-philippinus* (Weaver bird)	*P. pipistrellus*	
19	*Neopsylla secura* (ii)	–	Rodent, Shrew	Rodent, Shrew
20	*Neopsylla secura kashmirensis* (ii)	Rodent, Shrew	*Apodemes flavicollis, Mus musculatus, Rattus rattoides, Suncus murinus*	*A. flavicollis, M. musculatus, R. rattoides, S. murinus*
21	*Neopsylla* sp.-I(=*Setosa group*) (ii)	–	–	Rodent

Sl. No.	Fleas Species (Family)	Distribution of Host/s of the Fleas-Parasite		
		Jammu	Kashmir	Ladakh
22	*Neosopsyllus fasciatus* (ii)		*Alticola roylei –montosa*	–
23	*Nosopsyllus nilgirionsis* (i)	Cattle and small mammals	Cattle and small mammals	Cattle and small mammals
24	*Nosopsyllus punjabensis* (i)	Dog and small mammals	Dog and small mammal, *Rattus rattus ganguntrianus*	Dog, and small mammals
25	*Nosophyllus simla* (i)	Small mammals	*A. roylei montosa* and small mammals	Small mammals
26	*Paradoxopsylla custodis* (v)	Rodent	Rodent	Rodent
27	*Peromyscopsylla himala-yica* (v)	*Herpestes-edwardsii* (Mongoose)	–	*Alticola roylei* (Vole)
28	*Pulex irritants* (vi)	Dog, Goat, Sheep, Rodent, and Man	Dog, Goat, Sheep, Rodent, Man	Dog, Goat, Sheep, Rodent, Man
29	*Stivalius ferinus* (vii)	Rodent, Shrew	–	–
30	*Stivalius shalae* (vii)	Small mammals	–	–
31	*Thaumapsylla breviceps-oientalis* (vii)	Bat	–	–
32	*Xenodaeria telios* (iii)		Rodent, Shrew	–
33	*Xenopsylla astia* (vi)	–	Rat	–
34	*Xenopsylla cheopis* (vi)	Rodent, Shrew, Mongoose	Rat, Human, *Mus*	–

(i) Ceratophyllidae; (ii) Ctenophthalmidae; (iii) Hystrichopsyllidae; (iv) Ischnopsyllidae; (v) Leptopsyllidae; (vi) Plucidae; (vii) Stivaliidae; –: Absent.

Source: Jordhan (1937); Kalra and Roa (1951a); Das and Sharma (1970); Kulkarni and Bhat(1972); Rao *et al.* (1973); Smit (1977); Srivastava and Wattal (1981); Mir *et al.* (1993); Sharma and Sharma (1973); Sharma (1993); Wani *et al.* (2015); Yadav *et al.* (2006).

7.2 Lice (Sucking and Chewing)

Lice belong to insect order Phthiraptera, are small insect (5 mm to 8 mm), dorsoventrally flattened,wingless, blind, obligate ectoparasites of birds and mammals. They are highly host specific. The true lice of Jammu, Kashmir and Ladakh region are represented by two Sub-orders : Anopleura, including sucking lice; Amblycera and Ichnocera, chewing lice. In sucking lice, the mouth-parts are not visible, palps absent and are of piercing and sucking type. The sucking lice can cause and transmit diseases. The louse-borne infectious diseases reported from different parts of the world are epidemic typhus, relapsing fever, and trench fever, caused by *Rickettisia prowazekii*, *Borrelia recurrentis* and *Bartonella quintana* respectively. Besides these, human louse infestation or pediculosis, may cause serious skin irritations or secondary infections. Lice pierce the skin and suck blood of mammals and catte. Lice irritate cattle, causing them to bite, scratch and rub. In large number they can cause anaemia and secondary infection in mammals and cattle. In case of chewing lice, the antennae are visibe and mouth-parts are used for chewing. They spend their entire life generation and after generation on host. The chewing lice feed on feathers, hair, dermal material, fluid or blood from minor abrassions.

The Anopleurans (sucking lice) of Jammu, Kashmir and Ladakh, are represented by six families, *viz.* Haematopinidae (Ungulate lice), Polyplacidae (spiny rat lice), Hoplopleuridae (armoured lice), Linognathidae (pale lice), Pediculidae (primate body lice) and Pedicinidae. The chewing lice of ths region are represented by three families, under two sub-orders as: Menoponidae (bird/chicken lice) (Sub-order Amblycera); Trichodectidae (mammal chewing lice) and Philopteridae (sub -order Ischnocera).

A total of 32 spp., distributed over 9 genera of anopleurans, belonging to various families, is known from the UTs. The families of anopleurans, with total number in each family, are :Haematopinidae (2 spp.); Polyplacidae (7 spp.); Hoplopleuridae (9 spp.); Linognathidae (5 spp.); Pediculidae (5 spp.); Pedicinidae (4 spp.). Chewing lice (Amblycera and Ischnocera), include a total of 29 spp., belonging to 16 genera, under different families. The families, with total number of species each family, are : Menoponidae (5 spp.); Philopteridae (19 spp.) and Trichodecidae (5 spp.). For the preparation of updated checklist (Table 7.4 on lice (chewing and sucking), with their hosts and distribution in Jammu, Kashmir and Ladakh, Indian and the world Checklists on: chewing lice (Lakhshminarayana, 1979; Price *et al.*, 2003); sucking lice (Adhikary and Ghosh, 1994; Durden and Musser, 1994), have been consulted.

7.2.1 Systematic Survey, Diversity and Host Range of Louse-Fauna (Sucking and chewing)

The earliest information about the blood-sucking lice from J&K Himalayan region is available through the papers published by: Kalra and Rao (1949); Ferris (1951); Kalra (1952, 59); Narsimhan (1962); Das and Sharma (1970) and Mishra *et al.* (1973). Rao *et al.* (1973) gave an exhaustive survey report of anoplurans, found in the J&K Himalayan regions. They listed a total of 19 species of lice from various Districts of J&K State. These included 8 species from Ladakh, 11 species from Jammu and 9 species from Kashmir Valley, parasitizing hosts like rats, mice, goats, shrew, sheep, humans, cattle, buffaloes, dogs and monkeys.

Three species of lice, parasitic on rodents, mammals and buffaloes, occurring in the pastures of Gulmarg and Tangmarg localities of Kashmir Valley, have been reported by Shrivastava and Wattal (1981). Mishra (1981) published a monograph on the hoplopleurid lice of Indian sub-continent, therein he listed about 7 species of hoplopleurid lice, parasitizing mammal hosts, belonging to order Insectivora, Logomorpha and Rodentia. An updated checklist (Table 7.4) of lice (chewing and sucking), parasitic on birds and mammals of Jammu, Kashmir and Ladakh region, indicates that a total of 59 species of lice (Phthirapterans) under 23 genera are known to exist in this region. The sucking lice (anopleurans) included a total of 32 spp., under 7 genera and six families,whereas chewing lice incorporated a total of 27 spp., under 15 genera of 3 families and two sub-orders.

The number of species, under various genera, families of sub-order Anopleura (sucking lice) are:2spp. (genus *Haematopinus*), Haematopinidae; 7spp each (*Neohaematophinus*, *Polyplax*), Polyplacidae; 9 spp. (*Hoplopleura*), Hoplopleuridae; 5 spp. (*Linognathus*), Lingognathidae; 2 spp. (*Pediculus*),1sp. (*Pthirus*), Pediculidae and 4 spp. (*Pedicinus*), Pedicinidae. The number of species, under various genera of family Menoponidae of sub-order Amblycera (chewing lice) are: 1 sp. (*Actornithophilus*); 2 spp. (*Menocanthus*); 1 sp. (*Menopon*); 1 sp. (*Myrsidea*). The number species of various genera and families of sub-order Ichnocera (chewing lice) are: 5 spp. (*Brueelia*), 1 sp. (*Cuclotogaster*), 1sp. (*Degeeriella*), 6 spp. (*Goniodes*), 1 sp. (*Goniocotes*), 1 sp. (*Lipeurus*), 2 spp. (*Philopterus*), 1 sp. (*Reticulipeurus*), belonging to family Philopteridae; 2 spp. (*Bovicola*), 1 sp. (*Damalinia*), 1 sp. (*Suricatoecus*), 1 sp. (*Syrrhaptoecus*), pertaining to family Trichodectidae.

The dominant number of sucking lice species is as 16, found to be parasitizing rodents, small and burrowing mammals in different areas and localities of Jammu and Kashmir State. This is followed by Primates- *Presbytis* (monkeys) and *Macaca* (macaque), infested with 6 spp. The sheep and goat infested with 5 spp., humans with 2 species and 4 sub-species of *Pediculus humanus*, cattle with 3 spp.and, dog, camel and poultry birds with 1 sp. each. The highest number of chewing lice species is found in wild bird species. This is followed by poultry/domestic fowl and wild mammals, having 7 spp. and 4 spp. respectively (Table 7.3).

Table 7.3: Species Richness and Host Diversity of Louse-Fauna of Vertebrates (Birds, Mammals, including humans) of J&K, Ladakh (UTs)

Sl.No.	*Host-type*	*Sucking-Lice (No. of Species)*	*Chewing-Lice (No. of Species)*
1	Small and burrowing mammals	16	—
2	Sheep and Goat	05	—
3	Cattle	03	—
4	Monkeys and Macaque	06	—
5	Wild birds	—	16
6	Wildlife (mammals)	—	4
7	Humans	02 (4)*	—
8	Camel	01	—
9	Dogs	01	—
11	Equines	—	01
12	Poultry (domestic fowl)	01	07

* Indicates 4 sub-species of *Pediculus humanus.*

7.2.2 Lice of Veterinary Importance

(a) Lice Affecting Poultry and Domestic Animals

The sucking lice of veterinary importance are *Haematopinus eurysternus* and *H. tuberculatus*, affecting ducks, poultry, goat, cattle, usually causing skin irritation, allergic reactions, and general feeling of malaise. Besides these, the sucking lice of cattle in Jammu, Kashmir and Ladakh region are *Linognathus* spp. (5). These infest sheep, goat, cattle and dogs. These sucking lice species cause stress and anemia, leading to loss of production. Chewing lice of domestic animals do not usually spread disease pathogens, however, heavy infestation can cause skin irritation, weight loss and reduced egg production. The chewing lice prevalent in Kashmir and Ladakh region are : The Chicken head-louse (*Cuclotogaster heterrographus*), found on the head and neck region of the host; fluff-louse (*Goniocotes gallinae*), remain attach to down features and causes little irritation; *G. gigas*, largest louse on chicken, occur any where on the body and *Lipeurus caponis* occurs on the underside of the wing and tail feathers of domestic fowl (indigenous).

Table 7.4: Lice (Sucking and chewing) (Phthiraptera): Checklist and Vertebrates Host Diversity (Birds, Mammals, including humans) of Jammu, Kashmir and, Ladakh (UTs)

Scientific Name of Louse	*Host/s*	*Locality/ Localities*
**Haematopinus eurysternus*	*Bos indicus* Duck, hens, cocks, goat, cattle, musk shrew	Kmr. (Srg.)
Haematopinus tuberculatus	Cattle (calf), *Bubalus bubalis, Camelus dromedus*	Gul., Udh., Raj, Poo., Kat., Akh
**Neohaematopinus palearctus*	*Marmota caudata*	Ldk
Neohaematopinus petaurista	*Petaurista petaurista* albiventer (Asian giant, flying squard)	J&K
**Polyplax indica*	*Glunda ellioti*	Dod., Udh., Raj
Polyplax reclinata	Shrew, Mice – *Crocidura* sp., *Suncus murinus*	Dod., Bar., Ach.
Polyplax serrata	Mouse, *A. roylei*	Ldk., Bar
Polyplax stephensi	Rat	Udh., Raj.
Polyplax spinulosa	Rat, *M. musculus, Bendicot bengalensis*	Gul., Dod., Udh.
**Hoplopleura acanthopus*	*Mus musculus*	Gul., Tan
Hoplopleura captiosa	Rat, Shrew, *Mus* sp., *Mus booduga, Apodema flavicollis*	Ldk., Sgr., Ana., Udh., Bar, Kar.
Hoplopleura himalyana	*A.flavicolis*	J&K State (earstwhile)
Hoplopleura kondana	*Milardia mettada*	J&K (earstwhile)
Hoplopleura phaiomydis	*Microtus, Phaiomydis* sp.	Ldk.
Hoplopleura pacifica	Rat, Mice, *Suncus murinus tytleri*	Sgr., Ant., Ldk.Bar., Udh.
Hoplopleura ramgarh	*Mus.* Sp.	Udh.
Hoplopleura. sicata	Rodents, *Rattus fulvescens*	Agr., Ana., Ldk.Bar., J&K
Hoplopleura sp.	Rat	Udh.
**Linognathus apicanus*	Goat	Ldk., Bar., Udh.
Linognathus ovillus	Sheep, Goat	Sgr.

Scientific Name of Louse	Host/s	Locality/ Localities
**Linognathus setosus*	Dog	Dod., Udh. Poo, Raj.
Linognathus stenopsis	Sheep, Goat	Ldk.
Linognathus vituli	Cattle	Ant., Ldk., Bar., Khi.
**Pediculus humanus humanus*	Goat, Human	Jmu, Kmr., Ldk
Pediculus humanus capitis	Human	Jmu, Kmr, Ldk
Pediculus humanus corporis	Goat, Man, Monkey	Sgr.
Pediculus sp.	Monkey	Udh.
**Pthirus pubis*	Human	Jmu, Kmr, Ldk
**Pedicinus ancoartus*	*Presbytis*	Kmr.
Pedicinus (P.) eurygaster	*Macaca mulatta*	Kmr.
Pedicinus obtusus	*Macaca*	Kmr.
Pedicinus sp.	*Presbytis entellus*	Poo.
Actornithophilus grandiceps	*Ibidorhyncha struthersii*	Kmr.
Menocanthus cornutus	Domestic chicken	Kmr
Menocanthus stamineus (=Eomenacanthus stramineus)	Domestic chicken	Kmr.
Menopon gallinae	Domestic chicken	Kmr.
Myrsidea cornicis	*Corvus cornix*	Jmu, Kmr, Ldk
Brueelia bicellata	*Pica pica bacterian*	Ldk
Brueelia husaini	*Kitta flavirostris cuculatta*	Ldk
Brueelia multipunctate	*Nucifraga caryocatacte multipunctate*	Ldk
Brueelia myiophoneae	*Myiophoneus coeruleus temminckii*	Ldk
Brueelia varia	*Corvus corax*	Ldk
Campanulotes bidentus	*Columba livia*	Jmu
Cuclotogaster heterographus	Chicken	Ldk
Degeeriella rufa	*Falco tinnunculus*	Kmr.
Goniodes costatus	*Tetraogallus- himalayensis*	Kmr.
Goniodes dispar	*Alectoris graeca chukar*	Kmr.
Goniodes gigas (= Goniocotes gigas)	*Tetraogallus*	Kmr.
Goniodes intermedius	*Pucrasia macrolpha- biddulphi*	Kmr.
Goniodes ocellatus	*Liphura leucomelana hameltonii*	Kmr.
Goniodes tragopan	*Tragopan- melanocephalus*	Kmr.
Gonicotes gallinae	*Tetraogallus*	Kmr.
Lipeurus caponis	Gallinaceous Birds/Chicken	Ldk
Philopterus atratus	*Corvus corax*	Kmr.
Philopterus garruli	*Kitta flavirostris- flavirostris, Pica*	Ldk
Reticulipeurus himalayensis	*Tragopan himalayaensis*	Ldk
Bovicola fulva	*Hemitragus jemlaticus*	Kmr.

Scientific Name of Louse	*Host/s*	*Locality/ Localities*
Bovicola thompsoni	*Capricornis sumatrensis*	Kmr.
Damalinia equi	Equines	Kmr.
Suricatoecus vulpes	*Vulpes vulpes*	Kmr.
Syrrhaptoecus tibetanus	*Syrrhaptes tibetanus*	Ldk

Ach: Achabal; Ana- Anantnag; Akh: Akhnoor; Bar: Baramulla; Dod: Doda; Gul: Gulmarg; Jmu: Jammu; J&K: Jammu and Kashmir State; Kar: Kargil; Kat: Kathua; Khi: Khilanmarg; Kmr: Kashmir; Ldk: Ladakh; Poo: Poonch; Raj: Rajouri; Srg: Srinagar; Tan: Tangmarg; Udh: Udhampur; *: Sucking lice (genera); Rest listed genera as Chewing lice.

Source: Narsimhan (1962); Das and Sharma (1970); Mishra *et al.* (1973); Lakhshminarayan (1979); Mehmood *et al.* (2019); Mir *et al.* (1993); Sharma (1993a); Price and Graham (1997); Adhikary and Ghosh (1994); Das *et al.* (2000); Saleem *et al.* (2009); Tramboo *et al.* (2019). website: www. Phthiraptera.info.

Mir *et al.* (1993) studied the prevalence of chewing-lice as ectoparasites of indigenous fowl in Kashmir and recorded 1 sp. under family Menoponidae and 4 spp. belonging to family Philopteridae. The prevalence and seasonal variations of Philoperids (3 spp.) and Menoponids (3 spp.) the free range chicken of Kashmir showed *Lipeurus capornis* as most prevalent parasite throughout the year (Saleem *et al.*, 2009). These authors also reported that the parasite load was higher during summer for *L. capornis, Goniodes gigas* and *Menocanthus cornutus,* and during authumn for other parasites, *viz. Goniocotes gallinae, Menocanthus stramnius* (= *Eomenacanthus stramineus*) and *Menopan gallinae*. Prevalence of chewing lice, *viz. M. gallinae, M. stramineus, Goniocotes gallinae, Goniodes gigas* and *Lipeurus caponis,* infesting backyard poultry in Jammu region, showed *M. gallinae* as predominant in all the seasons, excepting during pre-monsoon season, whereas *M. stramineus* had highest prevalence during monsoon (Bhat *et al.*, 2014). Recently, prevalence of trichodectid, *Damalinia equi,* as ectoparasite of equines of Kashmir valley, has been reported (Tramboo *et al.*, 2019).

7.2.3 Lice of Medical Importance

The medically important family of this region is Pediculidae, having genus *Pediculus* and *Phthirus,* parasitic on humans and monkeys. The disease caused by *Pediculus humanus* is infection, known as pediculosis, responsible for causing victim to become highly sensitive, resulting in tissue reactions that is extremely irritating. Sometimes due to louse bites over a long period, the human skin often becomes deeply pigmented. The condition is known as "vagabond's disease". The irritability is not found in the patients suffering from this disease and victim shows immunity. The human body – louse (*P. humanus*) is responsible for transmitting fatal disease of humans known as epidemic typhus, caused by a rickettsian (*Rickettsia prowazekii*). Besides this, louse-borne human diseases include trench fever (five-day fever) caused by rickettsia, *Rochalimaea quintona* and epidemic relapsing fever, caused by a spirochaete, *Borrelia recurrentis.*

The earliest outbreak of typhus in earstwhile J&K State was in the year 1894, which followed the next in 1903 and from 1916 onwards the reports were more frequent. The major outbreak was in the year 1903 and extended to 1944, resulting in 2,481 cases and 730 deaths. This outbreak covered whole of Baramula district and also northern part of Anantnag, including Srinagar. Besides Kashmir Valley, outbreaks of typhus disease frequently appeared in hilly tracts of Jammu and Ladakh at refugee camps. The causative agent of this diseases was a rickettsian. The epidemic typhus disease showed high fever, eruptive rashes, purplish spotting of the skin, severe nervous and gastro-intestinal symptoms, occurrence of microscopic nodules around arterioles and great reduction of leucocytes. The transmission of typhus in man results due to contamination of bites or scratches by faeces or from inhalation of dried faeces of lice.

The first documented report, pertaining to typhus outbreak in Banihal (Kashmir) has been provided by Kaul (1944). Later, during the winter of the year 1947-48, there were two outbreaks of typhus, one started in December, 1947 in refugee camp at Baramulla and Srinagar, and other in March, 1948 at village Brari Pathri (Budgam). The symptoms of epidemic typhus of limited outbreaks were of typical epidemic typhus. The cause was confirmed to contact and the strains isolated from patients and lice, were similar to *Rictettsia prowazekii* (Kalra and Rao, 1949).

The spread of epidemic typhus in earlier days in earstwhile J&K State, was prevented by de-lousing. However, in the year 1958 and 1960, Srinagar area was surveyed, with the aim to determine the insecticide susceptibility status of the "Hariganiwan strain", "Wasan-Anderwan" and "Malaru" – local strains of the body louse (*Pediculus humanus corporis*). It was Narsimhan (1962) who determined the status of insecticide susceptibility of local strains of body-louse. He found that with the withdrawal of control measures at Hariganiwan, the local strain reverted to normal susceptibility to DDT and BHC and retained its susceptibility to an indigenous make of insecticide, *viz.* Pyrethrin. It was also observed by him that the strains at Wosan-Anderwan and Malaru, showed susceptibility to all the three insecticides.

Descriptions of an epidemic of typhus, which occurred for the first time in Ladakh in July 1961, was given by Patheja (1963). It was confirmed in Tibetan refugees at Leh. The infection was brought in by refugees, when they migrated to Leh in 1960 from eastern Tibet. It was also established on the basis of available data that the epidemic was louse-borne.

7.3 Parasitic Bugs

A rare bug (Hemiptera),*viz. Psammotettix striatus* (Linn.) has been found to parasitize the high-population of flying fox (bat) (Chiroptera), *Pterops,* which were hanging on the trees: *Plantinus orientalis* (Chinar), *Cedrella toona* (toon) and *Eucalyptus lanceolatus* (eucalyptus) in Poonch Valley, Jammu (Sharma,1973).

In earstwhile J&K State, two species of bed bugs, belonging to the family Cimicidae (Hemiptera), *viz. Cimex hemipterus* and *C. lectularius,* have been reported. Bed bugs are cosmopolitan, nocturnal, usually found in large numbers and are known to suck blood of animals, poultry (chicken) and man.They can be extremely annoying, persons who are victim of biting of bed bugs, suffer loss of sleep, sores from infected bites and iron/haemoglobin deficiency. Rao (1973), reported the existence of *C. hemipterus* and *C. lectularius* from Ladakh region, *C. hemipterus* from Jammu (Udhampur and Rajouri), from indoor places. The prevalence of bed bug, *C. lectularius* as an ectoparasite of indigenous fowls in Kashmir Valley, has been reported by Mir *et al.* (1992).

Chapter 8

Other Parasitic Insects: Mosquitoes and Flies (Diptera)

8.1 Mosquitoes

Mosquitoes belong to insect Order Diptera, Sub-order Nematocera, Infra-order Culicimorpha and Family Culicidae. The adult mosquitoes are parasitic, whereas the larval stages (immature stages) are aquatic. The adult female mosquitoes are blood-sucking parasites. The blood meal is required by many species of mosquito for their eggs maturation or to increase the ratio of egg production. The adult male mosquitoes feed on vegetation/plant material. The adult mosquitoes are elongated, long-legged, size ranges from 3-6 mm, having long forwardly directed proboscis, equal in lengths to head, thorax combined, wings show scales along the veins and forewing fringe, and second vein not arched but running parallel to vein I and II. The mosquitoes have medical and veterinary importance becacuse females not only annoy human and other animals but also act as potential biological vectors of important human, domestic animals and poultry diseases. The pathogens transmitted by mosquitoes include viruses (arboviruses), filarial worms (helminths) and protozoans.

In Jammu, Kashmir, and Ladakh UTs, a total of 65 species, under 14 genera and 17 sub-genera are known to occur, in the vast and diverse localities, and water bodies of these regions. These mosquitoe species are belonging to the subfamily Anophelinae and Culicinae. In this region, the mosquito-borne diseases like malaria and dengue are posing threat and are diseases of great public health importance. Malaria disease is transmitted by Anophelines(*Anopheles* spp.) and dengues by Culicines (*Stegomyia* spp.). The adults of living Anophelinae and Culicinae can be recognized by the posture they attain when resting on a surface. The adult *Anopheles* rest with proboscis, head, thorax and abdomen in one straight line, making an acute angle with the surface. The Culicine adults rest with their body angled and abdomen directed backwards the surface on which it is resting. *Aedes* and *Anopheles* can also be recognized by colour and spots. The colour of *Stegomyia* (= *Aedes*) mosquito is greyish black, with white stripes. In case of *Anopheles,* the wings have dark spots. The aquatic larvae of mosquito, *viz. Lutzia fuscana* has predatory potential on *Stegomya,* having importance for biological control of dengue vector, through the augmentation and release, particularly in smaller mosquito larval habitat (Pramanik *et al.*, 2016).

8.1.1 Systematic Survey, Diversity and Species Richness

The earliest studies on mosquito-fauna of earstwhile J&K State, including Ladakh region, have been done by: Gill (1920), Barraud (1924 a,1924 b,1924 c, 1934), Christopher (1933), Covell (1927,1931), Puri (1948) and James and Liston (1904). Bhat and Kulkarni (1971), for the first time reported the occurrence of *Culex* (*Neoculex*) *hortensis* from Ladakh region. Rao *et al.* (1973), during the course of their extensive surveys, conducted between the year 1966-70, in various districts and provinces of this State, established the existence of about 30 Species of mosquito. They showed the district-wise break of various species as: Doda, 5 spp.; Udhampur, 17 spp.; Rajouri, 3 spp.; Srinagar and Anantnag, 3 spp. each; Baramulla, 6 spp. and Ladakh (Province), 9 spp. Bhat and Kulkarni (1983) gave the records of 27 spp. of mosquito, collected from various regions and localities of Jammu, Kashmir and Ladakh, with ecological notes. These included 7 species, belonging to genus *Anopheles,* 19 spp., pertaining to genus *Culex* and 1 sp. of the genus *Culiseta.*

In the recent years, the research papers by various authors, *viz.* Nair (1973), Rao (1981), Buhroo and Chishti (1999), Kaur (1992, 2003) and Jabansen *et al.* (2012), have added to the already recorded mosquito species from Jammu, Kashmir regions. A detailed monitoring of mosquito-fauna of Jammu Province, covering District Udhampur and Jammu, has been accomplished by Dar *et al.* (2016). They have documented a total of 47 mosquito species, with new records, belonging to subfamily Anophelinae and Culicinae, surveyed in diverse localities and fifteen larval habitats in two Districts of Jammu. Now, a total of 32 Species of mosquito are known to occur in various localities of Jammu region.

Till -to- date, a total of 65 spp. of mosquito, belonging to 14 genera and 17 subgenera, is known to occur in diverse areas, localities and habitats of Jammu, Kashmir and Ladakh Himalyan regions. These species pertain to two main subfamilies of family Culicidae, *viz.* Anophelinae and Culicinae (tribe, Aedini, Culicini, Culisetini). The Jammu Divisiuon has shown the prevalece highest number of mosquito species *i.e.* 48 (15 spp.of anophelines, 33 spp. of culicines). This is followed by Kashmir region, having occurrence of of a total of 29 spp. (18 anophelines, 11 culicines). The Ladakh region, was found to be showing the existence of 10 spp. of culicines.

8.1.2 Mosquitoes of Medical Importance

(a) Vectors of Human Malaria

Six species of mosquitoes belonging to genus *Anopheles, viz., An. culicifacies, An. dirus, An. fluviatilis, An. minimus, An. sundaicus, An. stephensi,* are major human malaria vectors in India (Dash *et al.*, 2007; Das and Sharma, 2013). The malaria vector mosquito species reported from Jammu and Kashmir regions include *An. annularis,*

An. culicifacies, *An. fluviatilis*, *An. maculatus* and *An. stephensi*. Two important vectors of Dengue/Dengue haemorrhagic fever, *viz*. *Stegomyia* (*S.*) *aegypti* (= *Aedes aegypti*) and *St. albopicta*, also as invasive species, are known to be prevalent in many areas of Jammu and Kashmir (Buhroo and Chishti,1999; Gookhale and Mourya, 2003; Jabanesan *et al.*, 2012), however, absent from Ladakh. The prevalence of *St. albotica* was shown by Kalra *et al.* (1997) as scanty in Kashmir (South) and widely prevalent in north Kashmir (Jabanesan *et al.*, 2012). Malaria disease is a life threatening disease, transmitted through the bite by *Anopheles* spp., caused by protozoan parasites, *Plasmodium* spp. (*P. falciparum*, *P. malariae*, *P. vivax*) in Jammu and Kashmir. The details on signs and symptoms, and epidemiology of different types of human malaria in J&K (UT), are given in Chapter 1.4.5, pp. 15-17.

Dengue is also mosquito-borne, caused by dengue virus and dengue fever can lead to life-theating dengue haemorrhagic fever or leading to dengue shock syndrome. Dengue is posing threat to public health of Jammu region and the disease is endemic to this region, occurring annually. The outbreaks of dengue/dengue haemorrhagic fever have been reported from Jammu (Pedbidri *et al.*, 1996; Cecilla *et al.*, 2003; Sidhu *et al.*, 2015). There is no case of dengue reported from Kashmir region, depite the occurrence of mosquito vectors in the region. From Jammu region, on the basis of laboratory records of clinically suspected Dengue patients from the year 2011-2015, the analysis made retrospectively for the results of Ig M anti-dengue antibodies tested by Dengue Ig M capture (ELISA), 31.78 per cent (1891/5950) samples were serologically positive for dengue infection (Sudan *et al.*, 2016).

8.1.3 Systematic Checklist and Distribution of Mosquitoes of Jammu, Kashmir and Ladakh

Order : Diptera

Family : Culicidae

Subfamily 1: Anophelinae

Genus: *Anopheles*

Subgenus: *Anopheles*

1. *Anopheles* (*Anopheles*) *barianensis* James [11]
   (= *Anopheles barianensis* James) [8, 9, 10]
   Distribution : K (Nas, Sri)
2. *Anopheles* (*Anopheles*) *barberi* Coquillett [9]
   Distribution : K (Gul, Hok, Kha)
3. *Anopheles* (*A.*) *gigas* Giles [5, 8, 9,11, 18]
   (= *Anopheles* (*Anopheles*) *gigas simlensis*) [5, 15]
   (= *Anopheles simelensis* Giles) Gill (10)
   Distribution: K (Bar, Gul)
4. *Anopheles* (*A.*) *hyrcanus* (Pallas) [18]
   Distribution : K (Karn)

5. *Anopheles* (*A.*) *lindesayi* Giles [11, 19]
   (= *Anopheles lindesayi* Giles) [5, 9, 10, 14]
   Distribution : K (Gul), J (Jmu, Udh)
6. *Anopheles* (*Anopheles*) *maculipalpis* Giles [11]
   (= *Anopheles maculipalpis* Giles) [9, 10]
   Distribution : K
7. *Anopheles* (*Anopleles*) *peditaeniatus* Leicester[19]
   Distribution : J (Jmu, Udh)
8. *Anopheles* (*Anopheles*) *plumbeus* Stephens [11]
   (= *Anopheles plumbeus* Stephens) [10]
   Distribution: K (Nas, Sri)
9. *Anopheles* (Anopheles) sinensis Wiedemann [19]
   Distribution: J (Jmu, Udh)

Subgenus: *Cellia*

10. *Anopheles* (*Cellia*) *annularis* vander Wulp] [11, 19]
    (= *Anopheles annularis* vander Wulp) [5]
    Distribution : J&K, J (Jmu, Udh)
11. *Anopheles* (*C.*) *culicifacies* Giles [5, 9,18,19]
    Distribution : J (Jmu,Udh), K (Karn)
12. *Anopheles* (*C.*) *dthali* Patton [13, 19]
    Distribution : J (Jmu, Udh), K
13. *Anopheles* (*C.*) *dravidicus* Christophers [19]
    Distribution : J (Jmu, Udh)
14. *Anopheles* (*C.*) *fluviatilis* James [5, 17, 18, 19]
    Distribution : J (Jmu, Udh), K (Han, Karn, Uri)
15. *Anopheles* (*C.*) *karwari* (James) [11]
    (= *Anopheles karwari* James) [14]
    (= *Anopheles* (*Myzomyia*) *karwari* James (7)
    Distribution : K
16. *Anopheles* (*C.*) *leucosphyrus* Donitz [17, 18]
    Distribution : K (Karn)
17. *Anopheles* (*C.*) *maculatus* Theobald [11, 19]
    (=*Anopheles maculatus willmorei* James [7]
    (= *Anopheles maculatus* James) [5, 9]
    Distribution : J (Jmu, Udh), K (Dal)

18. *Anopheles* (*C.*) *pseudowillmori* [19]
    Distribution : J (Jmu, Udh)
19. *Anopheles* (*C.*) *moghulensis* Christopher [13]
    Distribution : K (Gan)
20. *Anopheles* (*C.*) *splendidus* Koidzumi [17, 19]
    Distribution : J (Jmu, Udh), K (Han)
21. *Anopheles* (*C.*) *stephensi* Liston [11, 19]
    (=*Anopheles stephensi* Liston) [9, 10, 12]
    Distribution : K (Gul, Hok, Kha), J(Jmu, Udh)
22. *Anopheles* (*C.*) *subpictus* Grassi [18, 19]
    Distribution : J (Jmu, Udh)
23. *Anopheles* (*C.*) *turkhudi* Liston [19]
    Distribution : J (Jmu, Udh)
24. *Anopheles* (*C.*) *willmori* (James) [11, 19]
    (= *Anopheles* (*Myzomyia*) *wilmori* James) [8]
    (= *Anopheles wilmori* James) [10]
    Distribution :J (Jmu, Udh), K (Dal, Gul, Sin, Sri)

Subfamily 2 : Culicinae

Tribe 1.: Aedini

Genus: *Aedimorphus*

25. *Aedimorphus vexans* (Meigen) [11]
    (= *Aedes* (*Aedimorphus*) *vexans* (Meigen) [4, 5, 16]
    Distribution : L (Kar, Leh)

Genus : *Armigeres*

Subgenus: *Armigeres*

26. *Armigeres* (*A.*) *subalbatus* (Coquillett) [19]
    Distribution : J (Jmu, Udh)

Genus : *Collessius*

Subgenus : *Alloeomyia*

27. *Collessius* (*A.*) *pseudotaeniatus* (Giles) [19]
    Distribution : J (Jmu, Udh)

Subgenus : *Collessius*

28. *Collessius* (*C.*) *elsiae* (Barraud) [11, 19]
    (= *Aedes* (*Finlaya*) *elsiae* (Barraud) [5, 18]
    Distribution : J (Jmu, Raj, Udh)

Genus: *Fredwardsius*

29. *Fredwardsius vittatus* (Bignot) [11, 19]
    (= *Aedes (Fredwardsius) vittatus* (Bignot) [16]
    (= *Aedes (Stegomyia) vittatus* (Bignot) [18]
    Distribution : J (Raj, Udh)

Genus : *Gilesius*

30. *Gilesius pulchriventer* Gilet[19]
    Distribution : J (Jmu, Udh)

Genus : *Neomelaniconion*

31. *Neomelaniconion lineatopenne* (Ludlow) [19]
    Distribution : J (Jmu, Udh)

Genus : *Ochlerotatus*

Subgenus : *Culicelsa*

32. *Ochlerotatus* (*C.*) *taeniorhynchus* (Wiedemann) [11]
    (= *Culex taeniorhynchus* Wiedemann) [12]
    Distribution : K (Gul, Hok, Kha)

Subgenus : *Finlaya*

33. *Ochlerotatus* (*F.*) *oreophilus* Edwards [11]
    (= *Aedes* (*F.*) *oreophilus* Edwards) [4, 5, 16]
    Distribution : L (Kar, Leh)
34. *Ochlerotatus* (*F.*) *sintoni* (Barraud) [11]
    (= *Finlaya sintoni* Barraud) [1]
    (= *Aedes* (*F.*) *sintoni* (Barraud) [5, 15]
    Distribution : K (Tan), L
35. *Ochlerotatus* (*F.*) *versicolor* (Barraud) [11, 16]
    (= *Aedes (Ochlerotatus) versicolor* Barraud) [2]
    (= *Finlaya versicolor* Barraud) [18]
    Distribution : K (Yus)

Genus : *Stegomyia*

Subgenus : *Stegomyia*

36. *Stegomyia* (*S.*) *aegypti* (Linnaeus) [11, 19]
    (= *Aedes aegypti* (Linnaeus) [6, 12]
    Distribution : J (Jmu, Udh), K (Gul, Hok, Kha)

37. *Stegomyia* (subgenus uncertain) *albopicta* (Skuss) [11, 19]
(=*Aedes albopictus* (Skuss) [12]
(= *Aedes* (*Stegomyia*) *albopictus* (Skuss) [5, 18]
Distribution : J (Jmu,Udh), K (Gul, Hok, Kha), L

38. *Stegomyia* (uncertain subgenus) *w-alba* Theobald [11, 19]
(= *Aedes* (*S.*) *w-albus* (Theobld) [5, 15, 17]
Distribution : J (Jmu, Udh)

Tribe 2: *Culicini*

Genus : *Culex*

Subgenus : *Barraudius*

39. *Culex* (*Barraudius*) *modestus* Fiscalbi [11]
(= *Culex eadithae* Barraud) [3]
Distribution : K (Sop, Wul)

Subgenus : *Culex*

40. *Culex* (C.) *barraudi* Edwards [5, 18,19]
Distribution : J (Jmu, Udh)

41. *Culex* (C.) *fuscocephala* Theobald [5, 18, 19]
Distribution : J (Jmu, Udh)

42. *Culex* (C.) *mimeticus* Noe [5, 18, 19]
Distribution : J (Dod, Jmu, Udh), K (Sri), L

43. *Culex* (C.) *mimulus* Edwards [19]
Distribution : J (Jmu, Udh)

44. *Culex* (C.) *murrelli* Lien [19]
Distribution : J (Jmu, Udh)

45. *Culex* (C.) *perexiguus* Theobald [19]
Distribution : J (Jmu, Udh)

46. *Culex* (C.) *perplexus* Leicester[19]
Distribution : J (Jmu, Udh)

47. *Culex* (C.) *pipiens* Linnaeus [11]
(= *Culex pipiens* Linnaeus) [6]
Distribution : J&K

48. *Culex* (C.) *pseudovishnui* Colless [18, 19]
Distribution :J (Jmu, Udh), L

49. *Culex* (C.) *quinquefasciatus* Say [5, 18, 19]
(= *Culex fatigans*) [12]
Distribution : J (Jmu, Udh), K (Ban, Bar, Gul, Hok, Kha)

50. *Culex* (C.) *theileri* Theobald [4, 5, 18, 19]
Distribution : J (Jmu, Udh), L (Kar, Leh)

51. *Culex* (C.) *tritaeniorhynchus* Giles[19]
Distribution : J (Jmu, Udh)

52. *Culex* (C.) *vagans* Wiedermann [4, 5, 18, 19]
Distribution : J (Jmu, Udh), K (Ana, Bar), L (Kar, Leh)

53. *Culex* (C.) *vishnui* Theobald [5, 18,19]
Distribution : J(Jmu. Udh)

Subgenus : *Culiciomyia*

54. *Culex* (C.) *pallidothorax* Theobald [5, 18,19]
Distribution : J, (Jmu, Udh)

55. *Culex* (C.) *viridiventer* Giles [5, 18, 19]
Distribution : J (Jmu, Udh), K (Ana, Pah)

Subgenus : *Eumelanomyia*

56. *Culex* (*E.*) *brevipalpis* (Giles) [19]
Distribution : J (Jmu, Udh)

57. *Culex* (*E.*) *malayi* (Leicester) [19]
Distribution : J (Jmu, Udh)

Subgenus : *Lophoceromyia*

58. *Culex* (*L.*) *minutissimus* (Theobald) [19]
Distribution : J (Jmu, Udh)

Subgenus : *Oculeomyia*

59. *Culex* (*O.*) *bitaeniorhynchus* Giles [11, 19]
(= *Culex* (*Culex*) *bitaeniorhynchus* Giles) [5,18]
Distribution : J (Jmu, Udh)

Subgenus : *Maillotia*

60. *Culex* (*M.*) *hortensis* Ficalbi [11]
(= *Culex* (*Neoculex*) *hortensis* Facalbi) [4, 5]
Distribution: L (Kar, Leh)

Subtribe : *Culicina*

Genus: *Lutzia*

Subgenus: *Metalutzia*

61. *Lutzia (M.) fuscana* (Wiedemann[19]

Distribution: J (Jmu, Udh)

62. *Lutzia (M.) halifaxii* [19]

Distribution : J (Jmu, Udh)

Tribe 3 : *Culisetini*

Genus: *Culiseta*

Subgenus : *Allotheobaldia*

63. *Culiseta (A.) longiareolata* (Macquart) [4, 18]

Distribution : L (Kar, Leh)

Tribe 4: Sabethini

Genus : *Malaya*

64. *Malaya genurostris* Leicestes [19]

Distribution: J (Jmu,Udh)

Genus : *Toxorhynchites*

65. *Toxorhynchites (Toxorhynchites) splendens* (Wiedmann) [19]

Distribution : J (Jmu, Udh)

Source: The above given Catalogue is modified after Bhagat (2014). For the purpose of updation *of nomenclatural and systematic changes of mosquito species, online databases as well as catalogues* on mosquitoes of the world, given by Knight and Stone (1977), Gaffigan *et al.* (2013) and Harbach (2014), Website, mosquito-taxonomic Inventory. Info., have been consulted.

Keys to Author (s) and Distribution in Table 8.1.3

Key to the author (s), given as numerals in long brackets in the above given systematic Catalogue

1: Barraud (1924a); 2: Barraud (1924b); 3: Barraud (1924 c); 4: Bhat and Kulkarni (1971); 5: Bhat and Kulkarni (1983:); 6: Buhroo *et al.* (1997); 7: Christopher (1933; 8: Covell (1927); 9: Covell (1931); 10 : Gill (1920); 11: Harbach (2014); 12: Jabanesan *et al.* (2012); 13: Jacob (1950); 14: James and Liston (1904); 15: Kaur (1992); 16: Kaur (2003); 17: Nair (1973); 18 : Rao *et al.* (1973); 19: Dar *et al.* (2016).

Key to the localities of mosquito species, distributed in Jammu, Kashmir and Ladakh region, given in the form of abbreviations in parentheses in above Checklist :

Ana: Anantnag; Ban: Bandipore; Bar: Baramula; Dal: Dal Lake Dod: Doda; Gan: Ganderbal; Gul: Gulmarg; Han: Handwara; Hok: Hokersar; J: Jammu (Jammu Province); Ja: (Jammu District) J&K: Jammu and Kashmir; Jeh: Jehlum River; K:

Kashmir; Kar: Kargil; Karn: Karnah; Kha: Khag; L: Ladakh; Leh: Leh; Nas: Naeem Bagh; Pah: Pahalgam; Raj: Rajouri; Sop: Sopore; Sin: Sind Valley; Sri: Srinagar; Tan: Tangmarg; Udh: Udhampur; Ur: Uri; Wul.: Wular Lake; Yus: Yusmarg.

8.2 Sandflies

Sandflies belong to order Diptera of class Insecta, family Psychodidae, a well defined group of sandflies species, are sometimes accorded family rank as the Phlebotomidae, however, more often regarded as sub-family Phlebotominae. Sandflies resemble tiny moth because of their hairy bodies and moth like appearance. Sandflies are small in size and usually yellowish or buff coloured. The female sandfly is haematophagous and nocturnal in habit, whereas the males are non-parasitic, feeding on moisture. Jacob and Kalra (1951) and Mitra and Roy (1953), were first research workers to detect few cases of visceral as well as cutaneous leishmaniasis in south-eastern part of Kashmir. Sandflies are vector of various types of leishmaniasis (visceral and cutaneous) disease in humans. In Jammu and Kashmir region, visceral leishmaniasis and cutaneous leishmaniasis, caused by protozoans, *viz., Leishmania donovani* and *L. tropica* respectively, have been reported during early fifties. The visceral leishmaniasis is also known as Kalazar. The synonyms of Kalazar are Black disease, Dum-dum fever and Tropical splenomegaly. This disease is mainly characterized by weight loss, hepato-splenomegaly and anaemia. Death can occur in untreated cases of visceral leishmaniasis. In cutaneous leishmaniasis disease of human, the patients develop nodular and ulcerative skin lesions. This infection, if treated properly may heal completely in a few months to few years

8.2.1 Systematic Survey and Species Richness of Sandflies

Earlier surveys on sandfly-fauna of Jammu and Kashmir regions have been made by: Sinton (1932, 1933, 33 a), Theodar (1948), Jacob and Kalra (1951), Mitra and Roy (1953) and Mitra (1959). A total of 9 different spp. of sandflies were observed by Jacob and Kalra (1951), occurring in different localities of Jammu, Kashmir and Ladakh region. They reported 7 spp. from Jammu, 5 spp. from Kashmir and 3 spp. from Ladakh, including one new species. In Riasi locality of Jammu, they found great variety of sandfly, covering four different species. Sandfly, *viz. Phlebotonus minutus* was found to be most abundant in Jammu (Riasi). In Kashmir, *P. major* was reported to be uniformly distributed and *P. chinensis*, occurring in villages in Zanaskar range (Jacob and Kalra,1951).

Rao *et al.* (1973) in his detailed paper on haematophagous insects of Western Himalayan region, reported the existence of only one species of sandfly, *viz. Sergentomyia baghdadis* from Rajouri district of Jammu, distributed at 500 to 750 mts. They have reported the mode of collection as indoor. More information has been provided regarding the existence of sandflies in J&K, with 8 spp. from Jammu and 5 spp. from Kashmir, and 4 spp. as common for both areas (Lewis, 1978). The occurrence of 6 spp. of genus *Sergentomyia* were reported by Kaul and Shetty (1983), from three Districts of Jammu of region.

Kumar *et al.* (1984), during the course of his extensive survey for the sandflies in the Jammu area in district Kathua and Udhampur in the month of August

1982, collected 5 spp. both from indoor situation (houses, cattle shed, *etc.*) and as well as outdoor (under stones, culverts, tree-holes and burrows). Various sandfly species reported by these authors are : *Sergentomyia* (*Parrotomyia*) *baghdadis*, *S.* (*Neophlebotomus*) *zeylanica*, *Phlebotomus* (*Phlebotomus*) *papatasi*, *Sergentomyia* (*Grassomyia*) *indica* and *S.* (*Parrotomyia*) *kauli*. According to these authors, the highly abundant species was *S.* (*P.*) *baghdadis* (46.86 per cent) and least prevalent *S.* (*P.*) *kauli* (0.62 per cent) of the total collection made. The sandflies, *viz.*, *S. zeylanica* and *S. kauli* were encountered by these authors for the first time from Jammu, further according to them a total of 11 spp. of sandflies are known to occur in Jammu and Kashmir region.

Table 8.1: Checklist of Sandflies (Diptera: Psychodidae: Phlebotominae) of Jammu and Kashmir, Ladakh (UTs)

Sl.No.	*Name of Species*	*Distribution*		
		Jammu	*Kashmir*	*Ladakh*
1	*Phlebotomus (Adlerius) chinensis chinensis*	+	+	+
2	*P. (A.) longiductus*	+	+	–
3	*P. (Larroussius) major major*	+	+	–
4	*P. (Paraphlebotomus) sergenti*	–	+	+
5	*P. (Phlebotomus)papatasi*	+	+	–
6	*P. minutus* var. *antennatus*	+	–	–
7	*P. montanus*	–	+	–
8	*P. squamipleuris*	+	–	–
9	*Sergentomyia bailyi*	+	–	–
10	*Sergentomyia (Grassomyia) indica*	+	–	–
11	*S. montana*	+	–	–
12	*S. (Neophlebotmus) zeylanica*	+	–	–
13	*S. (Parrotomyia)babu*	+	–	–
14	*S. (Parrotomyia) baghdadis*	+	–	–
15	*S. (P.) kauli*	+	–	–
16	*S. (Sergentomyia) punjabensis*	+	–	–

+: The existence of sandfly species; –: Not recorded in any locality of the Province.

Source: Jacob and Kalra (1951), Mitra and Roy (1953), Rao *et al.* (1973), Lewis(1978), Kumar *et al.* (1983), Koul and Shetty (1983).

It is evident from the Table 8.1 that a total of 16 spp. of sandflies belonging to family Psychodidae (sub family Phlebotominae) of insect order Diptera, are known to occur in J&K, Ladakh (UTs). Of these species, about 14 spp. have been recorded from Jammu, 6 spp. from Kashmir and 2 spp. from Ladakh. The species known to exist in Ladakh region are, *viz. Phlebotomus* (*Adlerius*) *chinensis chinensis* and *P.* (*Paraphlebotomus*) *sergenti*. The species common for both Jammu and Kashmir areas included *P.* (*A.*) *c. chinensis*, *P.* (*Larroussius*) *major major* and *P.* (*P.*) *papatasi*. *P*

(*A.*) *c. chinensis* showed wide distribution in many areas and localities of Jammu, Kashmir and Ladakh.

8.2.2 Leishmansiasis (Visceral and Cutaneous) in Jammu and Kashmir

Visceral leishmaniasis (VL) or Kala azar disease in human is caused by the parasite, *Leishmania donovani*, is transmitted by the bite of infected sandfly, *Phlebatomus argentipes*, the only known vector of this disease in India (Anon, 2016). The synonyms of Kala azar are Black disease, Dumdum fever and Tropical splenomegaly. This disease is mainly characterized by weight loss, hepatosphenomegaly and anaemia. Death can occur in untreated cases of visceral leishmaniasis. *Cutaneous leishmaniasis* (CL), also known as Oriental sore or tropical sore, Chiclero ulcer or Aleppo boils, affecting human. CL usually produces ulcers on the exposed parts of the body like face, arms and legs. This disease is caused by protozoan parasite, *Leishmania tropica*, transmitted by the bite of an infected sandfly, *Phelabotomus major*. The earliest evidence of human visceral leishmaniasis cases in Kashmir and Jammu region, has been provided by Jacob and Kalra (1951) and Khan *et al.* (1975) respectively.

(a) Visceral Leishmaniasis

The earlier definite cases of Kala azar patients in thie earstwhile Jammu and Kashmir State, have been detected in local inhabitants of Kashmir Valley, who had acquired infection inside this State. Jacob and Kalra (1951) confirmed this disease in two Kashmiris, one belonging to the district Anantnag and other to district Srinagar, who had never gone outside the State. According to these authors, 12 cases were reported from J&K and these included 7 Gurkhas, 2 Kashmiris and 3 Dogras. The 2 Kashmiri had never been outside State and 3 Dogras (2 from Samba and 1 from Poonch) has also never left the State in their life. The village of Budil was incriminated epidemiologically, as likely endemic area, which had only one species of sandfly *i.e. P. major*, as the most abundant species. in whole of Jammu and Kashmir. Further, the authors did not find parasites in tissues of the sandflies. The indigenous cases of visceral leishmaniasis have been reported during 1975 from Jammu region: village Dugan in Kathu district and in Kishtwar region of district Doda.

In the recent years, sporadic cases of VL have been reported not only from Jammu but also from Kashmir (Naik *et al.*, 1979; Anon, 2016). The occurrence of this disease in Jammu and Kahmir region, is posing threat to the public health and there is urgent need of continuous surveillance, detailed epidemiological and control measures. Recently, a case of Kala azar was found in 44 year male from Anantnag (South Kashmir), suffered hepatomegaly and massive splenomegaly (Nabi *et al.*, 2015). From Uri area in North Kashmir, Baramula District, a five year old child was reported to be suffered from VL, diagonsed on the basis of bone marrow examination (Mahajan *et al.*, 2009). In Bani region of District Kathua of Jammu, Bhat *et al.* (2014), reported a rare indigenous case of VL.

(b) Cutaneous Leishmaniasis

During recent past years, there have been increasing number of cases of human cutaneous leishmaniasis (CL), reported from Jammu as well as from Kashmir region.

The reports of cases of CL, were from different non-endemic geographical regions Kashmir : Karnah and Uri belt in South-west Kashmir(Leherwal, 2004; Masood *et al.*, 2009; Wani *et al.*, 2015); Kupwara and Barmulla (Northern-western frontiers) (Rather and Yaseen, 2017; Sajad *et al.*, 2017).The CL patients from these areas showed the symptoms in the form of lesions of noduloulceration kind on the face, including lip and nose. According to the observation made by Rather and Yaseen (2017), the patients with CL from two districts, *viz.* Kupwara and Baramulla, had "Tin Tack" sign in localized lesion, reported for the first time from these areas of Kashmir Valley.

A new focus of CL has been identified in Jammu Division. The disease affected more than hundred patients (67 females, 53 males), from November, 2012 to October, 2013 and the patients belonged to three Districts of Jammu, *viz.* Doda, Kishtwar and Poonch (Kaul *et al.*, 2016). Recently, more cases of CL have been reported from Jammu, with CL in lower eye lid of 6-year old girl, a rare entity(Arora *et al.*, 2017; Gupta, 2017). The emergence of CL in various areas of Jammu and Kashmir in the recent years is of great concern and the need of the hour is to conduct detailed epidemiological and molecular investigations to ascertain the vector, reservoir of infection and causative agent disease in Jammu and Kashmir. Dermoscopy of two cases of CL in Kashmir, has shown CL in 6-old boy on the nasal aspect of the right eye, with asymptomatic swelling and 15-year old girl showed nodoulcerative lesion on the right forearm (Bhat *et al.*, 2020).

8.3 Blackflies

Blackflies or buffalo gnats are important pests and vectors of veterinary and medical importances, belonging to Order Diptera, Infra-order Culiomorpha of Sub-order Nematacera, Family Simuliidae, Subfamily Simuliinae and Tribe Simuliini. Blackflies are characterized by small-sized, dark coloured, with short horn-like antenna, robust and hump-backed body. The mouth parts are provided with teeth adapted for sucking blood. The females are haematophagous and feed on blood of human and animals, whereas the males feed on nectar. Simuliids breed in aquatic habitats. Simuliids are vectors of nematode, *Onchocerca volvulus*, causing onchocerciasis or River blindness, transmitted through the bites of blackflies of the genus *Simulium*. River blindness disease in man is prevalent Afrotropical and Neotropical regions of the world, and also vectors of *Onchocerca* spp. in domestic animals. Besides, these transmit blood protozoan, *Leucocytozoon* among birds and domestic poultry in different regions of the world.

Brunetti (1911) was first worker to study blackflies of earstwhile J&K State. Therafter, records of blackflies from Kashmir, belonging to two species of genus, *Simulium*, have been given by Edwards (1927). Later, Puri (1932, 32a) described two more species from this region. Datta (1983, 86) studied the blackfly-fauna of Kashmir and recorded few more species from this region. A total of nine species of blackflies are known to occur in various regions of Kashmir. The checklist of Indian blackflies, including earstwhile J&K State, has been given by Mitra and Sharma (2010). A total of 69 species of black flies are reported from India (Mitra *et al.*, 2017). The percentage of Simuliid- fauna from J&K (UT) is 13.04 per cent. The various species of black flies prevalent in different localities of Kashmir region are: (i) *Simulium*

(Odagmia) adventicium Datta; (ii) *S. (Eusimulium) aureum* Fries; (iii) *S. (Simulium) alajense* Rubtsov; (iv) *S. (Simulium) barraudi* Puri; (v). *S. (S.) chrystophersi* Puri; (vi) *S. (S.) griseifrons* Brunetti; (vii) *S. (Himalayanum) indicum* Becher; viii) *S. (Wilhelmia) pseudequinum* Puri (= *S. (Wilhelmia) mediterranea*) and (ix) *S. (Testisimulium) stevansoni* Edwards.

8.4 Horseflies, Deerflies and Clegs

Tabanids include Horseflies (*Tabanus*), deerflies (*Chrysops*), or Clegs (*Haematopota*), belonging to the family Tabanidae of order Diptera, sub-order Brachycera, infra-order Tabanomorpha and Superfamily Tabanoidea. Tabanids are robust flies, body size ranging from 5-33mm. The male adults feed on nectar,where as females are haematophagous as well as nectrophagous. Rao *et al.* (1973) reported the prevalence of horsefly, *viz. Haematopota faciata*, as human pest in Ladakh. Tabanids are known to transmit different types of viral, bacterial and protozoan diseases. The important disease, transmitted by tabanids in Jammu and Kashmir is as trypanosomiasis," Surra" in equids, caused by parasite, *Trypanosoma evansi*, reported by Kumar *et al.*, 2013. In this disease the protozoan infect the blood of vertebrate hosts like cattle or horses, causing fever, weakness, which leads to weight loss and anaemia. The untreated cases can lead to fatality. In the updated Checklist of Tabanidae of India, Maity *et al.* (2016), have listed 247 species of tabanids from various regions of Indian sub-continent. The tabanid-fauna of earstwhile J&K State, has been studied by Recardo (1911), Stone and Philip (1974), Philip (1979) and Burger (1984).

In Jammu and Kashmir region, the tabanid-fauna is represented by : Genus *Chrysops* (2 spp.), belonging to subfamily Chrysopinae; *Haematopota* (3 spp.), under subfamily Tabaninae and tribe Haematopotini; *Hyobomitra* (3 spp.) and *Tabanus* (4 spp.), pertaining to subfamily Tabaninae, tribe Tabanini. The species checklist is as: *Chrysops designates* Ricardo, *Chrysops flavipes* Meigen, *Haematopota alticola* (Philip) (= *Chrysozona alticola* Philip), *Haematopota fasciata* Ricardo, *Haematopota kashmirensis* Stone and Philip, *Hybomitra tatarica* (Portschinsky) (= *Tabanus tataricus* Portschinsky), *Hybomitra tatrica tridentate*, *Hybomitra pecularis* Szilady (= *Hybomitra pecularis kashmirensis* Szilady), *Tabanus laetetinctus* Becker, *Tabanus leleani*, *Tabanus striatus* Fabricius and *Tabanus sufis* Jaennicke.

8.5 Parasitic Muscids

Many species of muscids in Jammu and Kashmir, bite and suck blood of vertebrates and these belong to genera : *Hydrotaea* (Azellinae, Azeliini); *Morellia* (Muscinae, Muscini); *Haematobia* (= *Lyperosia*) (Muscinae, Stomoxyini); *Musca* (Muscinae, Muscini); *Stomoxys* (Muscinae, Stomoxyini); *Helina* (Phaoniinae) and Muscina (Muscinae). Various species of the genus *Hydrotaea* recorded from Kashmir include *H. armipes* (Fallen), *H. dentipes* (Fabricius), *H. himalayensis* Pont, *H. kashmiriana* Pont, *H. floccose* Macquart, *H. ignava* (Harris) and *Hydrotaea* sp. (Hennig, 1955-64; Pont, 1975). The adult of these flies feed on blood of mammals. *Hydrotaea* spp. also also attack eyes, mouth, nose, ear, feed on the secretions. *H. armipes* is of medical and veterinary importance as it carries and transmits diseases to animals and humans. It is also associated with spread of mastitis disease, symptoms show as inflammation of breast or udder and fever. *Stomoxys calcitrans* (Stable fly) is a blood sucking muscid of mammals and cattles, reported earlier by Brunnetti (1920) from different areas of Jammu and Kashmir. *Muscina stabulans* is also a parasitic muscid, reported from this region by the same author. *M. stabulans* also causes intestinal myiasis.

According to Sewell and Roonwal (1965), four species of dipterous flies, belonging to the family Muscidae (Muscoidea), as haematophagous, like blood of animals and man, are known to occur in many areas of Kashmir Himalayan regions. These species are: *Helina* (*s.str*) *cashmirensis* (Mallock), often haematophagous in man, *Haematobia thirouxi titillans* (Bezzi)(= *Lyperosia* (*Glossinella*) *titillans* Bezzi, *Musca* (*Eumusca*) *autumnalis* De Geer, *M.* (*Viviparomusca*) *bezzi* Patton and Cragg and *Morellia hortorum cashmiranas* van Emden, found to bite cattle/horses and on cow/horse dung, distributed in various localities like Dal Lake, Ganderbal, Nigeen, Kokernag and high-altitude areas of Gulmarg. *Haematobia* sp. (Hornfly) is a blood sucking muscid and transmit disease like infectious keratoconjunctivitis, caused by *Moraxella bovis*, a case of this disease in dzos in Jammu region has been reported by Bhardwaj *et al.* (2012),

8.6 Pupipara Parasites

The dipteran pupipara parasites include ked/louseflies and batfly, belonging to the family Hippoboscidae and Nycteribiidae respectively. These are specialized ectoparasites that live on the body of birds and mammals, and both sexes feed on blood from host. In pupiparid insects, the female flies retain egg inside the uterus, giving nourishment to the developing larva and do not extrude the young until they are ready to pupate

In J&K, the main species of Hippoboscid of economic importance is sheep ked, *Melophagus ovinus*, parasitizing sheep and cattle. This parasite has completely lost its wings and spends entire life-cycle, sheltered in the fleece of its host. Four Species of hippoboscid, are reported to parasitize birds in Jammu and Ladakh region, these include, *Crataerina pallida, Lynchia albipennis, Ornithophila metallica,* and *Pseudolynchia canariensis. P. canariensis* (Pigeon lousefly), attacks nestling birds, causing much irritation to them and also acts as the intermediate host of common pigeon protozoan parasite. (*Haemoproteus columbae*), in other parts of the world (Table 8.2).

Hippobosca longipennis and *H.variegata,* have been found to parasitize goat, cattle and dog in Jammu, Kashmir and Ladakh region. Dermatitis in canines of Jammu due to *H. longipennis* (= *H. capensis*), has been observed by Sharma *et al.* (2009). *H. longipennis* is a blood sucking ectoparasites of dogs in Ladakh region and is a potential intermediate host a filiarid nematode parasite, *viz. Acanthocheilonemma* (Rani- Puteri *et al.,* 2011). Apart from these, lousefly parasites of cattle and domestic animals, goat is also attacked by a hippoboscid,*viz. Lipoptena careoli*, in Doda, Rajouri, Srinagar and Udhampur region (Table 8.2).

The batflies are having atrophied wings and remains continously attached to their hosts. The pupae of these parasite are found at the walls of caves, *etc.*, inhabited by bats. The batfly, *Eucampsida hyrtii orientalis*, has been reported to parasitize bat in

Udhampur. Besides this, two species of batflies, *viz. Nycteribia* (*Listropodia*) *pedicularis* Latreille and *Penicillidia dufouri* (Westwood), parasitic on Kashmir cave bat, *Myotis longipes* (= *Leuconoe longipes*), observed to be prevalent in caves in Mattan/Pahalgam localities in Kashmir. In earstwhile J&K State, about 11 spp. of pupiparous parasites, have so far been recorded. These cover 8 spp. of louseflies, spread over 7 genera, parasitizing birds, dogs, sheep, goats and cattle, and 3 spp. of batfly parasitic on bat. Of these 11 spp., 7 spp. are occurring in Jammu, 5 spp. in Kashmir and 3 spp. in Ladakh. The pupipara parasites- ked/louseflies and batflies recorded from Ladakh and J&K, UTs, are given in Table 8.2.

Table 8.2: Checklist and Diversity of Pupipara Parasites of Jammu, Kashmir and Ladakh

Sl.No.	Pupipara Species (Family)	Host	Distribution
1	*Crataerina pallida* (Hip.)	Bird	Ladakh
2	*Hippobosca longipennis* (Hip) (= Hippobosca capensis	Dog, goat	Ladakh, Baramulla, Udhampur, Rajouri
3	*Hippobosca variegata* (Hip.)	Cattle	Udhampur
4	*Lipoptena careoli* (Hip)	Goat	Srinagar, Doda, Udhampur, Rjouri
5	*Lynchia albipennis* (Hip)	Bird	Rajouri
6	*Melophagus ovinus* (Hip)	Sheep	Srinagar, Ladakh, Baramulla, Jammu, Poonch
7	*Ornithophila metallica* (Hip)	Bird	Doda
8	*Pseudolynchia canariensis* (Hip)	Bird, Pigeon	Udhampur, Rajouri, Jammu
9	*Eucampsida hyrtlii orientalis* (Nyc)	Bat	Udhampur
10	*Nycteribia* (*Listropodia*) - *pedicularis (Nyc)*	Bat	Kashmir (Pahalgam)
11	*Penicillidia dufouri* (Nyc)	Bat	Kashmir (Pahalgam)

Hip: Family Hippoboscidae (Ked/louseflies); Nyc: Family Nycteribiidae (Batflies).

Source: Mehmood *et al.* (2019); Scott (1925); Rao (1973); Sharma and Das (1971); Yadav *et al.* (2009).

8.7 Parasitic Ceratopogonids (Biting Midges)

Biting midges are haematophagous, belonging to the family Ceratopogonidae. Female biting midges feed on hosts either individually or in groups, by using their cutting teeth of mouth-parts, to puncture to host's skin to expose blood that is sucked into their digestive system. According to Bhagat (2014), 2 species of biting midges are reported from Jammu and Kashmir. These species are : *Culicoides* sp. and *Bezzia* sp. *Culicoides* sp. transmits blue tongue virus (BTV) disease, which is endemic to many States, including earstwhile J&K State (Ilango, 2006). BTV disease causes morbidity and mortality in sheep and cattle. Though this disease has been reported from this region, but its vector has not been established with authencity and further investigation is required on vector of BTV disease in this region.

8.8 Dipterous Flies of Myiasis–Causing Importance in Animals and Humans

In Jammu, Kashmir and Ladakh region, 43 species, under 28 genera of Dipterous flies, are having myiasis importance in animals (vertebrates and invertebrates) and humans. These flies belong to 9 nine families, *viz.* Acroceridae, Anisopodidae, Calliphoridae (blow flies/blue green bottle flis), Drosophilidae (fruit flies), Fannidae (latrine flies), Muscidae, Oestridae (bot flies/warble flies/sheep nostril flies/sheep head maggot), Sarcophagidae (flesh flies) and Syrphidae (flower flies) (Bhagat, 2016). The dominant family of mysiasis-causing, has been found to be Calliphoridae, having 16 species, followed by Sarcophagidae (9 spp.), Oesrtridae (7 spp.), Muscidae (5 spp.). Rest of the four families include single species each. (see Chapter 8.8.1, pp. 204-207).

According to Zumpt (1965), the myiasis is defined as the infestation of vertebrate (human and/or animals) with dipterous larvae, which at least for a certain period feed host's dead or living tissues, liquid body substances or ingested food and cause wide range of infestations, according to body location and the relationships of larvae with host. Different types of myiasis have been reported from different parts of the world in human and animals (Patton,1922; James, 1947; Kettle, 1984; Francesconi and Omar, 2012). These types of myiasis are: cutaneous myiasis (dermal, subdermal, creeping, furuncular); wound myiasis; nasopharyngeal myiasis (nose, sinuses), ophthalmomyiasis (ocular); oral (mouth); aural myiasis (ear); tracheopulmonary myiasis; enteric myiasis (gastric, rectal, intestinal myiasis); urogenital myiasis (bladder, urinary passages, anal, vaginal); cerebral myiasis; nosocomial myiasis (myiasis in hospital setting) (Table 8.3).

The myiasis may be: i) obligatory or specific, caused by flies dependent on host for larval development (*e.g.*, *Chrysomya* spp., *Dermatobia* spp., *Gasterophilus* spp., *Hypoderma* spp., *Oestrus* sp., *Przhevalskiana* spp.); ii) facultative of semi- specific, caused by flies, normally free living, usually lay eggs in decomposing animal or vegetable matters, but can initiate myiasis in host, if open wound or sore is available (*e.g.*, *Calliphora* spp., *Phormia* spp., *Lucilia* spp. Sarcophaga spp.); iii) accidental or pseudomyiasis, the larvae are normally free living, and the fly eggs or larvae may accidently ingested by host that are on food items or enter through host's anatomical openings like oral or genito-urinary (*e.g.*, *Drosophila*, *Eristalis tenax*. *Fannia* spp., *Musca domestica*,, *Muscina stabulans*).

Some myiasis- causing flies (*Calliphora* spp., *Lucilia* spp., *Sarcophaga* spp.) are of potential forensic importance. Forensic entomology helps to determine the amount of time a victim has been exposed to environment, facilitating the estimation of the post mortem interval (PMI) and the medicolegal forensic entomology is useful in providing important data during criminal investigation on corpse (Smith, 1986, Byrd and Castner, 2009). Calliphorid, *Lucilia sericata* finds common use in maggot debridement therapy (MDT) or biosurgery, for healing of non-healing wounds (Reames *et al.*, 1988; Sherman, 2014). The maggots of *L. sericata* are placed on wounds and made it to feed on necrotic tissues and bacteria. The maggots also secretes some secretions, having anti-microbial activity. Medicinal properties in wound healing potential of maggots of *L. sericata*, has been studies in experimentally infected cutaneous wound model in Wister rat, maggots possessing a definite antibacterial action (Borkataki *et al.*, 2018)

8.8.1 Systematic and Annotated Checklist of Dipterous Flies of Myiasis–Causing Importance in Animals and Humans

About 43 species, under 28 genera of Dipterous flies of myiasis–causing importance in animals and human, belonging to 9 families, pertaining to Section Aschiza (= Cyclorrhapa), Schizophora (Sub -section Calyptratae, Acalyptratae) and Sub-order Nematocera, are found to be distributed over vast and diverse localities of Jammu, Kashmir and Ladakh Himalayan regions of Indian sub-continent of great zoogeographical significance. The dipterous families with total number of species are :Acroceridae (1 sp.), Anisopodidae (1 sp.), Calliphoridae (16 spp.), Drosophilidae (1 sp.), Fannidae (1 sp.), Muscidae (5 spp.), Oestridae (7 spp.), Sarcophagidae (9 spp.) and Syrphidae (2 spp.). The systematic checklist of flies, belonging to these myiasis causing dipterous families, is presented as under:

(a) Systematic and Annotated Checklist

Order: Diptera

Section: Aschiza

Superfamily: Syrphoidea

Family: Syrphidae

1. *Eristalis* (*Eristalis*) *tenax* (Linnaeus) (6)[K]
2. *Tubifera arbustorum* (Linnaeus) (7) [K]
   (= *Eristalis arbustorum* Linnaaeus) (5) [K]

Section: Schizophora

Subsection: Calyptratae

Superfamily Muscoidea

Family: Fannidae

3. *Fannia lepida* (Wiedmann) (15)[K]

Family: Muscidae

4. **Musca domestica* Linnaeus (3, 9) [J, K, L]
5. *Musca* (*Musca*) *domestica nebula* (Fabricius) (29) [K]
6. *Musca domestica vicina* Macquart
   (=*Musca* (*Musca*) *vicina* Macquart)(29) [K]
7. *Muscina stabulans* (Fallen) (15, 29) [K]

8. *Stomoxys calcitrans* (Linnaeus) (9,29) [J, K]

Superfamily Oesteroidea

Family: Calliphoridae

Subfamily: Calliphorinae

Tribe: Calliphorini

9. *Calliphora pattoni* Aubertin (20, 27) [G]
10. *Calliphora vicina* Robineau- Desvoidy (4) [J, S]
(=*Calliphora erythrocephala* Meigen) (9, 25) [J,K]

Subfamily: Chrysominae

Tribe: Chrysomini

11. * *Achoetandrus rufifacies* (Macquart) (4, 27) [K]
12. **Chrysomya albiceps* (Wiedemann) (4)[S]
13. *Chrysomya bezziana* Villeneure (2, 8) [J, K]
14. * *Chrysomya megacephala* (Fabricius) (4, 9) [J, S]
15 *Chrysomya pinguis* (Walker) (4) [S]
16. **Chrysomya rufifacies* (Macquart) (4) [S]
17. *Chrysomya villeneuvi* Patton (4) [S]

Tribe: Phormiini

18. **Phormia regina* Meigen (9) [J]

Subfamily: Helicoboscinae

19. *Gulmargia angustisquama* Rognes (18,27) [G]

Subfamily: Lucillinae

Tribe: Luciliini

20. **Lucilia cuprina* (Wiedmann) (4, 9, 27) [J, S]
21. * *Lucilia porphyrina* (Walker) 4,20, 27) [G]
22. ***Lucilia sericata* (Meigen) (4) [S]

Subfamily: Polleniinae

Tribe: Polleniini

23. *Pollenia rudis* (Fabricius) (20, 27) [J, K]
24. *Polleniopsis kashmirensis* Kurahashi et Okadome (10) [K]

Family: Oestridae

Subfamily: Cuterebrinae

25. *Dermatobia* sp. (21) [K]

Subfamily: Gasterophilinae

26. *Gasterophilus intestinalis* (De Geer) (26) [J, K]
27. *Hypoderma lineatum* (Villers) (13, 17, 32)[J]
28. *Hypoderma sinense* (Pleske) (16) [L]
29. *Hypoderma* sp. (9) [J]
30. *Przhevalskiana silenus* (Brauer) (31, 33, 34) [J, K]
    (= *Hypoderma crossi* Patton) (23) [K]

Family: Oesterinae

Tribe: Oestrini

31. *Oestrus ovis* (Linnaeus) (1, 22, 24) [K]

Family: Sarcophagidae

Subfamily: Sarcophaginae

Tribe: Sarcophagini

32. *Rosella khasiensis* Senior-White (30) [K]
    (=*Sarcophaga khasiensis* Senior- White) (20) [K]
33. *Sarcophaga* (*Liosarcophaga*) *dux* Thomson
    (=*Sarcophaga dux* Thomson) (20) [K]
34. *Sarcophaga* (*Bellieria*) *melanura* Meigen (20, 35a, 35b) [K]
35. *Sarcophaga* (*Bercaea*) *Africa* Wiedmann (35a) [K]
36. *Sarcophaga* (*Parasarcophaga*) *albiceps* (Meigen) (35a, 35b)[K]
37. *Sarcophaga* (*Parasarcophaga*) *Misra* (Walker) (20, 35a, 35b)[K]
38. *Sarcophaga striata* (Fabricius) (7) [K]
    (= *Sarcophaga haematodes* Meigen) (20) [K]
39. *Thyroscnema kentejana* Rohdendro (19) [K]
40. *Thyroscnema kashmirensis* Lehrer (11) [K]

Section: Acalyptratae

Infra-order: Muscomorpha

Superfamily: Nemestrinoidea

Family: Acroceridae

41. *Pterodontia kashmirensis* Lichtwardt (12, 5) [K]

Superfamily: Ephydroidea

Family: Drosophilidae

Subfamily: Drosophilinae

Tribe: Drosophilini

42. *Drosophila melanogaster* Meigen (14) [J, K]

Sub-order: Nematocera

Infra-order: Psychodomorpha

Superfamily: Anisopodoidea

Family: Anisopodidae

Sub-family: Anisopodinae

43. *Sylvicola nigroclavatus* (Edwards) (28) [G]

* indicates, forensically important dipterous flies.

** indicate, fly finds use in human wound debridement therapy (maggot therapy)

Key to Abbreviations

G: Gulmarg (Kashmir); J: Jammu; K: Kashmir; L: Ladakh; S: Shopian (Kashmir).

Key to References of Author/s Recording and Describing Taxa

1: Allaie *et al.* (2015; 2: Baba *et al.*, 2015; 3: Bharti (2008); 4: Bharti (2012); 5: Brunetti (1920); 6: Datta and Chakraborti (1983); 7: James (1947); 8: Katoch *et al.*, 2014); 9: Kour *et al.* (2015); 10: Kurahashi and Okadome (1976); 11: Lehrer (2008; 12: Lichtwardt (1909); 13: Logar *al et.* (2008); 14: Prashad and Duggal (1966); 15: Pont (1986); 16: Puente *et al.* (2010); 17: Ranga- Rao (1994; 18: Rognes (1993); 19: Rohdendrof (1937); 20: Senior-White (1940); 21: Shah and Andrabi (2009); 22: Shahardar *et al.* (2001); 23: Shahardar *et al.* (1993); 24: Shahardar and Ahmad (1998); 25: Sharma (1993); 26: Sharma *et al.* (2005); 27: Singh and Sidhu (2004); 28: Stone (1973); 29: van- Emden (1965); 30: Wan-Qixue *et al.* (2011); 31: Yadav *et al.* (2014); 32: Yadav *et al.* (2013); 33: Yadav *et al.* (2011); 34: Yadav *et al.* (2006)

Addendum: 35a: Chakraborty *et al.*, 2017*a*; 35b: Chakraborty *et al.*, 2017 *b*; 36: Singh and Sindhu (2004)

8.8.2 Biodiversity of Flies and Myiasis of Veterinary and Medical Importance

From the systematic checklist (Chapter 8.8.1), it is evident that the dominant family of Diptera of myiasis- causing importance, is found to be Calliphoridae, with five subfamilies- Calliphorinae, Chrysominae, Helicoboscinae, Lucillinae, Polleniinae, having 2 spp. (2 genn.), 8 spp. (4 genn.), 1sp. (1 gen.), 3 spp. (1 gen.) and 2 spp. (2 genn.) respectively. This is followed by family Oestridae, with four sub-families, Cuterberinae, Gasterophilinae, Hypodertmatinae and Oesterinae, incorporating 1 sp., 1 sp., 4 spp. and 1 sp. respectively. Family Oestridae in dominance, is followed by the family Sarcophagidae, Muscidae and Syrphidae, including 6 spp., 5 spp. and 2 spp. respectively. Rest of the families, *viz.* Acroceridae, Anisopodidae, Drosophilidae and Fannidae, having 1 species each. The species-rich dipterous genera of myiasis importance are: *Chrysomya*, having 6 spp., followed by *Hypoderma, Lucilia, Sarcophaga*, having 3 spp. each. The genera, *Calliphora* and *Thyrosocnema,* include 2 spp. each. Rest of the genera (19), having 1 sp. each.

Of the nine dipterous families of myiasis-causing importance in Jammu and Kashmir, four families such as Calliphoridae, Oestridae, Sarcophagidae and Muscidae, are economically important in causing different types of myiasis in domestic animals and humans. Calliphorids (blow flies), are medium sized flies and cause myiasis in animals and human, besides playing an important role in transmitting pathogenic organisms. In this region, *Calliphora pattoni* is a larvivorous species, *Lucilia poryphyrina* breeds on carcases of birds and animals, cause dermal/ cutaneous myiasis and *Pollenia rudis* is reported to be parasitic on earthworm, *Allolobophora chlorotica* Savigay (White *et al.*, 1940; Rognes, 1972).

Oestrid, sheep nostril fly/head maggot, is moderately large bee-like fly, greyish brown in colour. It is represented in Kashmir by an important species, *viz. Oestrus ovis*, parasitizing sheep and goat. The other Oestrids are : *Dermatobia* (botfly), larger densely haired fly, look like bumble bee; *Gasterophilus*, horse botfly, is bee-like flies, with hairy head and thorax; *Hypoderma* spp. (Warble flies)., are large in size, hairy, bee-like and brown, orange or yellow in colour, parasitic on cattle; *Przhevalskiana silenus*, goat warble fly, adult flies are 8-14 mm. in length, has large eyes, grey thorax, cause a myiasis known as " goat warble fly infestation" (GWFI).

Sarcophagids (flesh flies) of UT include *Rosella khasiensis*, *Sarcophaga* spp. and *Thyroscnema* spp. Adults of Sarcophaga have red eyes,clear wings and, thorax with alternating black and grey stripes. They also act as myiasis-causing agents by invading wounds, responsible for cutaneous furnucular myiasis, characterized by inflammatory nodules or pus pockets. Sarcophagids are also known to peneterate into living tissues,causing severe condition or even death. The hosts of flesh flies are mammals and even insects.

Muscids of myiasis (accidental) importance in this region are *Musca* spp. (house flies), *Muscina stabulans* (false stable fly) and *Stomoxys calcitrans* (stable fly). Excepting, *S. calcitrans*, all muscids are non-biting. The adults of Stable flies, both male and female feed on blood of warm-blooded animals.The members of muscids are known to produce different types of accidental myiasis. Stable flies cause local cutaneous infections or wounds or secondary infection in case of myiasis.

Syrphid flies or flowerflies belong to the family Syphidae, are free-living, with dark bodies ornamented with brightly colour spots and bands, hovering over flowers. Datta and Chakrabarti (1983), during the course of survey on flowerflies of Kashmir, recorded the existence of syrphid fly, *viz. Eristalis*, (*Erastalis*) *tenax* (Linnaeus), (Milesinae: Volucellini), from number of localities of Kashmir, *viz.* Gulmarg, Phalgam., Hazratbal, Kabhtarkhana, Rainawari and Lal Mandi.According to Cheng (1973) the larvae of *E. tenax* are most common myiasis causing species (rectal, gastero-intestinal myiasis) in different parts of the world. The flies lay eggs over cesspools, liquid manur, open privies and waterbodies, contaminated with rotting animals matter. The eggs develops into larvae bearing long-breathing siphon, resembling a tail, hence commonly known as rat-tailed maggot.

8.8.3 Economically Important Myiasis in J&K and Ladakh (UTs)

(a) Cutaneous Myiasis (Dermal and Subdermal)

Different types of myiasis, on the basis anatomical position in or on the host animals and human, produced by 21 spp., under 12 genera, belonging to dipterous families, Calliphoridae, Muscidae, Oesteridae and Sarcophagidae, have been found in Jammu, Kashmir and Ladakh region. These flies of medical and veterinary importance include 11 spp. of Calliphoridae, 7 spp. of Oesteridae, 2 spp. of Muscidae and 1 sp. of Sarcophagidae. As many as 19 spp. of flies belonging to families, Calliphoridae, Muscidae and Oesteridae, have been reported to cause different types of myiasis in domestic animals- bovines, horse, yak, small ruminants (sheep and goat) and cattle carcasses (Table 8.3).

Cutaneous myiasis (dermal and subdermal) is seen in cow, buffalo, cow carcasses, yak, sheep, goat and human, in diverse areas and localities of Jammu, Kashmir and Ladakh, is caused by flies, *viz. Calliphora vicina, Chrysomya megacephala, Dermatobia, Hypoderma* spp., *Lucilia cuprina, Phormia regina, Przhevalskiana silensus.* In Ladakh region, Ranga Roa (1994) showed the prevalence of parasitic infection due to cattle warble fly, *Hypoderma lineatum* in Indian yak, *Bos (Poephagus) grunniens.*The incidence of *Hypoderma lineatum* in cattle, has been found in Jammu region, with 9.73 per cent prevalence, however, no infestation of hypodermosis found (Yadav *et al.*, 2013). The eggs of *H. lineatum* hatch with in a week and these penetrate the skin and migrate to oesophagus, cause swellings,"warbles". Large punctures in skin render cattle hide valueless. *H. lineatum* and *H. sinense*, have been observed to cause dermal/subdermal myiasis in human case in Ladakh region (Logar *et al.*, 2008; Puent *et al.*, 2010). Cutaneous myiasis, with traumatic wounds in tail region due to Calliphordid, *Chrysomya bezziana* in a buffalo, aged 5 years in Jammu region, has been recorded by Katoch *et al.*, 2014.

Goat warble fly, *Przhevalskiana silenus*, also causes myiasis, depending upon the location of the larvae. The goat warble fly larvae reduce the host physiological functions, destroy host tissues and cause significance loss of quality of animal hides. Shahardar *et al.* (1993), reported the occrurrence of *Przhevalskiana silenus* (= *Hypoderma crossi*), in goats for the first time from Kashmir, causing skin lesions and loss of hairs in these domestic animals. A sub-cutaneous myiasis due to Cuterebine (Osterid) parasite, *Dermatobia* sp. (robust bot fly) in cattle of Kashmir was observed, with treatment done by Shah and Andrabi (2009).

A study on the prevalence and biology of *P. silenus* in Jammu region has been conducted by Yadav *et al.* (2011). They found high prevalence of goat warble fly infestation in Udhampur District (56.44 per cent) and lowest in Doda District (13.49 per cent), based on observation made on larval stages in the slaughter house. In Samba and R. S. Pora areas of Jammu, *P. silenus* was reported in goat from late September to early March, highest in in winter (80.45 per cent), followed by autumn (65.43 per cent) and Spring (11.90 per cent), no infestation in summer and rainy season (Yadav *et al.*, 2006).

(b) Other Types of Myiasis (Gastro-intestinal, Nasopharyngeal, Tracheopulmonary, Wound and Oral)

Gastro-intestinal type of myiasis has been found in horses in Jammu regon, due to bot fly, *Gastrophilus intestinalis*. The bot fly larvae cause chronic gastritis ulcerations. These larvae attach to cardiac and non grandular portion of the stomach by hooks and penetrate digestive system. Gastro-intestinal myiasis in human in Kashmir region has been reported by James (1947).

Nasopharyngeal and tracheopulmonary myisis in sheep and goats in Jammu and Kashmr, have been observed by; Shahardar and Ahmad (1998); Shahardhar (2001) and Allaie *et al.* (2016). These types of myiasis are reported to be caused by *Oestrus ovis*, sheep nostril fly/sheep head maggot, parasitizing sheep and goat of this region. The female fly is ovoviviparous which deposits larvae in the nares and long maggots find their way back into nostrils. The larvae cause a profuse discharge of mucous (snotty nose). The larvae develop in the naso-pharyngeal cavities of sheep and goat, causing myiasis. Sheep nasal/nostril fly, causing nasal-sinusal myiasis, called oestrosis, has shown overall prevalence rate 99.16 per cent in ovine population in Plain and Kandi areas of Jammu, resulting in huge economic loss to the meat Industry (Saleem *et al.*, 2017)

Table 8.3: Different Types of Myiasis, Caused by Dipterous Flies in Animals and Humans in Jammu, Kashmir and Ladakh Region (After Bhagat, 2016)

Dipterous Fly Species.	*Host(s)*	*Myiasis Types (Depending on Host Body's Location)*						*References, Host Type, Distribution (Case Reports)*
		CU	*GI*	*NA*	*OR*	*TR*	*WO*	
Calliphora -vicina	Cow carcass, sheep	+	–	–	–	–	–	3- A, K; 6- A, J; 14- A, J.
Chrysomya-albiceps	Cow carcass	–	–	–	–	–	+	3- A, K.
Chrysomya -bezziana	Buffalo, Humans	–	–	–	+	–	+	5- A, J; 2-H, K.
Chrysomya -megacephala	Small ruminants, Cow -Carcass	+	–	–	–	–	+	6- A, J;3- A, K.
Chrysomya-pinguis	Cow carcass	–	–	–	–	–	+	3- A, K.
Chrysomya-rufifacies	Cow carcass	–	–	–	–	–	+	3- A, K.
Chrysomya-villeneuvi	Cow Carcass	–	–	–	–	–	+	3- A, K.
Dermatobia sp.	Cattle	+	–	–	–	–	–	10- A, K
Gasterophilus-intestinalis	Horses	–	+	–	–	–	–	15- A, JK.
Hypoderma-lineatum	Yak, Cattle, Human	+	–	–	–	–	–	7- H, L; 9- A, L; 18- A, J.

Dipterous Fly Species.	*Host(s)*	*Myiasis Types (Depending on Host Body's Location)*						*References, Host Type, Distribution (Case Reports)*
		CU	*GI*	*NA*	*OR*	*TR*	*WO*	
Hypoderma-sinense	Human	+	–	–	–	–	–	8- H, L.
Hypoderma sp.	Goat	+	–	–	–	–	–	6- A, J.
Lucilia cuprina	Cow, Sheep, Goat	+	–	–	–	–	–	3- A, K; 6- A, J.
Lucilia-porphyrina	Carcass of Birds, animals	–	–	–	–	–	+	16- A, K.
Lucilia sericata	Cow (carcass)	–	–	–	–	–	+	3- A, K.
Musca domestica	Small ruminants	+	–	–	–	–	+	6- A, J
Oestrus ovis	Sheep, Goat	–	–	+	–	+	–	1- A, K; 11- A, J;13- A, K
Phormia regina	Small ruminants	+	–	–	–	–	–	6-A,J.
Przhevalskiana-silensus	Sheep, Goat	+	–	–	–	–	–	12- A, K; 17- A, J; 19- A, J; 20- A, J.
Sarcophaga-striata	Human	–	+	–	–	–	–	4- H, K.
Stomoxys-calcitrans	Small ruminants	–	–	–	–	–	+	6- A, J.

J: Jammu; **K**: Kashmir, **L**: Ladakh; **JK**: Jammu and Kashmir; **+**: Present; **–**: Absent; **A**: Animal; **H**: Human.

Keys to Abbreviations and code numbers: CU: cutaneous myiasis (dermal, sub-dermal); **GI**: gastro intestinal/intestinal myiasis; **OR**: oral myiasis; **NA**: nasal/nasopharyngeal myiasis; **WO**: wound myiasis; **TR:** tracheopulmonary myiasis.

1: Allaie *et al.* (2015); **2**. Baba *et al.* (2015); **3**. Bharti (2012); **4.** James (1947); **5**: Katoch *et al.* (2014); **6**: Kour *et al.* (2015); **7**: Logar *et al.* (2008); **8** Puent *et al.* (2010); **9**: Ranga- Rao (1994); **10**: Shah and Andrabi (2009); **11**: Shahardar *et al.* (2001); **12**: Shahardar *et al.* (1993); **13**: Shahardar and Ahmad (1998); **14**: Sharma (1993); **15**: Sharma *et al.* (2005); **16**: Senior- White (1940); **17**: Yadav *et al.* (2014); **18**: Yadav *et al.* (2013); **19**: Yadav *et al.* (2011); **20:** Yadav *et al.* (2006)Above Table adopted after Bhagat (2016)

Wound myiasis in livestock (cow, buffalo and small ruminans) of J&K (UT), has been observed to be caused by: Calliphorids-*Chrysomya* spp., *Lucilia* spp.; Muscid-*Musca domesticata* (Sharma, 1993; Bharti, 2012; Katoch, 2014 and Kour *et al.*, 2015). In wound myiasis, the larvae (maggots) of Calliphoirids and common housefly invade the healthy as well as dead tissues. A case of oral myiasis has been detected in a patient (12-year old) in Kashmir region due to a Calliphorid, *Chrysomya bezziana* (Baba *et al.*, 2015).

Chapter 9

Entomophagous Parasitic Insects

9.1 Entomophagous Insects (Parasites, Parasitoids)

The entomophagous parasitic insects include parasites and parasitoids, belonging to the order Hymemoptera and Diptera, and involve feeding upon insects. Most insects, parasitic upon other insects are protelean *i.e.*, they are parasitic only in immature stages (larvae) and adults as free living. They devour all or most of the hosts' body and then pupate, either within or external to the host. The adult parasite emerge from the pupa and start the next generation. The adult parasite feeds on honeydew, nectar, pollen and many feed on host's body fluid. In general, parasitic insects are distinguished from entomophagous parasitic insect in that their hosts are typically vertebrates, and parasitic as adults and larvae/nymphs; smaller than their hosts, feed upon all stages of their hosts and do not kill their hosts, for example Siphonaptera (fleas), Anoplura, Mallophaga, dealt with separately (*see* Chapter 7, p. 173). Typically parasitoids attack one stage of host *i.e.* egg/larva/nymph or pupa. The parasitoid larvae kill their host. The parasitoids are of two types, *viz.* ectoparasitoids and endoparasitoids. In ectoparasitoids, larval stages feed and develop on the external surface of the host's body, whereas in case of endoparasitoids, larval stages feed and develop inside the body of the host. The parasitoids are common among some groups of hymenoptera and diptera.

Now-a-days, entomophagous parasitic insects (parasites/parasitoids) are being successfully utilized for suppression/control of insect pest populations by adopting various methods of applied biological control. Some parasites are non-beneficial, known as hyperparasites, as parasite of parasitoids. The original parasite of insect/pestes known as primary parasite and the hyperparasite as secondary parasite. When hyperparasites are present, they play a negative role in biological control programme.Various entomophagous insects (parasites/parasitoids) known to exist in Jammu, Kashmir and Ladakh Himalayan regions, belong to the insect orders, superfamilies and families as: Diptera- Bombyloidea (Bombyliidae), Conopoidea (Conopidae), Muscoidea (Muscidae, Sarcophagidae, Tachnidae), Syrphoidea

(Pipunculidae); Hymenoptera-Ichneumonoidea (Braconidae, Ichneumonidae), Ceraphronoidea (Megaspilidae), Chalcidoidea (Aphelinidae, Chalcididae, Encyrtidae, Eulophidae, Eupelmidae, Eurytomidae, Mymaridae, Pteromalidae, Torymidae and Trichogrammatidae) Cynipoidea (Cynipidae, Figitidae) and Platygastroidea (Scelionidae). The diversity, host range and economic importance of these parasitic groups, are discussed under the following heads:

9.2 Dipterans (Pipunculids, Muscids, Sarcophagid, Conopids, Tachinids and Bombyliids)

In Kashmir Valley, various dipteran parasites reported so far belong to the families,*viz.* Conopidae, Bombyliidae, Muscidae, Pipunculidae, Sarcophagidae and Tachinidae. Conopids are also known as tickheaded flies. These are solitary, internal parasites of hymenoptra (bumble bees and wasps), dipteran and also orthoptera. The adults of conopids are free living, however, larvae are parasitic on insect hosts. Females deposits eggs on host's body while in flight, into the abdominal cavity of host, mostly parasitizing crickets and cockroaches. From Kashmir Himalayan region, 6 species of conopids are known to occur, these are, *viz. Myopa confusa, Myopa nigriventris, M. testacea, Physocephalus rufescens, Sicus ferrugineus* and *Zodion griseum* (Brunetti, 1920, 23) (*see* Checklist 9.4.2, p. 230-231). In India, a total of 52 species of conopids are known to occur and, Jammu and Kashmir accounts for 11.54 per cent of conopid- fauna (Mitra *et al.*, 2015). The muscids (2 spp.), parasitizing some unknown species of insects, occurring in Dal lake, Ganerbal and Gulmarg localities (Sewel and Roonwal, 1965), include *Graphomyia maculate* and *Pogonomyia fumipennis.* Besides these, a muscid, *Muslina stabulanus*, has been found to parasitize forest/fruit tree pest *Lymantria obfuscata* on *Salix* spp. and *Populus* spp. (Rishi and Shah,1985) (*see* Checklist 9.4.3, p. 231).

A sarcophagid (*Sarcophaga* sp.), commonly known as fleshfly, usually myiasis- causing, however, at times acts as an internal parasite of other insects. In this region, *Sarcophaga* sp. has been found to parasitize *L. obfuscata,* infesting *Salix* spp. (Rishi and Shah,1985). This dipteran parasite is blackish in colour, with grey thoracic stripes. The larval stages feed on animal/insects and adults feed on sugar containing material. About 8 species of Tachinids, belonging to 5 genera are reported as endoparasitoids of larval lepidopterans (*L. obfuscata, Pieris brassicae* and *Plutella xylostella*) on Salix, poplars and brassicas vegetables from Kashmir region (Anon,1970; Bhat *et al.*, 2013), Rao,1960; Zutshi,1967, Simmond and Rao, 1960) (*see* Checklist 9.4.3, p. 231). Tachinid flies are covered with stout bristles or setae, usually drab coloured. An insect species when attacked by the tachinid, is always killed (*see* Checklist 9.4.3, p. 231).

Bombylids are having stout and hairy- body, commonly known as bee flies. The adults of these flies feed on nectar and pollens. Larvae are generally parasitoids of other insects, especially bees and wasps. Eight species of bombyliid larvae have been found to exist in Kashmir as *Anastoechus kashmirenis* Zaitzer, *Phthina gracilis* Walker and *Systoechus socius* Walker, besides *Bombylius* sp. and 2 species each of genera, *Anthrax* and *Villa* (Brunetti, 1920; Zaitzev,1988; Pal,1991; Evenhuis and Greathead, 1999; Banerjee and Mitra, 2006 and Najar and Bashir, 2016). Pipunculids or big-headed flies, are small in size (20-11.5 mm), dark in colour, wings long, narrow and hyaline. The females of most pipunculids deposit their eggs in early instar nymphs or adults of their hosts. The larval stages of pipunculids are endoparasaitoids of hemipterans. From Kashmir region, five species of pipunculids, under two genera are reported, these are : *Tomosvaryella curvipatis, Tomosvaryella longipygianus, T.epichalea, T.eusylvatica* and *Cephalops kashmirensis* (= *Pipunculus cephalops*) (Kapoor *et al.*, 1987) (*see* Checklists 9.4.1, p. 230 and 9.4.4, p. 231-232).

9.3 Hymenopteran Parasites/Parasitoids

The hymemopteran parasites/parasitoids, belonging to different families, covering systematic survey, diversity, species richness and as bio-control agents, are discussed under the following heads:

9.3.1 Aphelinids

The aphelinids (Aphelinidae) are minute insects measuring 0.6 to 1.4 mm in length and their bodies are not matelic, soft-bodied, showing yellow or brown colour. These are biologically diverse parasitoids and several species are important in biological control of scale insects and aphids. These parasites show wide prevalence in Jammu and Kashmir region, found to be parasitizing economically important insects like aphids and scale insects on plants of horticultural and forestry importance, such as *Prunus* spp., *Pyrus malus, Populus* spp., *Salix* spp., *Robinia pseudoacacia, Acer caesium, Celtis australis,* besides other economically important plant species.

In Jammu and Kashmir region, a total of 27 species of aphelinid parasites belonging to 11 genera, are known to be prevalent, parasitizing number host insects, infesting diverse species of host plants. The various genera, including total number of species are*viz. Ablerus* (2 sp.), *Aphelinus* (9 spp.), *Aphytis* (4), *Azotus* (1),*Centrodora*(1), *Coccophagus* (1), *Encarsia* (3) *Marietta* (1), *Promuscidea* (1), *Protaphelinus* (2) and *Pteroptrix* (2) (*see* Checklist 9.4.6, pp. 232-234).

11 different genera, covering 20 spp of aphids on various host plants are known to be parasitized by different species of aphelinid parasites and these include *Aphis* (6 spp.) *Brevicoryne* (1), *Drepanosiphum* (1), *Epipemphigus* (1sp.) *Eriosoma* (1), *Lipaphis* (1) *Macrosiphum* (1), *Myzus* (2), *Pemphigus* (5) and *Rhopalosiphum* (1). 6 spp belonging to 6 genera of scale insects, attacked by aphelinid parasites in this region are: *Ceroplastodes, Eulecanium coryli Lepidosaphes kirgisica, Parthenolecanium corni, Pulvinaria inconspiqua* and *Diaspidiotus perniciosus* (=*Quadraspidiotus perniciosus*) (San Jose Scale). The checklist of aphelinidae of India, has been provided by Manickavasagan and Menakaevi (2014). The list of various species of aphelinids occurring in J&K region and their host insect and plant species, with their distribution and authors reporting,are encapsulated in Checklist 9.4.6.

(a) Aphelinids as Biological Control Agents of Insect Pests

The earliest reports of *Aphelinus mali* as an effective parasite of woolly apple aphid, *Eriosona lanigerum,* damaging apple trees in Kashmir Valley have been provided by Narayanan (1957) and Rao *et al.* (1971). In India, *A. mali,* in the year

1937, was first released in woolly apple aphid pest infested orchards in Raison (Kulu) Himachal Pradesh for biological control of this pest, in later years, the parasite (*A. mali*) spread to Kashmir and became active in various areas of this region. In Kashmir valley, for the biological control of San Jose scale, a serious insect pest of apple deciduous fruit trees, the Russain and Chinese strains of an aphelinid parasite, *Aphytis diaspidis*, has been utilized in the year 1964/1965. Later, a Japanese strain of *Aphytis* sp. nr. d*iaspidis* from California (USA) was introduced to Kashmir. The recovery of this parasite was reported from Sumbal and Wadora region. The average percentage of parasitism in these localities during the period August, 1967 to July, 1968 were 68.6 per cent and 61.5 per cent respectively (Rao *et al.*, 1971). Masoodi *et al.* (1989) released aphelinid, *viz. Aphytis* sp. *proclia* group to control San Jose scale in three apple orchards at Humuhama, Awantipora and Narbal in Kashmir, during the year 1985 and 1986. The activity of this parasite was monitored through sticky traps at Narbal. These authors reported the establishment of parasites in these areas.

The effective parasites of San Jose Scale as *Encarsia perniciosi* (= *Prospaltella parniciosi*) has been found to play a significant role in suppressing this pest in Kashmir orchards. This specific parasite of San Jose scale, has established in many areas, showing high- population density during summer (Masoodi *et al.*, 1989a and Bhagat *et al.*, 1992). Singh *et al.* (1984), for the first time showed the natural parasitization in the eggs of gypsymoth (*L. obfuscata*), a major pest of silvicultural importance, due to parasite, *viz. Encarsia* (= *Prospaltella*) *kashmirensis* at Fruit Research Station, Shalimar, Srinagar. The parasitization of eggs of *L. obfuscata* was shown to be ranging from 15.78-20.89 per cent.

9.3.2 Aphidiines

Aphidiines (aphid parasitoids), under sub-family Aphidiinae, belong to the family Braconidae of super-Family Ichnneumonoidea, are exclusively parasitoids of aphids as solitary koinobiont endoparssitoids. Aphidiines are relatively small and fragile wasp, having real joint or artriculation below second and third abdominal segment, having reduced venation of forewing. Aphidiines are of great importance in checking natural population of aphids and are being utilized for biological control of aphid pests of agricultural and forestry importance, and other economically important plants. Aphids are responsible for causing different types of damages to host plants by sucking sap and also being vectors of virus diseases.

The female aphidiine parasitoid deposits eggs into the host aphid, in a few days develops into larva and eats out the inside of aphid. It cuts a hole in the bottom of aphid's empty skin and through this glues the carcass to the substratum. The full grown larva of most genera spin a cocoon inside empty skin of the victim. In case of genus, *Praon*, a peculiar tent like cocoon beneath the victim (host aphid) is made. The adult parasitoid emerges by cutting a circular lid. On t he basis of recent detailed morphological studies of aphidiines, genera *viz., Binodoxys* and *Trioxys* are, separate genera (Lazarevic *et al.*, 2017), besides genus *Betuloxys*. The molecular phylogeny of Aphidiines favours either three tribes as Ephedrini, Praini, Aphidiini or five tribes as Ephedrini, Praini, Monoctoni, Trioxini and Aphidiini (Sanchis, 2000). The three tribes of Aphidiinae- Ephedrini, Praini and Aphidiini, with Aphidiini further divided into three sub-tribes as Monoctonina, Trioixina and Aphidiina, are proposed (Shi and Chen, 2005).

The earliest descriptions of new species of aphidiines from Jammu and Kashmir region, have been given by: Dharmadhikari and Ramaseshia (1970), Stary and Bhagat (1978), Shuja-uddin (1970,74,78,82,83), Takada and Rishi (1980) and Bhagat (1981c 82a, 82c 84). Rakhshani *et al.* (2012) have given description of a new species and new records of aphidiines from Jammu and Rajouri areas. Studies on aphid-galls and their parasitoids in Kashmir, were done by Bhagat (1982). The records and host- range of hyperparasitiods of aphids from Kashmir, have been given by Bhagat (1983,87). New records, host range and biological notes on aphidiines of Jammu region, have been given by Bhagat and Ahmad (1995). In a recent detailed survey of aphidiine parasitoids of aphids, occurring in the agro-ecosystem, with special reference of horticultural ecosystem of Kashmir valley, for the first time an aphidiine, *viz. Lysaphidius erysimi* and also number of new host plant species of already known aphidiines, have been recorded (Khan and Shah, 2017).

In Kashmir Valley, detailed investigation on aphidiid parasitoids, controlling aphids, damaging valuable agricultural crops and other economically important plants, have also been carried out. In this direction, observations have been made on the seasonal occurrence of aphids, abundance of parasitoids and their role as natural biological control agents. The various papers encompassing such studies are on: medicinal plants (Bhagat, 1987 *a*); cereals, pulses and vegetables, fodder crops and ornamental plants (Bhagat, 1986 *b*); fruit trees (Bhagat, 1986 *a*); trees of forestry importance (Bhagat, 1985 *a*). The wide aphidiid parasitoid spectrum of aphids of Kashmir, is having mixed fauna of Palaearctic and Oriental region, and might be useful for exporting of some of the efficient parasitoids to other regions of our country and to other areas of the world for the bio-control of aphid pests. In India, a total of 125 spp. of aphidiines belonging to 22 genera, under 4 tribes have been reported (Akhter *et al.*, 2011). However, from Jammu and Kahmir region till date a total of 80 spp. of aphidiines, belonging to 17 genera are known to occur (*see* Checklist 9.4.17, pp. 243-250).

In Jammu and Kashmir, the various genera, with their total number of species prevalent are: *Adialytus* (3 spp.), *Aphidius* (18), *Areopraon* (1), *Betuloxys* (1), *Binodoxys* (12), *Diaeretiella* (1), *Ephedrus* (11), *Kashmiria* (1), *Lipolexis* (2), *Lysaphidius* (2), *Lysiphlebia* (1), *Lysiphlebus* (3), *Monoctonus* (3), *Pauesia* (4), *Praon* (8), *Toxares* (4) and *Trioxys* (6). These parasitoids are known to attack a total of 97 host aphid species, distributed over 49 genera, infesting/damaging more than 112 host plant species, distributed over 74 genera, covering 35 plant families (Table 9.1).

The various host aphid genera, with their total number of species, attacked by parasitoids are: *Acyrthosiphon* (3 spp.), *Amphicercidus* (1), *Aphis* (12), *Aulocarthum* (2), *Betacallis* (1), *Brachycaudus* (2), *Brevicoryne* (1), *Capitophorus* (2), *Cavariella* (3), *Chaitophorus* (5), *Chaetosiphon* (5), *Chromaphis* (1), *Cinara* (2), *Coloradoa* (1), *Duraphis* (1), *Eriosoma* (2), *Eumyzus* (1), *Hyalopterus* (1), *Impatientinum* (3), *Lachnus* (2), *Liosomaphis* (3), *Lipaphis* (1), *Macrosiphoniella* (6), *Macrosiphum* (3), *Melanaphis* (2), *Metalophilum* (1), *Microlophium* (1), *Mindarus* (1), *Myzaphis* (1), *Nosanovia* (1), *Myzus*

(4), *Ovatus* (1), *Panaphois* (1), *Phorodon* (2), *Prociphilus* (1), *Periphyllus* (2), *Pterocomma* (1), *Rhopalosiphum* (1), *Schizoneurella* (1), *Semiaphis* (1), *Shivaphis* (1), *Sipha* (1), *Sitobion* (3), *Stagona* (1), *Tetraneura* (1), *Therioaphis* (1), *Tuberlachnus* (1), and *Uroleucon* (2) (Table 9.1).

The different host plant families (33), incorporating various genera, with total number of species in case of each genus, damaged by above given aphid genera (49), with 97 species, parasitized by 17 aphidiid genera (81 spp.), are given as under:

Family Aceraceae – *Acer* (1 sp.); Apiaceae – *Conium* (1); *Daucus* (1), *Ferula* (1); Asclepiadaceae – *Calotropis* (1); Asteraceae – *Achillea* (1); *Chrysanthemum* (3), *Cichonum* (1), *Senecio* (1), *Sonchus* (3), *Zinnia* (1); Balsaminaceae – *Impatiens* (1); Berberidaceae – *Berberis* (2); Boraginaceae – *Cyanoglossun* (1); Brassicaceae – *Brassica* (2), *Capsella* (1), *Descuraina* (1), *Sisymbrium* (1); Cannabaceae – *Cannabis* (1); Caprifoliaceae – *Lonicera* (1); Cupressaceae – *Thuja* (1); Dioscoreaceae – *Doscorea* (1); Euphorbiaceae – *Euphorbia* (1); Fabaceae – *Lathyrus* (1), *Lespedeza* (1), *Medicago* (3), *Pisum* (1), *Robinia* (1), *Trifolium* (1), *Vicia* (1); Hippocastanaceae – *Aesculus* (1); Fagaceae – *Quercus* (1); Lamiaceae – *Mentha* (3), *Nepeta* (2), *Salvia* (1); Meliaceae – *Melia* (1); Nymphaeaceae – *Nymphaea* (2); Ongraceae – *Epilobium* (1), Pinaceae – *Abies* (1); *Pinus* (3); Poaceae – *Arundo* (1); *Hordeum* (1), *Pennisetum* (1); *Phragmites* (1), *Poa* (1), *Sorghum* (1); *Triticum* (1); Polemoniaceae – *Phlox* (1); Polygonaceae – *Rumex* (2); Punicaceae – *Punica* (1); Ranunculaceae – *Ranunculus* (2); Rosaceae – *Cotoneaster* (1), *Malus* (1) *Prunus* (3); *Rosa* (6), *Rubus* (2), *Spirea* (1); Rubiaceae – *Rubia* (1); Salicaceae – *Populus* (3), *Salix* (7); Scrophulariaceae – *Verbascum* (1), *Veronica* (1); Simaroubaceae – *Ailantus* (1); Solanaceae – *Lycopersicon* (1); Ulmaceae – *Celtis* (1); *Ulmus* (2); Urticaeae – *Urtica* (1).

The highest number of plant species *i.e* 14, is found to be pertaining to plant family Rosaceae, infesting aphids and their aphidiine parasitoids. This is followed by family Asteriaceae and Salicaceae, including 10 species each. The host aphid genus *Aphis*, with 12 species, infesting highest number of plant species *i.e.*, 29, has yielded 22 aphidiines, belonging to *Aphidius* (7 spp.), *Binodoxys* (6), *Lipolexis* (2), *Lysiphlebus* (2), *Toxares* (2), *Diaeretiella* (1), *Ephedrus* (1) and *Trioxys* (1). *Aphis* genus is followed by genus *Macrosiphoniella* and *Macrosiphlum* parasitized by 11 spp. and 8 spp. respectively. The number of aphidiine species of other genera, parasitizing rest of aphid genera, are presented in Table 9.1. The aphidiine genus *viz. Aphidius* and its species, is responsible for parasitizing highest number of aphid genera *i.e.*, 26 and its species. This genus is followed by genus *Praon* and *Binodoxys*, parasitizing number of genera of 6 aphid as 13 and 10 respecively. For rest of aphidiine genera, parasitizing total number of host aphid genera (*see* Table 9.1).

9.3.3 Braconids, Chalcidids and Figitids

(a) Braconids

Braconids (Braconidae: Ichneumonoidea) are relatively small-sized and stout-bodied wasps, with abdomen about as long as the head and thorax combined. The adults visit flowers and are often attracted to light. In majority of braconids, pupation takes place outside of the body of the host or spin cocoons away from the host. In *Apanteles* spp., pupation takes place externally and the pupa remains

Table 9.1: Species Richness and Diversity of Aphidiine Parasitoids (Hymenoptera), Attacking Host Aphid Genera and their Total Number of Species

Aphidiine Genera (Species)	*Aphidiine Genera, with Total Number of Species, Parasitizing Aphid Genera (Species)*																
Host Aphid Genera (Species)	*Areopraonn (1)*	*Adialytus (3)*	*Aphidius (18)*	*Betuloxys (1)*	*Binodoxys (1)*	*Diaeretiella (1)*	*Ephedrus (11)*	*Kashmiria (1)*	*Lipolexis (2)*	*Lysaphidius (2)*	*Lysephlebia (1)*	*Lysephlebus (3)*	*Monoctonus (3)*	*Pauesia (4)*	*Praon (8)*	*Toxares (4)*	*Trioxys (6)*
Acyrthosiphon (3)	–	–	2	–	–	–	1	–	1	–	–	–	–	–	–	–	1
Amphicercidus (1)	–	–	–	–	–	–	–	–	–	–	–	–	–	–	1	–	
Aphis (12)	–	–	7	–	6	1	1	–	2	–	–	2	–	–	3	2	1
Aulacorthum (2)	–	–	–	–	–	–	1	–	–	–	–	–	–	–	–	–	–
Betacallis (1)	–	–	–	–	–	–	–	–	–	–	–	–	–	–	–	1	–
Brachycaudus (2)	–	1	–	–	–	–	–	–	–	–	–	–	–	–	–	1	–
Brevicoryne (1)	–	–	–	–	–	1	–	–	–	–	–	–	–	–	–	–	–
Capitophorus (2)	–	–	2	–	–	–	–	–	–	–	–	–	–	–	–	–	1
Cavariella (3)	–	–	1	–	–	–	–	–	–	–	–	–	–	–	–	–	–
Chaitophorus (5)	–	1	2	–	1	–	–	–	–	–	–	–	–	–	1	–	–
Chaetosiphon (5)	–	–	3	–	–	–	–	–	–	–	–	–	–	–	–	–	–
Chromaphis (1)	–	–	–	–	–	–	–	–	–	–	–	–	–	–	–	–	1
Cinara (2)	–	–	–	–	–	–	–	–	–	–	–	–	–	2	–	–	–
Colorado (1)	–	–	1	–	–	–	–	–	–	–	–	–	–	–	–	–	–
Duraphis (1)	–	–	–	–	–	–	–	–	–	–	–	–	–	–	–	–	1
Dysaphis (1)	–	–	1	–	–	–	–	–	–	–	–	–	–	–	–	–	–
Eriosoma (2)	1	–	1	–	1	–	–	–	–	–	–	–	–	–	–	–	–
Eumyzus (1)	–	–	1	–	–	–	1	–	–	–	–	–	–	–	–	1	–

Aphidiine Genera (Species)	*Aphidiine Genera, with Total Number of Species, Parasitizing Aphid Genera (Species)*																
Host Aphid Genera (Species)	*Areopraonn* (1)	*Adialytus* (3)	*Aphidius* (18)	*Betuloxys* (1)	*Binodoxys* (1)	*Diaeretiella* (1)	*Ephedrus* (11)	*Kashmiria* (1)	*Lipolexis* (2)	*Lysaphidius* (2)	*Lysephlebia* (1)	*Lysephlebus* (3)	*Monoctonus* (3)	*Pauesia* (4)	*Praon* (8)	*Toxares* (4)	*Trioxys* (6)
Hyalopterus (1)	–	–	–	–	–	–	1	–	–	–	–	–	–	–	1	1	1
Impatientium (1)	–	–	1	–	–	–	–	–	–	–	–	–	–	–	–	–	–
Lachnus (2)	–	–	1	–	–	–	–	–	–	–	–	–	–	2	–	–	–
Liosompahis (3)	–	–	1	–	–	–	–	–	1–	–	–	–	1	–	–	–	–
Lipaphis (1)	–	–	–	–	–	1	–	–	–	–	–	–	–	–	–	–	–
Macrosiphoniella (6)	–	–	2	–	2	–	2	–	–	–	–	–	1	–	2	1	1
Macrosiphum (3)	–	1	2	–	3	–	–	–	–	1	–	–	–	–	1	–	–
Melanaphis (2)	–	–	1	–	–	–	1	–	–	–	–	1	–	–	1	–	–
Metaplophium (1)	–	–	–	–	–	–	–	–	–	–	–	–	–	–	–	1	–
Microlophium (1)	–	–	2	–	–	–	1	–	–	–	–	–	–	–	–	–	–
Mindarus (1)	–	–	–	–	2	–	–	–	–	–	–	–	1	–	1	–	–
Myzaphis (1)	–	–	1	–	1	–	–	1	–	–	–	–	1	–	1	–	–
Myzus (4)	–	–	2	–	–	–	–	1	–	–	–	–	–	–	1	1	–
Nosonovia (1)	–	–	–	–	–	–	–	–	–	–	–	–	–	–	1	–	–
Ovatus (1)	–	–	–	–	–	–	–	–	–	–	–	–	–	–	–	–	–
Panaphis (1)	–	–	–	–	–	–	–	–	–	–	–	–	–	–	–	–	1
Periphyllus (2)	–	–	1	–	–	–	–	–	–	–	–	–	–	–		–	–
Phorodon (2)	–	–	1	–	1	–	–	–	–	–	–	–	–	–		–	–
Prociphilus (2)	–	–	–	–	–	–	–	–	–	–	–	–	–	–	–	–	–
Pterocomma (1)	–	–	1	–	–	–	–	–	–	–	–	–	–	–	–	–	–

Aphidiine Genera (Species)	Aphidiine Genera, with Total Number of Species, Parasitizing Aphid Genera (Species)																
Host Aphid Genera (Species)	*Areopraonn* (1)	*Adialytus* (3)	*Aphidius* (18)	*Betuloxys* (1)	*Binodoxys* (1)	*Diaeretiella* (1)	*Ephedrus* (11)	*Kashmiria* (1)	*Lipolexis* (2)	*Lysaphidius* (2)	*Lysephlebia* (1)	*Lysephlebus* (3)	*Monoctonus* (3)	*Pauesia* (4)	*Praon* (8)	*Toxares* (4)	*Trioxys* (6)
Rhopalosiphum (1)	–	–	–	–	–	–	–	–	–	–	–	–	–	–	1	–	–
Schizoneurella (1)	1	–	–	–	–	–	–	–	–	–	–	–	–	–	–	–	–
Semiaphis (1)	–	–	1	–	–	–	–	–	–	–	–	–	–	–	–	–	–
Shivaphis (1)	–	–	–	–	1	–	–	–	–	–	–	–	–	–	–	–	2
Sipha (1)	–	–	–	–	–	–	–	–	–	–	–	2	–	–	–	–	–
Sitobion (3)	–	–	3	–	–	–	–	–	–	–	1	–	–	–	–	–	–
Stagona (1)	–	–	–	–	–	–	–	–	–	–	–	–	–	–	–	–	–
Tetraneura (1)	–	–	–	–	–	–	1	–	–	–	–	–	–	–	–	–	–
Therioaphis (1)	–	–	–	–	–	–	–	–	–	–	–	–	–	–	–	–	1
Tuberlachnus (1)	–	–	1	–	–	–	–	–	–	–	–	–	–	–	–	–	–
Uroleucon (2)	–	–	–	–	2	1	1	–	–	–	–	–	–	–	1	–	–
Undetermined aphid	–	–	–	1	–	–	–	–	–	–	–	–	–	–	–	–	–

enclosed in a cocoon, formed of fine threads of white, yellow or cream coloured silk. From various regions of Jammu, Kashmir and Ladakh, a total of 106 spp., distributed over 9 sub-families has so far been reported. These sub-families are,*viz.* Alysiinae, Aphidiinae, Blaciinae, Braconinae, Cheloninae, Doryctinae, Euphorinae, Microgastrinae and Opiinae. The Aphidiines include 80 spp. distributed over 80 genera, followed by Microgastrines, with 13 spp., Euphorinae (3 spp.), Opiinines (4 spp). and Alysiines (2 spp.). Rest of the subfamilies,are represented by single species each. (*see* Checklist 9.4.17, pp. 243-252).

Previous studies on braconid parasites in Kashmir Valley, attacking gypsy moth (*L. obfuscata*) and maize borer (*O. kashmirica*), for biological control of these forestry and agricultural pests in U.S.A, have been carried out by the entomologists of Common wealth Institute of Biological control during the year 1968 and 1969. It was reported that a total of 3 braconids, *viz. Apanteles conspersae, A. liparidis* and *Apanteles* sp., suppressed the population of *L. obfuscata.* Thereafter, during the course of extensive surveys for the natural enemies of *L.obfuscata,* Zutshi (1967) and Rishi and Shah (1985), recorded more microgastrine parasites, belonging to the genus *Apanteles, viz., A. flavicoxis, A. indiansis, A. melanoscelus, A. porthetriae,* and *A.* nr. *solitarius.* Four new species belonging to sub-family Opiinae from Poonch and Rajouri area of Jammu, have been described by Sami-uddin (2009). The behavioural study in laboratory on field (Jammu) collected solitary microgastrine and Euphorine endoparasitoids – *Glyptapanteles agamemnonis* and *Meteorus pulchicornis* respectively towards virus infected and parasitized *Spilarctia obliqua* larvae, has been made by Gupta (2019)

Two species, including one new species, pertaining to genus *Phaenocarpa* (Allysiinae), recorded by Bhat (1979) from Gulmarg, Pahalgam, Khilanmarg and Sonamarg localities of Kashmir. Description of a new species and first generic record of Blaciine from India (Kashmir), *viz. Blacus* (*B*) *imitator,* has been given by by Papp (1985). Pawar and Parry (1989) recorded the occurrence of *Apanteles* (sub-family Microgastrinae), suppressing the population of *Archips subsidiaria,* a major pest of apple in Nimmu region in district Leh. Buhroo *et al.* (2001) has found *Spathius* sp. (Doryctinae), parasitizing larvae/pupae of shot hole borer, *Scolytus nitidus* in fruit-growing areas of Kashmir valley. A detailed survey in six districts of Kashmir, *viz.* Anantnag, Badgam, Ganderbal, Pulwama and Srinagar, on *Spodoptera litura,*Noctuid Lepidopteran pest, damaging vegatables (cole crops)- cabbage, cauliflower, knol-khol and kale, has revealed the prevalence of a braconid parasite, *Zele chloropthalma* (Zelinae) and the highest parasitization reported to be 9.75 per cent (Bhat *et al.*, 2011).

(b) Chalcidids

Chalcidids have been placed in the family Chalcididae under Super-family Chalcidoidea. Theses are primary parasites of Lepidoptera, Diptera, Hymenoptera Coleoptera besides, bugs are also attacked. Some are also hyperparasites, attacking tachinids and ichneumonids. In Jammu and Kashmir, a total of 11 spp. belonging to 3 genera of chalcidid parasite as *Brachymeria* (8 spp.), *Bruchophagus* (1 sp.) and *Epitranus* (2 spp.), is known to occur and these attack pests of agricultural and forestry importance, besides medicinal plants (Checklist 9.4.7, pp. 234-235).

Earliest report of chalcidid, *viz. Brachymeria euloeae*, attacking *L. obfuscata* on poplars from Kashmir, has been given by Zutshi (1967. Later, chalcidid, *B. intermedia*, as pupal parasite of *L. obfuscata* on salix and poplar from Srinagar, has been reported by CIBC scientists, during the course of their project for evaluation of hymenopteran parasites for biological control of pests in U.S.A. (Anon., 1969, 70). Gaffar and Punjabi (1990) reported the prevalence of chalcidid, *Bruchophagus samsonowi* (= *Eurytoma samsonowi*) on apricot in Ladakh region. The natural parasitism of pests of brassicas vegetables, *viz. Pieris rapae, P. brassicae* and *Pontia daplidice* in Kashmir valley due to chalcidid parasites- *Brachymeria femorata, Brachymeria* sp. has been found (Bhat and Bhagat, 2009; Bhat *et al.*, 2010). Recently, Ahmad *et al.* (2012 a), described a new species of genus *Epitranus* as *E. husaini* from Srinagar.

Rishi and Shah (1985) and Masoodi *et al.* (1986), reported the existence of *B.lasus*, parasitizing pupae of *L. obfuscata* on salix and poplars in Kashmir. In addition to the above mentioned three species genus *Brachymeria*, 3 more species of chalcidid, have been reported from Kashmir region, these included *Brachymeria tachardia, Brachymeria ornatipes* (=*Chalcis ornatipes*) and *Epitrans kashmirensis* by Subba- Rao and Hyat (1986), Narendran (1986), (1906) and Husain and Agarwal (1981) respectively.

(c) Figitids

Figitids are included in the Super-family Cynipoidea, commonly called as gall-wasps, because majority of the species produce galls in plants (sub-family cynipinae).The larvae of figitds develop as internal parasitoids of endopterygote insects, especially dipterans (syrphids) and are classified as koinobionts (parasitoids) However, many species belonging to sub-family Eucolinae and Charipinae are parasitic. Charipinae includes species which are mainly hyperparasites of aphids through primary parasites (aphidiines) or other parasites. The figitids are minute insects, black or brown, with abdomen globose or compressed, shiny and macropterous.

In Kashmir Valley, two species of cynipids/figitds, *viz. Alloxysta* sp. and *Phaenoglyphis*, have been recognized as hyperparasites attacking aphidiines of different aphid hosts (*see* Checklist 9.4.16, pp. 242-243). Bhagat (1983,87), gave records and host range (parasites, aphid and plants), from different ecological habitats of these two species, with special emphasis on occurrence of hyperparasites of primary parasities of aphids infesting *Rosa* spp. *Alloxysta* spp. have been found in various localities *ca* 1500m to 2,500 m, parasitizing the primary parasite, *viz., Praon volucre, Aphidius eglanteriae* on aphid, *Myzus rosarum,* attacking mainly host plant, *Rosa, webbiane* (wild rose). Besides, it also attacked *Aphidius* spp. (4), *Lysiphlebus ambiguous, Toxares* spp. (2) on 8 spp. each of host aphid and plant. *Phaenoglyphis* sp. was abserved as hyperparasite of aphids *Microlophium carnosum* and *Aphis* spp, attacked by primary aphidiid parasitoids, *Aphidius* spp. (2) and *Praon abjectum* on host plants, *Urtica dioica* and *Nepeta salvifolia.* From Srinagar, a new figitid species, *Phaenoglyphis indica* Ferre-Suay and Pujade-villar has been reported (Ferre-Sway and Pujade-Villar, 2013). Rather *et al.* (2019) made observation on the foraging behavior of a newly recorded figitid, *Aspicera hartigi* on *Ephedra* plant in Srinagar (Shankrachary hill).

9.3.4 Encyrtids, Eulophids and Eupelmids

(a) Encyrtids

Encyrtids (Encyrtidae: Chalcidoidea) are endoparasites as well as hyperparasites of aphids,scale insects and whiteflies. A total of nine species of encyrtid, under nine genera are known to occur Jammu and Kashmir (Checklist 9.4.8, pp. 235-236). They have robust body, showing broad convex mesopleura. From Kashmir, encyrtid, *viz. Microterys chalcostomus,* has been reported as a endoparasite of scale insect, *Lecanium* (Rao and Hayat,1986). *Aphidencyrtus aphidivorus* and *Prionomitus* sp. are observed to be as hyperparasites of aphids in this region (Bhagat, 1982, 87). Encyrtid, *Hunterellus sagarensis* has been found to parasitize tick, *Haemaphylis bispinosa* in Kashmir (Kaul *et al.*, 1981).

In Kashmir region, studies have been conducted on host range of encrytids – *A. aphidivorus* and *Prionomitus* sp., with reference to gall-forming aphids and aphids infesting *Rosa* spp. *A. aphidivorus* has been found to parasitize 3 spp. of primary aphidiid parasitoid, *viz. Aphidius* sp. *Praon* sp. *Protaphelinus nickolskajae* on different host aphids, infesting 3 host plant species. In case of *Prionomitus,* primary aphidiid parasitoid species included *Aeropraon lepelleyi, Aphidius* spp. and *Trioxys indicus* on four different aphid and host plant species (Bhagat, 1982, 83, 87).

In addition to above mentioned encyrtid parasitoids, five more species have been found to be prevalent in Jammu and Kashmir region, *viz. Anagyrus* sp., *Cheiloneurus pyrillae, Cerchysiella kathuaensis, Copidosoma floridanum* and *Syrophagus* (Kazmi and Hayat, 1998; Anis and Hayat,; Singh, 2007 and Ahmad *et al.* (2013 b).

(b) Eulophids

Eulophids belong to the family Eulophidae under Superfamily Chalcidoidea, parasitic warps, small to very small size (16 mm). Some are as hyperparasite and a few phytophagous Species. The eulophids attack scale insects, concealed Lepidopterans and Dipterans (leaf-miners). Earlier, from Srinagar, Eulophid, *viz. Chrysocharis nitetis* has been reported by Hansson (1985). Ahmad (2005) gave description of new genus of Eulophid, *Farooquiella kashmirensis,* parsitizing pupal cocoon of *Plutella xylostella* on *Brassica campestris* in Khudwani area of district Anantnag. However, Narendran (2007) has raised doubt regarding validity of the genus. Narendran (2011) described new eulophids from Kashmir region belonging to subfamily Eulophinae, these include *Elachertus kashmiricus, Eulophus razaki* and *Sympiesis acicus.* In addition to these, Eulophids parasitizing insects like Coccinellid beetle, *Chilo* and leaf-miners of Kashmir region are : *Dighyphus guptai, Diglyphus kashmirensis, Euplectrus coimatorensis, Aprostocetus neglectus* (Narendran, 2007; Ahmad *et al.*, 2013; Gupta and Poorani, 2009). A taxonomic review of *Chrysocharis* of India, including Kashmir and Ladakh, has been provided by Narendran *et al.* (2013). This review includes 5 spp. of genus *Chrysocharis,* with descriptions of 4 new species.

Incidence of eulophid, *viz. Compsilura* sp. and *Tetrasticus* sp. parasites of *Lymantria obfuscata* (Gipsy moth) in Kashmir has been reported by Masoodi *et al.* (1986). Also, *Parasserodella* sp. (Entiinae), has shown occurrence in Kashmir region.

A detailed survey conducted on eulophid parasites of insect pests (*Liriomyza trifolii, Chromatomyia horticola, Thysanoplusia orichalcea, Helicoverpa armigera* of vegetable crops in Kashmir, has revealed as many as 12 spp. of eulophids. These are: *Asecodes*, sp., *Chrysocharis horticola, Chrysonotomyia* sp., *Closterocerus indica, Diglyphus horticola Diglyphus* sp., *Euplecrtus ceylonensis, E.euplexae, Pediobius imdicus, Hemiptarsenus varicornis, Euderus agromyzae* and *Quadrastichus* sp. (Bhat *et al.*, 2009 a, 2009 b; Bhat and Bhagat, 2009 b; Bhat and Bhagat, 2010; Bhat *et al.*, 2009, 2010).

Occurrence of ten species of eulophid, identiified up to genera level, with prevalence percentage, parasitizing apple leaf miner (Lyonetia clerkella) in Kashmir region, has been provided (Rather *et al.*, 2017). These eulophids, with prevalence percentage include *Chrysocharis* sp. (47.8 per cent), *Chrysonotomyia* sp. (19.65 per cent), *Minotetrastichus* sp. (8.18 per cent), *Pnigalio* sp. (8.8 per cent), *Sympiesis* sp. (5.81 per cent), *Stenonomesius* sp. (4.4 per cent), *Closterocerus* sp. (2.98 per cent), *Quadrastichus* sp. (1.7 per cent), *Achrysocharoides* (0.78 per cent) and *Baryscapus* sp. (0.15 per cent). For annotated checklist of eulophids covering as many as 37 spp. of Jammu and Kashmir (*see* Checklist 9.4.9, pp. 236-238).

(c) Eupelmids

Eupelmids have been placed under the family Eupelmidae and Superfamily Chalcidoidea. The female body is elongate, frequently metallic coloured and often "u" shaped in dried specimen. The larvae of eupelmids feed ectoparasitically and also endoparastically, showing wide host spectrum.The genus *Anastatus*, having four species (*A. bifasciatus, A.japonicus, A. kashmirensis* and *Anastatus* sp.), reported from Kashmir, have been bred from eggs of Lepidoptera: *L. obfuscata* on *Salix* spp. and *Populus* spp. and *Porthetria dispar* infesting *Salix* spp. and *Populus* spp. Apart from these three species from Kashmir, 3 more species belonging to the genus *Eupelmus* (*E.kashmiricus, E.valsus* and *E. vindox*) have been found to parasitize the larvae of coleopterans, *Scolytus* spp. (shot hole borer) on apple trees. (Islam and Hayat, 1986; Mathur, 1956; Masoodi *et al.*, 1986; Narendran, 2001).

Earlier, the programme for the evaluation of hymenopteran parasites of *L. obfuscata* by CIBC scientists for bio-control of pest in U.S.A, revealed the prevalence of *Anastatus* spp. (*A. kashmirensis* and *A. bifasciatus*) from the eggs of *L. obfuscata* (Rao, 1966 and Anon, 1969). Later surveys by Rishi and Shah (1985) and Masoodi *et al.* (1986) confirmed the wide prevalence of these effective egg parasites of gypsy moth from different parts of the Valley. Recently, Narendran *et al.* (2001) reported the existence of three Species of *Eupelmus*, given above as effective parasites of shot-hole borer beetles, damaging apple trees in many orchards of Kashmir region. For annotated checklist of eupelmids of this region (*see* Checklist 9.4.10, p. 238).

9.3.5 Eurytomids, Ichneumonids and Megaspilid

(a) Eurytomids

Eurytomids belong to the family Eurytomidae under Super-family Chalcidoidea, commonly known as seed chalcids usually black in colour, with thorax, head and antennae often hairy. From Kashmir, two species of eurytomids, *viz. Eurytoma morio* and *Eurytoma* sp., as larval endoparasites of *Scolytus* sp. on apple tree and *L.*

obfuscata on *Salix* spp., have been reported by Narendran *et al.* (2001) and Masoodi *et al.* (1986) respectively (*see* Checklist 9.4.11, pp. 238-239).

(b) Ichneumonids

The ichneumonids are mostly slender wasps, belonging to the family Ichneumonidae and superfamily Ichneunoidea, and is the largest family of parasitic hymenoptera, characterized by different size, form, colouration, presence of long antenna and in many ichneumonids, the ovipositor is long, often longer than the body. All species are parasitic on larval and pupal holometabolous insects. The parasite may complete its development in the stage of host in which the egg is laid or in later stage. Most species spin cocoon and pupate, where the host died or pupate within the cocoon or pupae of host. In Jammu, Kashmir and Ladakh region, ichneumonids are represented by about 69 species, belonging to 37 genera, parasitizing larval, prepupal stages of gypsy moth (*Lymantria obfuscata*),wood-boring coleopterans, sawflies and insects of agricultural and forestry importance such as *Plutella xylostella* (= *maculipennnis*), *Heliothis armigera*, *Pieris* spp. and *Ostrinia kashmiric; Glyphodes pyloalis* and pest of *Morus* spp.), besides other unknown insect species pertaining to orders like Lepidoptera, Hymenoptera and Coleoptera (*see* Checklist 9.4.18, pp. 252-257).

The rich ichneumonid- fauna is distributed among as many as 10 subfamilies. The various sub-families incorporating different genera, with total number of species are; Banchinae- *Anilastus/Eriborus* (1 sp.) *Exetastes* (1 sp.); Campopleginae- *Campoletis* (2 spp), *Charops* (1), *Diadegma* (1), *Hyposoter* (1); Cremastinae- *Eucremastus* (1), *Pristomerus* (2); *Cryptinae- Buathra* (5), *Caenocryptus* (1), *Glabridorsum* (1), *Itamoplex* (3), *Myrmetonostenus* (1); Ichneumoninae- *Agloajoppa* (1), *Amblyaenus* (1), *Ctenichneumon* (2), *Ichneumon* (1), *Protichneumon* (1), *Pterocomus* (1), *Triptognathus* (1); Ophioninae- *Enicosphilus* (1); Pimplinae- *Ecthrus* (1), *Exeristes* (1), *Itoplectis* (3), *Pimpla* (11), *Scambus* (6), *Theronia* (2), *Townesia* (1), *Xanthopimpla* (2); Poemeniinae- *Deuteroxorides* (1), *Eugalta* (1), *Poemenia* (1); Tryphoninae- *Eridolius* (1), *Netelia* (3), *Platylabus* (2), *Polyblastus* (1); Xoridinae- *Xorides* (2).

Subfamily Pimplinae is found to be dominant with respect to species richness, having 27 species, spread over 8 genera. The pimpline genus, *viz. Pimpla* contains highest number of species i.e 11. These are idiobiont endoparasite of lepidopteran pupae (pupivorus), present in concealed places like soil, leaf-litters, leaf-rolled, *etc.*A total of 7 spp. of ichneumonds, including 4 species of pimpline, belonging to genus *Pimpla*, are known to parasitize gipsy moth (*Lymentria obfuscata*) on forest trees like salix and poplars. More than 9 species of ichneumonids, have been found to be associated with insect pests of vegetable crops in Kashmir region. An annotated checklist of ichneumonids, is provided, giving their host range, distribution, besides authors recording and describing new species from Jammu, Kashmir and Ladakh region, is provided (*see* Checklist 9.4.18, pp. 252-257).

During the year 1968 and 1969, the evaluation of ichneumonid parasites as biological control agents of *L. obfuscata* was carried out by the scientists of Commonwealth Institute of Biological Control (CIBC) London. The parasites recorded and reared in Kashmir (for U.S.A) included *Anilastus* sp., *Pimpla* (= *Coccygominus*) and *Theronia atalantae atalantae.* (Anon., 1970). Later on, Rishi and Shah (1985), and Masoodi *et al.* (1986), have made detailed field observations on the ichneumonid parasites of gypsy moth in Kashmir. In Kashmir, the scientists of CIBC, during 1968-69 made thorough field studies on the parasites of *Ostrinia kashmirica* (Pyralidae: Lepidoptera), for biological control of *Ostrinia* major pest of agricultural importance in U.S.A. The samples were collected from localities, *viz.* Larkipora and Achabal. The larvae parasites of *O. kashmirica,* reported from the field samples included *Exeristes roborator, Pristomerus* and *Scambus flaricrus* (Anon, 1970).

(c) Megaspilids

Megaspilids belong to the family Megaspilidae and superfamily Ceraphronoidea, are small insects and mostly hyperparasites or secondary parasites of aphids through primary parasites as aphidiines. In Kashmir Valley, Bhagat (1983) reported the existence of Megaspilid, *viz. Dendrocerus carpenteri* and studied its host range. *D. carpenteri* was found to develop as hyperparasites of aphidiines *viz. Aphidius* spp. on aphids (*Macrosiphum rosae* and *Uroleucon sonchi*) and *Praon* spp. on aphids (*Aphis grossulariae* and *Amphicercidus tuberculatus*), infesting four species of host plants (*Rosa webbiana, Sonchus* sp., *Lonicera quinquelocularis* and *Epilobium hirsutum*) (*see* Checklist 9.4.5, pp. 232).

9.3.6 Mymarids, Pteromalids, Scelionid, Torymids and Trichogrammatids

(a) Mymarids

Mymaridae: Chalcidoidea) are commonly known as fairflies, extremely minute in size (0.2, preferring host eggs in concealed habitat, such as plant tissues under bark and in soil. Some species have been found as successful agents in biological control programme. In Jammu and Kashmir region, seven species of mymarids, belonging to genus *Erythemelus* (1 sp.), *Gonatocerus* (2 spp.) and *Polynema* (=*Paralletaptera*) (1 sp.), *Polaeoneura* (1 sp.) and *Lymaenon* (2 spp.) have so far been reported (Zeya and Hayat, 1995; Subba Rao and Hayat, 1986). These species are known to be as egg parasites of various unknown insect species on unknown host plant species, from various regions like Jammu and Kashmir (Pahalgam and Anantnag) (*see* Checklist 9.4.12, p. 239).

(b) Pteromalids

Pteromalids are minute-sized, black or metallic green or bronze coloured insects, belonging to the family Pteromalidae under super-family Chalcidoidea. These are important entomophagous form of speices, attacking wide variety of insects of different orders. Most of these are primary parasites and many are valuable in the biological control of crop pests. Many species are also secondary/hyperparasites. The most common species in Kashmir are: *Asaphes* spp., *Coruna* spp. and *Pachyneuron* spp. The female of hyperparasite species parasitized aphid that include last instar larva or pupa of the aphidiines (primary parasitoid).

In Jammu and Kashmir, about 31 species of pteromalids, spread over 20 genera have been reported as primary parasites and hyperparasites of insect hosts, such as hemipterans (aphids), coleopterans (scolytus, chrysomelids), dipterans (agromyzids, leaf-miners, cecidomyiids, chlorophrods, syrphids) and lepidopterans (pupae), on wide range of host plants. Various pteromalid genera, with number of species include *Asaphes* (2 spp.), *Coruna* (2 spp.). *Cheiropachus* (1 sp.), *Dibrachys* (1 sp.), *Halticoptera* (3 spp.), *Herbertia* (1 sp.), *Heydenia* (1 sp.), *Lamprotatus* (1 sp.), *Macroglenes* (1 sp.) *Metacolus* (4 spp.), *Macromesus* (1 sp.), *Pachyneuron* (4 spp.), *Pteromalus* (2 spp.), *Rhaphitelus* (1 sp.), *Roptrocerus* (1 sp), *Schizonotus* (1 sp.), *Sphaeripalpus* (1 sp.), *Spalangia* (1 sp.), *Shegigaster* (1 sp.) and *Systasis* (1 sp.).The species of pteromalid parasites are listed, with their host insects (primary/secondary) and host plant Species along with information on authority reporting the species (*see* Checklist 9.4.13, pp. 239-241).

Bhagat (1983, 87) studied the host range of pteromalids as hyperparasitoids of aphids on wide range of host plants, these include *Asaphes suspensus* (Nees) *Coruna* sp., *Pachyneuron aphidis* and *P. groenlandicum* (Holmgren). *A. suspenscus* was found to be hyperparasitoid of 7 spp. of primary parasitoids *Aphidius* (spp. 4), *Diaertiella rapae, Proan volucre* and *Trioxys* sp. on 8 species of aphids, infesting 8 host plant species. *Coruna* sp. attacking 5 species of primary aphidiid parasitoids of 6 species of aphids on host plants, *viz. Rosa* spp., *Aesculus indicus, Lonicera quinquelaculans* and *Prunus domestica insititia.* In case of *Pachyneuron* spp. (*P. aphidis, P. groenlandicum*), 7 species each having of host aphidiid parastoid and aphids, on 6 host plant species. Apart from this, Boucek *et al.* (1978) reviewed the pteromalidae of this region and reported *Coruna clavata* Walker as hyperparasitoid of aphidiid parasitoid on aphid, *viz. Macrosiphum* sp.

For the first time record of two species of pteromalids (*Cheiropachus quadrum, Rhaphitelus maculatus*) and also descriptions of two new Species- *Heydenia indica* and *Macromesus harithus,* as economically important parasite of shot-hole borer, *Scolytus nitidus* (Scolytidae: Coleoptera), damaging apple trees and other fruit trees in many orchards of Kashmir valley, have been provided (Buhroo *et al.*, 2001; Narendran, 2001).

The areas of Trans-Himalyan Ladakh region, *viz.* Leh, Nemu, Hemis and Changthang, have shown the prevalence of rich ptermalid-fauna, representing as many as 10 spp., under 8 genera, parasitizing aphids, lepidopterans pupae, chrysomalids, dipterans (Sureshan, 2012). Detailed survey has been made in Kashmir region for pteromalid parasites of insect pests like *Plutella xylostella, Liriomyza trifolii* and *Pieris brassicae,* damaging vegetables (cabbage, cauliflower and, bottle and sponge-gourd) (Bhat and Bhagat, 2008; Bhat *et al.,* 2009a, 10b). Besides, these authors have also found wide prevalence of pteromalids such as *Herbertia indica, Pteromalus puparium* and *Tetraticus sokolowskii,* in vegetable field eco-system of Kashmir region. Two new pteromalids, *viz. Sphegigaster agarwali* and *Systasis shalimarensis,* have been described from Kashmir region (Ahmad *et al.*, 2012b, 2012c).

(c) Scelionid

Scelionid wasps belong to family Scelionidae and Superfamily Proctotrupoidea. Adult is non-metallic,sclerotized, solitary, internal egg parasite of hemiptera, prefers eggs in aggregation clumps. Eariler, from Kashmir region, Scelionid, *viz.*, *Hadronotus*, has been recorded as a parasite of egg of *Cletus bipunctatus*(Coreidae:Hemiptera) (Mathur and Sharma,1962). Another scelionid parasite known to be prevalent in Srinagar, is *Telenomus pauriensis* Mukerjee (Rajmohan, 2015) (*see* checklist 9.4.19, p. 257).

(d) Torymid

Torymid wasps belong to the family Torymidae under superfamily Chalcidoidea their body is moderate seized and typically metallic blue or green, are ectoparasites of hymenoptera and coleoptera, some as phytophagous. In Kashmir Valley, torymid, *Monodontomerus acreus* has been found suppressing the population of *L. obfuscata*, a major pest of forestry importance and was reared in Srinagar by CIBC scientists during 1968, for use as bio-control agent against gypsy moth in U.S.A (Anon., 1970). In addition to *M. acreus*, five more species of tormid parasites known to occur in Kashmir and Ladakh region are : *Torymus barsulicus* from Baramulla and Srinagar areas of Kashmir (Narendran and Razak, 2010); *Megastigamus kashmiricus* and *Megastigmus* sp., associated with wild rose in Ladakh (Suresh, 2009); *Ecdamua lehensis* from Ladakh (Suresh, 2010). Besides these, *Ecdamua bouceki*, has been found to be prevalent in south Kashmir (Anantnag) (Ahmad *et al.*, 2012) (*see* Checklist 9.4.19, p. 242).

(e) Trichogrammatids

Trichogrammatid wasps have been placed under family Trichogrammatidae and Super- family Chalcidoidea, are smallest insects (0.2 to 1.0 mm), weakly sclerotized, without metallic colouration, showing 3 segmented tarsi and presence of microscopic rows of hairs on the wings. These are parasites of eggs of insects and genus *Trichogramma*, usually parasitize Lepidopterans. Trichogrammatids are being extensively utilized in biological control programmes in different parts of the world.

Pawar *et al.* (1981) conducted field study in the apple orchards in Ladakh region to determine the effectiveness of two species of trichogrammatids (exotic) pertaining to the genus *Trichogramma*, *viz. T. cacoeciae pallidium* and *T. embryophagum*, against codling moth (*Cydia pomonella*), damaging apple. With the introduction of these two exotic egg parasites of *C. pomonella*, reduction in the incidence of codling moth has been achieved from 17.5-37.5 per cent. The scientists have confirmed the effectiveness of the egg- parasitoids in Ladakh region against insect pest infestations and they further suggested that it would be useful if weekly releases of these bio-control agents were made against the moth (pest) from the month of June to July.

Host range of two species of *Trichogramma* (*T. flandersi*, *T. japonicum*) have been studied in Jammu and Kashmir, parasitizing economically important pests *viz.*, *Chilo infuscatellus*, *Agrius convolruli*, *Chilo polychrysa*, *Scirophaga incertulas*, *S. nivella* and *Trichoplusia ni* (Ahmad *et al.*, 2002). These parasitoids have shown wide distribution and potential for biological control of these economically important pests. *Trichogramma* sp. has been encountered, after rearing of pest, *viz. Helicoverpa armigera* on vegetables like *Solanum tuberosum* and *Lycospercicon esculentum* in Kashmir region (Bhat *et al.*, 2009) (*see* Checklist 9.4.15, p. 242).

9.4 Entomophagous Parasitic Insects, Host Insects and Plant Species Annotated Checklists

(Key to the abbreviations, used are given at the end of all the checklists

9.4.1 Bombyliidae (Bombyloidea: Diptera)

(1) *Anastoechus kashmirensis* [Kmr; 1, 3, 6,]

(2) *Anthrax keiseri* [Kmr; 3]

Host: Unknown

(3) *Anthrax orientalis* [Kmr; 3, 6]

Host: Unknown

(4) *Bombylius* sp. [Kmr; 4]

Host: Unknown

(5) *Phtina gracilis* [Kmr; 1, 3]

Host: Unknown

(6) *Systoechus socius* [Kmr; 1, 3, 5,]

Host: Unknown

(7) *Villa baluchianus* (= *Anthrax baluchianus*) [Kmr; 3, 5,]

Host: Unknown

(8) *Villa manifesta* (= *Anthrax manifesta*) [Kmr; 2]

Key to numerals in above given long brackets:

1 = Banejee and Mitra (2006); 2 = Brunetti (1920); 3 = Evenhius and Greathead (1999); 4 = Najar and Bashir (2016); 5 = Pal (1991); 6 = Zaitzev (1988).

9.4.2 Conopidae (Conopoidea : Diptera)

(1) *Myopa confusa* [Kmr; 1, 2]

Host: Unknown

(2) *Myopa nigriventris* [Kmr; 1, 2]

Host: Unknown

(3) *Myopa testacea* [Kmr; 1, 2]

Host: Unknown

(4) *Physocephala rufescens* [1, 2]

Host: Unknown

(5) *Sicus ferrugineus* [2]

Host: Unknown

(6) *Zodion griseum* [1, 2]

Host: Unknown

Key to the numerals given in above given long brackets: 1 = Brunetti (1923); 2 = Mitra *et al.* (2015).

9.4.3. Muscidae, Sarcophagidae, Tachinidae (Muscoidea : Diptera)

(a) Muscidae

(1) *Graphomyia maculata* [Kmr.; Sewel and Roonwal, 1965]
Host: Unknown

(2) *Pogonomyia fumipennis* [Kmr.; Sewel and Roonwal, 1965]
Host: Unknown

(3) *Muslina stabulanus* (Fab.) [Rishi and Shah,1985; Kmr].
Host: *Lymantria obfuscata* on *Salix* spp. and Poplars.

(b) Sarcophagidae

(1) *Sarcophaga* sp. [Rishi and Shah,1985 Kmr.]
Host: *L. obfuscata* on *Salix* sp.

(c) Tachinidae

(1) *Corcelia* sp. [Srg; Anonymous,1970; Srg]
Host: *L. obfuscata* on *Salix* and Poplars

(2) *Compsilura coccinnata* [Kmr; Bhat *et al.*, 2013]
Host: *Pieris brassicae* on Brassicas vegetables

(3) *Compsilura* sp.[Rao, 1960; Masoodi *et al.*, 1989a; Kmr.]
Host: *L. obfuscate* on *Salix* spp.

(4) *Drino discreta*, [Anonymous, Srg.)
Host: *L. obfuscata* on *Salix* spp. and *Populus* spp.

(5) *Drino inconspicuoides* [Zutshi, 1967; Kmr].
Host: *L. obfuscata* on *Salix*.

(6) *Exorista larvarum* [Kmr; Bhat *et al.*, 2013]
Host: Pieris brassicae on Brassicas vegetables

(7) *Exorista rossica Mesnil* [Rishi and Shah,1985; Zutshi,1967, Kmr]
Host: *L. obfuscata* on *Salix* spp.

(8) *Voria ruralis* spp. [Kmr; Anonymous,1970; Simmond and Rao, 1960 -Srg.]
Host: *Plutella xylostella* (= *maculipennis*) on unknown host plant species).

9.4.4 Pipunculidae (Syrphoidea : Diptera)

(1) *Cephalops kashmirensis* (=*Pipunculus kashmirensis*) [Kmr; Kapoor *et al.*, 1987)
Host: Unknown

(2) *Tomosvaryella curvbipatis* [Srg.; Kapoor *et al.*, 1987]

Host: Unknown

(3) *Tomosvaryella epichalea* [Srg; Kapoor *et al.*, 1987]

Host: Unknown

(4) *Tomosvaryella eusylvatica* [Srg; Kapoor *et al.*, 1987

Host: Unknown

(5) *Tomosvarella longipygianus* [Srg.; Kapoor *et al.*, 1987]

Host: Unknown

9.4.5 Megaspilidae (Ceraphronoidea : Hymenoptera)

(1) *Dendrocerus carpenteri* (Curtis) [Bhagat, 1983; Kmr.]

Hosts: Aphidiid parasitoids: *Aphidus avenae – Macrosipham rosae* on *R. webbiana; Aphidius* sp. – *Uroleucon sonchi* on *Sonchus* sp., *Praon dorsale – Amphicerchidus tuberculatus* on *Lonicera quinquelocularis; Praon – objectum – Aphis grossulariae nr. pollinosa* on *Epilobium hirsutum.*

9.4.6 Aphelinidae (Chalcidoidea : Hymenoptera)

(1) *Ablerus* sp. [Kmr; 21]

Host: Unknown

(2) *Ablerus perspeciosus* Girault [Kmr; 9]

Host: Unknown

(3) *Aphelinus albipodus* Hayat and Fatima [Kmr; 10]

Host: Unknown

(4) *Aphelinus (Aphelinus) abdominalis* (Dalman)[J&K; 7, 9]

Host: *Aphis craccivora* on unknown host plant

(5) *Aphelinus (Aphelinus) asychis* Walker [Kmr; 7, 16]

Host: *Brevicorynae brassicae, Lipaphis erysimi, Myzus persicae* on unidentified host plants.

(6) *Aphelinus (Aphelinus) flavipes* (Forester)[J&K; 7, 16]

Host: *Aphis gossypii, A. spiraecola, L. erysimi, M. persicae, Rhopalosiphum* sp.

(7) *Aphelinus* nr. *fulvus* Jashnosh[Duc., Khi., Kmr; 2]

Host: *Drepanosiphum plantantoids* (Schrauk) on *Acer caesium; M. cerasi* on *Spirea* sp.

(8) *Aphelinus kashmiricus* Hayat [Srg; 2, 4, 5]

Host: *Aphis gossypii* on *Zinnia elegans*

(9) *Aphelinus gossypii* Timber [Kmr; 7]

Host: Unknown

(10) *Aphelinus mali* Haldman[Kmr; 2, 7,14]

Host: *Aphis affinis* on *Epilobium hirsutum; A. cracciura* on *R. pseudoacacia; A. kurosawai* on *Dalphne oleoides; A. nerii* on *Verbascum thapsus; E. lanigerum* on *M. pumila; Sitobion avenae* (=*Macrosiphum avenae*) on *Pennisetum flaccidum.*

(11) *Aphelinus* sp.[Kmr; 3]

Host: *Eulecanium coryli* on *Prunus armeniaca, Pulvinaria inconspiqna* on *C. australis*

(12) *Aphytis diaspidis* How [Kmr; 9, 17]

Host: *Diasidiotus perniciosus* (=*Quadraspidiotus perniciosus*) on deciduous fruit trees, *Prunus* spp. and *Salix* spp.

(13) *Aphytis lingnanensis* Compere [Kmr., Srg., Pat., Han; 3]

Host: *Parthenolecanium* Corni (Bouche) on *Viburnum foetans, D. perniciosus on M. pumila.*

(14) *Aphytis paramaculicornis* DeBach and Rosen (Han., Pat., Srg., Ana. 3,12,13)

Host: *Lepidosaphes kirgisica* on *Salix fragilis, D. perniciosus* on *M. pumila*

(15 *Aphytis* sp. *proclia* group [Kmr;12, 13]

Host: *D. perniciosus* on *M. pumila.*

(16) *Azotus kashmirensis* Narayanan [Kmr; 7, 15]

Host: *D. perniciosus.*

(17) *Centrodora idioceri* Ferriere [Kmr; 9]

Host: Unknown

(18) *Coccophagus tschirchii* Mahdhihassan [Kmr; 7, 9]

Host: *Ceroplastodes* sp. on undertermined plant species

(19) *Encarsia centrina* Craw [Kmr; 18]

Host: *D. perniciosus*

(20) *Encarsia kashmirensis* Singh *et al.* [Sha; 20]

(= *Prospaltella perniciosi* Tower) [J&K;7, 9, 15]

Host: Egg of *L. obfuscata* on fruit tress (*Prunus* spp.)

(21) *Encarsia perniciosi* (Tower)

(= *Prospaltella perniciosi* Tower) [Kmr; 3, 6, 7, 8, 9, 13, 18]

Host: *D. perniciosus* on deciduous fruit trees, *Prunus* spp., *M. pumila* and *Salix* spp.

(22) *Marietta carnesi* (Howard)

(= *Marietta indica*) [J&K;7, 9, 15]

Host: *D. perniciosus*

(23) *Promuscidea ufasciadiventris* Giralt [Kmr; 9]

Host: Unknown

(24) *Protaphelinus dargahae* Rishi and Lone [Srg;7, 19]

Host: *Pemiphigus bursarius* on *Populus nigra* var. *italica*.

(25) *Protaphelinus nickolskajae* Joshnosh [Kmr;7, 9, 11]

Host: *Epipemphigus imaicus* on *Populus ciliata, Pemphigus aedificator* on *P. ciliata, P. immunis* on *Populus caspica, P. ciliata* and *P. nigra, P. mordvilkoi* on *P. caspica, P. napeus* on *P. alba*.

(26) *Pteroptrix chinensis* (Howard) (J&K 1, 7,18)

Host: *D. perniciosus*

(27) *Pteroptrix koebelie* Howard (9, Kmr)

Host: Unknown

Key to numerals in long brackets, in front of above listed aphelinid species: 1: Bhagat (1982); 2: Bhagat (1985); 3: Bhagat *et al.* (1992); 4: Bhat (1984); 5: Hayat (1972); 6: Hayat (1980); 7: Hayat (1986); 8: Hayat(1989); 9: Hayat (1998); 10: Hayat and Fatima (1992); 11: Mackauer (1972); 12: Masoodi and Trali (1987); 13: Masoodi *et al.* (1986a); 14: Narayanan (1957); 15: Narayanan (1961); 16: Ramaseshia and Dharmadhikari (1969); 17: Rao *et al.* (1971); 18 : Rehmat *et al.* (2011); 19: Rishi and Lone (1984); 20: Singh *et al.* (1984); 21: Ahmad *et al.* (2013 a).

9.4.7 Chalcididae (Chalcidoidea : Hymenoptera)

(1) *Brachymeria euloeae* Wester [Kmr; Zutshi,1967]

Host: *L. obfuscata* on *Salix* sp.

(2) *Brachymeria femorata* Panzer [Kmr.; Bhat and Bhagat, 2009]

Host: *Pieris rapae* and *Pontia daplidace* on *Brassica oleracea capitata, B.o.gongloides*

(3) *Brachymeria intermedia* (Nees)[Kmr; Anonymous,1969.)

Host: *L. obfuscata* on *Salix* spp. and *Populus* spp.

(4) *Brachymeria lasus* Walker [Kmr; Narendran, 1986; Rishi and Shah, 1983; Masoodi *et al.*, 1986; Sheela *et al.*, 2003.)]

Host: *L. obfuscata* on *Salix* spp.

(5) *Brachymeria ornatipes*(= *Chalcis ornatipes* Cameron) [Kmr; Cameron 1906]

Host: Unknown plant host and insect species.

(6) *Brachymeria tachardia* (Cameron)[Kmr; Subba Rao and Hayat,1986]

Host: Unknown

(7) *Brachymeria tibialis* [Kmr; Masoodi *et al.*, 1986]

Host: *Lymentria obfuscata*

(8) *Brachymeria* sp. [Bhat *et al.*, 2010 b; Ahmad and Zaki, 2008; Kmr]

Host: *Pieris brassicae* on *Brassica oleracea; Brassica campestris, B.o.gongylides; Parnara guttata* on paddy

(9) *Bruchophagus samsonowi* [Kmr; Narendaran,1994]

(=*Eurytoma samsonowi* Vassiliev [Kmr; Gaffar and Punjabi, 1990]

Host: Unknown insect on Apricot

(10) *Epitranus husaini* Ahmad, Khurshid and Azim [Kmr;Ahmad *et al.*, 2012 a]

Host: Unknown

(11) *Epitranus kashmirensis* Hussain and Aggarwal [Kmr; Hussain and Aggarwal,1982; Srg.]

Host :Unknown

9.4.8 Encyrtidae (Chalcidoidea : Hymenoptera)

(1) *Anagyrus* sp.[Ahmad *et al.*, 2013a]

Host: Unknown

(2) *Aphidencyrtus aphidivorus* (Mayr)[Kmr; Bhagat, 1982, 87)

Host: Aphid parasitoids*: Aphidius* sp. – *Ovatus* nr.*crataegarius* on *Mentha longifolia; Praon* sp. – *Chaetosiphon* sp. on *Rosa webbiana; Protaphelinus nickolskajae* – *Pemphigus* on *P. ciliata.*

(3) *Cerchysiella kathuaensis* (Singh, 2007; Jmu)

Host: Unknown

(4) *Cheiloneurus pyrillae* [J&K; Anis and Hayat, 2002]

Host: Eggs of *Pyrilla*

(5) *Copidosoma floridanum* [J&K; Pam; Kazmi and Hayat,1998]

Host: Unknown

(6) *Hunterellus sagarensis* [Kmr; Kaul *et al.*, 1978]

Host: Tick (*Haemophysalis bispinosa*)

(7) *Microterys chalcostomus*[Kmr; Rao and Hayat,1986]

Host: *Lecanium* on unknown plant species

(8) *Prionomitus* sp.[Kmr; Bhagat,1982, 1987]

Host*:* Aphidiid parasitoids: *Aeropraon lepelleyi* on aphid,*viz. Schizoneurella indica* on *Ulmus wallichiana; Aphidius* spp. – on *Aphis farinosa* on Salix alba, *Chaetosiphon tetrarhodus* on *Rosa canina* and *Rosa* sp.*; Trioxys indicus* on *Aphis affnis* on *M. longifolia* (= *M. sylvestris*).

(9) *Syrophagus* sp.[J&K; Ahmad *et al.*, 2013 a]

Host: Unknown

9.4.9 Eulophidae (Chalcidoidea : Hymenoptera)

(1) *Aprostocetus neglectus* [Kmr; 12]

Host: Unknown

(2) *Asecodes* sp. [Kmr, 4]

Host: *Liriomyza trifolii* on *Lagenaria siceraria* and *Cucurbita maxima*

(3) *Chrysocharis cuticisi* [Ldk, 13a]

Host: Unknown

(4) *Chrysocharis differentis* [Ldk, 13a]

Host: Unknown

(5) *Chrysocharis harithi* [Kmr., 13a]

Host: Agromyziid

(6) *Chrysocharis neoviridis* [Kmr., 13a]

Host: Agromyzid

(7) *Chrysocharis horticola* [Kmr; 7, 8]

Host: *Chromatomyia hortricola* on *Allium cepa, Brassica oleraceous acephala, B.o.gonglodes, Pisum sativum*

(8) *Chrysocharis nitetis* [Srg., 10, 13a]

Host: Coleopterans, Lepidopteran and secondary parasite of Hymenoptera

(9) *Chrysonotomyia* sp. [Kmr, 7, 8, 15]

Host: *L. trifolii* on *L. siceraria* and *Luffa cylindrica, Lyonetia clerkella* on *Malus domestica*

(10) *Closterocerus indica* [Kmr, 4]

Host: *Liriomyza trifolli* on *L. sinceraria, L.cylindrica, C.maxima, Lycopercicon esculentum, Cucumis melo, Cucumis sativus, Solanum tuberosum*

(11) *Compsilura* sp. [Kmr, 1]

Host: *Lymentria obfuscata*

(12) *Diglyphus guptai* [Kmr, 13]

Host: Unknown

(13) *Diglyphus horticola* [Kmr, 7,8]

Host: *C.horticola* on *Allium cepa, Brassica o.* and *varities, Malva sylvestris, P.sativum*

(14) *Diglyphus kashmirensis* [Kmr,2]

Host: *C.horticola*

(15) *Diglyphus* sp. [Kmr, 4]

Host: *C. horticola on Lagenaria siceraria, Luffa cylindrica, Cucubita maxima, Cucumis melo, Cucumis sativus, Solanum tuberosum.*

(16) *Elchertus kashmiricus* [Kmr, 13]

Host: Unknown

(17) *Euplectrus coimbatorensis* [Kmr,9, 2a]

Host: *Chilo partellus, Trichoplusis orechala*

(18) *Euplectrus ceylonensis* [Kmr, 5]

Host: *Thysanoplusia orechalcea on B.o.gonglodes*

(19) *Eulophus razakai* [Kmr, 13]

Host: Unknown

(20) *Euplectrus euplexae* [Kmr, 3]

Host: *Helicoverpa armiger* on *B.o.capitata*

(21) *Farooqiella kashmirensis* [Kmr, 1]

Host: *Plutella xylostella on Brassica campestris*

(22) *Parasserodella* sp. [Kmr, 2a]

Host: Unknown

(23) *Pediobius indicus* [Kmr, 7, 8]

Host: *C.horticola* on *A.cepa, P. sativum, B.o.acephala*

(24) *Hemiptarsenus varicornis* [Kmr, 4]

Host: *L.trifolii on L. siceraria, L.cylindrica*

(25) *Euderus agromyzae* [Kmr,7, 8]

Host: *C.horticola* on *A.cepa, B.campestis, B.o.acephala, P.sativum*

(26) *Quadrastichus* sp. [Kmr, 4, 15]

Host: Unknown insect host on *L. siceraria, L.cylindrica* and *Lyonetia clerkella* on *Malus domestica*

(27) *Sympiesis acicus* [Kmr,13]

Host: Unknown

(28) *Sympiesis* sp. [Kmr; 2a, 15]

Host: Unknown and *Lyonetia clerkella* on *Malus domestica*

(29) *Tetrastichus* sp. [Kmr, 11]

Host: Gipsy moth

(30) *Tetrastichus sokolwskii* [Kmr; 14]

Host: *Plutella xylostella* on *Brassica o. capitata* and *Brassica o.botrytis*

(31) *Chrysocharis* sp. [Kmr., 15]

Host: *Lyonetia clerkella* on *Malus domestica*

(32) *Minotetrasticus* sp. [Kmr., 15]

Host: *Lyonetia clerkella* on *Malus domestica*

(33) *Pnigalio* sp. [Kmr., 15]

Host: *Lyonetia clerkella* on *Malus domestica*

(34) *Stenomesius* sp. [Kmr., 15]

Host: *Lyonetia clerkella* on *Malus domestica*

(35) *Closterocerus* sp. [Kmr., 15]

Host: *Lyonetia clerkella* on *Malus domestica*

(36) *Achrysocharoides* sp. [Kmr,15.]

Host: *Lyonetia clerkella* on *Malus domestica*

(37) *Baryscapus* sp. [Kmr., 15]

Host: *Lyonetia clerkella* on *Malus domestica*

Key to numerals given in long brackets in front of the Eulophid species, listed above: 1: Ahmad (2005); 2: Ahmad *et al.* (2013); 2 a: Ahmad *et al.* (2013 a); 3: Bhat *et al.* (2009); 4: Bhat *et al.* (2009a); 5: Bhat *et al.* (2010); 6: Bhat and Bhagat (2009); 7: Bhat and Bhagat (2009 b); 8: Bhat and Bhagat (2010); 9: Gupta and Poorani (2009);10: Hanson (1985); 11: Masoodi *et al.* (1986a); 12: Narendran (2007); 13: Narendran (2011); 13a: Narendran *et al.* (2013);14: Bhat and Bhagat (2008); 15: Rather *et al.* (2017).

9.4.10 Eupelmidae (Chalcidoidea : Hymenoptera)

(1) *Anastatus bifasicatus* [Kmr; Rao, 1960; Masoodi *et al.*, 1986]

Host: Eggs of *L. obfuscata* on *Salix* spp. *Populus* spp.

(2) *Anastatus japonicus* [Kmr, Islam *et al.*, 1986; Narendran, 2009]

Host: Eggs of *Lymentria obfuscata*

(3) *Anastatus kashmirensis* - [Kmr; Mathur,1956,Hayat,1975; Rao and Hayat, 1986]

Host: *L. obfuscata* on *Salix* and *Populus*, *Porthetria dispar.*

(4) *Anastatus* sp. [Kmr, Rishi and Shah,1985; Singh and Udaya,1987]

Host: Eggs of *L. obfuscata* on *Salix* spp. and *Populus* spp., *P. dispar.*

(5) *Eupelmus kashmiricus* [Kmr, Narendran *et al.*, 2001]

Host: *Scolytus* sp. on *M. pumila* (= *Malus sylvestris*)

(6) *Eupelmus valsus* [Kmr, Narendran *et al.*, 2001]

Host: *Scolytus* sp. on *M. pumila*

(7) *Eupelmus vindox* [Kmr; Buhroo *et al.*, 2001]

Host: *Scolytus* sp. on *M. pumila.*

9.4.11 Eurytomidae (Chalcidoidea : Hymenoptera)

(1) *Eurytoma morio* [Kmr, Buhroo *et al.*, 2001; Narendran *et al.*, 20001]

Host: *Scolytus* sp. on *Malus pumila* (=*M. sylvestris*)

(2) *Eurytoma* sp.[Kmr, Masoodi *et al.*, 1986]
Host: *L. obfuscata* on *Salix* sp.

9.4.12 Mymaridae (Chalcidoidea : Hymenoptera)

(1) *Erythemelus* sp .[Kmr; 6]
Host: Unknown

(2) *Gonatocerus aegypticus* (= *Gonatocerus tarae*) [Kmr; 5, 6]
Host: Eggs of unknown insect

(3) *Gonatocerus longicornis* [Jmu; 6, 2]
Host: Eggs of *Cofana spectra, Nephotettix* sp.

(4) *Lymaenon munnarus* (= *Gonatocerus munnarus*) [J&K; 6]
Host: Eggs of *C.spectra* and *N. nigropictus*

(5) *Lymaenon aureus* (= *Gonatocerus pahalgamensis*) [Pah; 3, 5, 6]
Host: Eggs of unknown insects

(6) *Palaeoneura unimaculatum* [J&K; 1, 3]
Host: Unknown

(7) *Polynema anantanagona* (= *Paralletoptera anantanagona*) [Ant; 1,3,4]
Host: Unknown

Key to numerals in long brackets (1-6): 1: SubbaRao and Hayat (1986); 2: Mani and Saraswat (1973); 3: Manickava *et al.* (2018); 4: Hyat and Anis (1999); 5: Triapitsym (2013); 6: Zeya and Hyat (1995).

9.4.13 Pteromalidae (Chalcidoidea : Hymenoptera)

(1) *Asaphes fletacheri* [Kmr;14)
Host: *Eriosoma lanigerum* (aphid) on *Malus pumila.*

(2) *Asaphes suspensus* [Kmr, Ldk; 7, 9, 11, 15, 16)
Host: Aphidiid parasitoids: *Aphidius areolatus – Periphyllus aesculi* on *Aesculus indicus; A. avenae – Chaetosiphan chaetosiphon* on *R. macrophylla, Rosa indica; matricariae – Capitophorus elaeagni* on *Carduus edelbergi; A. rosae – C. chaetosiphon* on *R. macrophylla; Dieretiella rapae – Breviconyne brassicae* on *Brassica oleracea; Proan volucre – Macrosiphum* sp. on *Rosa* sp; *Binodoxys indicus – Aphis fabae solanella* on *Solanum nigrum* (= *miniatum*), *Aphis affinis* group on *M. longifolia;* unknown aphid species in Ladakh region.

(3) *Coruna clavate* [Kmr; 10, 11, 15]
Host: Aphidiid parasitoids on aphid (*Macrosiphum*); *Binodoxys indicus* (=*Trioxys indicus*) on aphids

(4) *Coruna* sp. [Kmr; 6, 7, 8]
Host: Aphid parasitoids: *A. areolatus – P. aesculi* on *A. indicus; Aphidius* spp. – *Chaetosiphon* sp. on *Rosa* sp., *M. rosae* on *Rosa* sp., *Macrosiphun* sp. on *Rosa indica; Ephedrus srinagarensis – Prociphilus* sp. on *L. quinquelocularis; P. volcure – Hyalopterus pruni* on *Prunus domestica insititia; Praon* sp. – *Chaetosiphon* sp. on *R. macrophylla.*

(5) *Cheiropachus quadrum* [Kmr, Ant; 13, 15, 17]

Host: *Scolytus* sp. on *M. pumila; Ips stebbingi, Polygrapha major* and *Pityogenes scitus* on *Pinuis wallichiana*

(6) *Dibrachys microgastri* (= *Dibrachys carvus*) [Ldk; 12, 16]

Host: Polyphagous parasite of lepidopteran pupae

(7) *Halticoptera budgami* [Kmr (Bud); 18]

Host: Unknown

(8) *Halticoptera circulus* [Ldk;16]

Host: Parasitoid of Diptera (Agromyzidae, cecidomyiidae and chlorophoridae)

(9) *Halticoptera smaragdina* [Ldk;16]

Host: Parasitoid of unknown insect host

(10) *Herbertia indica* [Kmr, 4]

Host: *Liriomyza trifolii* on *L. siceraria* and *L. cylindrica*

(11) *Heydenia indica* [Kmr; 13]

Host: *Scolytus* sp. on *M.pumila*

(12) *Lamprotatus splendens* [Ldk;16]

Host: Parasitoid of leaf-mining insects

(13) *Macroglenes conjungens* [Kmr.; 2a]

Host: Unknown

(14) *Macromesus harithus* [Kmr (Ant); 3, 15, 17]

Host: *Scolytus* sp. on *M. pumila; Ips stebbings, Polygraphus major* and *Pityogenes scitus* on *Pinus wallichiana*

(15) *Metacolus beesoni* (= *Zapachiua beesoni*) [Kmr; 14, 15]

Host: Unknown

(16) *Metacolus parasinicus* [Ant (Kmr); 17]

Host: *Ips stebbingi, Polygraphus major* and *Pityogenes scitus* on *Pinus wallichiana*

(17) *Metacolus sinicus* [Ant (Kmr); 17]

Host: *I. stebbingi, P. major* and *P. scitus* on *P.wallichiana*

(18) *Metacolus unifasciatus* [Kmr; 12]

Host: Unknown

(19) *Pachyneuron aphidis* [Kmr; 7, 8, 10, 11]

Host: Aphid parasitoids: *Aphidius gifuensis – A. fabae* on *Capsella bursa pestoris; A. matricariae- Capitophorus elaeagni* on *C. edelbergi; A. uzbekistanicus – Macrosiphum ? avenae* on *Sorghum halepense; Aphidius* sp. – *Aphis farinosa* on *Salix aegyptica; P. volucre – H. pruni* on *P.d. insititia; Trioxys acalephae – M. rosae* on *R. webbiana; Trioxys rishii – C. elaeagni* on *C. edelbergi*

(20) *Pachyneuron formosa* [Kmr.; 2a]

Host: Unknown

(21) *Pachyneuron groenlandicum* (=*P.karnalensis*) [Kmr; 7, 10, 15]

Host: Aphid parasitoid: *Trioxys rishii – C. elaeagni* on *C. edellergi; Brachycaudus cardui* (aphid)

(22) *Pachyneuron nelson* [Ldk; 16]

Hosts: Pasitoid of Diptera (syrphidae), Hemiptera (aphididae)

(23) *Pteromalus semotus* [Ldk; 16]

Host: Primary parasite or hyperparasites of lepidoptera and some coleopterans

(24) *Pteromalus puparium* [Kmr, Ldk; 5, 16]

Host: *Pieris brassica* on *Brassica oleraceous capitata*, parasitoid of pupae of butterflies, hyperparsasitoid

(25) *Rhaphitelus maculatus* [Kmr; 13, 15]

Host: *Scolytus* sp. on *M. pumila*

(26) *Roptrocerus mirus* [Ant (Kmr); 17]

Host: *Ips stebbingi, Polygraphus major* and *Pityogenes scitus* on *Pinus wallichiana*

(27) *Schizonotus latus* (= *Pteromalus latus*) [Ldk;16]

Host: Unknown

(28) *Sphaeripalpus lacunosus* [Ldk; 16]

Host: Unknown

(29) *Spalangia nigra* [Kmr; 2a]

Host: Unknown

(30) *Sphegigaster agarwali* [Kmr; 1]

Host: Unknown

(31) *Systasis shalimarensis* [Kmr; 2]

Host: Unknown

Key to numerals given in long brackets in front of above listed pterolamid species: 1: Ahmad *et al.* (2012 c); 2: Ahmad *et al.* (2012); 2a: Ahmad *et al.* (2013a); 3: Bhat and Bhagat (2008); 4: Bhat *et al.* (2009); 5: Bhat *et al.* (2010); 6: Bhagat (1981); 7: Bhat(1983); 8: Bhat (1986); 9: Bhat (1987); 10: Boucek *et al.* (1978); 11: Farooqi and Subba Rao (1986); 12: Gupta (2016); 13: Narendra *et al.* (2001); 14: Subba Rao and Hayat (1986); 15: Sureshan and Narendran (2003); 16: Sureshan (2012); 17: Khanday *et al.* (2019); Khurshid *et al.* (2018)

9.4.14 Torymidae (Chalcidoidea : Hymenoptera)

(1) *Torymus barsulicus*[Kmr; Narendran and Razak, 2010]

Host: Unknown

(2) *Ecdamua lehensis* [Ldk; Sureshan, 2010]

Host: Unknown

(3) *Ecdamua bouceki* [Ana; Ahmad *et al.*, 2012]

Host: Unknown

(4) *Megastigmus kashmiricus* [Ldk; Sureshan, 2009]

Host:Unknown

(5) *Megstigmus* sp. [Leh, Sureshan,2009]

(6) *Monodontomerus acreus*[Kmr; Annymous,1970

Host: *L. obfuscata* on unknown host plant.

9.4.15 Trichogrammatidae (Chalcidoidea : Hymenoptera)

(1) *Trichogramma cacoeciae pallidium* (Mayer) [Ldk; Hayat and Viggiani, 1984]

Host: *Cydia pomonella* on *M. pumila* (apple cultivated)

(2) *Trichogramma embryophagum*(= *Encyrtus embryophagum* Hartig) [Kmr, Ldk; Hayat and Viggiani, 1984]

Host: *C. pomenella* on *M. pumila*

(3) *Trichogramma flandersi* [Kmr; Ahmad *et al.*, 2002]

Hosts: *Agrius convolvuli* (Hawk moth), *Chilo fuscatellus* (yellow top Borer)

(4) *Trichogramma japonicum* [J&K; Ahmad *et al.*, 2002]

Host: *Chilo polychrysa, Trichoplusia*

(5) *Trichogramma kashmiricus* [Kmr; Nagaraja *et al.*, 2007]

Host: Sciomyzid on Paddy leaf

(6) *Trichogramma* sp. [Kmr; Bhat *et al.*, 2009]

Host: *Helicoverpa armigera* on *Solanum tuberosum* and *Lycopersicon esculentum*

9.4.16 Figitidae (Cynipoidea : Hymenoptera)

(1) *Alloxysta* sp. [Kmr; Bhagat,1983, 1987]

Host: Aphidiid parasitoid: *Aphidius uzbekistanicus –Impatientinum impatiens dalhousiensis* on *Impatiens, Macrosiphum avenae* sp., *Sitobion avenae* (=*Macrosiphum avenae*) on *Rosa* sp., *Macrosiphum* sp. on *Rosa brunoni; Lysiphlelus ambiguus – Sipha maydis* on *Poa angustifolia; P. volucre – Myzaphis rosarum* on *Rosa* sp., *Toxares deltiger – Betacallis* sp. on *Conium maculatum; Toxares zakai – Microlophium* sp. on *urtica dioica.*

(2) *Aspicera hartigi* [Srg; Rather *et al.*, 2019]

Host: Diptera (Syrphid fly) on *Ephedra* sp.

(3) *Phaenoglyphis indica* [Kmr; Ferre-Suay and Pujade-villar,2013)

Host: Unkown

(4) *Phaenoglyphis* sp.[Kmr; Bhagat,1983]

Host: Aphidiid parasitoid: *Aphidius ervi – Microlophium carnosum; Aphidius gifuensis- Aphis* sp. on *Nepeta salvifolia*

9.4.17 Braconidae (Ichneumonoidea : Hymenoptera)

(a) Aphidiinae

(1) *Adialytus ambiguus* (Haliday) [16, Bat, Raj]

Host: *Macrosiphum* sp.on *Sorghum*

(2) *Adialytus arvicola* (Stary) [25, Sop]

Host: Aphid undermined species on unknown host plant

(3) *Adialytus salicaphis* (Fitch) [19, 25, 31, Sop, Kmr, Srg]

Host: *Chaitophorus kapuri* on *Populus* sp., *Chaitophorus* sp. on *Populus* sp.

(4) *Areopraon lepelleyi* [2, 4, 10, 31, Kmr]

Host: *Eriosoma* (*Schizoneura*) *ulmi* on *Ulmus villosa* (=*laevigata*); *Schizoneurella indica* on *Ulmus wallichiana, Eriosoma lanigerum on Malus domestica*

(5) *Aphidius absinthi* Marshall [16, 31, Jmu, Kmr]

Host: *Macrosiphoniella* sp.on *Chrysanthemum; Macrosiphoniella sanbornii* on *Chrysanthemum cinerarifolium*

(6) *Aphidius areolatus* Ashmead [8, Kmr]

Host: *Periphyllus aesculi* Hille Ris Lambers on *Aesculus indicus; P.vandenboschii* on *Acer caesium*

(7) *Aphidius avenae* Haliday [9, Kmr]

Host: *Chaetosiphon glaber* on *Rosa* sp.; *Macrosiphum* sp. on *Rosa macrophylla; Myzaphis turanica* on *Rosa* sp.

(8) *Aphidius bifurcatus* Rakhshani and Tomanovic [20, Raj]

Host: Unknown aphids, collected in *Zea mays* and wild grass field by sweeping

(9) *Aphidius cingulatus* Rathe [30, Dac]

Host: *Pterocomma populeum* on *Populus aurea*

(10) *Aphidius colemani* Viereck [11, 20, 30,31, Srg, Gul, Raj]

Host: *Aphis nerii* on *Calotropis* sp.; *Melanaphis donacis* on *Arundo donax; Dysaphis pyri* on *Prunus persica* and *Malus domestica*

(11) *Aphidius eglanteriae* Haliday [9, Kmr]

Host: *Chaetosiphon tetrarhodum* on *Rosa macrophylla; Myzaphis rosarum* on *Rosa webbiana*

(12) *Aphidius ervi (Haliday)* [12, 13, 16, Kmr]

Host: *Acyrthosiphon pisum* on *Medicago* sp., *Pisum sativum* and *Trifolium* sp.; *Acyrthosiphon* sp. on *Vicia sativa; M. carnosum* (= *evansi*) on *Urtica dioica*

(13) Aphidius *gifuensis* Ashmead [9, 15, Kmr]

Host: *Aphus affinis* on *Mentha longifolia; Aphis* sp. on *Veronica persicae* and *Vicia sativa*

(14) *Aphidius hortensis* Marshall [30, 31, Kmr]

Host: *Liosomaphis himalayensis* on *Berberis pseudoumbulata*

(15) *Aphidius matricariae* Haliday [9, Jmu, Kmr- Man, Kup]

Host: *Aphis craccivora* on *Solanum nigrum; A. fabae on Rumex nepalensis;Aphis* sp. on *Cynoglossum glochidatum: Brachycaudus helichrysi* on*Ageratim conyzoides,Brassica oleraceous, Notonia grandifolia, Erigeron canadensis, Prunus persica, Prunus domestica* and *Malus pumila;Capitophorus elaeagni* (common oleaster aphid) on *Carduus edelbergi; Chaitophorus kapuri* on *Populus* sp; *Phorodon cannais* (*Diphorodon cannabis*) on *Cannabis sativa; Microlophium carnosum* on *Urtica dioica*; n *Myzus persicae* on *Descurainia sophia, Lycopersicon esculentum, Phlox dummondi; Myzus umecola* on *Rubia cardifilia*

(16) *Aphidius rosae* Haliday [18, 19, 31 Kmr]

Host: *Macrosiphum* on *Rosa macrophylla; Sitobion rosaeiformis* on *Rosa damascena*

(17) *Aphidius salicis* Haliday [9, Kmr- Srg]

Host: *Aphis* sp. on *Dioscorea deltoidei, Senecio chrysanthemoides; Cavariella biswasi* on *Ferula jaeschkeana; Semiaphis heraclei* on *Daucus carota*

(18) *Aphidius salignae* Watanabe [16, Ban]

Host: *Tuberolachnus salignus* on *Salix alba*

(19) *Aphidius smithi* Subba Rao and Sharma [9, 11,12, 20, Raj]

Host: *Acyrthosiphon pisum* (pea aphid) on *Lathyrus aphaca, Medicago* sp., *Ranunculus muricatus* and *Ranunculus* sp.; unknown aphid host, collected on wild grass by sweeping

(20) *Aphidius transcaspicus* Telenga [16, 20, Poo, Raj]

Host: *Aphis* sp. on *Prunus domestica;* unknown aphid,collected in grass field by sweeping

(21) *Aphidius uzbekistanicus* Luzhetzki (8, 20, 31, Kmr, Raj]

Host: *Impatientinum impatiens dalhousiensis* on *Impatiens; Sitobion* sp. on *Pennistum flaccidum;* unknown aphid, collected in wild grass field by sweeping

(22) *Aphidius* spp. [9, 10, 15, 20, 30, 31, Jmu, Kmr- Gul, Zab, Srg, Band, Bee]

Host: *Aphis farinosa* on *Salix aegyptica, S.alba* and *Salix* sp.; *Capitophorus* sp.on *Carduus* sp.; *Chaetosiphon gracilicorne* on *Rosa webbiana;Chaetosiphon* sp. on *Rosa* sp.; *Chaitophorus himalayaensis* on *Salox flabellaris; Coloradoa rufamaculata* on *Chrysanthemum; Eriosoma* sp. on *Ulmus* sp.; *Eumyzus* sp. on *Nepeta cataria; Lachnus tropicalis* on *Pinus wallichiana; Macrosiphoniella sanborni* on *Chrysanthemum morifolium* and *Chrysanthemum* sp.; *Myzus* sp. on *Euphorbia helicoscopia* and *Sisymbrium; Rhopalosiphum nymphaeae* on *Nymphaea alba*; unknown aphid, collection through sweeing in wild grass field

(23) *Betuloxys intermediatus* Shujauddin

(= *Trioxys* (*Betuloxys*) *intermediatus*) [22, Sop]

Host: Unknown aphid species on *Centaurea iberica*

(24) *Binodoxys acaplephae* (Marshall

(= *Trioxys* (*Binodoxys*) *acalephae* Marshall) [9, 15, 29, Kmr- Srg, Bur, Zab]

Host: *Aphis affinis* on *Mentha longifolia; Macrosiphum* spp on *Rosa* sp.; *Phorodon cannabis* on *Cannabis sativa*

(25) *Binodoxys basicurvus* Shuja-uddin

(= *Trioxys (Binodoxys) basicurvus* Shuja-uddin) [23, Sop]

Host: *Aphis gossypii* on Rubus sp.

(26) *Binodoxys brevicornis* (Haliday)

(= *Trioxys* (*Binodoxys*) *brevicornis* Haliday) [9, Srg]

Host: *Chaitophorus pakistanicus* on *Salix babylonica*

(27) *Binodoxys centaureae* Haliday

(= *Trioxys* (*Binodoxys*) *centaureae* Haliday) [9, 13, 14, 29, 31; Kmr- Srg]

Host: *Mcrosiphoniella artimisae* and *M. pseudoartimisae* on *Artemisia indica; Macrosiphoniella sanborni* on *Chrysanthemum morifolium; Macrosiphoniella* sp. on *Achiellia millefolium* and *Chrysanthemum* sp.; *Uroleucon compositae* on *Chrysanthemum cinaerifolium*

(28) *Bionodoxys indicus* Subba Rao and Sharma [20, Jmu- Raj]

(= *Trioxys* (*Binodoxys*) *indicus* Subba Rao and Sharma) [9, 10, 12, 15, 16,29, Kmr- Khi]

Host: *Aphis affinis* on *Mentha longifolia; Aphis craccivora* on *Robinia pseudoacacia, Rumex nepalensis, Solanum miniatum* and *S. nigrum; Aphis fabae solanella* on *R. nepalensis; Aphis gossypii* on *Melia azedarach* and *Urtica dioica; Aphis pomi* on *Malus pumila; Eriosoma* sp. on *Cotoneaster aitkinsoni; M. pseudoartemisae* on *Artemisia absinthium; Macrosiphum* sp on *Rosa brunoni; Mindarus japonicus* on *Abies pindrow; Phorodon* sp. on *Punica granata;* unknown aphid, collection through sweeping in *Zea mays* field

(29) *Binodoxys jaii* Bhagat

(= *Trioxys* (*Binodoxys*) *jaii* Bhagat) [6, 10, Srg]

Host: *Aphis craccivora* on *Robinia pseudoacacia; Shivaphis celti* on *Celtis australis*

(30) *Binodoxys kashmirensis* Takada and Rishi [30, 31, Kmr]

Host: *Macrosiphum euphorbiae* on *Solanum melongena; Macrosiphum* sp. on *Rosa brunonii*

(31) *Binodoxys longispinus* Shuja-uddin

(= *Trioxys* (*Binodoxus*) *longispinus* Shuja-uddin) [28, Kmr]

Host: *Aphis farinosa* on *Salix caprea*

(32) *Binodoxys rosaecola* Bhagat

(= *Trioxys* (*Binodoxys*) *rosaecola* Bhagat) [6, Kmr]

Host: *M. rosarum* on *Rosa webbiana*

(33) *Binodoxys rubicola* Shuja-uddin

(= *Trioxys* (*Binodoxys*) *rubicola* Shuja-uddin) [23, Sop]

Host: *Aphis gossypii* on *Rubus* sp.

(34) *Binodoxys uroleucon* Takada and Rishi [30, Kmr]

(= *Trioxys uroleucon*)

Host: *Uroleucon compositae* on *Chrysanthemum cinaerifolium; Uroleucon* sp. on *Chrysanthemum morifolium*

(35) *Binodoxys* sp.

(= *Trioxys* (*Binodoxys*) sp.) [10, 16, Kmr- Ban]

Host: *Aphis craccivora* on *Robinia; Mindrus japoinicus* on *Abies pindrow*

(36) *Diaeretiella rapae* (M'Intosh) [13, 16, 20, 30, Kmr- Srg, Jmu- Raj]

Host: *Aphis craccivora* on *Solanum tuberosum; Brevicoryne brassicae* on *Brassica oleracea* var. *acephala, Descuraini sophia; Lipaphis erysimi* on *Brassica napus* (= *Brassica campestris*); *Uroleucon sonchi* on *Sonchus oleraceous*; unknown aphid, collected through sweeping in wild grass and *Zea mays* field

(37) *Ephedrus* (*Ephedrus*) *campestris* Stary [26, Kmr]

Host: *Macrosiphoniella sanborni* on

(38) *Ephedrus* (*Ephedrus*) *dioscoreae* Bhagat [7, 15, Kmr]

Host: *Aphis* sp. and *Aulacorthum* sp. on *Dioscorea deltoidea*

(39) *Ephedrus laevicollis* Dalla

(= *Ephedrus minor* Stelfox) [30, 31, Kmr- Har]

Host: *Cavariella aegopodii* on *Salix alba; Cavariella* sp. on *Salix alba*

(40) *Ephedrus niger* Gautier, Bonnamour and Gaumont [13, 30, Kmr]

Host: *Macrosiphoniella* sp. on *Chrysanthemum* sp.; *Uroleucon* sp. on *Chrysanthemum morifolium*

(41) *Ephedrus longiarticulatus* Rakhshani and Tomanovic [20, Raj]

Host: Unknown aphid, collection through sweeping in wild grass and *Zea mays* field

(42) *Ephedrus persicae* Froggatt [9, 10, 12, 29 Kmr]

Host: *Aphis craccivora* on *Rumex acetosella; Chaitophorus populeti* on *Populus capsica; Hyalopterus pruni* on *Prunus domestica; Melanaphis donacis* on *Arundo donax* on *Arundo donax*

(43) *Ephedrus plagiator* (Nees) [4, 9, 10, 20, 31; Kmr- Duc, Jmu- Raj]

Host: *Michelia champa* on *Prociphilus micheliae; Prociphilus* sp. on *Lonicera quinquelocularis; Tetraneura* on *Ailantus altissima;* aphid species not known, collected through sweeping in wild grass field

(44) *Ephedrus salicicola* Takada [4, 9,10, Kmr- Kan]

Host: *Cavariella aegopodi* on *Salix aegyptica; Cavariella* sp. on *Salix babylonica; Eumyzus* on *Nepeta cataria*

(45) *Ephedrus srinagarensis* Stary and Bhagat [4, 9, 29, Kmr]

Host: *Prociphilus* sp. on *Lonicera quinquelocularis*

(46) *Ephedrus (Ephedrus) urticae* Bhagat [7, Bee]

Host: *Microlophium carnosum* (Common nettle aphid) on *Urtica dioica*

(47) *Ephedrus (Ephedrus)* sp.[16,20, Kmr- Ban, Jmu- Raj]

Host: *Acyrthosiphon* sp. on *Quercus elex; Aulacorthum* sp. on *Spirea* sp.; unknown aphid, collection by sweeping in *Zea mays* field

(48) *Kashmiria aphidis* Stary and Bhagat [29, Kmr- Peri.m]

Host: *Myzus* sp. and *Prociphilus* sp.on *L. quinquelocularis*

(49) *Lipolexis gracilis* Forester [9, 10, 13, 16, 29, 30, Jmu, Kmr]

Host: *Acyrthosiphon rubi* on *Rubus fruticosus; Aphis affinis* on *Mentha aquatica* and *M. longifilia* (= *sylvestris*); *Aphis craccivora* on *Robinia pseudoacacia* and *Solanum nigrum; Aphis gossypii* on *Zinnia elegans; Aphis nerii* on *Verbascum thapsus, Liosomaphis himalayensis* on *Berberis pseudoumbulata*

(50) *Lipolexis oregmae*(Gahan)

(= *Lipolexis scutellaris* Mackauer)[29, Kmr]

Host: *Aphis craccivora* on *Robinia pseudoacacia*

(51) *Lysaphidius erysimi* [Kmr, 31]

Host: *Lipaphis erysimi* on *Brassica oleraceous capitata, Brassica oleraceous, Brassica juncea, Brassica rapa* and *Raphanus sativus*

(52) *Lysaphidius staryi* Bhagat [5, Srg]

Host: *Sitobion avenae* (= *Macrosiphum* (*Sitobion*) *avenae*) on *Pennisetum flaccidum*

(53) *Lysiphlebia mirazai* Shuja-uddin [17, 20, Srg, Raj]

Host: *Sitobion avenae* (= *Macrosiphum avenae*) on *Triticum vulgare*; aphid not known, collection made through sweeping in wild grass field

(54) *Lysiphlebus ambiguus* Haliday [13, Kmr]

Host: *Sipha* (*Rungsia*) *maydis* on *Sorghum halepense*

(55) *Lysiphlebus confusus* Trembly and Eady [4, 13, Kmr]

Host: *Aphis* nr.*intybi* on *Cichorium intybus*; *Chaitophorus niger* on *Populus capsica*; *Sipha maydis* on *Hordeum vulgare*, *Sorghum halepense* and *S.vulgare*; *Melanaphis sacchari* on *Sorghum* sp.

(56) *Lysiphlebus fabarum* (Marshall) [9, 10, Kmr]

Hosts: *Acrythosiphon rubi* on *Rubus fruticosus*; *Aphis craccovora* on *Rumex nepalensis*; *Aphis* nr. *intybi* on *Cichorium intybus*; *Aphis* nr. *salviae* on *Salvia moorcraftiana*

(57) *Monoctonus crepidis* Haliday [9, Srg]

Host: *Liosomaphis atra* on *Berveris pseudoumballatus*; *Liosomaphis* sp. on *Berberis zebbiliana*

(58) *Monoctonus* (*Monoctonus*) *fotedari* Bhagat [1, Kmr]

Host: *Myzaphis rosarum* on Rosa sp.

(59) *Monoctonus* sp. [30, Kmr]

Host: *Macrosiphoniella oblonga* on *Artemisia vestita*

(60) *Puesia gulmargensis* Bhagat

(= *Pauesia* (*Paraphidius*) *gulmargensis* Bhagat)[3, Gul]

Host: *Cinara lachnirostris* on *Pinus* sp.; *Cinara* sp. on *Abies pindrow*

(61) *Pauesia hazratbalensis* Bhagat

(= *Pauesia* (*Kashmirpauesia*) *hazratbalensis* Bhagat) [3, Srg]

Host: *Lachnus* sp. on *Thuja orientalis* (=*Biota orientalis*)

(62) *Pauesia himalayensis* Bhagat

(=*Pauesia* (*Paraphidius*) *himalayensis* Bhagat [3, Srg- Zab]

Host: *Lachnus tropicalis* on *Pinus wallichiana*

(63) *Pauesia* (*Paraphidius*) *pini* (Haliday) [3,10, Gul, Tang, Bar]

Host: *Cinara lachinirostris* and *Cinara* sp. on *Pinus wallichiana*

(64) *Praon abjectum* Haliday [9, 29, Kmr- Kok]

Hosts *Aphis farinosa* (Sallow aphid/Small willow aphid) on *Salix alba* and *S.babylonica; Aphis grossulariae* nr. *pollinosa* on *Epilobium hirsutum; Melanaphis donacis* on *Arundo donax; Myzaphis rosarum* on *Rosa* sp.

(65) *Praon dorsale* Haliday [9,13, 17, Kmr]

Host: *Macrosiponiella sanborni* on *Chrysanthemum morifolium; Macrosiphoniella* sp. on *Chrysanthemum* sp.

(66) *Praon kashmirensis* Bhagat [7,10, Kmr]

Host: *Aphis farinosa* on *Salix alba; Chaitophorus pakistanicus* on *Salix* sp.

(67) *Praon nymphaeae* Subba Rao and Sharma [9,13, Srg]

Host: *Rhopalosiphum nymphaeae* on *Nymphaea* spp.

(68) *Praon orientale* Stary and Schlinger [20, Raj]

Host: Unknown aphid, collected through sweeping in wild grass field

(69) *Praon stagona* Takada and Rishi [30, Zab]

Host: *Amphicercidus tuberculatus* and *Stagona* sp. on *Lonicera quinquelocularis*

(70) *Praon volucre* (Haliday) [8, 9, Srg- Nag, Nis,]

Host: *Aphis* sp. *Lespadeza* sp.; *Hyalopterus pruni* (= *arundis*) on *Prunus domestica* and *Berberis pseudoumbellatus; Macrosiphum* sp. on *Rosa* sp.; *Myzus turanica* on *Rosa* sp.; *Nosonovia (Hypermyzus) lactucae* (= *Hypermyzus lactucae*) on *Sonchus oleraceous; Uroleucon* sp. on *Sonchus asper.*

(71) *Praon* sp. [9, 13, Kmr]

Host: *Macrosiphoniella* sp. on *Chrysanthemum* sp.

(72) *Toxares deltiger* (Haliday) [9,10,12,15, 31 Kmr- Srg, Gul]

Host: *Aphis craccivora* on *Robinia pseudoacacia; Betacallis* on *Conium maculatum; Eumyzus sp. on Nepeta cataria; Hyalopterus pruni* on *Prunus domestica insititia; Macrosiphoniella yomogifolia* on *Artemisia absinthium; Metopolophium montanum* on *Rosa indica; Ovatus* nr. *crataegarius* on *Mentha longifolia.*

(73) *Toxares macrosiphophagum* Shuja-uddin [20, 24, Kmr, Raj]

Host: *Macrosiphoniella sanborni* on undetermined host plant; collected through trapping in wild grass field

(74) *Toxares shigai* Takada [18, 30, Kmr- Har]

Host: *Aphis farinose* on *Salix alba*

(75) *Toxares zakai* Shuja-uddin [24, Kmr]

Host: *Myzus persicae* on *Solanum nigrum*

(76) *Trioxys complanatus* Quilis

(= *Trioxys (Trioxys) complanatus* Quilis) [9, Srg]

Host: *Acrythosiphin pisum* on *Lythrus aphaca; Therioaphis* on *Medicago polymorpha, M. sativa* and *Vicia sativa*

(77) *Trioxys pallidus* (Haliday)

(=*Trioxys* (*Trioxys*) *pallidus* Haliday) [27, 31, Kmr]

Host: *Chromaphis juglandicola* (small walnut aphid) and *Panaphis juglandis* (Large walnut aphid on *Juglans regia*

(78) *Trioxys pannonicus* Stary

(= *Trioxys* (*Trioxys*) *pannonicus* Stary) [9, Kmr]

Host: *Macrosiphoniella pseudoartemisae* on *Artemisia* sp., *A. absinthium* and A.*indica*

(79) *Trioxys rishi* Stary and Bhagat

(=*Trioxys* (*Trioxys*) *rishii* Stary and Bhagat) [9,10,12, 29,32 Srg]

Host: *Aphis craccivora* on *Robinia pseudoacacia*; *Brachycaudus* sp. on *Carduus edelbergi*; *Capitophorus* sp. on *Carduus* sp.; *Hyalopterus pruni* on *Prunus domestica insititia*; *Diuraphis* (*Holcaphis*) *calamagrostis* (= *Hyalopterus arundinis*) on plum tree.

(80) *Trioxys shivaphis* Takada

(=*Trioxy* (*Trioxys*) *shivaphis* Takada) [6,10, Srg]

Host: *Shivaphis celti* on *Celtis australis*

(81) *Trioxys sporensis* Shuja-uddin

(= *Trioxys* (*Trioxys*) *sporensis* Shuja-uddin)[4, Sop]

Host: Unknown

Key to the numerals, given in the long brackets, in front of above listed aphidiine species: 1: Bhagat (1981a); 2: Bhagat (1981b); 3: Bhagat (1981c); 4: Bhagat (1982); 5: (Bhagat 1982a); 6: Bhagat (1982b); 7: Bhagat (1982c); 8: Bhagat (1983); 9: Bhagat (1984); 10: (1985); 11: Bhagat (1986); 12: Bhagat (1986a); 13: Bhagat (1986b); 14: Bhagat (1987); 15: Bhagat (1987a); 16: Bhagat and Ahmad (1995); 17: Bhagat (2000); 18: Dharmadhikari and Ramaseshia (1970); 19: Pawar and Parry (1989); 20: Rakhshani *et al.* (2012); 21: Rishi (1976); 22: Shuja-uddin (1970); 23: Shuja-uddin (1974); 24: Shuja-uddin (1976);25: Shuja-uddin (1978); 26: Shuja-uddin (1978a); 27: Shuja-uddin (1982); 28: Shuja-uddin (1983); 29: Stary and Bhagat (1978); 30: Takada and Rishi (1980); 31: Khan and Shah (2017); 32: Bhagat and Masoodi (1988)(addendum).

(b) Alysiinae

(1) *Phaenocarpa kashmirensis*[Khi, Pah, Son; Bhat,1979]

Host: Unknown insect and host plant species

(2) *Phaenocarpa tridentata* [Gul, Son, Pah; Bhat,1979]

(c) Blaciinae

(1) *Blacus* (*Blacus*) *imitator* [Srg;Papp, 1985]

Host: unknown insect and plant species

(d) Braconinae

(1) *Bracon brevicornis* [Ach, Lark; Anonymous,1970]
Host: *O. kashmirica*

(e) Cheloninae

(1) *Chelonus* sp.[Kmr.; Nighat *et al.*, 2002]
Host: *Diaphania pyloalis* (= *Glyphodes pyloalis*) on *Morus* spp.

(f) Doryctinae

(1) *Spathius* sp.[Kmr; Buhroo *et al.*, 2001; Narendran *et al.*, 2001]
Host: *Scolytus* sp. on *M.pumila*

(g) Euphorinae

(1) *Dinocampus coccinellae* [Kmr.; Bhat *et al.*, 2010)
Host: *Coccinella septempunctata* on vegetable crop

(2) *Meteorus pulchricornis* [Jmu; Gupta *et al.*, 2019]
Host: *Spilarctia obliqua* on *Ricinus communis*

(3) *Peristernus nitidus* [J&K; Shamim *et al.*, 2008]
Host: Unknown

(h) Hormiinae

(1) *Pambolus ruficeps* [J&K; Ahmad *et al.*, 2019]
Host: Unknown

(i) Microgastrinae

(1) *Apanteles conspersae* [Kmr; Anonymous,1969,1970]
Host: *L. obfuscata* on *populus* spp. and *salix* spp.

(2) *Apanteles flavicoxis* [Kmr; Rishi and Shah,1985]
Host: *L. obfuscata* on *Salix* spp.

(3) *Apanteles indiensis* [Kmr; Rishi and Shah,1985]
Host: *L. obfuscata* on *salix* sp.

(4) *Apanteles liparidis* [Kmr, Anonymous, 1969,70; Rishi and Shah,1985]
Host: *L. obfuscata* on *Salix* spp. and *Populus* spp.

(5) *Apanteles melanoscelus* [Kmr; Rishi and Shah,1985]
Host: *L. obfuscata* on *Salix* spp. and *Populus* spp.

(6) *Apanteles porthetriae* [Kmr; Zutshi, 1967; Rishi and Shah,1985)
Host: *L. obfuscata* on *Salix* spp. and *Populus* spp.

(7) *Apanteles nr. solitarius* [Kmr; Zutshi, 1967]
Host: *L. obfuscata* on unidentified plant

(8) *Apanteles* sp.[Ldk; Anonymous,1970; Pawar and Parry,1989; Ahamad and Zaki,2008]

Host: *Archips subsidiaria* on *M. pumila, Ostrinia kashmirica* (Corn borer) on *Cirsium* (= *Cnicus* sp.), *L. obfuscata* on *Salix* spp. *Populus* spp.; *Parnara guttata* (Rice Skipper) on paddy

(9) *Cotesia glomerata* [Kmr., Nighat *et al.*, 2002)

(= *Apanteles glomeratus*)

Host: *Diaphania pyloalis* (= *Glyphodes pyloalis*) on *Morus* spp.

(10) *Glyptapanteles agamemnonis* [Jmu; Gupta *et al.*, 2019]

Host: *Spilarctia obliqua* (hairy caterpillar) on *Ricinus communis*

(11) *Microplitis manilae* [Kmr; Gupta,2013]

Host: Unknown

(12) *Pholetesor hayati* [Srg; Akhter *et al.*, 2010]

Host: Unknown

(13) *Protomicroplitis indicus* [Poo; Ahmad and Usmani,2016]

Host: Unknown

(j) Opiinae

(1) *Opius* (*Opiothorax*) *baraelliensis* [Raj; Ahmad *et al.*, 2009]

Host: Unknown

(2) *Opius* (*Nosopaeopius*) *longiterebrata* [Raj; Ahmad *et al.*, 2009

Host: Unknown

(3) *Opius* (*Apodesmia*) *poonchensis* [Poo, Ahmad *et al.*, 2009]

Host: Unknown

(4) *Opius* (*Phaedrotoma*) *patnaensis* [Darhal, Ahmad *et al.*, 2009]

Host: Unknown

(k) Zelinae

(1) *Zele chloropthalma* Nees [Kmr, Bhat *et al.*, 2011]

Host: *Spodoptera litura* on vegetables (cole crops)- cabbage, cauliflower, Knol-khol and kale

9.4.18 Ichneumonidae (Ichneumonoidea : Hymenoptera)

(1) *Anilastus* sp./*Eriborus* sp. [Kmr., Anon., 1970]

Host: *Lymentria obfuscata*

(2) *Aglaojoppa flavomaculata* [Pha; Gupta, 1987,88]

Host: Unknown

(3) *Amblyaeneus malaise* [Pha; Gupta,1987,88]
Host: Unknown

(4) *Buathra excavate* [Gul,Khi; Jonathan,2006]
Host: Unknown

(5) *Buathra insidiator* [Kmr; Jonathan,2006]
Host: Unknown

(6) *Buathra luculenta* [Pha, Gul, Srg, Ldk; Jonathan,2006]
Host: Unknown

(7) *Buathra nigritergia* [Kmr; Jonathan,2006]
Host: Unknown

(8) *Buathra rufitergia* [Ldk (Dra); Jonathan,2006]
Host: Unknown

(9) *Caenocryptus rugosus* [Gul; Jonathan,1999,2006]
Host: Unknown

(10) *Campoletis chloridae* [Kmr; Bhat *et al.*, 2009,2010; Kapoor and Shankar, 2019]
Host: *T. orichelasa; B.rapae* on *A. cepa, P.sativum; H.armigera* on *S.tuberosum, L. esculentum*, Soya bean; *H. armigera* on *Vigna mongo*

(11) *Campoletis longipes* [Kmr; Ahmad and Dar,1978; Hortmann,1992]
Host: Unknown

(12) *Charops bicolor* [Kmr; Bhat *et al.*, 2009]
Host: *H.armigera* on *L.esculentum*

(13) *Ctenichneumon panzer* [Ldk; Gupta,1987, 88]
Host: Unknown

(14) *Ctenichneumon victoriae* [Ldk, Gupta,1987,88]
Host: Unknown

(15) *Deuteroxorides indicus* [Pah; Gupta,1980, 85]
Host: Coleopteran larvae on hardwood coniferous tree

(16) *Diadegma fenestral* [Kmr; Bhat *et al.*, 2009; Bhat and Bhagat, 2009]
Host: *H.armigera* on *L. esculentum*

(17) *Ecthrus coracinus* [Gul; Gupta,1978]
Host: Coleopteran larvae

(18) *Enicosphilus striolatus* [Bha; Nikam,1980]
Host: Unknown

(19) *Eridolius aithogaster* [Pah; Rao and Hayat,1986; Gupta,1991]
Host: Larvae of sawflies (Tenthredinidae)

(20) *Eucremastus parvipes* (=*Cephalobolus parvipes*) [Kmr; Ahmad and Dar, 1978)

Host: Unknown

(21) *Eugalta leucopus* [Pha; Gupta,1980]

Host: Unknown

(22) *Exeristes roborator*[Lar, Ach; Anonymous,1970]

Host: *O. kashmirica* on *Cirsium sp*

(23) *Exetastes fornicator* (=*Ichneumon fornicator*)[Kmr; Ahmad and Dar, 1978]

Host: Unknown

(24) *Glabridorsum semilunatum* [Gul; Jonathan,2006]

Host: Unknown

(25) *Hyposoter ebeninus* [Kmr; Bhat and Bhagat, 2009 a; Bhagat *et al.*, 2010b]

Host: *Pieris deplidice, P. braassicae, P.rapae* on *Brassica* spp. and its varities

(26) *Ichneumon rufofemoratus* [Ldk; Gupta,1987]

Host: Unknown

(27) *Itamoplex carnifrons* [Khi; Jonathan,2000,2006]

Host: Unknown

(28) *Itamoplex lanceolatum* [Gul. Jonathan, 2000,2006]

Host: Unknown

(29) *Itamoplex kashmirensis* [Gul; Jonathan,2000]

Host: Unknown

(30) *Itoplectis himalayensis* [Pah, Bat, Srg; Gupota,1967; Rishi and Shah, 1985; Bhat and Bhagat, 2008]

Host: *L. obfuscate* on *Salix* spp and *Populus* spp.; *Plutella xlostella* on *rassica* spp. (*B.o., B.o.campestris*)

(31) *Itoplectis tibetensis* [Gul; Gupta, 1967]

Host: Unknown

(32) *Itoplectis* sp.[Sgr; Simmonds and Rao, 1960]

Host: *P. xylostella* on unknown crop

(33) *Myrmeteonostenus nigrimaculatus* [Pah; Jonathan,2000; Jonathan.2006]

Host: Unknown

(34) *Netelia montanes* (=*Paniscus montanes*) [Kmr; Cameron,1906]

Host: Unknown

(35) *Netelia ocellaris*(=*Paniscus ocellaris*) [Srg; Ahmad and Dar,1978]

Host: Unknown

(36) *Netelia kashmirensis* (= *Paniscus kashmkirensis*) [Kmr; Cameron,1906]

Host: Unknown

(37) *Pimpla arctica* [Kmr; Ahmad and Dar,1978]

Host: Unknown

(38) *Pimpla alternans* [Kmr; Annonymous, 1970

Host: Unknown

(39) *Pimpla cyanator* (= *Coccygominus cyanator*)[Kmr; Pha; Gupta and Saxena,1987]

Host: Unknown

(40) *Pimpla laothoe* (= *Coccygominus laothoe*) [Kmr; Rao, 1972; Gupta and Saxena, 1987]

Host: *L. obfuscata*

(41) *Pimpla instigator* (= *Coccygominus instigator*) [Ldk; Gupta and Saxena,1987]

Host: *L. obfuscate*

(42) *Pimpla indra* (= *Coccygominus indra*)[Srg; Gupta and Saxena, 1987]

Host: Unknown

(43) *Pimpla turionellae* (= *Coccygominus turionellae*) [Srg; Gupta and Saxena, 1987]

Host: *L. obfuscata*

(44) *Pimpla semitibialis* (= *Coccygominus semitibialis*) [Dra (Ldk); Gupta and Saxena, 1987]

Host: *Unknown*

(45) *Pimpla russula* (= *Coccygominus russula*) [Veri; Gupta and Saxena, 1987]

Host: Unknown

(46) *Pimpla shiva* (= *Coccygominus shiva*) [Pah; Gupta and Saxena, 1987]

Host: Unknown

(47) *Pimpla* sp. [Kmr, Anonymous, 1969; Rao, 1966; Bhat *et al.*, 2010b]

Host: Pupae of *L.obfuscata* on *Salix* spp. and *Populus* spp.; *Pieris brassicae* on *Brassica oleracea gongyloides*

(48) *Platylabus taiwanus*[Ldk, Pah; Gupta,1987]

Host: Unknown

(49) *Platylabus takenchii* [Gul; Gupta, 1987]

Host: Larvae of wood boring (hard wood) coleopterans

(50) *Poemenia kashmirica* [Pah; Gupta,1980, 85, 99]

Host: Larvae of coleropteran (wood boring conifers)

(51) *Polyblastus kashmirensis* [Pah; Gupta, 1984]
Host: Larvae of sawfly

(52) *Pristomerus sulci* [Kmr; Mittal *et al.*, 2004]
Host: *Diaphania pyloalis* (= *Glyphodes pyloalis*) on *Morus* sp.

(53) *Pristomerus* sp.[Lar, Ach; Anonymous,1970]
Host: *Ostrinia kashmirica* on *Crisium*

(54) *Protichneumon pisorius* [Kmr.; Wachkoo *et al., 2019*]
Host: Unknown

(55) *Pterocormus americanicolor* [Kmr; Gupta, 1987]
Host: Unknown

(56) *Scambus aithomelus* [Srg; Gupta and Tikar, 1967]
Host: *O.kashmirica* on *Crisium wallichi*

(57) *Scambus flavicrus* [Lar, Ach; Annonymous, 1970]
Host: *O. kashmirica* on *Crisium* sp. (*Crinus* sp)

(58) *Scambus indicus* [Gul; Gupta and Tikar, 1967]
Host: Larvae of lepidopterans

(59) *Scambus kashmiricus* [Sho; Gupta and Tikar, 1967]
Host: Lepidopteran larvae and spider eggs

(60) *Scambus nursei* [Kmr; Narayanan *et al.*, 1955]
Host: Larvae of lepidopterans and coleopterans

(61) *Scambus* sp.[Kmr; Bhat *et al.*, 2010]
Host: *Thysoplusia orichalcea* on *Phaseolus vulgaris*

(62) *Theronia atalantae* [Kmr. Anonymous,1970; Rishi and Shah,1985]
Host: *L. obfuscata* on *Salix* spp. and *Populus* spp.

(63) *Theronia* sp. { Kmr.; Zutshi,1962; Masoodi *et al.*, 1986]
Host: *L. obfuscata* (pupa) on *Salix* and *Populus* spp.

(64) *Townesia exilis* [Gul; Shu-Ping *et al.*, 2019]
Host: Unknown

(65) *Triptognathus subalpinus* [Ldk; Gupta,1987]
Host: Unknown

(66) *Xanthopimpla nana* [Kmr; Akhter *et al.*, 2010]
Host: Unknown

(67) *Xanthopimpla* sp. [Kmr.; Mittal *et al.*, 2011]
Host: *Diaphania pyloalis* (= *Glyphodes pyloalis*) on *Morus* sp.

(68) *Xorides anthracinus* [Pah; Gupta and Chandra,1974]

Host: Unknown

(69) *Xorides minimus* [Gul; Gupta and Chandra,1974)

Host: Unknown

9.4.19 Scelionidae (Platygastroidea : Hymenoptera)

(1) *Hadronotus* sp. [Kmr.; Mathur and Sharma, 1962]

Host: Eggs of *Cletus bipunctatus*(Coreid)

(2) *Telenomus pauriensis* [Srg; Rajmohan, 2015]

Host: Eggs of unknown insects

Key to the abbreviations used in Chapter 9.4, pp. 230-257.

Ach: Achabal; Ana: Anantnag; Ban: Banihal; Band: Bandipore; Bar: Baramulla; Bat: Batote; Bee: Beerwah; Bha: Bhadrawah; Bud: Budgam; Bur: Burzhama; Che: Chesmashahi; Dac: Dachigam; Dra: Dras; Duc: Ducksum; Gan: Ganderbal; Gul: Gulmarg; Han: Handwara; Har: Harwan; Hari: Hariparvat; Hee: Heevan; J&K: Jammu and Kashmir; Jmu: Jammu; Kan: Kangan; Khi: Khilanmarg; Kis: Kishtwar; Kmr: Kashmir; Kok: Kokernag; Kup: Kupwara; Lar: Larkipora; Ldk: Ladakh; Man: Manasbal; Nag: Nageen; Nis: Nishat; Pah: Pahalgam; Pat: Pattan; Peri.m: Peri Mahal; Poo: Poonch; Raj: Rajouri; Sho: Shopian; Son: Sonamarg; Sop: Sopore; Srg: Srinagar; Tan: Tangmarg; Veri: Verinag; Zab: Zabervan.

Chapter 10

A Catalogue of Host-Plant Species, Host-Insect Species and Entomophagous Parasitic Insect Species (Parasites/Parasitoids) of Jammu and Kashmir (UT), Ladakh (UT)

(1) *Abelmoschus esculentus* (= *Hibiscus esculentus*) (Okara) (Malvaceae)

Aph – *Aulacorthum solani*

Apd- *Aphidius* sp.

(2) *Abies pindrow* (The Himalayan Silver Fir) (Pinaceae)

Aph. - *Cinara* sp.

Apd. - *Pauesia. (P.) gulmargensis*

Aph. - *Mindrus. japonicus* (Balsam aphid)

Apd. –*Binodoxys indicus, Binodoxys* sp.

(3) *Acer caesium* (Mapple) (Aceraceae)

Aph. – *D. plantanoids* (Sycamose aphid)

Apl. – *Aphelinus* nr. *fulvus*

Aph. – *P. vandenboschii*

Apd. – *A. areolatus*

(4) *Achillea millefolium* (Milfoil, Yarrow, Sanguinary) (Asteraceae)

Aph. – *Macrosiphoniella* sp.

Apd. –*Binodoxys centaureae*

(5) *Aesculus indicus* (Indian Horse Cheshnut) (Hippocastanaceae)

Aph. – *Periphyllus aesculi*

Apd. – *A. areolatus*, Pte. – *Asaphes suspensus* (hyperparaste)

(6) *Ageratum conyzoide* (Goat weed) (Asteraceae)

Aph. – *Brachycaudus helichrysi* (Leaf-curling plum aphid)

Apd. – *Aphidius matricariae*

(7) *Ailantus altissma* (Ailanto, Tree of Heaven) (Simaroubaceae)

Aph. – *Tetraneura* sp.

Apd. – *Ephedrus plagiator*

(8) *Alium cepa* (Onion)(Amaryllidaceae)

Agr – *Chromatomyia horticola* (Pea leaf-miner)

Eul. – *Chrysocharis horticola, Diglyphus horticola, Pediobius indicus, Euderus agromyza*

Agr.- Undetermined

Eul.- *Chrysocharis harithi, Chrysocharis neoviridis*

Lep- *Trichoplusia orichalcea*

Ich.- *Campoletis chloridae*

Lep.- *Brassica rapae*

Ich. *Campoletis chloridae*

***Artemisia* spp. (Asteraceae)**

(9) *Artemisia absinthium* (Absinthe, Wormwood, Madderwood)

Aph. – *Macrosiphoniella. nr. artemisae*

Apd. –*Trioxys pannonicus*

Aph. – *M. pseudoartemisae*

Apd. –*Binodoxys indicus*

Aph. – *M.* nr. *yomogifoliae*

Apd. – *T. deltiger*

(10) *Artemisia indica*

Aph. – *Macrosiphoniella artemisiae* (Artemisia aphid)

Apd. –*B. centaureae*

Aph. – *M. pseudoartemisae*

Apd. –*B. centaureae, B. indicus, Trioxys pannonicus*

(11) *Artemisia vestita*

Aph. – *Macrosiphoniella oblonga*

Apd. – *Monoctonus* sp.

(12) *Artemisia* sp.

Aphd.- *Macrosiphoniella pseudoartimisae*

Apd.- *Trioxys pannonicus*

(13) *Arundo donax* (Great Reed, Spanish Cane) (Poaceae)

Aph. – *Melanaphis donacis*(Giant reed aphid)

Apd. – *Aphidius colemani, Ephedrus persicae, Praon objectum*

***Berberis* spp. (Berberidaceae)**

(14) *Berberis pseudoumbellatus* (Barberry)

Aphd.-Hyalopterus pruni (Mealy plum aphid)

Apd.- P.volcure

Aph. – *Liosomaphis atra*

Apd. – *Monoctonus crepidis, P. volucre*

Aph. – *Liosomaphis himalayensis*

Apd. *A. hortensis, L. gracilis*

Aph. – *Liosomaphis* sp.

Apd.- *Monoctonus crepidis*

(15) *Berberis zebbiliana* (Barberry)

Aph. – *Liosomaphis* sp.

Apd.- *M. crepidis*

***Brassica* spp. (Brassicaceae)**

(16) *Brassica campestris* (Field mustard)

Lep.-*Pieris brassicae*

Cha.- *Brachymeria sp.*

Lep.- *Plutella xylostella*

Eul.- *Farooquiella kashmiricus*

(17) *Brassica napus* (Rape) *B.* var. *glavica* (Indian Colza) (=*Brassica campestris*) (Field mustard)

Lep.- *Pieris brassicae* (cabbage butterfly, Large white)

Cha.- *Brachymeria* sp.

Lep.- *Plutella xylostella* (Diamondback moth)

Eul.- *Farooquiella kashmiricus*

Aph. – *Lipaphis erysimi* (Turnip/Mustard aphid)

Apd. – *Diaeretiella rapae*

Aph.- *Lipaphis erysimi*

Apd.- Lysaphidius erysimi (= pseudoartimisae)

(18) *Brassica oleracea*

Aph.- *Brevicoryne brassicae*

Apd.- *Diaeretiella rapae*

Pte.- *Asaphes suspensus* (Hyperparasite)

Aph.- *Brachycaudus helichrysi*

Apd.- *Aphidius matricariae*

(19) *Brassica olearcea* var. *acephala* (Knol- Khol, Turnip Kale)

Agr.-*Chromatomyza horticola*

Eul- *Chrysocharis horticola, Pediobius indicus, Euderus agromyzae*

Aph. – *B. brassicae*

Apd. – *Diareretiella rapae,* Pte.- *A. suspensus* (hyperparasite)

Lep.- *Spodoptera litura*

Bra.-*Zele chloropthalma*

(20) *Brassica oleracea botrytis* (Cauliflower)

Lep.- *Plutella xylostella*

Eul.- *Tetrasticus sokolwskii*

Lep.- *Spodoptera litura*

Bra.-*Zele chloropthalma*

(21) *Brassica oleracea capitata* (Cabbage)

Aph.-*Lipaphis erysimi*

Apd.- *Lysaphidius erysimi*

Lep.- *Pieris brassicae*

Pte.- *Pteromalus puparium*

Lep.- *Pieris rapae* (Cabbage small white)

Cha.- *Brachymeria femorata*

Lep.- *Pontia daplidice* (The whitebath)

Cha.- *B. femorata*

Lep.- *Plutella xylostella*

Eul.- *Tetrasticus sokolwskii*

Lep.- *Helicoverpa armigera*

Eul.- *Euplectrus euplexa*

Lep.- *Spodoptera litura*

Bra.- *Zele chloropthalma*

(22) *Brassica gongylodes* (Turnip)

Agr.- *Chromatomyia horticola*

Eul.- *Chrysocharis horticola*

Lep.- *Pieris brassicae*

Cha- *Brachymeria femorata, Brachymeria* sp.

Lep.- *Pontia daplidice*

Cha.- *B. femorata*

Lep.-*Trichoplusia orechalacea*

Eul.- *Euplectrus ceylonensis*

(23) *Brassica juncea* (Mustard)

Aph.- *Lipaphis erysimi* (= *pseudobrassicae*)

Apd.- *Lysaphidius erysimi*

(24) *Brassica* sp.

Agr.- *Chromatomyia horticola*

Eul.- *Diglyphus horticola, Euderus agromyzae*

Lep.- *Pieris brassicae*

Tach.- *Compsilura concinnata, Exorista larvarum*

Lep.- *Plutella xylostella*

Ich.- *Itoplectis himalayensis*

(25) *Calotropis* sp. (Milkweed) (Asclepiadaceae)

Aph. – *Aphis nerii* (Oleander aphid)

Apd. – *Aphidius colemani*

(26) *Cannabis sativa* (Hemp.) (Marijuana) (Cannabaceae)

Aph. – *Phorodon cannabis*

Apd. – *A. matricariae, T. acalephae*

(27) *Capsella bursa pastoris* (Shepherd's purse) (Brassicaceae)

Aph. – *Aphis fabae* (Bean aphid)

Apd. – *A. gifuensis*, Pte.- *Pachyneuron aphidis* (hyperparasite)

(28) *Carduus edelbergii* (Thistle) (Asteraceae)

Aph.- *Brachycaudus* sp.

Apd.- *Trioxys rishii*

Aph. – *Capitophorus elaeagni*

Apd. – *A. matricariae, Pte. – A. suspensus,P. aphidis* (hyperparasite);

Apd. – *T. rishii, Pte.- P. groenlandicum* (hyperparasitoid)

Aph. – *Capitophorous* sp.

Apd. – *Aphidius* sp., *T. rishii*

Aph. – *Brachycaudus* sp. (Thistle aphid)

Apd. – *T. rishi*

(29) *Carduus* sp.

Aph.- *Capitophorus* sp.

Apd.- *Aphidius* sp., *Trioxys rishii*

(30) *Celtis australis* (Ulmaceae)

Sc-i. – *Pulvinaria inconspiqua* (Woolly/Cottony Scale insect)

Apl. – *A. mali*

Aph. – *Shivaphis celti*

Apd. –*Binodoxys jaii, Trioxys shivaphis*

(31) *Celtis* sp.

Aph.- *Shivaphis* sp.

Apd.- *Trioxys sporensis*

(32) *Centaurea iberica* (Asteraceae)

Aph. – Undetermined aphid species

Apd.- *Betuloxys intermediatus*

***Chrysanthemum* spp. (Asteraceae)**

(33) *Chrysanthemum cinerariifolium* (Pyrethum)

Aph.- *Mcrosiphoniella sanborni* (Chrysanthemum aphid)

Apd.- *Aphidius absinthii*

Aph.- *Uroleucon compositae*

Apd.- *Binodoxys centaureae, Binodoxys uroleucon*

(34) *Chrysanthemum morifolium*

Aph. – *Macrosiphonoiella sanborni*

Apd. – *Aphidius* sp., *P. dorsale, Trioxys centaureae*

Aph. – *Uroleucon* sp.

Apd. – *Binodoxys uroleucon, Ephedrus niger*

(35) *Chrysanthemum leucanthemum* (Ox-eye-Daisy)

Aph. – *Uroleucon tanaceti*

Apd. – *Trioxys shigai*

(36) *Chrysanthemum* sp.

Aph. – *Coloradoa rufamaculata*

Apd. – *Aphidius* sp.

Aph. – *M. sanborni*

Apd. – *Aphidius* sp.

Aph. – *Macrosiphoniella* sp.

Apd. – *Aphidius. absinthi, Ephedrus niger, Proan dorsale, Proan* sp., *B. centureae*

(37) *Cichorium intybus* (Chicory/Wild Endive) (Asteraceae)

Aph. – *Aphis* nr. *intybi*

Apd. – *L. fabarum, L. (P.) confuscus*

(38) *Cirsium wallichi* (Thistle Plant)

(=*Cnicus wallichi*)

Pyr. –*Ostrinia kashmirica* (Corn Borer)

Ich. – *Exeristes roborator, Scambus (A.) aithomelus*

(39) *Cirsium* sp.

(=*Cnicus* sp.)

Pyr. – *O. kashmirica*

Bra. – *Apanteles* sp., *Bracon brericornis*; Ich. - *Pristomerus* sp., *Scambus flavicrus*

(40) *Conium maculatum* (Poison Hemlock) (Apiaceae)

Aph. – *Betacallis* sp.

Apd. – *Toxares. deltiger*, Cyn.– *Alloxysta* sp. (hyperparasite)

(41) *Cotoneaster aitkinsonii* (Rosaceae)

Aph. – *Eriosoma* sp.

Apd. –*Binodoxys indicus*

***Cucumis* spp. (Cucurbitaceae)**

(42) *Cucumis melo* (Musk melon)

Agr.- *Liriomyza trifolii*(American serpentine leaf-miner)

Eul.- *Closterocerus indica*

Agr.- *Chromatomyia horticola* (Pea leaf-miner)

Eul. – *Diglyphus* sp.

(43) *Cucumis sativus* (Cucumber)

Agr.-*Chromatomyia horticola*

Eul.- *Diglyphus* sp.

Agr.- *Liriomyza trifolii*

Eul.- *C. indica, Diglyphus* sp.

(44) *Cucubita maxima* (Squash) (Cucubitaceae)

Agr.- *Chromatomyia horticola*

Eul.- *Diglyphus* sp.

Agr.- *Liriomyza trifolii*

Eul.- *Asecodes* sp., *Closteroceus indica*

(45) *Cynoglossum wallichii* (Boraginaceae)

(=*Cynoglossum glochidatum*)

Aph. – *Aphis* sp.

Apd.- *A. matricariae*

(46) *Daucus carota* (Wild carrot) (Apiaceae)

Aph. – *Semiaphis heraclei*

Apd. – *Aphidius salicis*

(47) *Daphne oleoides* (Spurge) (Thymelaceaceae)

Aph. – *A. kurusawai*

Apl. – A *mali*

(48) *Descurainia sophia* (Flax/Flixweed) (Brassicaceae)

Aph. – *B. brassicae*

Apd. – *Diaeretiella rapae*

Aph. – *Myzus persicae* (Green Peach-potato aphid)

Apd. – *A. matricariae*

(49) *Dioscorea deltoidea* (Dioscoreaceae)

Aph. – *Aphis* sp.

Apd. – *A. salicis, Ephedrus* (*E.*) *dioscoreae*

Aph. – *Aulacorthum* sp.

Apd. – *E.* (*E.*) *dioscoreae*

(50) *Ephedra* sp. (Mormon or Brigham tea) (Ephedraceae)

Dip.- Syrphid fly

Fig.- *Aspicera hartigi*

(51) *Erigeron canadensis* (Asteraceae)

Aph.- *Brachycaudus helichrysi*

Apd.- *Aphidius matricariae*

(52) *Epilobium hirsutum* (Hainy willow weed) (Onagraceae)

Aph. – *Aphis affinis*

Apl. – *Aphelinus mali*

Aph. – *Aphis grossulariae* nr. *pollinosa*

Apd. – *Praon abjectum*, Meg. – *Dendroceros carpentri* (hyperparasite)

Aph. – *Rhopalosiphum nymphaeae*

Apd. – *P. abjectum.*

(53) *Euphorbia helicoscopia* (Sun Spurge) (Euphorbiaceae)

Aph. – *Myzus* sp.

Apd. – *Aphidius*

(54) *Ferula jaesckeana* (Apiaceae)

Aph.- *Cavariella biswasi*

Apd.- *Aphidius salicis*

(55) *Glycine max* (Soya bean) (Fabaceae)

Lep.- *Helicoverpa armigera* (Cotton bollworm)

Ich.- *Campoletis longipes*

(56) *Hordeum vulgare* (Barley) (Poaceae)

Aph.- *Sipha maydis*

Apd.- *Lysephlebus confuscus*

(57) *Impatiens* sp. (Balsum plant) (Balsaminaceae)

Aph. – *Impatientinum impatiens dalhousiensis*

Apd. – *A. uzbekistanicus*, Fig. – *Alloxysta* sp. (hyperparasitoid)

(58) *Juglans regia* (Walnut) (Juglandaceae)

Aph.- *Chromaphis juglandicola*

Apd.-*Trioxys pallidus*

Aph.- *Panaphis juglandis* (Large walnut aphid)

Apd.- *Trioxys pallidus*

(59) *Lagenaria siceraria* (Calbash or bottle gourd) (Cucurbitaceae)

Agr.- *Chromatomyia horticola*

Eul.- *Diglyphus*

Agr.- *Liriomyza trifolii*

Eul.- *Ascodes* sp., *Chrysonotomyia* sp., *Hemiptarsenus varicornis*

Pte.- *Herbertis indica*

(60) *Lathyrus aphaca* (Yellow Yetchling) (Fabaceae)

Aph.- *Acyrthosiphon pisum* (Pea aphid)

Apd.- *Trioxys complanatus, Aphidius smithi*

(61) *Lespedeza* sp. (Lespedeza) (Fabaceae)

Aph - *Aphis* sp.

Apd. – *Praon volucre*

(62) *Lonicera quinquelocularis* (Honey suckle) (Caprifoliaceae)

Aph.- *Amphicercidus tuberculatus*

Apd.- *Praon dorsale, Meg. – D. carpentri* (hyperparasitoid)

Apd.- *Praon stagona.*

Aph. – *Prociphilus* sp.

Apd.- *Ephedrus plagiator, E. srinagarensis,* Pte.- *Coruna* sp. (hyperparasitoid)

Apd.- *Kashmiria aphidis*

Aph.- *Stagona* sp.

Apd. – *Praon stagona.*

Aph.- *Myzus* sp.

Apd.- *K. aphidis*

(63) *Lycopersicon esculentum* (Tomato) (Solanaceae)

Aph.- *M. persicae* (Green peach aphid or peach potato aphid)

Apd.- *A. matricariae*

Lep (Noc).- *Helicoverpa armigera*

Ich.-*Campoletis chlorides, Charops bicolor, Diadegma fenestralis*

Noc.- *H. armigera*

Tri.- *Trichogramma* sp.

(64) *Luffa cylindrica* (Sponge gourd) (Cucurbitaceae)

Agr.- *Liriomyza trifolii*

Eul.-*Chrysonotomyia* sp., *Diglyphus* sp., *Herbertia indica*

(65) *Magnolia champaca* (=*Michelia champa*) (Champa) (Magnoliaceae)

Aph.- *Prociphilus micheliae*

Apd.- *Ephedrus plagiator*

(66) *Malus* spp. (*M. pumila, M. sylvestris, M. domestica*) (Apple) (Rosaceae)

Aph.- *Aphis pomi* (green apple aphid)

Apd.- *Binodoxys indicus*

Aph. – *Eriosoma lanigerum* (Woolly apple aphid)

Apl. – *A. mali;* Pte. – *Asaphes fletacheri*

Sca-i.- *Diaspidiotus perniciosus* (= *Quadraspidiotus perniciosus*) (San Jose Scale)

Apl.- *Aphytis paramaculicornis, Aphytis* sp. (*proclia* group), *A. lingnanensis, Encarsia (Prospaltella) perniciosi*

Aph.- *Brachycaudus helichrysi*

Apd.- *Aphidius matricariae*

Aph.- *Dysaphis pyri*

Apd.- *Aphidius colemani*

Aph.- *Eriosoma lanigerum*

Apd.- *Areopraon lepelleyi*

Lep. – *Lyonetia clerkella* (Apple leaf-miner)

Eul.- *Achrysocharoides* sp., *Baryscapus* sp., *Chrysocharis* sp., *Chrysonotomyia* sp., *Closterocerus* sp., *Minotetrastichus* sp., *Pnigalio* sp., *Quadrasticus* sp., *Stenomesius* sp. and *Sympiesis* sp.

Sco.- *Scolytus* spp. (Shot hole borer)

Bra.- *Spathius* sp.

Eup.- *Eupelmus kashmiricus, E. valsus, E. vindox*

Eur.- *Eurytoma morio*

Pte. - *Cheiropachus quadrum, Heydenia indica, Macromeseus harithus, Rhaphitelus maculatus*

Tor. - *Archips subsidaria* (Fruit-tree tortricid)

Bra. – *Apanteles* sp.

Tor. - *Cydia pomonella* (Codling Moth)

Tri. - *Trichogramma embryophagum* and *T. cacoeciae pallidum*

(67) *Malva Sylvestris* (Common mallow) (Cucurbitaceae)

Agr.- *Chromatomyia horticola*

Eul.- *Diglyphus horticola*

***Medicago* spp. (Fabaceae)**

(68) *Medicago polymorpha*

Aph. - *Therioaphis* sp. (Clover aphid)

Apd. *Trioxys complanatus*

(69) *Medicago sativa* (Lucerne/Alfalfa)

Aph.- *Therioaphis* sp.

Apd.- *T. complanatus*

Aph.- *Acyrthosiphon* sp.

Apd.- *A. smithi*

(70) *Medicago* sp.

Aph.- *A. pisum*

Apd.- *Aphidius ervi, A. smithi*

(71) *Melia azedarach* (Persian Liliac/Bead tree (Meliaceae)

Aph.- *A. gossypii* (Melon/cotton aphid)

Apd. –*Binodoxys indicus*

***Mentha* species (Mint) (Lamiaceae)**

(72) *Mentha aquatica* (Marsh/water Mint)

Aph. - *Aphis affinis*

Apd. - *Lipolexis gracilis*

(73) *Mentha longifolia* (Horse mint)

(=*Mentha sylvestris*)

Aph.- *A. affinis*

Apd.- *A. gifuensis, L. gracilis, Binodoxys acalephae; Binodoxys indicus.*

Pte. – *A. suspensus* (hyperparasite)

Aph.- *Ovatus* nr. *crataegarius*

Apd.- *Aphidius* sp.,

Enc. – *Aphidencrytus aphidivorus* (hyperparasite)

Apd.- *Toxares deltiger*

(74) *Morus* spp. (Mulberry) (Moraceae)

Lep.- *Bombyx mori*

Bra.- *Pholetesor hayati*

Lep.- *Diaphania pyloalis* (= *Glyphodes pyloalis*) (Lesser mulberry snout moth)

Bra.- *Chelonus* sp., *Cotesia glomerata* (= *Apanteles glomeratus*)

Ich.- *Pristomerus sulci, Xanthopimpla*

***Nepeta* species. (Lamiaceae)**

(75) *Nepeta cataria* (Cat Mint)

Aph. - *Eumyzus* sp.

Apd. - *Aphidius* sp., *Ephedrus salicicola* and *Toxares deltiger*

(76) *Nepeta salvifolia*

Aph.- *Aphis* sp.

Apd.- *A. gifuensis,*

Fig.- *Phaenoglyphis* sp. (hyperparasitoid)

(77) *Notonia grandifolia* (Asteraceae)

Aph.- *Brachycaudus helicherysi*

Apd.- *Aphidius matricariae*

***Nymphaea* spp. (Nymphaeaceae)**

(78) *Nymphaea alba* (White water-lily)

Aph.- *Rhopalosiphum nymphaeae* (Water lily aphid)

Apd.- *Aphidius* sp., *Proan nymphaeae*

(79) *Nymphaea* sp.

Aph.- *R. nymphaeae*

Apd.- P. *nymphaeae*

(80) *Oryza sativa* (Paddy/Rice) (Poaceae or Graminae)

Lep- *Parnara guttata* (Rice skipper)

Cha.- *Brachymeria*

(81) *Pennisetum flaccidum* (Fountain grass) (Poaceae)

Aph.- *Sitobion avenae* (=*Macrosiphum avenae*) (Grain aphid)

Apl.- *Aphelinus mali*

Apd.- *A. uzbekistanicus,*

Fig.- *Alloxysta* - (hyperparasitoid);

Apd. - *Lysaphidius staryi*

Aph.- *Sitobion* sp.

Apd.- *A. uzbekistanicus.*

(82) *Phlox dummondi* (Polemoniacae)

Aph.- *M. persicae*

Apd.- *A. matricariae*

(83) *Phragmites australis* (= *communis*) (Common Reed) (Poaceae)

Aph.- *H. pruni*

Apd.- *P. volucre*

***Pinus* spp. (Pinaceae)**

(84) *Pinus roxburghii* (Chir Pine/The Himalayan long-leaved Pine)

Aph.- *Cinara lachnirostris* (Black Pine Aphid)

Apd.- *Pauesia pini*

(85) *Pinus wallichiana* (= *excelsa*) (Blue Pine/Bhutan Pine)

Aph.- *C. lachnirostris*

Apd.- *Pauseia pini*

Aph.- *Cinara* sp.

Apd.- *P. pini*

Aph.- *Lachnus tropicalis*

Apd.- *Aphidius* sp., *Pauesia himalayensis*

Col.- *Ips stebbing, Polygraphus major, Pityogenes scitus*

Pte.- *Cheiropachus quadrum, Macrmesus harithus, Metacolus parasinicus. Metacolus sinicus* and *Roptrocerus mirus*

(86) *Pinus* sp.

Aph.- *C. lachinirostris*

Apd.- *Pauesia gulmargensis, P. pini*

Aph.- *Cinara* sp.

Apd.- *P. pini*

(87) *Pisum sativum* (Pea) (Fabaceae)

Aph.- *Acyrthosiphon pisum* (Pea aphid)

Apd.- *Aphidius ervi*

Agr.- *Chromatomyia horticola*

Eul.-*Chrysocharis horticola,Diglyphus horticola, Euderus agromyzus, Pediobius indicus*

Lep.- *Thysanoplusia orichalcea* (=*Trichoplusia orichalcea*)

Ich.- *Campoletis chloridae*

(88) *Phaseolus vulgaris* (Common bean) (Fabaceae)

Lep.- *Thysoplusia orichelesa*

Ich.- *Scambus* sp.

(89) *Poa angustifolia* (Grass) (Poaceae) (Hedgehog grain aphid)

Aph.- *S. maydis*

Apd.- *Lysiphleibus ambiguus,*

Fig. – *Alloxysta* sp. (hyperparasite)

***Populus* species. (Salicaceae)**

(90) *Populus aurea*

Aph. – *Pterocomma populeum* (Poplar bark aphid)

Apd. – *Aphidius cingulatus*

(91) *Populus caspica* (white Poplar)

Aph.- *Chaitophorus niger* (Dark willow leaf aphid)

Apd.- *L. confusus*

Aph.- *Chaitophorus populeti* (Poplar shoot aphid)

Apd.- *Ephedrus persicae*

Aph.- *Pemphigus mordvilkoi*

Apl.- *Protaphelinus nickolskajae*

(92) *Populus ciliata* (The Himalayan Poplar)

Aph.- *Epipemphigus imaicus*

Apl.- *P. nickolskajae*

Aph.-*Pemphigus aedificator*

Apl.- *P. nickolskajae*

Aph.- *Pemphigus immunis*

Apl.- *P. nickolskajae*

Enc.- *A. aphidivorus* (hyperparasite)

(93) *Populus nigra* var. *italica* (Lombardy Poplar)

Aph.- *Pemphigus bursarius*

Apl.- *Protaphelinus dargahae*

Aph.- *P. immunis*

Apl. – *P. nickolskajae*

(94) *Populus* spp.

Aph.- *Chaitophorus kapuri*

Apd.- *Adialytus salicaphis, A. matricariae*

Aph.- *Chaitophorus* sp.

Apd.- *Adialytus salicaphis*

Aph.- *Pemphigus napeus*

Apl. – *P. nickolskajae*

Lym.- *Lymantria obfuscata* (Gypsymoth)

Bra.- *Apanteles conspersae, A. liparidis, A. melanoscelus, A. porthetriae, A. solitaria, Apanteles* sp;

Cha.- *Brachymeria intermedia, Brachymeria lasus.,*

Eup.- *Anastatus kashmirensis* and *Anastatus* sp.,

Ich.- *Itoplectis himalayensis,Pimpla* sp., *Theronia atlantae*

Mus.- *Muslina stabularis*

Sar.- *Sarcophaga* sp.

Tac.- *Corcelia* sp., *Drino discreta, Exorista rossica*

Torm.- *Monodontomerus acreus*

***Prunus* species. (Rosaceae)**

(95) *Prunus armeniaca* (Common Apricot)

Sci.- *Eulecanium coryli* (Lecanium scale insect)

Apl.- *Aphelinus* sp.

(96) *Prunus domestica* (Common plum)

Aph.- *Aphis* sp.

Apd.- *Aphidius transcaspicus*

Aph.- *Hyalopterus pruni* (=*arundinis*) (Mealy plum aphid)

Apd.- *P. volucre*

Aph.- *Brachycaudus helichrysi*

Apd.-*Aphidius matricariae*

Aph.- Duraphis (*Holcaphis*) *calamagrostis*) (= *Hyalopterus arundinis*)

Apd.- *Trioxys rishii*

(97) *Prunus domestica* var. *insititia* (Cultivated variety of plum)

Aph.- *Hyalopterus pruni*

Apd.- *E. persicae*

Pte. – Coruna, Pachyneuron aphidis (hyperparasites)

Apd.- *P. volucre*

Pte. – Coruna sp. and *Pachyneuron aphidis*

Apd.- *T. deltiger* and *T. rishi*

(98) *Prunus persica* (Peach)

Aph.- *Dysaphis pyri*

Apd.-*Aphidius colemani*

Apd.- *Brachycaudus helichrysi*

Apd.- *Aphidius matricariae*

(99) *Punica granatum* (Pomgernate) (Punicaceae)

Aph.- *Phorodon* (=*Paraphorodon*) sp.

Apd. –*Binodoxys indicus*

(100) *Quercus ilex* (Holi or Holm Oak) (Fagaceae)

Aph.- *Acyrthosiphon* sp.

Apd.- *Ephedrus* sp.

Ranunculus **species. (Ranunculaceae)**

(101) *Ranunculus muricatus* (Edible Buttercup)

Aph. - *Acyrthosiphon* sp.

Apd. – *Aphidius smithi*

(102) *Ranunculus* sp.

Aph. - *Acyrthospihon* sp.

Apd. - *A. smithi*

(103) *Raphanus sativus* (Brassicaceae)

Aph.- *Lipaphis erysimi*

Apd.- *Lysaphidius erysimi*

(104) *Ricinus communis* (Castor oil plant) (Euphorbiaceae)

Lep.- *Spilarctia obliqua* (Hairy caterpillar)

Bra.- *Glyptapanteles agamomnonis, Meteorus pulchricornis*

(105) *Robinia pseudoacacia* (False Acacia or black locust)

Aph.- *Aphis craccivora* (Cowpea aphid)

Apd.- *Lipolexis gracilis, Lipolexis oregmae (= scutellaris), Lysephlebus fabarum, Binodoxys jaii, Binodoxys indicus, Ephedrus persicae, Toxares deltiger, Trioxys rishii*

Apl.- *Aphelinus mali*

(106) *Robinia* sp.

Aph.- *Aphis craccivora*

Apd.- *Binodoxys* sp.

Rosa **species. (Rosaceae)**

(107) *Rosa brunonii* (Himalayan Musk Rose)

Aph.- *Macrosiphum* sp.

Apd.- *Aphidius* sp

Fig.- *Alloxysta* sp. (hyperparasitoid);

Apd.-*Binodoxys indicus, Binodoxys kashmirensis.*

(108) *Rosa canina* (Dog rose)

Aph.- *Chaetosiphon tetrarhodus*

Apd.- *Aphidius* sp.,

Euc.- *Prionomitus* sp. (hyperparasitoid)

(109) *Rosa chinensis* (=*Rosa indica*) (China/Bengal or Monthly Rose)

Aph. - *Macrosiphum rosae* (Rose aphid)

Apd. - *A. avenae, Pte. – A. suspensus, Coruna* sp. (hyperparasitoid)

(110) Rosa damascene (Damask rose)

Aph.- *Sitobion rosaeiformis*

Apd.- *Aphidius rosae, Aphidius uzbekistanicus*

(111) *Rosa indica* (Cyme Rose)

Aph.- *Macrosiphum rose*

Apd.-*Aphidius rosae*

Aph.- *Metopolophium montanum*

Apd.-*Toxares detiger*

(112) *Rosa macrophylla* (Wild Rose)

Aph.- *Chaetosiphon chaetosiphon*

Apd.- *Aphidius rosae,*

Pte.- *A. suspensus* (hyperparasitoid)

Apd.- *Aphidius avenae*

Pte.- *A. suspensus*

Aph.- *Chaetosiphon tetrarhodum* (Small green Rose aphid or Strawberry aphid)

Apd.- *Aphidius eglanteriae*

Aph.- *Chaetosiphon* sp.

Apd.- *Proan* sp.

Aph. - *Macrosiphun* sp.

Apd. - *Aphidius rosae*

Pte.- Asaphes suspensus (hyperparasitoid)

(113) *Rosa webbiana* (Wild Rose)

Aph.- *Chaetosiphon gracilicorne*

Apd.- *Aphidius* sp.

Aph.- *Chaetosiphon* sp.

Apd.- *Proan* sp.

Enc.- *A. aphidivorus* (hyperparasitoid)

Aph.- *M. rosae*

Apd.- *A. avenae*

Meg. – *Dendrocerus. carpenteri* (hyperparasitoid)

Apd.- *Binodoxys acalephae,*

Pte. – *P. aphidis* (hyperparasitoid)

Aph.- *Myzaphis rosarum* (Lesser Rose aphid)

Apd.- *A. eglanteriae, Binodoxys rosaecola*

(114) *Rosa* spp.

Aph.- *Chaetosiphon glaber*

Apd.- *Aphidius avenae*

Aph.- *C. tetrarhodum*

Apd.- *Aphidius* sp.

Euc.- *Prionomitus* sp. (hyperparasitoid)

Aph.- *Chaetosiphon* sp.

Apd.- *Aphidius* sp.

Fig.- *Alloxysta* sp. (hyperparasitoid)

Pte.- *Coruna* sp. (hyperparasitoid)

Aph.- *Microsiphum rosae*

Apd.- *Aphidius* sp.

Pte.- *Coruna* sp.

Aph.- *Macrosiphum* sp.

Apd.- *P. volucre*

Pte.- *A. suspensus* (hyperparasitoid)

Apd.- *Binodoxys acalephae*

Aph.- *Myzaphis rosarum*

Apd.- *Monactonus. (M.) fotedari*

Fig.- *Alloxysta sp.* (hyperparasitoid)

Apd.- *Proan abjectum*

Fig.- *Alloxysta sp.*

Apd.- *Proan volucre*

Fig.- *Alloxysta* sp. (hyperparasitoid)

Aph.- *Myzaphis turanica*

Apd.- *Aphidius avenae* and *P. volucre*

(115) *Rubia cardifolia* (Indian Madder) (Rubiacea)

Aph.- *Myzus umecola*

Apd.- *A. matricariae*

***Rubus* spp. (Rosaceae)**

(116) *Rubus fruticosus* (Alish, Akhi, Kanachi)

Aph.- *Acyrthosiphon rubi*

Apd.- *L. gracilis, Lysiphlebus fabarum*

(117) *Rubus* sp.

Aph.- *A. gossypii*

Apd.- *Binodoxys basicurvus, Binodoxys rubicola*

Rumex species (Polygonaceae)

(118) *Rumex acetosella* (Sheep Sorrel)

Aph.- *A. craccivora*

Apd.- *Ephedrus persicae*

(119) *Rumex nepalensis* (Sorrel/Broad- leaved Dock herb)

Aph.- *A. craccivora*

Apd.- *Binodoxys indicus, Lysiphlebus fabarum*

Aph. - *Aphis fabae* (The black bean aphid)

Apd. – *A. marticariae*

Aph.- *A. fabae solanella*

Apd.- *Lysiphlebus fabarum, B. indicus*

***Salix* spp. (Willows) (Salicaceae)**

(120) *Salix alba* (European willow/white Willow)

Aph.- *Aphis farinosa* (Small willow aphid)

Apd.- *Aphidius* sp.,

Enc- *Prionomitus* (hyperparasitoid)

Apd.-. *Proan objectum, Proan kashmirensis, Toxaresshiagi*

Aph.- *A. farinosa*

Apd.- *Toxares shigai*

Aph.- *Cavariella aegopodii*

Apd.- *Ephedrus laevicollis* (= *minor*)

Aph. – *Cavariella* sp.

Apd. – *E. laevicollis* (=. *minor*)

Aph.- *Tuberolachnus salignus* (Giant willow aphid)

Apd.-*Aphidius salignae*

(121) *Salix aegyptica*

Aph.- *A. farinosa*

Apd.- *Aphidius* sp.

Pte.- *P. aphidius* (hyperparasitoid)

Aph.- *Cavariella aegopodi* (Willow-Carrot aphid)

Apd.- *Ephedrus salicicola*

(122) *Salix babylonica* (Weeping Willow)

Aph.- *A. farinosa*

Apd.- *P. abjectum*

Aph.- *Chaitophorus pakistanicus*

Apd.- *Binodoxys brevicornis*

Aph.- *Cavariella* sp.

Apd.- *E. salicicola*

(123) *Salix caprea* (Sallow Goat willow)

Aph.- *A. farinosa*

Apd.- *Binodoxys longispinus*

(124) *Salix flabellaris* (Himalayan shrub willow)

Aph.- *Chaitophorus himalayensis*

Apd.- *Aphidius* sp.

(125) *Salix fragilis* (Crack Willow/Kashmir Willow/Red-wood willow)

Sca.- *Lepidosaphes kirgisica*

Apl.- *Aphytis paramaculicornis*

(126) *Salix* spp.

Aph.- *A. farinosa*

Apd.- *Aphidius* sp.

Aph.- *C. pakistanicus*

Apd.- *P. kashmirensis*

Lep.- *Porthetria dispar* (Gypsy/Tussock Moth)

Eup. - *Anastatus kashmirensis*

Lep.- *Lymantria obfuscata* (Gypsy Moth)

Bra. - *Apanteles conspersae, A. flavipes, A. indiansis, A. liparidis, A. porthetriae, A. melanoscelus, A.*nr. *solitarius, Apanteles* sp.

Cha.- *Brachymeria euloeae, B. intermedia, B. lasus*

Ich.-*Anilastus* sp., *Pimpla* sp. (*Coccygominus* sp.), *Itoplectis himalayensis, Theronia atlantae, Theronia* sp.

Eup.- *Anastatus bifasciatus, A. japonicus, A. kashmirensis, Anastatus* sp.

Eur.- *Eurytoma* sp.

Mus.- *Muslina stabulans*

Sar.- *Sarcophaga* sp.

Tac.- *Compsilura* sp.; *Corcelia* sp.; *Drino discreta, D. inconspicuoides, Exorista;*

Torm.- *Monodontomerus acreus.*

(127) *Salvia moorcroftiana* (Lamiaceae) (Moorcroft sage)

Aph.- *Aphis* nr. *salviae*

Apd.- *L. fabarum*

(128) *Senecio chrysanthemoides* (Asteraceae)

Aph.- *Aphis* sp.

Apd.- *A. salicis*

(129) *Sisymbrium* sp. (Brassicaceae)

Aph.- *Myzus* sp.

Apd.- *Aphidius* sp.

***Solanum* Species (Solanaceae)**

(130) *Solanum nigrum* (Black Nightshade)

Aph.- *A. craccivora*

Apd.- *A. matricariae, Binodoxys indicus, L. gracilis*

Aph.- *Aphis fabae solanella* (Bean aphid)

Apd.- *Binodoxys indicus*

Pte.- *A. suspeusus* (hyperparasitoid)

Aph.- *Myzus persicae*

Apd. – *Toxares zakai*

(131) *Solanum melongena* (Brinjal)

Aph.- *Macrosiphum euphorbiae*

Apd.- *Binodoxys kashmirensis*

(132) *Solanum tuberosum* (Potato)

Agr.- *Chromatomyia horticola*

Eul.- *Diglyphus* sp.

Agr.- *Liriomyza trifolii*

Eul. – *Closterocerus indica*

Aph.- *A. craccivora*

Apd. – *D. rapae.*

Lep.- *H. armigera*

Ich.- *Campoletis chloridae*

Noc.- *Helicoverpa armigera*

Tri.- *Trichogramma* sp.

***Sonchus* sp. (Asteraceae)**

(133) *Sonchus asper* (Spiny-leaved Sow Thistle)

Aph.- *Uroleucon* sp.

Apd.- *P. volucre*

(134) *Sonchus oleraceus* (Milk Thistle/Common Sow Thistle)

Aph.- *Nosonovia (Hypermyzus) lactucae* (=*Hypermyzus lactucae*)

Apd.- *P. volucre*

Aph.- *Uroleucon (Uromelan) sonchi*

Apd. – *D. rapae*

(135) *Sonchus* sp.

Aph.- *U. sonchi*

Apd.- *Aphidius* sp.

Meg. – *Dendrocerus carpentri* (hyperparasitoid)

Aph.-*Uroleucon* sp.

Apd.- *Proan volucre*

***Sorghum* species (Poaceae)**

(136) *Sorghum halepense* (Johnson grass/Millet)

Aph.- *Sitobion avenae* (= *Macrosiphum avenae*) (English grain aphid)

Apd. – *Lysiphlebus ambiguus, L. confuscus*

Aph.- *Macrosiphum avenae*

Apd.- *Aphidius uzbekistanicus*

Pte- *Pachyneuron aphidis* (hyperparasitoid)

(137) *Sorghum vulgare* (Sorghum or great millet)

Aph.- *Sipha (Rungsia) maydis*

Apd.- *L. confuscus*

(138) *Sorghum* sp.

Aph. – *Macrosiphum* sp.

Apd. – *A. ambiguus*

Aph. – *Melanaphis sacchari* (Sugar-cane Aphid)

Apd. – *L. confusus*

(139) *Spiraea* sp. (Rosaceae)

Aph. – *Aulacorthum* sp.

Apd. – *Ephedrus (E.)* sp.

Aph. – *Myzus cerasi* (Black Cherry aphid)

Apl.- *Aphelinus* nr. *fulvus*

(140) *Thuja orientalis* (Oriental Arbor vitae) (Cupressaceae)
(=*Biota orientalis*)
Aph.- *Lachnus* sp.
Apd.- *Pauesia hazartbalensis*

(141) *Trifolium* sp. (Clover) (Fabaceae)
Aph.- *Acyrthosiphon pisum*
Apd.- *Aphidius ervi*

(142) *Triticum aestivum* (=*vulgare*) (Poaceae)
Aph.-*Sitobion avenae* (=*Macrosiphum avenae*)
Apd.- *Lysiphlebia mirzai*
Unidentified Aphid Species
Apd.-*Toxares shigai*

***Ulmus* spp. (Ulmaceae)**

(143) *Ulmus villosa* (=*Ulmus laevigata*)
Aph.- *Eriosoma* (*Schizoneura*) *ulmi*
Apd.- *Areopraon lepelleyi*

(144) *Ulmus wallichiana* (The Himalayan Elm)
Aph.- *Schizonurella indica.*
Apd.- *Areopraon lepelleyi*

(145) *Ulmus* sp.
Aph.- *Eriosoma*
Apd.- *Aphidius*

(146) *Urtica dioica* (Stinging Nettle) (Urticaceae)
Aph.- *Aphis* sp.
Apd.- *Binodoxys indicus*
Aph.- *Microlophium carnosum*
Apd.- *A. ervi*
Fig.- *Phaenoglyphis* sp. (hyperparasitoid)
Apd.- *Aphidius matricariae, Ephedrus urticae*
Aph.- *Microlophium* sp.
Apd.- *T. zakai*
Fig. -*Alloxysta* sp. (hyperparasitoid)

(147) *Verbascum thapsus* (Cow's lungwort/Common Mullein) (Scrophulariaceae)
Aph.- *A. nerii* (Oleander aphid. Milkweed aphid)

Apd.- *L. gracilis,*

Apl.- *A. mali*

(148) *Veronica persicae* (Scrophulariaceae)

Aph.- *Aphis* sp.

Apd.- *A. gifuensis*

(149) *Viburnum grandiflorum* (Sheep berry, Stag Bush/Black Had) (Caprifoliaceae)

Sca.- *Parthenolecanium corni* (Plum scale or European Fruit lecanium)

Apl.- *A. lingnanensis*

(150) *Vicia sativa* (Common Vetch, Spring Vetch) (Fabaceae)

Aph.- *Acyrthosiphon* sp.

Apd.- *Aphidius. ervi*

Aph.- *Aphis* sp.

Apd. – *A. gifuensis*

Aph.- *Therioaphis* sp.

Apd.- *Trioxys complanatus*

(151) *Vigna mungo* (Black gram) (Fabaceae)

Aph.- *Helicoverpa armigera*

Ich.- *Campoletis chloridae*

(152) *Zinnia elegans* (Youth-and-old Age) (Asteraceae)

Aph.- *Aphis gossypii* (Cotton aphid or melon aphid)

Apl.- *A. kashmiricus*

Apd.- *L. gracilis*

Key to the Abbreviations

Agr: Agromyzid; Aph: Aphid; Apd: Aphidiine; Apl.-Aphelinid; Bra: Braconid; Cha: Chalcidid; Cic: Cicadellida; Col: Coleopteran; Con: Conopid, Cor: Coreid; Cra: Crambidae; Enc: Encyrtid; Ere: Erebidae; Eul: Eulophid; Eup: Eupelmid; Eur: Eurytomid; Fig: Figitidae; Hym: Hymenopteran; Ich: Ichneumonid; Lep: Lepidopteran; Lip: Liparid; Lym: Lymantriid; Meg: Megaspilid; Mus: Muscidid; Mym: Mymarid; Noc: Noctuidae; Pte: Pteromalid; Sar: Sarcophagid; Sca: Scale insect; Sce: Scelionid; Sci: Sciomyzid; Sco: Scotylid; Tac: Tachnid; Ten: Tenthredinid; Tor: Tortricid; Torm: Torymid and Tri: Trichogrammatid.

References and Bibliography

(* *Indicates references not cited in the text of the book, however, as recommended/ further reading*)

Abbas, Arshad and Wali-ullah, M. I. S. (2010). Prevalence of plant-parasitic nematodes of *Gladiolus hortulanus* L. in and around Srinagar, Kashmir. *Indian J. Nematol.*, 40 (2): 179-183.

Abdualkader, A. M, Ghawi, A. M., Alaama, M., Awang, M. and Merzouk, A. (2013). Leech therapeutic applications. *Indian J. Pharm. Sci.*, 75 (3): 127-137, Doi: 10. 4103/0250-474.115456.

Abdullah, Iram., Tak, Hidayat-ullah, Ahmad, F., Gul, Nazima, Nabi, S. and Sofi, T. A. (2016 a). Predominance of Gastro-intestinal protozoan parasites in children: A brief Review. *J. of Health Education Res. and Dev.*, 4: 194, doi: 10.4172/2380-5439.1000194.

Abdullah, I., Tak, Hidayat-ullah, Ahmad, F. and Gul, N. (2016 b).Prevalence and associated risk factors of giardiasis among children in District Anantnag of Kashmir valley, India. *J. Gastroenterology and Hepatobiliary disorders*, 2 (1): 1-4.

Abdullah, I., Tak, H. and Ahmad, F. (2017). Impact of giardiasis on Haematological Profile of infected children. *J. Medical Sciences*, 17 (3): 140- 143.

Abdullah, I., Tak, H. and Ganie, S. A. (2000). Prevalence of Faciolosis and Dicrocoeliosis in sheep slaughtered in Kashmir valley. *Intl. J. Scientific and Technol, Res.*, 9 (1): 570-573.

Abdullah, I., Tak, Hidayat-ullah, Ahmad, F. and Gul, N. (2016 b). Prevalence and associated risk factors of giardiasis among children in District Anantnag of Kashmir valley, India. *J. Gastroenterology and Hepatobiliary Disorders*, 2 (1): 1-4.

Abrol, D. P. (2013). *Beekeeping: A comprehensive guide to bees and beekeeping*. Scientific Publishers (India), Jodhpur, Rajasthan, 896 pp.

*Abrol, D. P. and Kakroo, S. K. (1999). Observation on concurrent parasitization by mites on four honey bee species in India. *Tropical Agriculture*, 74: 150- 156.

Abrol, D. P. and Kotwal, D. R. (1996). Occurrence of *Varroa jacobsoni* Oudemans in Jammu, India on honey bees. *Insect Env.*, **2**(2): 48.

Abrol, D. P. and Putatunda, B.N. (1995). Discovery of ectoparasite mite *Tropilaelaps koenigenus*. Dilfinado Baker and Baker (Acari: Laelapidae) on *Apis dorsata, A. mellifera* and *A. cerana* F. in Jammu and Kashmir, India. *Curr. Sci.*, **68**(1): 90.

Abrol, D. P. and Putatunda, B.N. (1996). New record of *Tropilaelaps clareae* (Acarina: Mesostigma) associated with *Xylocopa irridipennis* from India. *Sci. and Culture*, **62**: 59.

Acha, P. and Szyfres, B. (2003). Pentastomiasis, In: *Zoonoses and Communicable Diseases common to man and animals*. Vol.III, Third Edition, Washington: Pan American Health Organization, pp. 345-350.

Adhikary, C. C. and Ghosh, A. K. (1994). Anopluran- fauna of India: Sucking lice infesting domesticated and wild animals. *Occ. Paper No.164, Zool. Surv. of India*, Calcutta, 212 pp.

*Adie, J. R. and Adie, H. (1913). Note on an enquiry into Malaria and Mosquitoes in Kashmir Valley. *Indian Medical Gaz.*, **48**: 341-342.

Adl, Sina, M. *et al.* (2012). The revised classification of Eukaryotes. *J. of Eukaryotic Microbiol.*, **59** (5): 429-514. Doi: 10.1111/j 1550.7408.2012.00644.x. PMC 3483872.

Agarwal, B. L. (1951). Relapsing fever in Kashmir. *Indian Med. Gaz.*, **86**: 446-53.

Ahad, S., Tanveer, S. and Malik, T. A. (2015). Seasonal impact on the prevalence of coccidian infection in broiler chicks across poultry Farms in Kashmir Valley. *J. Parast. Dis.*, 39 (4): 736-740, doi: 10.1007/s 12639-014-0434-6.

*Ahad, S., Tanveer, S. and Malik, T. A. (2016). Anticoccidial activity of aqueous extract of wild mushroom (*Ganoderma applanatum*) during experimentally induced coccidial infection in broiler chicken, *J. Parasitic Dis.*, **40** (2): 408-414.

*Ahad, S., Tanveer, S., Malik, T. A. and Nawchoo, I. A. (2018). Anticoccidial activity of fruit peel of *Punica granatum* L. *Microbiol. Pathogenesis*, **116**: 78-83.

*Ahangar, A. A., Godara, R., Yadav, Bhantyal, D., S. and Katoch, M. (2015). Deltamethrin resistence in field population of *Rhipicephalus* (*Boophilus*) *microplus* (Acari: Ixodidae) in Jammu and Kashmir, India. *Experimental and Applied Acarology*, **67** (3): 467-475.

*Ahangar, M. A., Jan, N. A. and Chishti, M. Z. (2008). Histopathology of indigenous carp (*Schizothorax* spp.) infected with *Pomphorhynchus* species in River Jehlum, Kashmir pp. 490-492, ***In***: M. Z. Chishti and Fayaz Ahmed (eds.), *Science for Better Tomorrow*. Printed at Universal Printers, Khanyar, Srinagar.

Ahmad, B. D., Azmi, S., Katoch, R., Sood, S. and Nashiruddulah (2015). Pathological study on an outbreak of Hepatic coccidiosis in Rabbit Farm. *Shanlax Intl. J. Vet. Sci.*, **3** (2): 19-25.

Ahmad, Bashir, Bhatti, Gurjeet, Thokar, N. A. and Malla, Nancy (2002). Human Toxocariasis and ascariasis: concomitant parasitism in Srinagar, Kashmir, India. *Indian J. Pathology Microbiol.*, **45** (31): 315-318.

Ahmad, D. and Dar, H. U. 1975 (1978). Biological Control of potentially serious orchard pest in Kashmir and feasibility of integrated control. *J. Sci.Univ. Kashmir.*, **3** (2): 93-98.

Ahmad, Farooq, and Sharma, K. (2016): First record of *Dactylogyrus rectotrabus* (Monogenea) on *Garra gotyla* (Pisces, Cyprinidae) in India from Poonch River and its tributaries, District Poonch, Jammu and Kahmir. *Intl. J. Recent Scienntific Res.*, **7** (3): 9402-9405.

Ahmad, Fayaz (1993). *Trematode and Cestode Parasites of Pisces and Aquatic aves of Kashmir, with Histology and Histopathology of Genus Adenoscolex* Fotedar, 1958, Vol.I, 430 pp. Vol.II, 488 pp. + I-XCVII plates. Ph.D. Thesis, Univ. of Kashmir.

Ahmad, Fayaz (2000). Avian Trematode Parasites of Kashmir. Part VI. Genus *Paryphostomum* Dietz, 1909. *J. Parasitic Diseases,* **24**(2): 155-158.

*Ahmad, F. (2010). *Helminths of Medical Importance, with special reference to Kashmir*. Published by University of Kashmir, in Collaboration with J&K Council of Science and Technology.

*Ahmad, F., Fatima, M. and Chishti, M. Z. (2008). Ruminant Trematode parasite of Kashmir, *Dicrocoelium dendriticum* (Rudolphi, 1819, Looss, 1899), *In*: M. Z. Chishti and Fayaz Ahmad (eds.), *Science For Better Tomorrow*. Published by Universal Printers, Khanyar, Srinagar, Kashmir.

Ahmad, F. and Chishti, M. Z. (1994). Trematode parasites of fishes of Kashmir. Part I. *Diplozoon* (Monogenea: Polyopisthocotylea), with description of a new species, *Diplozoon guptai. J. Parasitic Diseases,* **19** (1): .

Ahmad, Fayaz and Chishti, M. Z. 1995). Avian Trematode parasites in Kashmir. Part.II. Genus *Notocotylus* Diesing, 1839. *Oriental Sci.* **1** (1-2): 59-65.

Ahmed, Fayaz and Chishti, M. Z. (1995 a). Avian trematode parasites of Kashmir Part I: The genus *Paramonostomum* Luhe, 1909, with description of a new species *P. zahoorensis* (*Proc.* 12th *Natl. Congr. Parasitol.*), held at Panaji, Goa, from 23-25th Jan., 1995). *J. Parasitic Diseases,* 1981-105.

Ahmad, Fayaz. and Chishti, M. Z. (1997). Avian trematode parasites of Kashmir. Part III. Genus *Echinoparyphium* Dietz. 1909. *Oriental Sci.*, **2** (1): 41-48.

Ahmad, Fayaz and Chishti, M. Z. (1997*a*). Avian trematode parasites of Kashmir. Part IV. *Echinostoma revolutum*, (Forelich, 1802) Looss, 1899. *Oriental Sci.*, **2** (2): 49-59.

*Ahmad, F. and Chishti, M. Z. (1998). Histological studies on a Caryophyllidean Cestode infecting fresh water fishes in Kashmir. *Oriental Sci.*, **3**(1): 39-50 (On microanatomy of cestode, *viz. Adenoscolex*).

*Ahmad, F. and Chishti, M. Z. (1999). Fish Trematode parasites of Kashmir, Part I. Genus *Diplozoon* Nordmann, 1832 (Monogenea: Polyopisthocotylea). *Oriental Sci.*, **4** (2): 79-91 (Three trematodes, *viz. Diplozoon aegyptensis, D. kashmirensis* and *D. guptai* n. sp., infecting fishes from Anchar and Dal lake, have been describrd).

Ahmad, Fayaz and Chishti, M. Z. (2000). Avian trematode parasites of Kashmir. Part V. Genus *Euparyphium Dietz* 1909). *J. Parasitic Diseases,* **24** (2): 219-220.

Ahmad, Fayaz and Chishti, M. Z. (2000 *a*). Fish trematode parasites of Kashmir. Part II. Genus *Clinostomum* Leidy, 1856 (Digenea: Clinostomatidae). *Oriental Sci.,* **5** (1): 13-22.

Ahmad, Fayaz and Chishti, M. Z. 1999 (2001). Fish trematode parasites of Kashmir. Part-I. Genus *Diplozoon* Nordmann, 1832 (Monogenea, Polyophisthocotylea) *Oriental Sci.,* **4** (2): 79-91.

Ahmad, Fayaz and Dhar, R. L. (1996). Avian trematode parasites of Kashmir. Part IV. Genus *Echinostoma* Rudolphi, 1809. *J. Parasitic Diseases,* **20** (1): 113.

*Ahmad, F., Fatima, M. and Chishti, M. Z. (2008). Ruminant Trematode Parasite of Kashmir, *Dicrocolelium dendriticum* (Rudolphi, 1819, Looss, 1899). **In**: M. Z. Chishti and Fayaz Ahmed (eds.), *Science For Better Tomorrow*, Printed at Universal Printers, Khanyar, Srinagar, Kashmir.

*Ahmad, F., Fazili, K.A., Sofi, T. A., Sheikh, B.A., Waza, A. A., Rashid, R. and Gani, T. T. (2015). Morphological and molecular characterization of *Diplozoon kashmirensis, D. aegyptensis* and *D. guptai,* collected from fishes of Kashmir Valley, India. *Fish Aqua J.* **6**: 147, doi: 10.4172/2150-35080.1000147.

*Ahmad, F., Hussain, A. and Khan, A. R. (2008). Host specificity of helminth parasites in fishes of Kashmir: Phylogenetic Prespective, *pp.* 483-489, **In:** M. Z. Chishti and Fayaz Ahmad (eds.), *Science For Better Tomorrow,* Printed at Universal Printer, Khanyar, Srinagar, Kashmir.

Ahmad, F., Hussain, A. and Khan, A. R. (2008). Comparartive analysis of *Adenoscolex* infection in indigenous and exotic fish species in River Jehlum, pp. 525-528, **In**: M. Z. Chishti and Fayaz Ahmad (eds.), *Science for Better Tomorrow,* Printed at Universal Printers, Khanyar, Srinagar, Kashmir.

Ahmad, Imtiaz, Ahmad, Ishtiyaq, Dar, S., Awas, S., Kaur, H., and Ganai, A. B. (2019). *Myxobolus himalayaensis* sp. nov. (Cnidaria: Myxozoa), Parasitizing *Schizothorax richardsonii* (Cyprinidae, Schizothporacinae)from River Poonch, in North Himalaya, India. *J. Aquaculture Report.* Doi: 10.1016/j.aqrep.2019.100192.

Ahmad, Imtiaz, Ahmad, Ishtiyaq, Dar, S., Awas, S., Kaur, H., and Ganai, A. B. (2019). *Myxobolus himalayaensis* sp. nov. (Cnidaria: Myxozoa), Parasitizing *Schizothorax richardsonii* (Cyprinidae, Schizothporacinae)from River Poonch, in North Himalaya, India. *J. Aquaculture Report.* Doi: 10.1016/j.aqrep.2019.100192.

Ahmad, F. Sheikh, B.A., Sofi, O. M. and Sofi, T. A. (2018). Prevalence of helminth parasites in fishes (*Salmo trutta* and *Schizothoraz plagiostomus*) in India. *Revista Veterinaria,* **29** (1): 45-51, dx Doi: org/10.30972/vet. 2912790.

*Ahmed, F., Tariq, K. A. and Chishti, M. Z. (2000). Epidemiological studies on *Haemonchus* of sheep in Kashmir Valley. *Oriental Sci.,* **12**: 43-47.

Ahmad, F., Tariq, K. A. and Chishti, M. Z. (2009). Seasonal prevalence of intestinal cestodes of goats, with respect to age, sex and breed in Kashmir valley, India. **In:** Veena Tandan, A. K. Yadav and Bishnupada Roy (eds.), *Current Trends in*

Parasitology, Proc. 20th National Congress of Parasitology, Shilong (Nov. 3-5, 2008), pp.147-152.

*Ahmad, F., Wani, S. A., Zargar, S. A., Ahmad, Z. and Tak, H. (2008). Prevalence of Gastro-intestinal Helminthic infection and associated Risk factors among school childrens in District Kupwara of Kashmir valley, ***In***: M.Z. Chishti and Fayaz Ahmad (eds.). *Science For Better Tomorrow*, printed at Universal Printers, Khanyar, Kashmir.

Ahmad, Ishtiaq, Ahmad, Farooq, Mir, A. H., Wani, N. A., Rifat and Mir. N. A. (2001). Solitary hydatid sigmoid mesocolon. *JK-Practioner*, **9** (1): 49-50.

*Ahmad, Tariq, Baqui, A., Chishti, M.Z., Tanveer, S., Ahmad, P. and Mir, S. A. (2008). Seasonal incidence of *Paramphistomum cervi* in sheep in Kashmir Valley, ***In***: M. Z. Chishti and Fayaz Ahmad (eds.), pp. 445-446, *Science For Better Tomorrow*, printed at Universal Printers, Khanyar, Srinagar, Kashmir.

Ahmad, I., Yadav, A., Katoch, R., Godara, R., Saleem, T. and Nisar, M.A. (2015). Prevalence and analysis of associated risk factors of *Cryptosporidium* infection in lambs in Jammu District. *J. Parasit. Dis*. **39**: 414-417.

Ahmad, M. J. (2005). A new genus and species of Tetrastichnae (Chalcidoidea: Eulophidae) from south Kashmir. *Shashpa*, **12** (2): 71-72.

Ahmad, M. J., Khursheed, I. and Azim, M. N. (2012). A new species of *Ecdamua* (Chalcidoidea: Torymidae) from South Kashmir, India. *Intl. J. Sci., Env. and Technol.*, **1**(5): 409-415.

Ahmad, M. J., Khursheed, I. and Azim, M. N. (2012 a) A new species of *Epitranus* Walker Chalcidoidea: Chalcididae from Kashmir (India). *J. Insect Science*, **25** (3): 316.

Ahmad, M. J., Khursheed, I. and Azim, M. N. (2012 b). Description of a new species of *Sphegigaster* Spinola (Chalcidoidea: Pteromalidae) from Kashmir. *Indian J. Entomol. Res.*, **36** (4): 388-389.

Ahmad, M. J., Khursheed, I. and Azim, M. N. (2012 c). Description of a new species of *Systasis* Walker (Chalcidoidea: Pteromalidae) from Kashmir, India. *Annals of Plant Protection Sciences*, **20** (2): 352-353.

Ahmad, M. J., Khursheed, I. and Azim, M.N. (2013). A new species of *Diglyphus* Walker (Chalcidoidea: Eulphidae) on mustard leaf-miner, *Chromatomyia horticola* (Gourea) from Kashmir. *J. Entomol. Res.*, **37**(2): 163-165.

Ahmad, M. J., Khursheed, I. and Azim, M. N. (2013 a). First records of Chalcids from Jammu and Kashmir, India. *Annals of Plant Protection Sciences*, **21**(1): 72-76.

Ahmad, M. J. and Ahmad, B. (2013). Dynamics of *Diaretiella rapae* (M'Intosh) (Braconidae: Aphidiidae) and its hyperparasitoid on mustard aphid, *Lipaphis erysimi* Kalt, infesting brown mustard, *Brassica campestris* in Kashmir, India. *J. Biol. Control*, **27**(4): 247-253.

Ahmad, M. J. and Zaki, F.A. (2008). Natural Parasitism of Rice Skipper, *Parnara guttata* (Bremer and Grey) (Lepidoptera: Hesperiidae), infesting Paddy at Anantnag, Kashmir. *J. Biol. Control*, **22** (1): 113-117.

Ahmad, M., Ahmad, J., Mishra, R. K. and Khan, H. R. (2002). Egg parasitoid, *Trichogramma* and its host range in India. *Indian Forester*, **128** (4): 412-.

*Ahmad, P. A., Sharma, R., Katoch, R., Anis, Yadav, A., Godara, R. and Rafiqui, S. I (2018). Incidence of gastro-intestinal helminths of stray dogs: A Post Mortum Study. *Vet Parasitol.*, **24**: 83-86.

Ahmad, Rayies, Wani, Z. A., Allaie, I. M., Bushra, M. S. and Hussain, H. A. (2016). *Toxocara vitulorum* in a sucking calf: A case study. *J. Parasitic Dis.*, **40**: 1330-1331.

Ahmad, S., Ahmad, Z and Shamim, M. (2009). Taxonomic studies of the genus *Opius* (Hymenoptera: Braconidae: Opiinae) from India. *Oriental Insects*, **43**: 159-199.

Ahmad, T., Reshi, M. L., Chishti, M. Z., Tanveer, S. and Shah, Z. A. (2013). A case report of *Gastrothylax crumenifer* incidence in sheep in Kashmir valley. *Iraqi J. Vet. Sci.*, **27** (1): 71-72.

*Ahmad, T. S. and Wani, Shabeer (2014). Aphid Parasitoids (Hymenoptera: Braconidae: Aphidiinae) of Kashmir: A Checklist. *Acta Zoologica Mexicana* (N. S.) 30 (3): 539- 552 (A total of 61 spp., under 15 genera of aphidiine, parasitizing diverse host aphids on plant species).

Ahmad, Tariq, Baqui, A., Chishti, M. Z., Tanvee, S., Ahmad, Pervaiz and Mir, S. A. (2008). Seasonal incidence of *Paramphistomum cervi* in sheep in Kashmir Valley, pp. 445-446, **In**: *Science For Better Tomorrow*, printed at Universal Printers, Khanyar, Kashmir.

Ahmad, Muzzafar (1978). *Incidence of Helminthic* and *protozoan infection in gut in rural population of Kashmir*. M.D. Medicine Thesis. Department of Medicine Government Medical College. The University of Kashmir, Srinagar. pp.111.

Ahmad, Ishtiaq and Usmani, M. K. (2016). A new species of the genus *Protomicroplitis* (Braconidae: Microgastrinae) from India. *J. Entomol. and Zool. Studies*, **4** (1): 540-42.

*Ahmad, J, Tanveer, S. and Zargar, B. A. (2013). In vitro anthelmintic activity of *Mentha longifolia* (L.) leaves against *Ascaridia galii*. *Global Veterinaria*, **1** (1): 112-117.

Ahmad, Q., I, . Ahmad, S., Choh, S, Ahmad, S. M., Ali, Wajid and Chaudhary, J. (2010). Clinico-epidemiological profile of maternal and congenital toxoplasmosis in Kashmir valley: A hospital based study. *J. Pediatric Infectious Diseases*, **5** (4): 333-337.

Ahmad, Z., Ahmad, Z., Ghramti, H. A. and Ansari, A. (2019). Two new species of braconid wasps (Hymenoptera: Braconidae) from India. *Zookeys*, 889 (22): 23-35.

Ahmad, Zubair, Tanveer, S., Yattoo, G. N., Chishti, M. Z., Sofi, B. A. and Wani, S. A. (2008). Seroprevalence of *Toxocara canis* infection among children population of Disrict Srinagar and Kupwara of Kashmir Valley. **In**: M. Z. Chishti and Fayaz

Ahmad (eds.), *Science For Better Tomorrow*, pp. 305- 308. Printed at Universal Printers, Khanyar, Srinagar, Kashmir.

*Ahmed, Muneer, Zaki, F. A. and Mohi-uddin, S. (2009). Surveillence of San Jose Scale and its parasitization in orchards of Kashmir valley. *Bull. Env. Sciences*, **28** (2): 121-124.

Akhtar, A. Khan, and Shah, M. A. (2017). Records of aphids and their natural enemies in agro-ecosystem with special reference to horticultural ecosystem of Kashmir. *J. Entomol. and Zool. Studies*, **5** (4): 189.

Akhter, Mir Samin, Zubair, A. and Ramamurthy, V. K. (2010). Descriptions of two new species of Microgastrini (Hymenoptera: Braconidae). *Zootaxa*, 2608: 57-62.

Akhter, M. S., Kumawat, M. M. and Ramamurthy, V. K (2010). An annotated checklist of *Xanthopimpla* Saussure (Hymenoptera: Ichneumonidae). *Oriental Insects*, 44: 243-469.

Akhter, M. S., De, Debjani and Usmani, M. K. (2011). A catalogue of aphid parasitoids (Hymenoptera: Braconidae) from India. *Insecta Mundi*, 670 http. digital commons.unl.edu/insectmundi/670.

*Akhter, S., Ahmed, F. and Chishti, M. Z. (2008). Infection Dynamics of *Bothriocephalus acheilognathi* in copepod intermediate Host. *Oriental Sci.*, **15**: 55-59.

*Altaf, Aqsa and Tanveer, S. (2018). Morphology and Presence of Helminth parasites, of *Anas platyrhynchos* (Duck) collected from Dal and Wular Lakes in Kashmir. *Recent Trends in Aqua., Food Science, Forestry, etc. AFHABEC- 2018*, pp. 1-3 (Parasites included *Echinostomum revolutum*, *Raillietina tetragona* and *Polymorphus minutus* (= *Polymorphis*).

Anis, S. B. and Hayat, M. (2002). A Revision of the Indian species of *Cheiloneurus* Westwood (Hymenoptera: Chalcidoidea: Encyrtidae). *Oriental Insects*, **36**: 129-209.

Ajjampur, S. S., Sankaran, P. and Kang, G. (2008). *Cryptosporidium* species in HIV infected individuals in India: an overview. *Natl. Med. J. India* (July-August), **21**(4): 178-184.

Akhtar, Yasmin, Ahmad, W. and Jairajpuri, S. (1994). Two new species of Dorylaimida (Nematoda) from Kashmir Valley, India. *Rec. Zool. Surv. India*, **94** (1): 87-92.

*Akhter, S., Ahmed, F. and Chishti, M. Z. (2008). Infection Dynamics of *Bothriocephalus acheilognathi* in copepod intermediate Host. *Oriental Sci.*, **15**: 55-59.

*Akhoon, Z. A. and Peer, F. U. (2004). Prevalence of Fascioliasis in District Srinagar of Kashmir Valley. *Indian J. Animal Health*, **53** (1): 35-40 (Investigation during 2001-2002 on a total 693 faecal sample, 361 were detected for liver flukes, prone to fascioliasis).

*Al- Mendalawi, M. Dhahir (2016). Clinical study of cutaneous leishmaniasis in Kashmir Valley. *Indian Dermatology Online Journal*, Nov. Dec., 7 (8): 538 (Author has made observation in response to the earlier research paper of Wani *et al.*, 2016, *Ind. Der. Online J.*, **6**: 387 92).

Akhtar, Yasmin, Ahmad, W. and Jairajpuri, S. (1994). Two new species of Dorylaimida (Nematoda) from Kashmir Valley, India. *Rec. Zool. Surv. India,* **94** (1): 87-92.

*Allaie, I. M. (2018). Studies on Benzimidazole Resistance in gastro-intestinal nematodes of small ruminants in Kashmir Valley. *SKAUST (Kashmir) publication,* (cf. http//scholar. google.com).

Allaie, I. M., Wani, Z. A., Malik, A. M., Shahardar, R. A. Zulhuma, M. (2016). *Oestrus ovis* larvae in nasal cavity of sheep: a case report. *J. Parasitic Diseases,* **40** (4): 1221-1222.

Allaie, I. M., Wani, Z. A., Shahardar, R. A., Maqbool, I., Malik, A.A. and Hussain, H.A. (2014). Parasitic-fauna of live stock in Gurez area of Kashmir Valley. *SKAUST J. Res,* **16** (2): 143-146.

Allaie, K. A., Shah, A.A., Wani, S. M., Bhat, K. A., Kouser, N. and Rawhat-ul- Nisa (2020). Diversity of Plant- parasitic nematodes associated with rhizosphere of maize from Different areas of District Rajouri, Jammu and Kashmir. *Plant Protection,* **4** (1) https//doi. org/10.33804/pp004.01.3211.

Allaie, I. M., Shahardar, R. A., Tramboo, S. R., Bulbul, K. H., Wani, Z. A. and Khan, A. A. (2018). Prevalence of Gastro-intestinal nematodes in small ruminants of Kashmir. *J. Ent. and Zool. Studies,* 6 (2): 2554-2559.

*Ali, Neelofar (1999). *Parasitism in Aquatic vertebrates of Anchar lake.* Ph. D. Thesis. P. G. Deptt. Of Zool., Faculty of Sci., Univ. of Kmr., Sgr., Kmr. (J&K).

Ali, N., Nazir, S., Ganai, H.A., Muzaffar, Gausia and Chishti, M.Z. (2012). Prevalence studies of parasite, *Pomphorhynchus kashmirensis* in local fish species, *Schizothorax. European Scientific J.,* **8** (30): 1857-1859.

Almeida, M. O. and Christffersen, M. I. (1999). A cladistic approach to relationships in Pentastomida. *J. of Parasitol.,* **85**: 695-704.

*Ambani, M. M. (2014). Fish Parasites (*Labeo rohita* and *Schizothorax niger*) of Jammu and Kashmir in India. *Advances in Aquaculture and Fisheries Management,* **2** (1): 203-205.

Ambasta, S. P. *et al.* (eds) (1986). *The useful plants of India.* Published by publication and Information Directorate of Scientific and Industrial Research, New Delhi (India).

Amin, O. M. (2002).Revision of Neoechinorhynchidae Stiles and Hassall, 1905(Acanthocephala: Neoechinorhynchidae), with keys to 88 species in two genera. *Systematic Parasitol.,* **53** (1): 1-18.

*Amin, Ayshia and Khan, A. R. (2008). Prevalence of Endohelminth infection in slaughtered sheep in Doda District of Jammu and Kashmir. **In**: M.Z. Chishti and Fayaz Ahmad (eds.), *Science For Better Tomorrow.,* Printed at Universal Printers, Khanyar, Srinagar, Kashmir. (On Cestode, Trematode and Nematode parasites).

*Amin, A. and Wani, S. A. (2012). Gastro-intestinal helminth parasites of slaughtered sheep, with special reference to age, sex and seasonal distribution in District Doda, Jammu and Kashmir State, India. *Indian J. Scientific Res. Technol.,* **1**: 25-28.

Amin, O. M. (2013). Classification of the Acanthocephala. *Folia Parasitologica*, **60** (4): 273-305.

Amin, O. M., Shamall, M. A. A. and Furhan, T. M. (2003). Description of *Pomphorhynchus spindletruncatus* from fresh water fishes in northern Iraq, with the erection of a new Pomphorhynchid genus *Pyriproboscis* n.g.and key to genera of Pomphorhynchidae and the species of Pomphorhynchus Monticelli, 1905. *Syst. Parasitol.*, **54**: 229-235.

Amin, O. M., Crompton, D.W.T. and Brent, B. N. (1985).*Classification and Biology of the Acanthocephala.* Cambridge University Press, Cambridge, Massachusetts.

Amin, O. M., Heckmann, R. A., Zargar, U.R. and Chishti, M.Z. (2012). The morphology of the long forgotten *Pomphorhynchus kashmirensis* (Acanthocephala: Pomphorhynchidae) from fresh water fish in Kashmir, using SEM, with notes on histomorphology. *Scientia Parasitologica.*, **13** (3): 93-99.

Amin, O. M., Heckmann, R. and Zargar, U. R. (2017). Description of a new Quadrigyrid Acanthocephalan from Kashmir, with notes on metal Analysis and Histology, and key to species of the subgenus.*Acanthosentis* from the Indian subcontinent. *J. Parasitol.*, **103** (5): 458-470.

Anderson, R. C. (2000). *Nematode parasites of Vertebrates: Their development and transmission* (2nd Edition).CBI Publishing, CAB International Wallingford, Oxan, U.K. 650 pp.

Andrabi, A., Tak, H., Lone, B. A. and Para, B. A. (2020). Seroprevalence of human cystic echinococcosis in South Kashmir, India. *Parasite Epidemiology and Control*, **e00172.**

Anjum, Aurangzeb (2013). A new species and first host record of *Rhabdochona* Ralliet, 1916 (Nematoda: Thelazoidea) in fresh water fish, *Tor putitora* from Poonch River of J&K State, India. *Intl. J. Adv. Biol. Res.*, **3** (1): 110-115.

Anonymous (1969). *Report of Work carried out during 1968.* Commonwealth Institute of Biological Control, Commw. Agri. Bureaux, London, England, 100 pp.

Anonymous (1970). *Report of Work carried out during 1969.* Commonwealth Institute of Biological Control, Commw. Agri. Burx., London, England, 183 pp.

Anonymous (1999). *National Family Health Survey (NFHS), India. Fact Sheet, Jammu and Kashmir National Family Health Survey* (*NFHS*).http: //rchiips.org/nfhs/data/jkfctsum.pdf.

Anonymous (2007). National Rural Health Mission (2005-2012). Goverenment of Jammu and Kashmir, District Health Action Plan, District Jammu, December, 2007, pp.1-205.

Anonymous (2009). Scrub Typhus and other Rickettsiosis: *CD Alert*, May to June 2009, 13(1): 1-8.

Anonymous (2016). *Zoonotic Diseases of Public Health Importance.* Published by Zoonosis Division, National Centre for Disease Control. Directorate General of Health Services, Ministry of Health and Family Welfare, New Delhi.

Anwar, P., Hassan, I. Yaseen, J. Imran, Majid, Qazi, M., Mubashir, S. Nisa, N., Rather, S., and Smeera, F. (2015). Pattern of sexually transmitted infections in muslim majority region of North India. *Indian J. Sexually transmitted Diseases and AIDS*, **36** (1): 30-34. Doi: 10.4103/0253-7184.156697.

Appannanavar, S.B. and Mishra, Baijayantimala (2011). An update on Crimean Congo Haemorrhagic Fever, *J. Global Infect. Dis.*, **3**(3): 285-292, doi: 10.41031/0974-777x.83537.

Ara, Jahan., Khan, A. R. and Ahmad, Fayaz (2000). First record of Pseudophyllidean cestode, *Bothriocephalus* (Rudolphi, 1808) from fishes of Kashmir. *Oriental Sci.*, **5**(1): 23-26.

Arora, R. R., Biswas, M., Mathur, K.K. Verghese, T. and Das, A. (1976). Prevalence of intestinal parasites in rural community in Jammu and Kashmir. *J. Communicable Diseases*, **8** (4): 302-315.

Arora, S., Mendonca, S., Malik, A., Ramesh, V, and Khandpal, R. (2017). Leishmanial abscess. *Indian J. Dermatol.*, **62** (4): 431-433, DOI: 10.4103/ijd.IJD_533_1.

*Ashraf, Aiman, Shahardar, R. A., Wani, Z. A. and Allaie, I. M. (2017). Prevalence of gastro-intestinal helminths in cattle of North Kashmir. *Indian J. Animal Sci.*, **87** (2): 147-152. (Corprological examination of cattle of Baramulla, Kupwara and Bandipore, showed prevalence of GI helminths as 59.86 per cent, covering trematodes: *Paramphistomes*, *Fasciola* spp., *Dicrocoelium*; Cestode - *Moniezia* spp. and Nematodes – *Strongyle, Strongyloides, Trichuris*).

*Ashraf, A., Shahardar, R. A., Wani, Z. A. and Allaie, I. M. (2018). Comparative efficacy of Fendendazole, Ivermectin and Oxyclozanide against gastro-intestinal of cattle in North Kashmir. *J. Pharmacognosy and Phytochemistry*, **7** (6): 2232-2235.

*Ashraf, A., Shahardar, R. A., Bulbul, K. H., Wani, Z. A., Allaie, I. M. and Mir, A. Q. (2020). Efficacy of amprolium against coccidiosis of goats in Central Kashmir. *J. Entomol. Zool. Studies*, **8** (3): 1667-1670.

*Askary, T. H. (2012). Managenent of root-knot nematode, *Meloidogyne javanica* in pigeon pea through seed treatment. *Indian J. Ecol.*, **39**: 151-152.

*Askary, T. H. and Haider, M. G. (2010). Parasitic nematodes associated with forest nurseries. Indian *J. Nematol.*, **40:** 239-240.

Askary, T. H., Khalil, A., Khan, A. A. and Nazir, N. (2018 a), Population Fluctuation of Plant-parasitic nematodes associated with grapevine under Hi-tech and Low-tech Polyhouse condition. *Intl. J. Curr. Microbiol. Appl. Sci.*, **7** (5): 2133-2140.

Askary, T. H., Khalil, A. and Nazir, Nowsheen (2018 b). Resistance of Grapevine cultivars to *Meloidogyne hapla* under polyhouse condition. *Intl. J. Curr. Biol. Appl. Sci.*, **7** (8): 2351-2359.

Askary, T. H., Shah, M. I., Wali-ullah, M. I. S. and Mir, M. M. (2014). Distribution of phytonematodes associated with stone and nut fruits in Kashmir valley, *India. J. Applied Horticulture*, **16** (2): 136- 142.

*Askary, T. H., Wali-ullah, M. I. S., Mir, M. M. and Banday, S. A. (2013). Population dynamics of plant-parasitic nematodes associated with pome fruits. *Annals of Plant Prot. Sci.,* **21** (1): 157-159 (Biodiversity of 8 spp. of plant parasitic nematodes of rhizosphere of two pome fruits- apple and pear, from three old nurseries of Kashmir, is studied).

Ayala, M.A. and Yencha, M.W. (2012). *Gongylema*: A Parasitic Nematode of the oral cavity. *Arch. Otolaryngol Head, Neck Surg.,* **138** (11): 1082- 1084. Doi: 10.1001/2013. Jamaoto.386.

Baba, A. A., Ahmad, S.M. and Sheikh, K. A. (2009). Intestinal ascariasis: The commonest cause of bowl obstruction in children at teritiary Care Centre in Kashmir. *Pediatric Surg. Intl.,* **25** (12): 1099-1102.

Baba, I. A., Shah, A. F., Yousuf, A. and Nabi, S. (2015). Mental deprivation and oral myiasis: A special case Report *Annals of Dental Speciality,* **3** (2): 69.

Baer, W. S. (1931). The treatment of chronic oesteomyletis with maggots (larva of blowfly). *J. Bone and Joint.,* **13**: 438-475.

*Bakshi, Shafqat (1999). *Fish Parasites of some Lakes of Kashmir, with an analysis of seasonality of incidence with regard to different ecological factors.* Ph. D. Thesis, Faculty of Sci., P. G. Deptt. of Zool., The Univ. Kmr., Sgr., Kmr. (J&K).

Bali, H. S. (1971). On the occurrence of *Calicophoron caliciphorum* (Fishoeder, 1901) Nasmark, 1937 in sheep in J&K State. *Indian Vet. J.,* **49** (4): 442-43.

Bali, H. S. (1972 *a*). On the morphology of a new trichostrongylid, *Nematodirus fotedari* n. sp. of sheep (*Ovis aries*) from Kashmir. *J. Res. Panjab Agri. Univ.,* **9**: 195-198.

Bali, H. S. (1972*b*). Study on a new trematode belonging to the genus *Paramphistomum* Fishoeder, 1901 from *Ovis aries* in J&K State. *Indian J. Anim. Sci.,* **42** (3): 231.

Bali, H. S. and Fotedar, D.N. (1972 *a*). On the morphology of a Paramphistome in sheep. *Ceylonocotyle scoliocoelium* from Jammu and Kashmir *J. Res. Punjab. Agri. Univ.,* **9** (1): 199-200.

Bali, H. S. and Fotedar, D.N. (1973). On a new Amiphistome, *Olveria thapari* n.sp. from rumen. of cross-bred sheep from Kashmir. *Indian J. Helminthol.,* **24** (1): 36-39.

Bali, H. S. and Fotedar, D. N. (1974). Effect of experimental infection of some bursate nematodes in sheep in Kashmir. *J. Res. Panjab. Agri. Univ.,* **11** (1): 114-119.

Bali, H. S. and Fotedar, D. N. (1974*a*). Studies on Juvenile stages of some helminth parasites of Kashmir sheep. *J. Res. Punjab Agric. Univ.,* **11** (8): 329-343.

Bali, H. S. and Fotedar, D. N. (1976). Studies on the pathology of some helminth-borne disease in sheep of Kashmir. *J. Anim. Health Prod.* P. A. U., **4** (1-2): 9-14.

Bambroo, Neena (1970*a*). On two species of *Gastrothylax* Poinier, 1883 from sheep in Kashmir. *Kashmir Sci.,* **7** (1-2): 147-155.

Bambroo, N. (1971). On a new bursate nematode from sheep in Kashmir. *Indian J. Helminthol.,* **23**(2): 86-94.

Banerjee, D. and Mitra, B. (2006).Diversity of bee-flies in India. *Records of the Zool. Surv.Ind., Occ. Pap.* No.252, Zoological Survey of India, Kolkata.

*Baneth, G. and Weigler, B. (1997). Retrospective case control study of Hepatozoonosis in dogs in Israel. *J. Vet. Internal Med.*, **11**: 365-370.

Bansal, G. C., Sharma, R. L. and Dhar, D. N. (1982). Lungworms-Threat to sheep and goats. *Farmers Journal*, **1**: 42-44.

Baqui, A. and Shah, Farooz, A. (2002). Alteration in haematological parameter in nematodiasis. *Oriental Sci.*, **7**(1): 25-34.

Bari, Shams- ul, Malik, A. A., Parry, F. Q., Sammon, H., Murifat, M. and Bakshi, H. (2008). Deleyed diagnosis of traumatic rupture of Hydatid cyst of liver – a case Report. *Intl. J. Surgery*, **6**: 18-19.

Barraud, P. J. (1924a). A new mosquito from Kashmir. *Indian J. Med. Res.*, **11**: 967-968.

Barraud, P. J. (1924 b). A new mosquito from Kashmir and the North-West Frontier province. *Indian J. Med. Res.*, **12**: 73-74.

Barraud, P. J. (1924c). Four new species of mosquitoes from Western Himalayas, India. *J. Medical Res.*, **11**: 999-1006.

Barraud, P. J. (1934). *The Fauna of British India Vol 6, Family Culicidae, Tribe Magarhinini and Culicini,* xxviii + 1-463 pp. Tylor and Francis, London, England.

Basir, M. A. (1941). Two new nematodes from an aquatic beetle. *Proc. Ind. Acad. Sci.*, **13**: Sec. B., 163-167.

*Beigh, A. B., Darzi, M. M., Bashir, S., Kashani, B., Shah, A. and Shah, S. A. (2017). Gross and histopathological alterations associated with cystic echinococosis in small ruminants. *J. Parasitic Dis.*, **41** (4): 1028-1033.

Bezerra, T. N. *et al.* (2020). *Nemys World database of nematodes.* http: / /nemys. Doi.10. 14284/366.

Bhadwal, M.S. and Raina, A.K. 1994: Thelaziasis in a buffalo. *Indian J. Vet. Med.*, **14**(2): 94.

Bhagat, K.C. and Masoodi, M. A. (1988). Natural enemies of mealy plum aphid, *Hyalopterus arundinis* Fabricius (Aphididae: Homptera) in Kashmir. *J. Advanced Zool.*, **9** (2): 145- 147.

Bhagat, K. C. (2000). Natural parasitism of wheat aphid, *Macrosipham* (*Sitobion*) *avenae* Fab. B. *Lysiphlebia mirzai.* Shuja-uddin (Hymenoptera: Braconidae). *Insect Envir.*, **6** (1): 9.

*Bhagat, M., Sood, S., Yadav, A., Katoch, R., Chakraborty, D., Godara, R., Sultana, M. and Sangha, Navrose (2018). Clinico-haematological studies on experimental *Cryptosporidium parvum* Jammu isolate infection in Swiss albino mice. *Indian J. Animal Res.*, **52** (7): 1025-1030.

Bhagat, R. C. (1981*a*). A new species of *Monoctonus* (Hymenoptera: Aphidiidae) from Kashmir, India. *Curr. Sci.*, h. **56** (24): 1087-1088.

Bhagat, R. C. (1981*b*). *Areopraon lepelleyi* (Hymenoptera: Aphidiidae): - a new record for India. *Geobios,* **8** (4): 177-178.

Bhagat, R. C. (1981*c*). Some new aphid parasitoids. (Hymenoptera: Aphidiidae: from Kashmir, India. *Orient. Insects,* **15** (2): 195-200.

Bhagat, R. C. (1982). Aphid-galls and their parasitoids from Kashmir. *Entomon,* **7** (1): 103-105.

Bhagat, R. C. (1982*a*). On a new aphid parasitoid (Hymenoptera: Aphidiidae) from Kashmir. *Indian J. Zool.,* **10** (1): 19-22.

Bhagat, R. C. (1982*b*). On two new aphid parasitoids of Genus *Trioxys* (Aphidiidae: Hymenoptera) from Kashmir, India. *Entomon,* **7** (3): 321 -324.

Bhagat, R. C. (1982*c*). New aphid parasitoid (Hymenoptera: Aphidiidae) from Kashmir, India. *Oriental. Insts.,* **16** (1): 113-117.

Bhagat, R. C. (1983). Records and host range of hyperparasitoids (Insects: Hymenoptera) of aphids (Homoptera: Aphidiidae) from Kashmir, India. *Sci. and Cult.,* **49** (3): 150-151.

Bhagat, R. C. (1984). New records and hosts of aphids parasitoids (Hymenptera: Aphidiidae) from Kashmir, India. *J. Bombay Hist. Soc.,* **81** (1): 93-98.

Bhagat, R. C. (1985). Host-aphid complex of Aphelinid parasites (Hymenoptera: Insecta) in Kashmir Valley, India. *Geobios new Reports,* **4**: 176-179.

Bhagat, R. C. (1985*a*). On Aphids and their parasitoids. (Aphidiidae: Hymenoptera), infesting forest-trees in Kashmir Valley, India. *Indian Forester,* **111** (6): 467-473.

Bhagat, R. C. (1986). Additional records and host aphids and aphidiid parasitoids from Kashmir Valley, India. *Geobios new Repts.,* **5**: 161-162.

Bhagat, R. C. (1986*a*). Aphid pests of fruit trees and their natural enemies in Kashmir Valley, India. *J. Agri. Sciences,* **56** (7): 532 -536.

Bhagat, R. C. (1986*b*). On aphid pests and their aphidiid parasitoids of agricultural importance. *Indian Agriculturist,* **30** (3): 229-235.

Bhagat, R. C. (1987). Field observations on hyperparasitoids of aphid pests, infesting *Rosa* spp. in Kashmir Valley. *Indian. J. Biol. Control,* **1** (2): 104 -105.

Bhagat, R. C. (1987*a*). On aphid damaging medicinal plants and their parasites from Kashmir Valley, India. *Geobios new Repts.,* **6**: 11-14.

*Bhagat, R. C. (2008). *Biodiversity of Parasite- fauna: Jammu, Kashmir and Ladakh Himalayas.* New Age International (P) Limited, Publishers. New Delhi (India) 211 pp.

*Bhagat, R. C. and Bala, P. (2002). *Biological Diversity of Jammu, Kashmir and Ladakh: A Classified and Comprehensive Bibliography.* A.P.H.Publishing Corporation. New Delhi, India 429 pp.

Bhagat, R. C. (2014). An update on Inventory and Biodiversity of Mosquito-fauna of Jammu, Kashmir and Ladakh Himalaya, India, *Cibtech Journal of Zoology,* **3** (3): 28-34.

Bhagat, R. C. (2014). Diversity and distribution of Midge-fauna (Insecta: Diptera) of Jammu, Kashmir and Ladakh Himalaya (India). *Research and Rev. Biosciences,* **9** (6): 199-206.

Bhagat, R. C. (2016).Biodiversity of Dipterous flies (Insecta) of myiasis causing importance in animals and human of Jammu, Kashmir and Ladakh Himalayas (India). *J. Global Biosciences,* **5** (7): 4341-4349.

Bhagat, R. C. and Ahmad, Nissar (1995). Aphidiid parasitoids (Hymenoptera) of aphids (Homoptera) of Jammu – New record, host range and biological notes. *J. Aphidology,* **5** (1 and 2): 90-96.

Bhagat, R. C., Farhan, N. and Ramzan, A. (1992). Natural enemies of scale insects in Kashmir Valley, India. *J. Biol. Control,* **6**(1): 38-39.

Bhalerao, G. D. (1973). Studies on the Helminths of India, Trematoda IV. *J. Helminthol.,* **15**: 97-124.

*Bhardwaj, R. K. (2013). Therapeutic management of Clinical Eaquine Trypanosomiasis. *Indian Vet. J.,* **90** (10): 42-44.

Bhardwaj, R. K., Singh, R., Himalini and Singh, V. (2015). Trypanosomiasis in Bakerwali Dog – A case report. Indian *J. Canine Practice,* **7** (2): 113-114.

Bhardwaj, R., Tiku, A. K. and Mir, I.A. (2012). Infectious Keratoconjunctivitis in dzos in Jammu. *Indian Vet. J.,* **89** (7): 101-102.

Bhat, A. M., Malik, H., Wani, N. H., Paul, S., Gupta, S., Dolma, T. and Singh, S.V. (2019). First Report of *Cryptosporidium* sp. infection in sheep population of Ladakh. *J. Parasitic Diseases,* **43** (3).

*Bhat, A. S., Rishi, A. A., Mir, R. M., Husain, I., Sheikh, B. A. and Khan, H. M. (2014). Assessment of prevalence of parasitic infection in sheep reared in Kashmir valley, *India. J. Anim. Prod. Adv.,* **4** (5): 416-421(On the prevalence of *Strongyle* spp., *Strongyloides* spp., *Eimeria* spp., *Nematodirus* spp. and *Moneizia* spp.).

Bhat, D. M. (2018). Incidence and Diversity of Lepidopterous insect pests and their parasitoids (Natural enemies) on cole crops at Danderkhah locality in Srinagar District (J&K), India. *Intl. J. Entomol. Res.* **3** (2): 107-113.

*Bhat, D. M. (2020a). Biodiversity of Lepidopterous pests (Insecta) and their Natural bio-control agents associated with vegetable crops in J&K. *J. Entomol. and Zool. Studies,* **8** (3): 31-42 (The present work includes 28 spp. of Lepidopterans- moths and butterflies on vegetable crops and their 40 spp. of natural biocontrol agents-parasitoids, predators and pathogens).

*Bhat, D. M. (2020 b). Lepidopterous insect pests and their natural enemies in vegetable ecosystem of Jammu, Kashmir and Ladakh, India. *J. Biodiversity and Environmental Sciences,* **16** (14): 150 -164.

Bhat, D. M. (2011), Bhagat, R. C. and Qureshi, A. A. (2011). Natural parasitization of *Spodoptera litura* (Lepidoptera: Noctuidae) by *Zele chloropthalma* Nees (Hymenoptera: Braconidae) in vegetable ecosystem of Kashmir Valley, India. *Halteres,* **3**: 88-90.

*Bhat, D. M. (2020). Lepidopterous insect pests and their natural enemies in vegetable ecosystem of Jammu, Kashmir and Ladakh, India – A Review. *J. Biodiversity and Env. Studies*, **15**(14): 150- 164.

Bhat, D. M. and Bhagat, R.C. (2008). Studies on parasitoids of cabbage diamond moth, *Plutella xylostella* (L) (Lepidoptera: Plutella) in Kashmir valley. *J. Entmol. Res.*, **32** (4): 303-308.

Bhat, D. M. and Bhagat, R. C. (2009). Natural Parasitism of *Pieris rapae* (L) (Lepidoptera: Pieridae) and *Pontia daplidace* (L) on cruciferous crops in Kashmir valley, India. *American-Eurasian J. of Agri. and Env. Sci.*, **5** (4): 591.

Bhat, D. M. and Bhagat, R.C. (2010). Record of some some hymenopterous parasitoids of leaf-miner on vegetables in Kashmir Valley. *Annals of Plant Protection Sciences*, **18** (1): 237-238.

Bhat, D. M., Bhagat, R. C. and Azim, M. N. (2009). Record of natural enemies of *Helicoverpa armigera* from Kashmir Valley. *Annals of Plant Protection Sciences*, **17** (1): 229-.

Bhat, D. M., Bhagat, R. C. and Qureshi, A. A. (2009 a). Record of some hymenopteran parasitoids of serpentine leaf-miners, *Liriomyza trifolii* in vegetable ecosystem in Kashmir. Indian *J. Plant Protection*, **37** (1 and 2): 188-189.

Bhat, D.M, Bhagat, R. C and Qureshi, A.A. (2010a). Newly recorded hymenopteran parasitoids of semilopper, *Thysanoplusia arichalcea* F. on some vegetable crops in Kashmir Valley (India). *J. Entomol. Res.*, **34** (1): 53-56.

Bhat, D. M., Bhagat, R. C. and Qureshi, A. A. (2010 b). Some natural enemies of *Pieris brassicae* in cruciferous crops in Kashmir valley. *Annals of Plant Protection Sciences*, **18** (2): 516-518.

Bhar, D.M., Bhagat, R. C. and Qureshi, A. A. (2011). Natural parasitization of *Spodoptera litura* (Lepidoptera: Noctuidae) by *Zele chloropthalma* Nees (Hymenoptera: Braconidae) in vegetable ecosystem of Kashmir Valley. *Halteres*, **3**: 88-90.

*Bhat, D. M., Bhagat, R. C. and Qureshi, A. A. (2017). Parasitoid-fauna associated with insect pests of vegetable crops of Kashmir Himalaya, India: Checklist and Biodiversity. *Munis Entomol. And Zool.*, **12** (1): 168-174.

Bhat, D. M., Bhagat, R. C. and Gulzar (2010). Report of natural parasitism of *Coccinella* septempunctata L. (Coleoptera: Coccinellidae) by *Dinocampus coccinellae* (Schrank) (Hymenoptera: Braconidae: Euphorinae) from Kashmir. *Annals of Entomology*, **28** (2): 112.

*Bhat, D. M., Khan, S. A., Ahangar, F. A. and Sheikh, M. A. (2020). Diversity of aphid Pests (Homoptera: Aphididae) and their natural bio-control agents in vegetable crop ecosystem of Jammu and Kashmir (India). *Intl. J. Curr. Microbiol. Appl. Sci.*, **9** (5): 2529-2546 (Present study included 18 spp. of aphids, under 12 genera, damaging as many as 25 species and 30 varieties of vegetable crops. 17 spp. of parasitoids found to parasitize the aphid species in vegetable ecosystem).

Bhat, D. M., Qureshi, A. A. and Bhagat, R.C. (2013). Natural parasitization of *Pieris brassicae* (Lepidoptera: Pieridae) by Tachinid flies (Diptera) on some vegetable crops in Kashmir. Indian *J. of Applied and Pure Biol.*, **28**: 31-34.

*Bhat, D. M., Sheikh, M. A. and Khan, S. A. (2020). The butterfly pests (Lepidoptera: Pieridae) of Brassicaceae crops and their biological control in J&K – A review. *Intl. J. Zool. Studies*, **5** (2): 5-8.

Bhat, H. R. and Kulkarni, S. M. (1971). A report on the occurrence of *Culex* (*Neoculax*) *hortensis* Fiscalbi, 1889, in Ladakh. *Oriental Insects*, **5**(4): 583-584.

Bhat, H. R. and Kulkarni, S. M. (1983). Records of mosquitoes collected from Jammu and Kashmir, with ecological notes. *Recd. Zool. Surv. India.* 81: (1-2): 192-203.

Bhat, K. J., Pandita, K.K., Khajuria, A. and Wani, S.M. (2014). Visceral leishmaniasis (Kalazar) migrating west: A new autochthonous case from sub-Himalaya. *Indian J. Med. Microbiology*, **32** (1): 94-95.

Bhat, O. K., and Kaul, V. (1994 *a*). Influence of soil temperature on the population of root knot nematode, *Meloidogyne icognita* in a tomato field. *Annl. Plant Prot.*, **2** (1): 28-29.

Bhat, O. K., Kaul, V. and Bhagat, K. C. (1994). Incidence of pests associated with the rhizosphera of tomato in Jammu. *Annls. Plant Protec. Soc.*, **2**(2).

Bhat, M. R. (1987). *Aphelinus kashmiricus* Hayat- a parasite of *Aphis gossypii* Glover (melon aphid) in Kashmir. *Fagopyrum*, **7**: 20.

Bhat, M. S., Sudhan, N. A., Shahardar and Mir, A.Q. (2007). Prevalence of gastro-intestinal nematodes in sheep in Kashmir Valley. *J. Vet. Parasitol.*, **21** (1): 89-91.

*Bhat, R. A., Tak, H., Bhat, B. A., Fazili, M. F., Wani, H. M., Haq, Iqram-ul and Ahmad, R. (2019). Live stock helminth infestations as a potential threat to wildlife ungulates in Hirpora Wildlife Santuary. *J. Himalayan Ecol. Sustain Dev.*, **14**: 71-78.

Bhat, Showkat A., Mir, Manzor -ur-Raham, Qadir, Sawleha, Allaie, I. M., Khan, H. M., Husain I. and Sheikh, B. A. (2012). Prevalence of gastro-intestinal parasitic infection in sheep of Kashmir valley, India. *Vet. World*, **5** (1): 667-671.

Bhat, S. (1979). The genus *Phaenocarpa* Forester (Hymenoptera: Braconidae: Alysiinae). *Oriental Ins.*, **13** (1-2): 1-27.

*Bhat, S. A., Mir, M., Qadri, S., Khan, M. H., Hussain, I. and Ali, A. (2011). Comparative resistance of sheep breeds to *Haemonchus controtus* in Pasture infection in Jammu and Kashmir. *J. Vet. Sci. and Technol.*, 2: 110, doi: 10.4172/21577579.1000110 (Sheep breeds are: Bakerwal, Corriediale and Kashmir Merino).

*Bhat, S. A., Manzoor, R. M., Qadri, S., Allaie, I. M., Khan, H. M. *et al.* (2011). Comparative Resistance of sheep Breeds to *Strongyle* spp. in Pasture infection in Jammu and Kashmir. *Intl. J. Vet Sci. and Tech.* 10.4172.2157-7579.1000110.

*Bhat, S. A., Mir, Manzoor-ur Rehman, Qadir, S., Allaie, R. M., Khan, M. H., Husain, I. and Sheikh, B. A. (2012). Prevalence of gastro-intestinal parasitic infection in sheep of Kashmir valley of India. *Vet. World*, **5** (11): 667-671.

Bhat, S. A., Rishi, A. A., Mir, H. M., Hussain, I., Sheikh, B. A. and Khan, H. M. (2014). Assessment of Prevalence of Parasitic infections in sheep reared in Kashmir Valley, India. *J. Animal Prod. Adv.*, **4** (5): 416- 421.

Bhat, S. A., Wani, M.Y., Khajuria, J. K., Katoch, R. and Dhama, K. (2014). A rare report of Ectoparasites in Backyard Poultry in Jammu region: Prevalence study and economic importance. *Asian J. Animal and Veterinary Advances*, 1-5 DOI: 10.3923/ajava.2014.

Bhat, Shama. (1979). The genus *Phaenocarpa* Foerster (Hymenoptera: Braconidae, Alysiinae). *Oriental Inst.*, **13**(1-2): 1-27.

Bhat, Yasmeen, J., Yaseen, A., Sheikh, and Hassan, I. (2020). Dermoscopy of two cases of cutaneous leishmaniasis. *Indian J. Dermatology*, **65** (3): 232-234.

Bhatia, B. L. (1936). *The Fauna British India, including Ceylon and Burma: Protozoa: Sporozoz.* Taylor and Francis, London.

Bhatia, M. L. (1930). Surune nouvella hirudinec thynchobdellida, *Glossiphonia cruciata* n. sp. provenant du vivier a truiteid Achhabal Kashmir. *Annls Parasit. Kum. Camp.*, **8**: 244-348.

Bhatia, M. L. (1934). Nouvella sangue rhynchobdellida *Glossiphonia lobata* n. sp. de Testablispiment de pisciculture Achabal. *Annls. Parasit. Kmr. Camp.*, **12**: 121-129.

Bhatia, M. L. (1939). On some leeches from Kashmir. *Bull. Punjab Univ. Zool.*, G **2**: 1-17.

Bhattacharya, S. B. (2007).*Handbook on Indian Acanthocephala*. Zoological Survey of India. Kolkatta., India.

Bihaqi, S.J., Allaie, I. M., Banday, M.A. and Wani, Z. A. (2013). Occurrence of Coccidial infection in rabbits of Kashmir valley. *J. Vet. Parasitol.*, **26** (2): 181-182.

Bihaqi, S. J., Allaie, I. M., Banday, M. A. A. and Shahardar, R. A. (2017). Prevalence of caprine GI helmints in temperate areas of Jammu and Kashmir. *J. Parasitic Dis.*, **41:** 843-849.

Bihaqi, S. J., Allaie, I. M., Banday, M. A. A., Sankar, M., Wani, Z. A. and Prasad, A. (2019). Multiple anhelmintic resistance in gastro-intestinal nematodes of Caprines on Mountain Research Centre for sheep and goat at Kashmir valley.

*Bihaqi, S. J., Allaie, I. M., Banday, M. A. A., Shanker, M., Wani, Z. A. and Prasad, L. (2020). Multiple anthelmintic resistance in gastro-intestinal nematodes of caprines on Mountain Research Centre for sheep and goat at Kashmir, India. Parasite *Epidemiology and Control* Doi: 10.1016/j.parepi.2020.e00163.

Borkataki, Sanku., Katoch, R., Goswami, P., Khajuria, Yadav, A., Kour, R., Godara, R. and Mir, I. (2013 a). Incidence of *Haemoproteus columbae* in pigeons in Jammu District. *J. Parasit. Dis.*, **39** (3): 426-428.

Borkataki, S., Katoch, R., Goswami, P., Godara, R., Khajuria, J. K., Yadav, A. and Kaur, R. (2013b). Prevalence of Parasitic infections of Stary cats in Jammu, India. *Sokoto J.Vet. Sciences*, **11**(1): 1-6, http//dx doi.org./10.4314/sokjvs.v ll i.

Borkataki, S., Katoch, R., Gosawami, P., Bhat, V., Bhardwaj, A., Chakraborty, D. and Chandrawanthani, P. (2018). Therapeutic uses of *Lucilia sericata* maggot in controlling bacterial bio-burden in Rat wound model. *Trop. Biomed*., **35**: 627-638.

Borkataki, S. Gosawmi, Katoch, R. and Raghuvanshi, P. (2017). Detection of *Artyfechinostomum surfrartyfex*- A zoonotic parasite from the small Indian Mangoose, *Herpestes auropunctatus* (Mammalia: Carnivora: Herpestidae) in Jammu and Kashmir, India. *J. of Threatened Taxa,* Vol **9** (11): 10980-10983: 10.11609/jot. 3219.9.11.10980-10982.

Brown M.J.E. (2017). Microspordia an emerging threat to bumble bees. *Trends Parasitol,* SI471-4922.

Boucek, Z., Subba – Rao, B. R. and Farooqi, S. I. (1978). A preliminary review of Pteromalidae (Hymenoptera) of India and adjacent countries. *Oriental Insects,* **12** (4): 433-468.

Boxshall, G. A. (1976). A new genus and two new species of copepod parasitic on fresh water fishes. *Bull. Br. Mus. Nat. Hist.* (Zool.), **30** (6): 209-215.

Brunetti, E. (1911). New Oriental Nematocera. *Rec. Indian Mus.*, **4**: 259-316.

Brunetti, E. (1920). *The Fauna of British India, including Ceylon and Burma. Diptera: Brachycera.* Vol. I. Taylor and Francis, Calcutta (Indian reprint Edition).

Brunetti, E. (1920). *The Fauna of British India, including Ceylon and Burma. Part-III. Pipunculidae, Syrphidae, Conopidae and Oestridae.* Taylor and Francis, London, England.

Brunetti, E. (1923). Diptera: Pipunculidae, Syrphidae, Conopide, Oestridae. In: Shipley, A.E. (ed) *Fauna of British India, including Ceylon and Burma* Vol III, Taylor and Francis, London, England.

Buhroo, A. A., Chishti, M.Z and Ahmad, D. (1999). Notes on some domestic pests of Kashmir. *Oriental Science,* **2** (2): 39-48.

Buhroo, A. A., Masoodi, M. A. and Chishti, M. Z. (2001). Description new species of chalcidoidae (Hymeroptera) from fruit orchard of Kashmir. *SKAUST J. Res.,* **3** (1): 14-20.

Buhroo, A. A., Chishti, M. Z. and Masoodi, M. A. (2002). Biocontrol agents of shot-hole Borer, *Scolytus nitidus* Schedl (Coleoptera: Scolytidae), infesting apple orchards. *Indian J. Plant Protection,* **30** (1): 71-73.

*Buhroo, A. A., Chishti, M. Z. and Masoodi, M. A. (2006). Parasitoid- complex and its efficacy in management of shot-hole borer, *Scolytus nitidus* (Coleoptera: Scolytidae) on apple trees. *J. Biol. Control,* **20** (1): 25- 32 (On 9 spp. of hymenopterous parsites of *S. nitidus*).

Bulbul, K.H., Shahardar, R. A., Allaie, I. M., Wani, Z. A., Tramboo, S. R., Asharaf, A., Maqbool, I, Ganaie, Z. A., Sheikh, I. U. and Sheikh, S. (2018). Avian trichomonosis, with special reference to pigs. *Intl J. Vet Sci.and Animal Husbandary,* **3** (5), Part A: 11-13.

Burger, J. (1984). Notes on Tabanidae (Diptera) of the Oriental Region II. Distribution records of Tabanidae from South-eastern Pakistan and adjacent areas. *Proc. Entomol. Soc. Washington*, **86**(3): 643-647.

Bushra, M., Shahardar, R. A. and Maria, A. (2013). Prevalence of gastro-intestinal helminth parasites of cattle in Central zone of Kashmir Valley. *J. Vet. Parasitol.*, **27**-33-36.

Byrd, J. H. and Castner, J. L. (2009). Insect of Forensic importance, In: Byrd, J.H. and Castner, J.L. (Eds.), *Forensic Entomology: The utility of Arthropods in Legal Investigations* (2nd Edition), pp. 43-45, CRC Press Boca Ranto, FL, 688 pp.

Cairns, S. D., Gershwin, L., Brook, F. J. *et al.* (2009). Phylum Cnidaria; Vorals, medusa, hydroids, Myxozoans, In: Gordon, D.P. (Ed.). *New Zealand Inventory of Biodiversity 1. Kingdom Animalia: Radiata, Lophotrochozoa, Deuterstomia*, pp.59-101.

Cameron, P. (1906). On the Tenthedinidae parasitic hymenoptera in Baluchistan by Major C. G. Nurse. *Bombay Nat. Hist. Soc.*, **17** (2): 274-280.

Cameron, P. (1906*a*). On the Tenthedinidae and parasitic hymenoptera collected by Major C. G. Nurse in Kashmir, *Bombay Nat. Hist. Soc.*, **17** (2): 289-292.

Cavalier-Smith, T. (2010). Kingdom Protozoa and Chromista and the eozoan root of the eukaryotic tree. *Biology Letters*, **6** (3): 342-345, doi: 10.10.98/rsbl.2009.0948.

*Cavalier, Smith, T., Chao, E. and Oates, B. (2004). Molecular phylogeny of Amoebozoa and the Evolutionary significance of unikont Phalansterium. *Eur. J. Protistol.*, 40: 21-48.

Cecilla, D., Shah, P. S. and Pawar, J. A. (2003). *Molecular Epidemiology of DEN viruses*, pp.38, Dengue, Annual Report, 2002-2003. National Institute of Virology, Pune, India.

Chakraborty, A., Hora, G., Pauri, P., Saha, G. K. and Benerjee, D. (2017 a). A biosystematics Species Inventory of Indian Sarcophagidae (Insecta: Diptera: Sarcophagidae). *J. Entomol. and Zool. Studies*, **5** (1): 465-473.

Chakraborty, A., Saha, G.A. and Banerjee, D. (2017 b). Biosystematic approach to inventorization of umbiquitious myiasis- causing flies of veterinary importance in India. *Res. In Agri. and Vet. Sci.*, 1 (2): 116-134.

Chakravorty, G. K. (1942). A new nematode *Camallanus salmonae* from Kashmir. *Curr. Sci.*, **11**: 441-442.

Chandler, A. C. and Read, C. P. (1961). *Introduction to Parasitology, with special reference to the parasites of man.* John Wiley and sons Inc., Indian Reprint Edition 1970, Wiley Eastern Private Ltd., New Delhi, 822 pp.

Chandniwala, K. M. (1996). *Plant Nematodes – Recent Advances in Plant Diseases.* Anmol Publication Pvt. Ltd. New Delhi, India.

Chandra, M. (1982). A checklist of leeches of India. *Recd. Zool. Surv. India.*, **80**: 265-270.

Cheng. C.T. (1973). *General Parasitology.* Academic Press Inc. U.S.A. pp.917.

Christoffersen, M.I. and Assis, J.E.D. (2013). A systematic monograph of the recent Pentastomida with compilation of their Hosts. *Zool. Med. Leiden*, **87** (1), 29 iii, 1-206.

Christopher, S. R. (1933). *The Fauna of British India, Diptera* Vol.IV, Family Culicidae, Tribe Anophelini. Taylor and Francis, London, England, pp. 1-371.

Chishti, M. Z. (1964). *Cestodes and Nematode Parasites of some Aves of Kashmir*. Ph. D. Thesis, Faculty of Science, P. G. Deptt. of Zool., The Uni. Kmr., Sgr., Kmr. (J&K), 319 pp.

Chishti, M. Z. (1973). On a new species of *Choanotaenia* Railliet., 1896 from *Acridotheres tristis* in Kashmir. *J. Sci. Kmr.*, **1** (1-2): 51-54.

Chishti, M. Z. 1975 (1978). A new record of *Acuaria skrjabini* Ozerskeya, 1926 from *Passer domesticus* in Kashmir. *J. Sci. Univ. Kashmir*, **1** (1-2): 109-116.

Chishti, M. Z. (1980). *Dilepsis fotedari* n. sp. (Dilepididae Furhmann, 1907: Cestoda) from *Anas platyrhynchos* in Kashmir. *Indian J. Helminthol*, ., **36** (1): 1-3.

Chishti, M. Z. (1981). On a new record of *Syngamus trachea* (Montagu, 1811) Chapin 1925 (Syngamidae, Leiper, 1912): Nematoda) from *Corvus splendens* in Kashmir. *Kashmir Univ. Res. J.*, **1** (1): 53-55.

Chishti, M. Z. (1982). *Pseudoschistotaenia pindchii* sp. nov. (Acoela: cestoda) from *Podiceps ruficollis capensis* in Kashmir. *Osnovy Cestolognii*, **11**: 69.

*Chishti, M. Z. (1998). Fish Parasitology in Kashmir: Effect of fish parasites on the fish population, their pathogenicity and control. *Parasitology Intl.*, **47**: 170.

.Chishti, M. Z. and Abdullah, R. (1986). On *Rusguniella* (*Rusgunioides*) *turdi* sp. nov. (Acuarioidea: Nematode) from *Turdus unicolor* in Kashmir. *Indian J. Helminthol.*, **38** (1): 79-82, (1986/87).

*Chishti, M. Z. and Ahanger, A. G. (1998). Epidemiology and control of Human hydatidosis in Kashmir, India. *Parasitol. Intl.*, **47** (Suppl.), 164 http: dx.doi. org./10.1016/s1383-5769 (98) 80399-1.

Chishti, M. Z., Tak, H. and Ahanger, A. G. (2000). Epidemiology of Hydatidosis in Human population of Kashmir valley. *Oriental Sci.*, **6** (1): 1-8.

Chishti, M. Z. and Ahmad, F. (1992). A new record of *Pseudoechinostomum indicum* Mehra, 1952 from *Anas platyrhynchos* in Kashmir. *Indian J. Helmintol.*, **44** (2): 101-104.

Chishti, M. Z. and Ahmad, F.1991. (1992a). On a new species of *Echinostoma kashmirensis* sp. n. (Echinostomatidae Poche, 1926) from *Anas platyrhynchos domesticus* in Kashmir. *Indian J. Helminthol.*, **43** (2): 153-155.

*Chishti, M.Z. and Ahmad, F. (2004). Seasonal occurrence of Helminth parasites in *Schizothorax* in Dal Lake, Kashmir. *J. Parasitic Dis.*, **28** (1): 23-28.

*Chishti, M. Z. and Ahmad, F. (2015). Protein profiling of *Haemonchus controtus* found in sheep of Kashmir Valley. *J. Parasitic Diseases*, **39** (4): 639- 644.

Chishti, M. Z., Ahmad, Fayaz and Bakshi, Shafqat (2002). Comparative Fish Parasite Diversity in two fresh water Lakes in Kashmir, in relation to ecological factors in different seasons. *Proc. 10th Intl. Congr. Parasitol.* ICOPA: X; Vancouver (Canada), August 4-9, 2002, pp.379-386.

Chishti, M. Z. and Bakshi, S. (1990). On *Rhabdochona guptai* n. sp. from the intestine of fish *Cyprinus carpio specularis,* fish from Hokersar Wetland in Kashmir. *Indian J. Parasitol.,* **14** (1): 45-49.

Chishti., M. Z. and Bakshi, Shafaqat 1992. New species of the genus *Pomphorhynchus* Monticelli, 1905 from *Cyprinus carpio communis,* a fresh water fish in Kashmir. *In:* Yousuf, A.R., Raina, M.K and Qadri, M.Y. (Eds.), pp. 95-102, *Curr. Trends in Fish and Fishery Biology and Aquatic Ecology* (*Proc. 3rd. Natl Ser. Fish* and their Env. field at P.G Deptt. of Zool. The Univ. of Kashmir, Srinagar, Kashmir, from 19-22. Sept., 1989.

Chishti, M. Z. and Khan, A. R. 1982 (1984). *Mayhewia kavini* sp. nov. (Hymenolepidae: Raillet et Henry 1909) from *Corvus monedula* in Kashmir. *Indian J. Helminthol.,* **34** (2): 107-111.

Chishti, M. Z. and Mir, A.A. 1989 (1990). On a new species of *Echinochasmus* (*Echinochasmus*) Dietz, 1909 (Echinostomatidae, Echinoschaminae Odhunner, 1911) from little grebe *Podiceps ruficollis capensis* in Kashmir, *Indian J. Helminthol.,* **41** (2): 141-145.

Chishti, M. Z., Mir, A. A. and Asmat Rasool 1 986 (1987). *Choanotaenia micracantha* sp.nov. (Dilepoidea: Cestoda) from *Corvus monedula* in Kashmir. *Indian J. Helminthol.,* **38** (2): 107-111.

Chishti, M. Z. and Naqash, A. R. (1982). *Varroa jacobsonii* – a serious menance to *Apis cerana* in Kashmir, *Kashmir Univ. Res. J.,* **2**: 43-45.

Chishti, M.Z. and Peerzada, M.Y., 1998. Host and seasonal occurrence of *Acanthocephala* in fishes of Wular Lake. *Oriental Sci.,* **3** (1): 31-38.

*Chishti, M. Z., Shah, F. A. and Mahmood, H. (2004). Study of humoral immune response to helminth infection in some fishes of Kashmir. *J. Parasitic Dis.,* **27** (2): 94-98.

Chishti, M. Z., Tak, Hidayat-ullah and Ahangar, A. G. (2001). Epidemiology of hydatid disease in Human Population of Kashmir Valley. *Oriental Sci.,* **6** (1): 1-8.

Chishti, M. Z. and. Tanveer, S. (1992). On a new species of trematode genus *Echinoparyphium Dietz* 1909 from domestic fowl in Kashmir. *Indian J. Parasitol.,* **16** (2): 155-157.

Chishti, M. Z. and Tanveer, Syed. (1993). On *Paryphostomum srivastavi* n.sp. from *Gallus gallus domesticus. J. Indian J. Helminthol.,* **45** (42): 152-54.

Chitwood, B. G. (1950). An outline classification of nematoda. pp.15-25, *In:* Chitwood, B.G. and Chitwood, M.B. (Eds.). *Introduction to Nematology.* University of Park Press, Baltimore, Maryland. U.S.A.

Christopher, S. R. (1933). *The Fauna of British India, Diptera,* Vol. IV, *Family Culicidae, Tribe Anophelini.* Tylor and Francis, London, England.

Covell, G. (1927). The distribution of anopheline mosquitoes in India and Ceylon. *Indian Med. Res. Memoirs,* **No.5**: 1-85.

Covell, G. (1931). The distribution of anopheline mosquitoes in India. Malaria Bureau No. 8, *Health Bulletin No.17, Malaria Survey of India, Kasauli, Govt. of India, Central Publication Branch,* Calcutta.

D'Hondt, J.L. (1997). Qu'est-il advanu des anciens-Aschelminthes. *Bulletin de la Societe. Biologique France,* **122:** 355-364.

Dar, A. M., Tanveer, S., Chishti, M.Z. and Shah, F. A. (2008).Prevalence and seasonal occurrence of Gastro-intestinal Helminths in indigenous domestic fowl, ***In***: M.Z. Chishti and Fayaz Ahmad (eds.). pp. 493-496, *Science For Better Tomorrow,* printed at Universal Printers, Khanyar, Srinagar, Kashmir.

*Dar, G. H., Bhagat, R. C. and Khan, M. A. (2002). Biodiversity of Kashmir Himalaya. Valley Books, Hazratbal, Srinagar (Kashmir). 399 pp.

Dar, J. A. and Tanveer, S. (2013). Prevalence of cestode parasites in free range backyard chicke (*Gallus gallus domesticus*) of Kashmir, India. *Agric and Biol. J. North America,* **4** (1): 67-70.

Dar, J. A., Tanveer, S., Dar, S. A. and Kuchai, J. A. (2010). First report of of *Capillaria anatis* Nematoda: Capillaridae) from *Corvus* species of Kashmir, India. *Global Veterinaria,* **10**(4): 467-471.

*Dar, J. A. and Tanveer, S. (2013). Prevalence of cestode parasites in free-range backyard chicken (*Gallus gallus domesticus*) of Kashmir, India. *Agri. and Biol. J. of North America,* **4**(1): 67-70 (Chickens collected from different localities of Kashmir, were found to be highly infected, with prevalence rate of 85. 83 per cent.

Dar, Javid, A., and Tanveer, S. (2013). Two avian cestodes parasitic to *Corvus* species of Kashmir. *J. of Parasitol. and Vector Biol.,* **5** (4): 46-52.

Dar, M. A., Shah, O. J. M., Wani, N. A., Khan, N. A. and Shah, P. (2002). Surgical Management of splenic Hydatidosis. *Surgery Today,* **32**: 224-229.

Dar, R. A., Mir, L. M., Darzi, M.M., Khan, H. M. and Kamil, S.A. (2014). Prevalence of *Cysticercus tenuicollis* infection in slaughtered sheep of Kashmir. *SKAUST J. Res.,* **10** (1): 62-65.

Dar, S. A., Ahmad, F., Mir, M. R., Dar, J. A., Tak, H. and Kuchai, J. A. (2012). Helminth infection in cold water fishes of Suru River, Ladakh, Jammu and Kashmir, India. *Recent Res. in Sci. and Techol.,* **4**(12): 1-4.

Dar, S. A., Ahmad, F., Mir, M. R., Dar, J. A., Kuchai, J.A., Lone, B. A. and Tak, H. (2013). Helminth infection in relation to size and sex of snow trout, *Schizothorax. Intl. J. Recent. Res.,* **4** (1)1-4.

*Dar, S. A., Chishti, M. Z., Kaur, H. and Dar, G. H. (2014). Parasites of some fishes (*Labeo rohita* and *Schizothorax niger* of Jammu and Kashmir in India. *Intl. J.*

Fisheries and Aquaculture, 6(9): 104- 107 (On the prevalence of parasites, *viz. Chilodonella and Argulus*).

Dar, S. A. Rashid, M. and Jan, U. (2015). Monogenean infestation of *Schizothorax* (Heckel, 1838). and its relation with size and sex of the host, with seasonal dynamics of parasitism. *European J. Biotech and Biosci*, **3** (1): 17-20.

* Dar, S. A. and Sheikh, B. (2013). Occurrence of nosemosis (*Nosema aphis*), affecting honey Bee (*Apis mellifera* L) colonies in Kashmir. *Applied Biol. Res.*, **15** (1): 53-56.

Dar, S. A., Tanveer, S., Kuchai, J. A. (2013). Study on cestode parasites of *Corvus* Species of Kashmir, India. *J. World's Poult. Res.*, **3:** 28-34.

Dar, S. H., Allaie, I. M. and Wani. Z. A. (2016). Surgical Management of *Coenurus* Cyst in Palpebral Sub-conjuntiva in sheep. *Intas Polyvet.*, **17** (2): 443-445.

Dar, S., Kaur, H., Chishti, M.Z., Ahmad, F., Tak, Irfan-ur-Rauf Dar, G. H. (2016 a). First record of Protozoa parasites in cyprinid fish, *Schizothorax niger* Heckel 1838 from Dal Lake in Kashmir Himalaya, with study on their pathogenesis. *Microbial Pathogenesis*, **93**: 100-104 (Elsevier Publisher).

Dar, S., Kaur, H., Chishti, M.Z. and Ahmad, F. (2016 b). Protozoan parasites in cold water cyprinid fish, *Schizothorax niger* from Kashmir Himalayas. *Bull. of Pure and Applied Sciences- Zoology*, July- December. 2016, **35** (2); 65-71.

Dar, S., Kaur, H., and Chishti, M. Z. (2017). *Myxobolus chushi* n. sp. (Myxozoa: Myxosporea). *Parasitol. Intl.*, **66** (3): 272-278, http//doi.org.10.1016/ jparint.2017.02.004.

Dar, T., Natarajan, R., Shanumuganthan, J. and Jambulingam, P. (2016).Mosquitoes (Diptera: Culicidae) of Jammu Division of Jammu and Kashmir State, India, with new records. *J. Entomol. and Zool. Studies*, **4** (1): 388-393.

*Dar, Z. A., Tanveer, S., Yattoo, G. N., Sofi, B. A., Wani, S. R., Dar and P. A. (2009). Alkaline phosphatise in serum as a marker of Human Toxocariasis. *Iranian J Blood and Cancer*, 1(2): 65-69.

*Dar, Z. A., Tanveer, S., Yattoo, G. N., Sofi, B. A., Dar, P. A. and Wani, S. A. (2008 a). Presence of anti-Toxocara antibodies in children population of district Anantnag and Pulwama of Kashmir valley. *Indian J. Medical Microbiol.*, **26** (4): 400.

Dar, Z. A., Tanveer, S., Yattoo, G. N., Sofi, B. A., Wani, S. A., Dar, P. A. (2008). Seroprevalence of Toxocariasis in children in Kashmir, J&K State, India. *Iranian J. Parasitol.*, **3**: 45-50.

Dar, Z. A., Tanveer, S., Yattoo, G. N., Sofi, B. A., Wani, S. A. and Dar, P. A. (2013). Haematological response in human toxocariasis patients. *Intl. J. Medicine and Medical Sciences*, **5** (7): 320-323.

Dar, Z. A., Tanveer, S., Yattoo, G. N., Sofi, B. A., Dar, P. A. and Wani, S. A. (2014). Prevalence of *Toxocara canis* in stray dogs of Kashmir valley. *Res. Rev.: J. Zool. Sci.*, **2** (3): 20.

Darzi, M. M., Pandit, B. A., Shahardar, R. A. and Mir, M. S. (2002). Pathology of *Taenia hydatigena* cysticercosis in naturally infected Corriedale lambs. *J. Vet. Parasitol.*, **16**: 173-174.

*Darzi, M. M., Mir, M. S., Kamil, S. A., Nashiruddulaha, N. and Munshi, Z. H. (2007). Pathological changes and local defense reaction occurring in spontaneous hepatic coccidiosis in rabbit (*Oryctolagus cuniculus*). *World Rabbit Science*, **15** (1): 23-28.

Darzi, M. M., Saleem, M., Shahardar, R. A. and Pandit, B. A. (2007). Clinicopathological, histochemical and therapeutic studies on concurrent sarcoptic and notoedric acrariosis in rabbits (*Oryctologus cuniculus*). *Veterinarski Archiv.*, **72** (2): 167-175.

Das, B.C., Chattopadaya, A. and Datta, A. L. (2000). A list of Menoponidae (Phthiraptera) from India. *Recd. Zool. Surv. India*, **98** (3): 111-137.

Das, Dev and Sharma, V.P. (2013). The dominant Mosquito vectors of Human Malaria, **In**: Sylvie Manguin (ed.), Chapter 9. *Anopheles* mosquitoes – a New insight into Malaria vector, In Tech.

Das, S. M. and Sharma, B. D. (1970). A new record of *Ctenocephalides felis orientis* (Jordon, 1925), (Siphonaptera; Pulicidae) from goats in Kashmir. *Kashmir Sci.*, **7**: 75-78.

Das, S. M. and Sharma, B. D. (1970 *a*). A survey of ticks of Kashmir. *Kashmir Sci.*, **7**: 111-118.

Das, S. M. and Sharma, B. D. (1970 *b*). On *Haematopinus eurysternus* Nitzsch (1818) (Siphunculata: Haematopinidae) from Kashmir. *Kashmir Sci.*, **7** (1-2): 79-81.

Das, S. M. and Shaw, A. A. (1968). Studies on Parasitic Protozoan of Domestic animals of Kashmir. I. On the occurrence of *Babesia argentina* Ligenieres (Protozoa, Haemosporida) in imported Australian Jersey Cows. *Kashmir Sci.*, **5** (1-2): 238-292.

Dasan, V., Kaur, Prabhdeep and Murhker, M. V. (2014). Rickettsial Disease outbreak in India: A Review. *Annl. Trop. Med. Public Health.***7** (6): 249-254.

Dash, A. P., Adak, T., Raghvendra, K. and Singh, O. P. (2007). The biology and Control of malaria vectors in India. *Current Science*, **92** (11): 1571-1578.

Datta, M. (1983). A review of the Simuliidae (Dipter) from Oriental Region. *Oriental Insects*, **17**: 215-267.

Datta, M. 1985 (1986). Notes on blackflies (Diptera: Simuliidae) from Kashmir, India. *Bull. Zool. Surv. India*, **7** (2-3): 231-236.

Datta, M. and Chakraborti, M. (1983). On a collection of flowerflies (Diptera: Syrphidae), with new records from Jammu and Kashmir. *Rec. Zool. Surv. India*, **81**: 237-253.

Datta, M.N. (1936). Scientific results of the Yale North India Expedition, Biological Rept. No. 20. Helminth parasites of fishes from North-India., with special reference to Acanthocephalans. *Rec. Indian Mus.*, **38**: 211-229.

Datta, M. N. (1937). Acanthocephala from India. Description of male of *Eosentis rigidus* van Cleave, occurring in the intestine of *Schizothorax zarudnyi* from Seistan *Rec. Ind. Mus.*, **39**: 303-304.

De Leon, Gerado, P., O, Launas-Calvo, Garcia- Prieto, L., Briosio-Aguilar, R. and Aguilar-Aaguilar, R. 2018). Update on the distribution of the co-invasive *Schyzocotyle acheilognathi* (= *Bothriocephalus acheilognathi*), the Asian fish tapeworm, in fresh water fishes of Mexico. *J. of Helminthol.*, **92** (3): 279-290.

De Ley, P. and Blaxter, M. (2004). A new system of Nematoda: combining Morphological characters molecular trees, and translating clades into ranks and taxa. *Nematology Monographs and Prespective*, 2004, 633 - 653.

Desquesnes, M., Holzmuller, P., Lai, DE-Hua, Dargantes, A., Lum, Zhao-Rong and Jittaplapong, S. (2013). *Trypanosoma evansi* and Surra: A Review and Perspective on origin, History, Distribution, Taxonomy, Morphology, and Pathological effects. *BioMed Res. Intl.*, **vol 2013** (2013), Article ID 194176, 22 pages. http// dx. Doi. org./10.11/2013/194176.

Dhanda, V. and Bhat, H.R. (1971). *Haemaphysalis* (*Herpetobia*) *sundrai* Sharif (Ixodoidea: Ixodidae), a tick parasitizing sheep in the Western Himalaya, redescription of female, description of male and ecological observation. *J. Parasitol.*, **57** (3): 646-650.

Dhanda, V. and Kulkarni, S. M. (1969). Immature stages of *Haemaphysalis cornupunctata* Hoogstraal and Varma, 1962 (Acarina: Ixodiadae), with new host and locality records and notes on its ecology. *Oriental Insects.*, **3** (1): 15-21.

Dhanda, V., Kulkarni, S. M. and Kaul, H. N. (1979). Record of *Nosomma monstrosum* (Acarina: Ixodidae) from Jammu and Kashmir. *Acarology Newsletter*, No. 8: 3-4.

Dhanda, V., Kulkarni, S. M. and Pratt, P. (1971). *Dermacentor raskemensis* (Ixodoidea: Ixodidae), redescription and notes on ecology. *J. Parasitol.*, **51** (6): 1324-29.

Dhar, D. N. (1980). Recent studies on lungworm infection of sheep and goats in India. *Indian J. Parasitol.*, **3**: 127-128.

Dhar, D. N. (1982). Recent studies on *Dictycaulus filaria* and other lungworm of sheep and goats in India. *J. Nuclear Agric. and Biol.*, **11**: 1-5.

Dhar, D. N. (1985). Bovine and ovine parasitic bronchitis, with particular reference to the work done in India. *In.* Sen, A. B., Katiyar, J. C. and Guru, P. Y. (Eds.), pp. 53-88, *Prespective in Parasitology.* Print House (India), Lucknow, India.

Dhar, D. N. and Sharma. R. L. 1978. *Dictyocaulus filaria* in lambs. The effect of varying single infection on subsequent production. *Vet. Parasitol.*, **4:** 221-229.

Dhar, D. N. and Sharma, . R. L. (1978*a*). Studies on the comparative susceptibility of sheep and goats to infection with *Dictyocaulus filaria. Ind. J. Anim. Sci.*, **48**: 29-31.

Dhar, D. N. and Sharma, R. L. (1979). A note on the prevalence of lungworm infections in sheep and goats in Tehsil Kargil of district Ladakh (Jammu and Kashmir). *Indian J. Anin. Sci.*, **49**: 585-588.

Dhar, D. N., Sharma, R. L. and Bansal, G. C. (1979). Occurrence of *Mullerius* sp. in sheep in India. *Indian J. Parasitol.*, **3**: 187.

Dhar, D. N., Sharma, R. L. and Bansal, G. C. (1982). Gastro-intestinal nematodes in sheep in Kashmir. *Vet. Parasitol.*, **10**: 91-95.

Dhar, D. N., Sharma, R. L. and Bansal, G. C. (1983). Ascarid infection in sheep in Kashmir Valley. *Indian J. Parasitol.*, **7**(1): 97-99.

Dhar, D. N., Sharma, R. L. and Raina, O. K. (1988). Fasioliasis in animals in Kashmir Valley. *J. Vet. Parasitol.*, **2** (1): 31- 35.

Dhar, D. N., Tewari, H. C. and Rao, Y.V. B.G. (1972). Prevalence of lungworm infections in sheep in Kashmir. *Indian. J. Animal Sci.*, **42**: 112-114.

*Dhar, R. L. (1973). *Studies on parasites of fishes of Jammu and Kashmir.* Ph.D. Thesis, P.G. Deptt. of Zool., Univ. Kmr., Srg., Kmr. (J&K).

Dhar. R. L. 1975 (1978). *Astiotrema fotedari* n. sp. (Trematoda: Plagiorchidae Luhe, 1901) from the intestine of *Laleo dero* (Gemilton). *J. Sci. Univ. Kashmir*, **3** (12): 99-104.

Dhar, R. L. and Ahmad, F. (1992). A new record of *Echinoparyphium clerci* Skrybin from *Anas platyrhynchos* in Kashmir. *Indian J.Helm.*, **44** (2): 93-96.

Dhar, R. L. and Ahmad, F. (1992a). First record of *Cotylurus cornutus* Rudolphi, 1887 Szidal, 1928, from *Anas platyrhynchos domesticus* in Kashmir. *Indian J. Helminthol.*, **44** (2): 105-108.

Dhar, R. L. and Ahmad, F. 1991(1992). *Pseudoechinostomum srinagarensis* sp. nov. (Trematoda: Echinostomatidae Looss, 1902) from a domestic duck *Anas platyrhynchos domesticus* from Kashmir. *Indian J. Helmintol.*, **43** (2): 147-49.

Dhar, R. L. and Anwari, M. D. (1982). On new host record of *Rhaldochona* Raillict, 1916. (Rhaldochonidae Skrjabin, 1946, Nematode). *Kashmir Univ. Res. J.*, **2**: 39-40.

Dhar, R. L. and Fotedar, D. N. (1979). On *Procammalanus*(*Monospiculus*) *kashmirensis* sp. nov from fresh water fish, *Wllago attu* from Jammu, India. *Indian J. Helminthol.*, **40** (2): 128-134.

Dhar, R.L. and Fotedar, D. N. 1979 (1980). Some protocephalid cestodes in fresh water fishes of Jammu and Kashmir. *Indian J. Helminthol.*, **31** (2): 111-127.

Dhar, R. L. and Kharoo, V. K. 1985. A new species of Acanthocephala, *Neoechinorhynchus glyptosternumi* n.sp. from Intestine of Kashmir Fish, *Glyptosternum* sp. *Indian. J. Helminthol.*, **36** (1): 36-39.

Dhar, R. L. and Kharoo, V. K. (1985 *a*). On a new trematode *Allocreadium fotedari* sp. nov. of the family Allocreadiidae Stossich, 1903, from the intestine of *Schizothorax niger*, a fresh water fish of Kashmir, India. *Indian J. Helminthol.*, **36** (1): 32-35.

Dhar, R.L. and Kharoo, V. K. 1986 (1987). Studies on trematode parasites of fishes, genus *Clinostomum* Keidy 1856 from fresh water fishes of Kashmir.*Indian. J. Helminthol*, **38** (1): 74-78.

Dhar, R. L. and Kharoo, V. K. (1991). Trematodes of aquatic and migratory birds in Kashmir. *Trudy Gel mintologicheskoi Laboratorii.* No.**38,** 28-32.

Dhar, R. L. and Majidah (1983). *Gangesia* (*Gangesia*) *fotedari* sp. nov. (Protocephalidae; Cestoda) from fresh water fish in Wular Lake, Kashmir. *Indian. J. Helminthol.,* **35** (1): 73-77.

Dhar, R. L. and Majidah Rasheed (1987). Fish Parasitization of Helminths in Wular Lake, Kashmir. India *Indian J. Helminthol.,* **34** (2): 143-152.

Dhar, R. L. and Peerzada, M. Y. (1992). Seasonal variation in the occurrence and maturation of *Adenoscolex oreini* Fotedar, 1958 (Cestoda: Caryophyllaeidae) infecting some cyprinid fishes of Wular Lake (Kashmir), *In*: Yousuf *et al.* (eds.), pp. 103-107. *Current trends in Fish and Fishery Biology and Aquatic Ecology* (*Proc. 3rd. Natl. Semr.* on *Fish and their Env.*), held at Srinagar, Kmr. on 19-22 September, 1989, at P. G. Deptt. of Zool., Univ. of Kmr., Sgr., Kmr. (J&K).

Dhar, R.L. and Peerzada, M.Y. 1993(1994). Seasonal variations in helminth population of common snow trout, *Schizothorax niger* Heckel of Wular Lake, Kashmir. *Indian. J. Helminthol.,* **45** (41/2): 86-91.

Dhar, R.L. and Raina, M. K. (1983). On a new cestode, *Protocephalus kashmirensis* sp. nov. from a fresh water fish, *Nemacheilus kashmirensis* (*Noemacheilus kashmirensis*) (Hora) from Kashmir, India. *Indian J. Helminthol.,* **35** (1): 30-34.

Dhar, S. N. and Kaul, B. L. (1988). Hydatid disease of Lung. *Sem Resir. Med.,* **9**: 409-413.

Dharmadhikari, P. R. and Ramaseshia, G. (1970). Recent records of aphidiids (Hym: Aphididae) in India. *Tec. Bull. Commonw. Inst. Biol. Control,* **13**: (83-89.

*Dharmadhikari, P.R., Ramasehiah, G. and Achan, P. D. (1985).Survey of *Lymentaria obfuscata* and its natural enemies in India. *Entomophaga,* **30** (4): 4-3.

Digra, S. K., Saini, G. S., Singh, V., Sharma, S.D. and Kaul, R. (2010). Scrub Typhus in Children: Jammu Experience. *JK Science,* **12** (2): April- June, 95-97.

Diller, E. (1984).Orientalische Taxa der Gattuug. *Promethes* Forester, (1989) (Hymenoptera, Chneumonidae, Diplzontinae). *Entomofauna,* **5** (6): 69-80.

Durden, L. A. and Musser, G. G. (1994). The sucking lice (Insecta: Anoplura) of the world: a taxonomic checklist with records of mammalian hosts and Geographical distribution. *Bull. American Mus. Nat. Hist.* **No. 218**, pp. 90.

Dutta, R. N., Mukerjee, D.G. and Bhattacharjee, S.N. (1963). Tick-borne relapsing fever. *Armed Forces Med. J. India.,* **19**: 125-33.

*Dutta, S., Godara, R., Katoch, R., Yadav, A., Katoch, M. and Singh, N. K. (2017). Detection of amitraz and malathion resistance in field populations of *Rhicephalus* (*Boophilus*) *microplus* (Acari: Ixodidae) in Jammu region of India. *Experimental and Applied Acarology,* **71** (3): 291-301.

Dwivedi, S. R., Sahu, H., Pattanayak, S and Roy, R. G. (1979). Prespective treatment of *P.vivax* malaria in Uttar Pradesh, Haryana, Himachal Pradesh and Jammu and Kashmir. *Indian J. Medical Res.,* **70**: 54-56.

Eastop, V. F. and Hille Ris Lambers (1976). *Survey of World Aphids*. Dr. W. Junk. b. v. Publishers. The Hague, Netherland.

Edwards, F. W. (1927). Two new *Simulium* from Kashmir. *Bull. Ent. Res.*, **18**: 169-170.

*Euzeby, J. (1988). *Comparartive Medical Zoology*, Vol. 3 Apicomplexa; Vol 2. Haemosporidiosis. Published by Fondation Marcel, Merieux, Lyon, France.

Evans, G. O. (1992). *Principles of Acarology*. CAB International, Willingford, Oxon, U.K.

Evenhuis, N. L. and Greathead, D.J. (1999). *World Catalaogue of Bee flies (Diptera: Bombyliidae)*. Backhuys Publishers, Leiden, xiviii, 756 pp.

Farhana, A., Bali, N., Kaul, F., Farooq, R., Inam-ul Haq and Shah, Parvaiz (2016). Serological evidence of Scrub Typhus among cases of PUO in Kashmir Valley-A Hospital based study. *J. Clinical and Diagnostic Res.*, **10** (5): DC24-DC26, doi10.7860/JCDR/2016/18471.7825.

*Farooq, T., Khan, I., Dar. S. A. and Yousuf, A. R. (2016). Endoparasites of some economically important food fishes of River Jehlum, Kashmir (India). *J. Parasitic Diseases*, **40** (3): 860-871 (On helminth parasites of *Schizothorax*, including *Adenoscolex kashmirensis, Bothriocephalus achliognathi, Echinorhynchus* sp. and *Pomphorhynchus kashmirensis*).

Farooqi, S. I. and Subba Roa, B.R. (1986). Family: Pteromalidae. In: Subba Rao. B.R. and Hayat, M. (eds), pp.279-306. The Chalcidoidea (Insecta: Hymenoptera) of India and adjacent countries. *Oriental Insects*, 20: 1-430.

*Fatima, Maqbool (2017). *Helminth Parasites of Bovines and Bubalines of Kashmir*. Published by Dilpreet Publishiung House.

*Fatima, M., Chishti, M.Z., Ahmed, F. and Lone, B. A. (2012). Epidemiological study of Fasciolosis in cattle in Kashmir. *Adv. In Biol. Res.*, **6** (3): 106-109.

*Fatima, M., Ahmed, F. and Chishti, M.Z. (2008). First Report of *Fasciola gigantica* Cobbold, 1855 in ruminants. *Proc. Of 2nd J&K Science Cong.*, pp. 446-450.

*Fatima, M., Ahmed, F., Chishti, M. Z. and Lone, B. A. (2012). A survey on the Bovine Amphistomiasis in Kashmir Valley. *Intl. J. Recent Sci. Res.*, **3** (1): 50-52.

* Fatima, M., Ahmed, F., Sofi, T. A., Chishti, M. Z. (2015). The microscopic and associated histological abnormalities in liver and bile duct of buffalo (*Bubalus bubalis*), naturally infected with amphistomes, slaughteres in different areas of Kashmir valley. *Agri. Adv.*, **4**: 108-114.

* Fatima, M., Ahmed, F., Chishti, M. Z. and Sofi, Tanveer, A. (2019). Gastro-intestinal helminth parasites of cattle and buffalo in Kashmir Valley, India. *Intl. J. Biol. and Medical Res.*, **10** (1): 6579-6586. (Authors have recorded 18 helminth species, with nematode and trematode as dominant, followed by cestodes. The overall prevalence has been 53.96 per cent and 49. 64 per cent in c attle and buffalo respectively. Besides, data on monthly/seasonal prevalence of parasites, has been presented).

Faust, E. C. ('1927). Studies on Asiatic Holostomes class Tre matoda. *Rec. Indian Mus.,* 215-218.

Faust, E. C., Russell, P. F. and Jung, R. C. (1974). *Craig and Fausts Clinical Parasitology* (8th Edition). Lea and Febiger Philadelphia, U.S.A., 890 pp + 340 illust, 8 col. Plts.

Fazili, A., Wani, N. A., Tahir, M. S., Khan, Saleem and Mir, A. R. (2002). Hydatidosis – rare presentation. *JK Practioner,* **9**(4): 252-253.

Ferre-Suay, Selfa, J. and Pujade-Villar, J. (2013). Charipinae fauna (Hymenoptera: Figitidae) from Asia, with descriptiona of 11 new species. *Zoological Studies,* **52**: 41.

Ferris, G. F. (1951). The sucking- lice. *Mem. Pacif. Coast. Ent. Soc.,* **1**: 1-320.

Ferris, H. (2020). Nemaplex website on soil and plant nematodes. *nemaplex.ucdavis.edu.*

Fomda, B.A., Khan, A., Thokar, M.A., Malik, A., Fazili, A., Dar, R.A., Sharma, M. and Malla, N. (2015). Sero- epidemiological survey of human cystic Echinococcosis in Kashmir, North India. *PLoSONE* **10**(4).eo124813. https//doi.org/10.1371/journal.pone.0124813.

Fomda, B. A., Sofi, B.A., Thokar, M. A. and Kakru, D. K. (2002). Human hydatid disease in Kashmir North India – A hospital based study. *J. Parasitic Diseases,* **26** (1): 34-37.

Fomda, B. A., Ahmad, Z., Khan, M. N., Tanveer, S. and Wani, S. A. (2007). Ocular toxocariasis in child: A case Report from Kashmir, North India. *Indian J. Medical Microbiology,* **25**: 411-412. dx.doi.org/10.4103/022-0857.37352.

Fotedar, D. N. (1958). On a new caryophyllaeid cestode, *Adenoscolex oreini* gen. et sp. nov. from fresh water fish in Kashmir and note on the related genera. *J. Helm.,* **32**: 1-16.

Fotedar, D. N. (1959). On a new species of the genus *Ganeo* Klein and some notes on the genus *J. Helminthol.,* **33** (2/3): 151-160.

Fotedar, D. N. (1960). On a new species of *Oxysomatium* Railliet and Henry, 1911 and some notes on the genera *Oxysomatium* and *Aplectana., J. Helminthol.,* **34**: 141-150.

Fotedar, D. N. (1964). A new species of the nematode genus, *Pseudonymus* Diesing 1857 from an aquatic beetle in Kashmir, *Kashmir Sci.,* **1** (1-2): 73-75.

Fotedar, D. N. (1964*a*). On a new species of the trematode genus *Typhlocoelum* Stossich, 1902 from mallard duck in Kashmir and a review of previous work on the genus. *Kashmir Sci.,* **1**(1 and 2): 44-53.

Fotedar, D. N. (1965). On the larval stages of a Brachylaemid trematode in the land snail, *Buliminus dextrosinister,* in Kashmir. *Kashmir sci.,* **2**: 101-113.

Fotedar, D. N. (1965 *a*). *Rhabdias bufonis* (Schrank, 1788) Stiles and Hassall, 1905 from the lungs of *Bufo viridis* in Kashmir and notes on some amphibian species of the genus. *Kashmir Sci.,* **2**(1-2): 20-26.

Fotedar, D. N. (1966). New Nematotaeniid Cestode from *Bufo viridis* in Kashmir. *Kashmir Sci.*, **3** (1-2): 77-82.

Fotedar, D. N. (1967). On the morphology of *Prostocus himalayai* Pande, 1937 and synonymy of two species of the genus. *Kashmir Sci.*, **4** (1-2): 227-234.

Fotedar, D.N. (1968). New species of *Neoechinorhynchus* Hamann, 1892 from *Oreinus sinuatus*, fresh water fish in Kashmir. *Kashmir Sci.*, **5** (1-2): 147-152.

Fotedar, D. N. (1968*a*). On two oxyurid nematodes of genus *Thelandros* Wedl, 1962 from some lizards of Jammu and Kashmir and notes on the genera, *Thelandros* and *Parapharyngodon* Chatterji, 1933. *Kashmir Sci.*, **5** (1-2): 208-221.

Fotedar, D. N. (1969). Note on some Georgoderid trematodes of fishes and amphibians in Kashmir and other parts of India, and notes on the classification of the family Gorgoderidae Loss, 1901. *Kashmir Sci.*, **6**(1-2): 89-100.

Fotedar, D. N. (1970).On a new trematode, *Pleurorchis nensie* gen et sp. Nov. from *Bufo viridis* in Kashmir. Proc. 57th Sess. Indian Sci. Congr., Part-III, pp454-455 (Abt.).

Fotedar, D. N. (1971). A review of the genera in Pleurogenetinae Looss, 1899 (Trematoda: Lecithodriidae). *Kashmir Sci.*, **8**(1-2): 127-142.

Fotedar, D. N. (1980). *Oswaldcruzia kashmirensis* n. sp. (Nematode: Trichostrongylidae) from *Bufo viridis* in Kashmir. *Indian J. Zool.*, **8**(2): 27-31.

Fotedar, D. N. and Bambroo, N. (1965). On a new Trichostrongylid Nematode from sheep in Kashmir. *Kashmir Sci.*, **2**(1-2): 136-141.

Fotedar, D. N. and Bambroo, N. 1967 (1968). On a new species of the nematode genus *Chabertia* Railliet and Henry: 1909 from sheep in Kashmir. *Indian J. Helminthol;* **19**(2): 192-196.

Fotedar, D. N. and Bambroo, N. 1975 (1978). On a new species of cestode genus *Thysaniezia* Skrjabin, 1926 from local sheep in Kashmir. *J. Sci. Univ. Kmr.*, **3**(1-2): 205-108).

*Fotedar, D. N. and Bhagat, R. C. (1985). *Jammu, Kashmir and Ladakh (Zoological Studies): An Annotated Bibliography (1848 – 1980)*. International Books and Periodical Supply Service, New Delhi (India), 133 pp.

Fotedar, D. N. and Chishti, M. Z. (1973). On a new species of *Anomotaenia kashmirensis* (Choanotaenidae Mathevossian, 1953) from *Sturnus vulgaris* in Kashmir. *J. Sci. Univ. Kmr.*, **1**(1-2): 48-50.

Fotedar, D. N. and Chishti, M. Z. (1974). On *Toxocara mystax* (Zeder, 1800) Stiles, 1907 from *Felis domestica* in Kashmir. *J. Sci. Univ. Kashmir.* **2**(1-2): 16-18.

Fotedar, D. N. and Chishti, M. Z. (1974 *a*). Redescription of *Choanotaenia orioli* Joyeux et Baer, 1955 and *C. infundibutum* (Bloch, 1779), with a note on the synonymy of *C. dutti* Mukherjii. 1964, *J. Sci. Univ. Kmr.*, **2** (1-2): 73-78.

Fotedar, D. N. and Chishti, M. Z. (1976). *Anomotaenia acrocephati* n. sp. and first record of *A. galbulae* (Gmelin, 1790) Fuhr., 1932 from some birds of Kashmir. *Rivista di Parasitologica*, **37** (2): 247-252.

Fotedar, D. N. and Chishti, M. Z. (1980). *Psendoschistotaenia indica*, new genus, new species (Amabilcidae, Cestoda) from *Podiceps ruficollis capensis* in Kashmir, India. *Riv. Parassitol.*, **41** (1): 39-44.

Fotedar, D. N. and Chishti, M. Z. (1980 *a*). *Streptocara indica* n. sp. (Acuarioidae Sobolev, 1949: Nematoda) from *Gallus domestica* in Kashmir, India. *Indian J. Helminthol.*, **32** (1): 4-7.

Fotedar, D. N., Chisthti, M.Z. and Khan, A. R. (1970). A new record of *Prosthogonimus (Macrogenotrema) cuneatus* (Rudolphi, 1809) from cloaca of *Passer domesticus* from Kashmir. *Kashmir Sci.*, **7** (1-2): 92-94.

Fotedar, D. N. and Dhar R. L. (1970). On the morphology of *Filochona kashmirensis*, (Thapar, 1950) from fresh water fishes of Kashmir. *Kashmir Sci.*, **7** (1-2): 107-110.

Fotedar, D. N. and Dhar R. L. 1973, (1975). On a species of nematode Genus *Spinitectus* Fourment from *Oreinus richardsonii*, a fresh water fish of Jammu, India. *J. Sci. Kash. Univ.*, **1**: 63-66.

Fotedar, D.N. and Dhar, R. L. 1973 (1975*a*). On the first record of a later juvenile form of an Acanthoccphalan Genus *Polymorphus* Luhe, *1911* from Rectum *of Rana cyanophlyctis* in Kashmir. *J.Sci. Univ. Kashmir.*, **1** (1-2): 55-56.

Fotedar, D. N. and Dhar, R. L. (1977). On five new species of genus *Pomphorhynchus* Monticelli 1905 from fresh water fishes of Jammu and Kashmir, India. *Indian J. Zool.*, **5** (1): 15-24.

*Fotedar, D. N. and Dhar, R. L. (1977). On the infection of *Rhabdochona* Raillet, 1916 (Nematoda) in fishes of J&K, **In**: pp. 34-35, *Proc. All India Symp. of Helminthol., University of Kashmir, Srinagar, Kashmir*, 8-11 August.

Fotedar, D.N., Duda, P. L. and Raina, M. K. (1970). a new species of the genus *Pomphorhynchus* Monticelli., 1905 from *Bufo viridis* in Kashmir. *Indian J. Helminthol.*, **22** (1): 41-45.

Fotedar, D. N. and Handoo, Z. A. (1974). Two new species of *Helicotylenchus* Steiner (Hopololaiminae Nematoda) from Kashmir. *J.Sci. Univ. Kashmir*, 2(1-2): 55-62.

Fotedar, D. N. and Handoo, Z. A. (1977). *Aerotylenchus safroni* n. gen., n. sp. (Nematoda: Tylenchida) from Kashmir. *Indian J. Nematol.*, **7** (2): 145-147.

Fotedar, D. N. and Handoo, Z. 1975 (1978). A revised scheme of classification to order Tylenchida Thorne, 1949. (Nematoda) *J. Sci. Univ. Kashmir*, **3** (1-2): 55-82.

Fotedar, D. N. and Kaul, V. (1982). Plant-parasitic nematodes associated with various crops in Kashmir. *Kashmir Univ. Res. J.*, **2**: 26-31.

Fotedar, D. N. and Kaul, V. 1983 (1984). *Tylenchus andrassyi* sp. nov. (Nematoda: Tylenchinae) from Kashmir. *J. Helminthol.* **35** (1): 43-46.

Fotedar, D. N. and Kaul, V. (1985). On some species of *Helicotylenchus* Steiner, 1945. (Hoplomaimidae: Nematoda) as common plant parasitic nematodes in Kashmir, India. *Indian J. Nematol.*, **15** (1): 9-13.

Fotedar, D. N. and Kaw, Lalita (1965). New Diplotriaenid nematode from the body cavity of *Acridotheres tristis* (Linnaeus). *Kashmir Sci.*, **2**: 125.

Fotedar, D. N. and Khan, A. R. (1974). *Paralecithodendrium indicum* n.sp. from the intestine of *Pipistrellus mimus*, common Bat in Kashmir. *J. Sci. Univ. Kashmir,* **2** (1-2): 63-65.

Fotedar, D. N. and Khan, A. R. (1974*a*). *Plagiorchis* (*Plagiorchis*) *pipistrellicola* n.sp. from Kashmir Bat, *Pipistrellus mimus. J. Sci. Univ. Kashmir.*, 2 (1-2): 66-69.

Fotedar, D. N. and Khateeb, N.G. 1986 (1987). Occurrence and seasonal variations of Helminth parasites of domestic fowl in Kashmir. *Indian. J. Helminthol.* **38** (1): 49-54.

Fotedar, D. N. and Mahajan, R. (1971). On a new species of genus *Tylenchorhynchus* Cobb. 1913 (Nematoda: Tylenchorhynchidae) from Kashmir, India. *Kashmir Sci.,* **8** (1): 120-122.

Fotedar, D. N. and Mahajan, R. (1972). A new species of the genus *Basiria* Siddiqi, 1954 from around the roots of *Brassica oleracea* in Kashmir *Proc. Indian Sci. Congr.,* Part III, p.430. (Abt.)

Fotedar, D. N. and Mahajan, R. (1972 a). Two new nematode species (Nothotylenchidae) from Kashmir. *Indian J.nematod.,* 2 (2): 169-172.

Fotedar, D. N. and Mahajan, R. (1973). On two species of the genus *Basiria* Siddiqi, 1959 (Nematoda: Psilenchinae) from Kashmir, India. *Indian J. Zool.,* **1** (1): 49-52.

Fotedar, D. N. and Mahajan, R. (1974). Four new species of genus *Helicotylenchus* Steiner, 1945 (Nematoda: Hoplolaiminae) from India. *Rivista di Parasitologica,* **35** (2): 119-124.

Fotedar, D.N., Mahajan, R., Dhar, R. L. and Chishti, M. Z. (1970). New variety of *Raillietina* from common Blue- Rock Pigeon in Kashmir. *Kashmir Sci.,* **7**: (1-2): 103-106.

Fotedar, D. N. and Parveen, Q. N. (1987). On two species of trematode genus *Diplozoon* Nordmann, 1832 from fishes of Kashmir and notes on some species of the genus. *Indian J. Helminthol.,* **39** (1): 56-76.

Fotedar, D. N. and Raina, M. K. (1964). On a new species of the trematode genus *Posthodiplostomum* Diabois, 1936 from *Milvus migrans lineatus* (Gray), common kite in Kashmir. *Kashmir Sci.,* **1** (1 and 2): 64-69.

Fotedar, D. N. and Raina, M. K. (1965). On a new species of the trematode Genus *Lyperosomum* Looss, 1899 from *Corvus monedula soemmeringii* common Jackdew in Kashmir. *Indian. J. Helminthol.,* 17 (1): 54-62.

Fotedar, D. N., Raina, M. K., Mahajan, R. and Dhar. R. L. (1972). Redescription of *Contracaecum milvi* Yamaguti; 1940 from *Milvus, migrans lineatus* from Kashmir. *Indian J. Helminthol.,* **24** (1-2): 1-4.

Fotedar, D. N., Raina, M. K. and Tikoo, Ratni (1966). On a new species of the nematode, Genus *Kalicephalus* Moliu, 1861, from a snake in Kashmir. *Kashmir Sci.,* **3** (1-2): 92-100.

Fotedar, D. N. and Wali, N. 1982 (1984). On the development of juvenile stages of some bursate nematode parasites of sheep and pathogenecity of *Oesophagostomum*. *Indian J. Helminthol.*, **34**(2): 123-135.

Francesconi, F. and Omar, L. (2012). Myiasis. *Clinical Microbiology Review*, **25** (1): 179- 105 ?.

Gaffar, S.A. and Punjabi (1990). Occurrence of apricot chalcidoid, *Eurytoma samsonovi* (Vasiljev) on apricots in Ladakh, Jammu and Kashmir. *Bull. Entomol. Soc. of India*, **31**(2): 231-232.

Gaffigan, T., Wilkerson, R.C., Pecor, J. E., Stoffen, A. and Anderson, T. (2013). *Systematic Catalogue of Culicidae*. Walter Reed Biosystematic Unit, U.S.A. website; http: / /www.mosquito- catalog.org.

Ganai, A., Parveen, S., Kaur, D., Katoch, R., Yadav, A., Godara, R. and Ahmad, I. (2015). Incidence of *Buxtonella sulcata* in bovines in R.S.Pora, Jammu. *J. Parasit. Dis.*, **39**(3): 446-447.

*Ganai, N. A., Khan, A., A., Mir, M. S. and Bhat, G. A. (2004). Effect of Levamisol on endoparasitic infestation of fowl in Kashmir Valley. *The Blue Cross Book*, **22**: 24-25.

Ganie, N. A., Sahaf, K. A., Munshi, N. A., Zargar, M.A. and Baba, Z.A. (2008): Incidence of Protozoan disease (Pebrine) of Silkworm, *Bombyx mori* L. in Kashmir. *SKAUST J. Res.*, **10**: *134-138.*

Ganai, T. A. S., Sheikh, F. D., Pandit, B.A., Misra, S.S. and Borkataki, Sanku (2008.). Prevalence of endoparasitic infection in Chiru (Tibetan antelope) of Ladakh. *J. Vet. Parasitol.*, **22** (1): 83.

Gantait, V. V. and Venkataraman, K. (2013). *Catalogue of Arthropod Parasitic Nematodes of India* Occ. Paper No. 345, pp.1-83, Published by Director, Zoological Survey of India, Calcutta.

Geraert, E. (2008). *The Tylenchida of the World: Identification of the family Tylenchidae* Academic Press, 540 pp.

Gauld, I. D., Wahl, D. B. and Broad, G. R. (2002). The Supra-generic group of the Pimplinae (Hymenoptera: Ichneumonidae) – a cladistic revaluation and evolutionary biological study. *Zoological J. of Linneaus Society*, **136**: 421-481.

*Ghosh, A. and Chakrabarti, S. (2013). *Checklist of Indian Trematode parasites from Amphibian Hosts. Occ. Pap. No. 346*: 1-36, Published by the Director, Zoological Surv. of India, Calcutta. (Also, includes trematode parasites from earstwhile Jammu and Kashmir State).

*Gibson, D. J., Joines, A. and Bray, R. A. (2002). Keys to the Trematodes Vol. 1-3, CABI Publishings, Oxfordshire, OX108DE. UK.

Gill, C. A. (1920). Notes regarding Malaria in Kashmir. *J. Medical Res.*, **7** (3): 610-617.

Global Invasion Species Database (2018). Species Profile, *Plasmodium relictum* Downloaded http/ /: www.lucngisd.org/gisd/species php?sc=32 on 17.04.2018.

*Godara, R., Parveen, S., Katoch *et al.* (2014). Acaricidal activity of extract of *Artemisia absinthium* against *Rhipicephalus sanguines* of dogs. *Parasitol. Res.*, **113:** 747-754.

*Godara, R., Parveen, S., Katoch, R., Yadav, A., Katoch, M., Khajuria, J. K., Kaur, D., Ganai, A., Verma, P. K., Khajuria, V. and Singh, N. K. (2015). Acaricidal activity of ethanolic extract of *Artemisia absinthium* against *Hyalomma anatolicum* ticks. *Experimental and Applied Acarology*, **65** (1): 141-148.

*Godara, R., Katoch, R., Yadav, A., Khajuria, J. K. and Borkataki, S. (2011). Coenurosis in small ruminants: an overview. *Vet. Practitioner*, **12** (1): 102-105.

Godara, R., Katoch, R., Yadav, A., Azmi, S. and Shafiqur- Rahman (2013). Incidence and histopathology of *Linguatula serrata* in goats of Jammu. *Veterinary Practioner*, **14** (2): 292-293.

Godara, R., Katoch, R., Yadav, A. and Borah, M. K. (2014 a). Dicrocoeliosis in goats in Jammu, India. *J. Parasitic Dis.*, **38** (2): 201-204.

*Godara, R., Katoch, R. and Yadav, A. (2014). Hydatidosis in goats in Jammu, India. *J. Parasitic Dis.*, **38** (1): 73-76.

*Godara, R., Katoch, R. Yadav, A. and Rastogi, A. (2014). Epidemiology of paramphistomosis in sheep and goats in Jammu, India. *J. Parasitic Dis.*, **38**: 423-428.

Godara, R., Katoch, R., and Yadav, A. (2014 b). Hydatidosis in goats in Jammu, India. *J. Parasitic Dis.*, **38** (1): 73-76.

Godara, R., Katoch, R. and Yadav, A. (2018). Prevalence of Ixodid ticks in cattle of Jammu. *Vet. Practitioner*, **19** (1): 27-28.

Gokhale, M. D. and Mourya, D. T. (2003). Studies on molecular aspects of transovarial transmission of Dengue in *Aedes aegypti* mosquitoes. *Dengue Annual Report-2002-2003.* National Institute of Virology, Pune, pp.51.

Gordh, G. and Headrick, D. H. (2001). *A Dictionary of Entomolgy.* CABI Publishing, CAB International, Wallingford, Oxon, U. K., 1032 pp.

Goswami, P., Mir, M. S., Qureshi, S. and Borkataki, S. (2015). Concomitant Demodicosis and *Microsporum canis* infection in dog. *Indian J. of Canine Practice*, **7** (2): 191-192.

*Gul, N., Hazari, Y., Fazali, K. M., Tak, H. and Sofi, T. A. (2019). Immunoprofiling of *Fasciola gigantica* infected adult bovines: 41K Da protein as potential immunodiagnostic target. *Intl. J. Biol. Med. Res.*, **10** (2): 6745-6750 (Study made for detection of *F. gigatica* soluble antigens readily diagnosed by *F. gigantica* infected bovine sera using western blot analysis).

Gul, Nazima, Tak, H. (2016). Abattoir survey of abamossal nematodes of sheep in Srinagar district (J&K). *J. Parasitic Dis.*, **40**: 377- 380.

Gul, N., Tak, H., Fazili, K. M., Abdullah, I. and Sofi, T. A. (2016). Prevalence of *Fasciola* infection in slaughtered animals in Kashmir. *Intl, J. Vet. Sci. and Animal Husbandary*, **1** (2): 30-36.

Gul, N. and Tak, H. (2016). Prevalence of *Trichuris* spp. in small ruminants, slaughtered in Srinagar District (J&K). *J. Parasitic Dis.*, **40:** 741-744.

Gupta, A. (2013) Revision of the Indian *Microplitis* Foerster (Hymenoptera: Braconidae : Microgastrinae) Description of new species. *Zootaxa,* **3620** (3): 429-452.

Gupta, A. (2016). *Checklist of Indian Pteromalidae (Hymenoptera- Chalcidoidea- Indian Fauna of Ptermalidae.* National Bureau of Agricultural Insect Resources, Banglore. http://wwww.nbair.res.in/Ankita.php.

Gupta, A. and Poorni, J. (2008). New distribution and host records of Chalcidoidea (Insecta : Hymenoptera) various parts of India. *Checklist,* **4** (4): 412.

Gupta, A. and Poorni, J. (2009). Taxonomic studies on a collection of Chalcidoidea (Hymenoptera) from India, with new distributional records. *J. Threatened Taxa,* **1** (5): 300-304.

Gupta, K., Gupta, N., Anjum, A. and Gupta, D. K. (2014). Taxometry and Ecology of a new species of *Diplozoon* Nordman 1832 from the gills of *Schizothorax richardsonii,* inhabitant of Poonch River, Jammu and Kashmir State, India. *The Bioscan,* **9** (3): 917-923.

Gupta, Marinal (2017). Cutaneous leishmaniasis of eyelid: An uncommon presentation of a common entity. *Indian J. Paediatric Dermatol.,* **18** (2): 116-118.

Gupta, N. K. and Uma (1980). *Quinisulcinus punici* (Tylenchorhynchidae: Tylenchida), a new nematode species from India. *Proc. Indian Acad. Sci.* (Animal Sci.), **89** (5): 415-418.

Gupta, N. K. and Uma (1981). Descriptions of two new species of genus *Tylenchorhynchus* Cobb, 1913, family Tylenchorhynchidae (Eliava, 1964). Golden, 1971 from India. *Helminthologica,* **18** (1): 53-59.

Gupta, R., Shah, Imtiaz and Gupta, S. L. (2002). Echinococceal Disease of Kidney, presenting as renal filling defect. *JK Practioner,* **9** (4): 254-55.

Gupta, R.K., Kour, R. and Gani, M. (2019).Behaviour of Solitary endoparasitoids *Glyptapanteles agamemnonis* and *Meteorus pulchricornis* towards virus infected and parasitized *Spilarctia obliqua* larvae and interaction there of. *J. Entomol. and Zool. Studies,* **7** (2) 1006-1014.

Gupta, R. K. and Narendran, T. C. (2007) Field parasitism and biological characteristics of potent larval endoparasites of *Spilosoma obliqua* Walker in Jammu and Kashmir. *J. Biological Control,* **21** (1): 43-52.

Gupta, S. (1978). Discovery of the genus *Ecthrus* in India. Hymenoptera: Ichneumonidae. *Oriental Insects,* **12:** 211-214.

Gupta, S. (1988). New Distributional Records for Ichneumonidae (Hymenoptera: Ichneumonidae) of the Indo- Austrialian Area. *Oriental Insects,* **22:** 301-359.

Gupta, S. C., Sehgal, H. and Jyoti, M.K. (2002). A new fish parasite, *Ergasilus putitori,* sp. nov. (Crustacea, Copepods, Ergasilidae)from the gills of *Tor putitora* (Hamilton) from River Tawi, J&K State, **In:** V.K.Gupta, Verma, A.K. and

Sharma, J.P (Eds.), pp. 131-140, *Perspective in Animal Ecology and Reproduction* Vol.1 Daya Publishing House, New Delhi.

Gupta, S.P. (1967). Helminthic-fauna of Kashmir. *Kashmir Sci.*, **4** (1-2): 56-61.

*Gupta, S. K. and Singh, A. (2013). Seasonal prevalence of gastro-intestinal helminths in sheep and goats in middle-climatic zone of Jammu Province. *J. Parasit. Dis.*, **37**: 21-25.

*Gupta, S.K., Bhardwaj, R. K., Singh, R., Raina, A. K. and Kumar, A. (2002). Management of Canine babesiosis: A case Report. *J. Vet. Med.*, **22** (2): 124.

Gupta, S. K., Yadav, A., Raina, A.K. and Singh, R. (2004). Theilerosis in seven-day old calf – a case Report. *Indian J. Vet. Med.*, **24** (1): 55.

Gupta, Santosh (1978). Discovery of genus *Echthrus* in India (Hymenoptera: Ichneumonidae). *Oriental Inst.*, **12** (2): 211-214.

Gupta, V. K. (1967). Indian species of *Itoplectis* Forester (Hymenoptera: Ichneumonidae). *Oriental Inst.*, **1** (1-2): 45-53.

Gupta, V. K. (1980). A revision of the Tribe Poemenini in the Oriental region (Hymenoptera: Ichneumonidae). *Oriental Inst.*, **14** (1): 73-130.

Gupta, V. K. (1984). The Tribe Tryphonini in India, with descriptions of new species (Hymenoptera: Ichneumonidae). *Oriental Inst.*, **18**: 173-186.

Gupta, V. K. (1985). A review of the tribe Neoxoridini of the world (Hymenoptera: Ichneumonidae Pimplinae). *Oriental Inst.*, **19**: 323-329.

Gupta, V. K. (1987). The Ichneumonidae of the Indo- Australian Area (Hymenoptera). *Mem. Amer. Ent. Inst.*, **41** (2): 895-1068.

Gupta, V. K. (1991). A review of the exenterine Genus *Eridolius* (Hymenoptera: Ichneumonidae) and description of new species from Oriental region. *Oriental Insects*, **25**: 435-446.

Gupta, V. K. and Chandra, Girish. (1974). Oriental species of *Xorides* (*Xorides*) (Hymenoptera: Ichneumonidae). *Oriental Inst.*, **8** (4): 395-411.

Gupta, V. K. and Saxena, K. (1987). A Revision of the Indo-Australian species of *Coccygominus* (Hymenoptera: Ichneumonidae). *Oriental Insects*, **2**: 263-431.

Gupta, V. K. and Tikar, D. T. (1967). Indian species of *Scambus* Hertig. (Hymenoptera: Ichneumonidae). *Oriental Inst.*, **1**(3-4): 215-237.

Gupta, V. K. and Tikar, D. T. (1976). Ichneumonologia Orientalis, Part I. The Tribe Pimplini. *Oriental Insects*, Monograph No. 1.

*Hafiz, A., Tufani, N. A. and Makhdoomi, D. M. (2011). Sesonal prevalence of Gastro-intestinal Helminthiasis in cattle from Kashmir valley. *The Indian J. of Field Veterinarians*, **6** (3): 50-52 (Prevalence percentage of helminths found as: *Strongyle* sp. (22.73 per cent). *Fasciola* sp. (6.18 per cent) and *Paramphistome* (4.04 per cent).

Hamdani, H. A., Khan, Z. A., Wani, H. Jalel, S. J. A., Bihaqi, Mir, M. S. (2017). Parasitic Profile of Domestic Geese of Kashmir. *Intl. J. Livestock Res.*, **7** (5): 129-133 doi: 10.5455/ijlr - 20170409094535.

*Handoo, Z. A. (1977). *Soil and Plant Parasitic nematodes of vegetables and fruit crops in Kashmir.* Ph. D. Thesis, Deptt. of Zool., Faculty of Science, University of Kashmir, Srinagar, 290 pp.

Hangloo, V. K., Koul, I. Dev, G., Dhar, U., Chrungoo, R. K. and Khan, M. (1989). *Biliary* ascariasis. *Indian J. Gastroenterol.*, **8** (4): 265-266.

Hansson, C. (1985). The entedontine genera – Achrysocharoides Girault, *Chrysocharis,* Foerster and *Kratoysma* Boucek (Hymenoptera: Eulophidae) in Oriental region. *Entomologica Scand.*, **16** (3): 217-226.

Haq, Abrar-ul, Tufani, N. A., Malik, H. U., Hussain, S. A., Rajaq, R., Bhat, U. A. and Nabi, S. U. (2017 a). Cross seasonal study on the prevalence of Ovine babesiosis in different breeds of Kashmir Valley. *J. Entomol. and Zool. Studies,* **5** (6): 1492-1496.

Haq, Abrarul, Tufani, N. A., Malik, H. U., Nabi, S. U., Hussain, S. A., Beigh, S. A. and Shahardar, R.A. (2017 b). Study on Ovine babesiosis in different age groups in Kashmir Valley. *J. Entomol. and Zool. Studies,* **5** (6): 2024-2027.

Haq, Abrarul, Tufani, N. A., Gugjoo, N. B., Nabi, S.U. and Malik, H. U. (2017c). Therapeutic amelioration of severly anaemic local Kashmiri goats affected with babesiosis. *Advances in Anim. and Vet. Sci.*, **5** (11): 463-467s.

Harbach, R. E. (2014). Valid mosquito Species, In: Mosquito Taxonomic Inventory. http: //www.mosquito taxonomic inventory-info/validspecies list.

Harish, Rekha and Gupta, Sanjeev (2009). *Plasmodium vivax* malaria presenting with severe Thrombocytopenia, cerebral complication and Hydrocephalus. *Indian J. Pediatr.,* **76**(5): 551-552.

Harwood, R. F. and Maurice, J. T. (1979). *Entomology in Human and Animal Health* (7th ed.). MacMillan Publishing Co. In., New York, U.S.A. 548 pp.

*Hassan, J. (2009). Seasonal nematode population density on maize and mustard. *World Appl. Sci. J.*, **6** (6): 734- 736.

*Hassan, J., Chishti, M. Z., Rashid, M. and Lone, B. A. (2009). Redescription of some nematodes associated *Zea mays. World Applied Sciences J.*, **6** (9): 1261- 1263 (On nematode species *viz., Tylenchorhynchus zeae, Pratylenchus sativus, Aphelenchoides srinagarensis*).

Hassan, J. and Ahangar, M. A. (2018 a). Dynamics of *Tylenchus arcuates* Siddiqi, 1963 on some crops of Kashmir and its morphological redescription. *Intl. J. Multidisciplinary Res. and Dev.*, **5** (3): 113-115.

Hassan, J. and Ahangar, M. A. (2018 b). Phytonematode infestations on *Brassica oleracea*, with special emphasis on *Boleodorus cylindricus* Dhanachand, Renubala and Anandi, 1993. *Intl. J. Adv. Educational Res.*, **3** (2): 157-159.

Hassan, J., Ahangar, M. A. and Iqbal, J. M. (2018). A management strategy for the significant reduction of *Filenchus cylindricaudus* (Wu, 1969) Siddiqi, 1986 on *Allium cepa* (onion). *Intl. J. Biol. Res.*, **3** (1): 324- 325.

Hassan, J., Chishti, M. Z., Rasheed, M. and Tak, H. (2008 a). A report of *Hirschmanniella* sp. and *Tylenchus* sp. on rice in Kashmir, with control strategy. *World Appl. Sci, J.* **5** (5): 543-545.

Hassan, J., Chishti, M.Z., Rasheed, M. and Bhat, F. A. (2008 b). Nematodes associated with *Brassica campestris* and their control. *World Appl. Sci. J.*, **5** (5): 543-545.

*Hassan, J., Chishti, M.Z., Ahmad, Imtiaz, Iqbal, M. and Lone, B.A. (2009). Seasonal nematode population density on maize and mustard. *World Appl. Sci. J.*, **6** (6): 734-736.

Hassan, J. Chishti, M. Z., Rasheed, M., Ahmad, F. and Lone, B. A. (2009). Nematodes associated with *Zea mays* and their control through organic soil amendments. *Intl. J. Plant Production*, **3** (4): 73-78.

Hassan, J., Malik, M, N., Bhat, M. A., Iqbal, J. M and Farhan, N. (2016 a). Nematode infestation on *Solanum tuberosum* and *Raphanus sativus* with generic build-up statistics- A first report from Kashmir. *Intl. J. of Innovative Res. and Adv. Studies*, **3** (5): 84-87.

Hassan, J. Chishti, M. Z., Rasheed, M., Tak, I., Dar, S. A., Lal, E. P. and Mohi-uddin, D. (2016 b). Population analysis of the genera build-up on some commercially important vegetable crops grown in Kashmir. J. *Parasitic Dis.*, **40** (3): 877- 880.

Hayat, M. (1972). The species of *Aphelinus* Dalman, 1820. (Hymenoptera: Aphelinidae) from India. *Entomophaga*, **17** (1): 49-58.

Hayat, M. (1975). Some Indian species of *Anastatus* (Hymenoptera: Chalcidoidea: Eupelmidae). *Oriental Insects*, **9** (3): 261-271.

Hayat, M. (1980). Taxonomic notes on some Oriental Aphelinidae, with new records. *Oriental Insects*, **14** (1): 461-472.

Hayat, M. (1986). Family Aphelinidae. ***In***: Subba Rao, B.R. and Hayat, M. (eds), pp.143-171 The Chalcidoidea (Insecta: Hymenoptera) India and adjacent countries. *Oriental Insects*, **20**: 1-430.

Hayat, M. (1989). A revision of the species of *Encarsia* Foerster (Hymenoptera: Aphelinidae) from India and adjacent countries. *Oriental Insects*, **23**: 54.

Hayat, M. (1998). Aphelinidae of India (Hymenoptera: Chalcidoidea)- A taxonomic Revision. *Mem. Entomol. Inst.*, **13:** 127.

Hayat, M. and Anis, S.B. (1999). The Indian species of *Polynema,* with notes on *Stephanodes reduvioli* (Hymenoptera: Mymaridae). *Oriental Insects*, **33**: 321.

Hayat, M. and Fatima, K. (1992). Taxonomic studies on *Aphelinus* (Hymenoptera: Aphelinidae), descriptions of new species and further record of *A. gossypii*, with new synonym. *Entomon*, **17**: 103-106.

Hayat, M. and Subba Rao, B.R. (1986). Family Trichogrammatidae, *In*: Subba Rao and Hayat, M. (eds), pp. 193-208, The Chalcidoidea (Insecta: Hymenoptera) of India and adjacent countries. *Oriental Insects*, 20: 1-430 pp.

Hayat, M. and Viggiani, G. (1984). Preliminary Catalogue of the Oriental Trichogrammatidae (Hymenoptera: Chalcidoidea). *Bollettino del Laboratorio di Entomologia Fillippo Silvestri Portia*, **41**: 34.

Hennig, W. (1955-1965). *Family Muscidae In*: E.Linder (Ed.) *Die Fliegen der Palaarktischen Region* 63b: 1110pp. Stuttart.

Hibbett, D.S. *et al.* (2007). A higher level of phylogenetic classification of fungi. *Mycological Res.*, **111**(5): 509-47.

Hoogstraal, H. and McCarthy, V.C. (1965). Host and distribution of *Haemaphysalis kashmirensis*, with description of immature stages and definition of the subgenus *Herpetobia canestrini* (resurrected). *J. Parasitol.*, **51**: 674-79.

Hoogstraal, H. and Verma, M.G. R. (1962). *Haemaphysalis cornupunctata* and *H. kashmirensis* sp. n. from Kashmir, with notes on *H. sundrai* Sharif and *H. swelli* Sharif of Indian and Pakistan (Ixodoidea: Ixodidae). *J. Parasitol.*, **48:** 185-194.

Hopkins, G. H. E. and Rothschild, M. (1953, 1962, 1966, 1966 *a*, 1971). *An illustrated Catalogue of the Rothschild Collection of fleas (Sphonaptera) in the British Museum (Natural History)*. Vol.I., 1953, 361 pp. *Tungidae and Pulicidae.*, Vol.II, (1962), 445 pp., *Vermipsyllidae to Xiphinopsyllidae;* Vol., III, 500 pp. (1966), *Histrichopsyllidae;* Vol.IV, pp.549 pp. (1966 *a*), *Histrichopsyllidae* and Vol.V, 530 pp. (1971), Leptopsyllidae. British Museum, National History, London, England.

Horstmann, K. (1992) Neue Taxa dev campopteginae aus deu Gattungen *Campoplex* Gravenhost *Daidegma* Forster and *Nemeritis* Holmgren (Hymenoptera Ichneumonidae). *Z. Arbeitsgem Oesterr. Entomol.*, **44** (3-4): 116-127.

*Hussain, A., Ahmad, F. and Khan, A. R. 2008. Studies on the influence of parasitic infection in relation to the decline in the population of the native fish of Kashmir, pp. 509-513, **In**: *Science of Better Tomorrow*, Printed at Universal Printers, Khanyar, Srinagar, Kashmir.

Husain Tasawer and Agarwal, M. M. (1981). Taxonomic studies on Indian Epitraninae (Hymenoptera: Chalcididae). *Oriental Insects.*, **15** (4): 413-432.

Hyman, L. B. (1951). *The Invertebrates*, Vol. III McGraw Hill. New York., U.S.A.

*Ilango, K. (2011). Spatial and temporal distribution of Phlebotomine sandflies in relation to Leishmaniasis in India, **Chap., 5**, pp. 99-104, **In**: Tyagi, B.K. and Veer, V. (eds.), *Entomology, Ecology and Biodiversity*. Scientific Publisher (India), Jodhpur, India. (In India sandfly-fauna represents 50 spp., 6 spp. are vectors and 3 spp. as species complexes. The causative agent of leishmaniasis, *Leishmania infantum*, transmitted by the bite of sandfly, *Phlebotomus major* affect children in Jammu and Kashmir).

Ilango, V. (2006). Bluetongue virus outbreak in Tamil Nadu, Southern India: Need to study the Indian biting midge vectors, Culicoides Latreille (Diptera: Ceratopogonidae). *Curr. Sci.*, **90**: 163-167.

*Iqbal, A., Tariq, K. A., Wazir, V.S. and Singh, R. (2013 a). Antiparasituc efficacy of *Artemisia absinthium* Tolfauril and Amprolium against intestinal coccidiosis in goats. *J. Parasitic Dis.*, **37** (1): 88- 93.

Iqbal, A., Wazir, V.S., Katoch, Singh, R., Suri, S., Yadav, A. and Wani, N. (2012). Prevalence of *Eimeria* spp. in goats of Jammu region of Jammu and Kashmir State. *J. Vet. Parasitol.*, **26** (1): 57-59.

Iqbal, A., Wazir, V. S., Malik, M.A., Tripathi, A.K., Najimanna, Tickoo, A. and Mir, I.A. (2013). Prevalence of caprine coccidiosis in Jammu, India. *Indian Vet. J.*, **90** (1): 17-19.

Islam, S. K and Hayat, M. (1986). Family Eupelmidae. *In*: Subba Rao, B.R. and Hayat, M. (eds.), pp.57-66. The Chalcidoidea (Insecta: Hymenoptera) of India and adjacent countries. *Oriental Insects*, **20**: 1-430 pp.

ITS Report (2014). Integrated Taxonomical Information System.

Ivanov, A. and Tsachev (2008). *Hepatozoon canis* and Hepatozoonosis in the dogs. *Trakia J. Sciences*, **6** (2): 27-35.

Iyenger, R. (1973). The Siphonaptera of Indians sub-region *Oriental Insects* (Suppl.), **3** (1-102).

Jabaneson, A., Rajasekar, and Paul, S. A. (2012). Influence of climatic factors on the distributioin of tree hole mosquitoes collected from Kashmir valley, India. International *J. of Env. Biol.*, 2 (2): 92-96.

Jacob, V. (1950). Some aspects of Malaria in Jammu and Kashmir State *Indian J. Malariology*. **4(3)**: 251-260.

Jacob, V. P. and Kalra, S. L. (1951). Kalazar in Kashmir. *Indian J. Med. Res.*, **39**: 323-327.

Jad, Beena, Dogra, S. and Mahajan, B. (2015). Significant decrease in prevalence of intestinal parasites among patients seeking treatment in Tertiary Care Hospital in Jammu: a Changing trend. *Intl. J. Curr. Microbiol. and Applied Sci.*, **4** (11): 659-664.

Jairajpuri, M. S. and Baqri, G. H. (1966). *Sectonema procta* n. sp. and *Pungenatus angulatus* n. sp., two new soil inhabiting nematodes. *Nematologica*, **12**(3): 396-402.

Jairajpuri, M. S. and Loof, P. L. A. (1976). On some species of *Euchodelus* Thorne, 1933. (Nematoda: Dorylaimidae) from India. *Nematol.*, **12** (3): 501-508.

Jairajpuri, M. S. and Siddiqi, A. H. (1963). A new and known species of the genus *Tylencholaimellus*, M. O. Cobb., 1915 (Nematoda: Dorylaimoidea) from India, with key to its species. *Zeitschrift fur Parasitenkunde*, **22** (6): 489- 494.

Jairajpuri, M. S. (1965). *Basiria kashmirensis* n. sp. (Nematoda: Tylenchida) from India. *Labdev J. Sci. Technol.*, **3**(1): 23-25.

James, M.T. (1947). *The Flies that cause Myiasis in Man*. U.S.Deptt. of Agriculture, Misc. Publ. No.631 (Diptera), pp 175.

James, S. P. and Liston, E.G. (1904). *A monograph of the anopheles mosquitoes of India*. Thacker Spink and Co, Calcutta, India.

*Javaid, Moien, Rashid, N., Nabi, B., Mir, M., Bhat, M. A., Wani, N. M., and Bhat, M. (2018). Prevalence of gastro-intestinal Nematodes in sheep and goats, reared by nomads of Jammu region of Jammu and Kashmir. *Intl. J. Livestock Res.* 8: 219- 225 (Prevalence of 8 genera of nematodes is given and genus *Haemonchus* was found to be predominant in faecal samples of animals).

Jehangir, M., Mariam, A., Singh, R. and Prabhakar (2020). Primary Alveolar Hydatid disease: A rare case series, with Review of Imaging findings. *Galician Medical J.*, **27** (1), F202019, doi: 10.21802/gmj.2020.1.9.

Jonathan, J. K. (1999). Four new species of *Coenocytus* Thomson from India Mynmar and Taiwan (Hymenoptera: Ichneumonidae). *Rec. Zool Surv. India*, **97** (Part 2): 1-13.

Jonathan, J. K. (2000). Five new species of *Itamoplex* Forester from India (Hymenoptera: Ichneumonidae). *Rec. Zool. Surv. India*, **98**(2): 173-184.

Jonathan, J. K. (2006). *Ichneumonologica Indica- Ichneumonidae*. Zoological Survey of India, Calcutta, e-book.

Jordan, E. L. (1930). On a new species of *Lepoderma* (Looss), *Lepoderma himalayii* n. sp. *Proc. Ind. Sci. Cong.*, **17**: 246.

Jordan, K. (1937).Three new Bird fleas from Kashmir. *Novitates Zoologicae*.**40**: 299-306.

Kalala, B. S. and Shivalli, S. (2017). Comment on serological evidence of scrub typhus among cases of PUO in Kashmir Valley- A Hospital based Study". *J. Clinical and Diagnostic Res.*, **11**(6): DL01-DL02, doi: 10.7860/JCDR/2017/25345.100038.

Kalha, C. S., Chib, H. S. and Tikoo, M. L. (1992). Earcockle of wheat, the first record from Jammu. *Plant Disease Res.*, **7** (2): 254.

*Kalha, R. K., Agarwal, N., Pande, Yadav, A. and Katoch, R. (2010). Prevalence of gastro-intestinal parasites in bovines of Jammu district. *Indian J. Vet. Med.*, **30**: 113-114.

Kalra, N. K., Kaul, S.M. and Rastogi, R. M. (1997). Prevalence of *Aedes aegypti* and *Aedes albopictus* vectors of Dengue and Dengue haemorrhagic fever in North, North-east and Central India. *Dengue Bulletin*, **21**: 84.

Kalra, S. L. (1952). Natural History of Typhus fever in India. *Indian J. Med. Sci.*, **6**: 569-575.

Kalra, S. L. (1959). Progress in the Knowledge of Rickettsiae Disease in India. *Indian J. Med. Res.*, **476**: 477-483.

Kalra, S. L. and Rao, R. N. A. (1949). Typhus fever in Kashmir State. Part I. Epidemic Typhus. *Indian J. Med. Res.*, **37** (4): 395-400.

Kalra, S. L. and Rao, R. N. A. (1951). Observation on the epidemiology of relapsing fever in Kashmir. *Indian J. Med. Res.*, **39**: 319-21.

Kalra, S. L. and Rao, R. N. A. (1951*a*). Typhus fever in Kashmir State Part II. Murine Typhus *Indian. J. Medical Res.*, **39** (3): 297-302.

Kamil, S. A., Darzi, M. M., Mir, M. S., Shah, S. A., and Khan, F. A. (2011). *Tetrameres fissipina* infection in ducks from Bandipore areas of Kashmir valley. *Israel J. Veterinary Medicines*, **66** (2): 43-47.

Kamil, S. A., Nashir-uddullah, N., Darzi, M. M. and Mir, M.S. (2006). Occurrence of histomoniasis (Enterhepatistis) in broiler breeder chickens by possible lateral transmission. *Indian J. Vet. Pathol.*, **30** (2): 14-17.

Kani, A., Nagase, K., Totani, Morishita, T. and Okamoto, K. (1981). An observation of parasitic and sanitary animals in Jammu and Kashmir State of India. 51 (Japanese) (cf. *Helminthol.* Abt. Vol.**50** (6): 325.

Kapoor, B and Shankar, Uma (2019). Trap catches of *Helicoverpa armigera* Hubner and Marnea, Vitrat Fabricius and their natural enemies on black gram, *Vigna mungo* (L.). *J. Entomol. and Zool. Studies*, **7** (6): 246-251.

Kapoor, V. C., Grewal, J. S. and Sharma, S. K. (1987). *Indian Pipunculids (Diptera: Pipunculidae): A comprehensive Monograph*. Atalantic Publishers and Distributors, New Delhi.

Katoch, R., Godara, Yadav, A., Sharma, S. and Ahmad, I. (2014). Occurrence of *Chrysomya bezziana* in a buffalo in Jammu. *J. Parasit. Diseases*, **38** (4): 420-422.

Katoch, R., Vohra, S., Yadav, A. Khajuria, J. K., Pandey, V. and Singh, Y. (2008). Prevalence of parasitic infection in Pigs of Jammu Region. *Vet. Practioner*, **9**: 102- 103.

*Katoch, R., Yadav, A., Vohra, S., Khajuria, J. K. and Singh, R. (2006). Effect of managemental practices of Ticks on equines in Jammu. *Vet. Prac.*, **7:** 19-20.

*Katoch, R., Yadav, A., Vohra, S., Taggar, R. K. and Pandey, V. (2007). Parasitic diseases of dairy animals in and around R. S. Pora of Jammu District. *Vet. Pract.*, **8**: 186-187.

*Katoch, R., Yadav, A. Godara, Khajuria, R., Borakataki, S. and Sodhi, S. S. (2012). Prevalence and impact of gastro-intestinal helminth on body weight gain in backyard chickens in subtropical and humid zone of Jammu, India. *J Parasitic Diseases*, **36** (1): 49-52.

Kaul, K. and Agarwal, M. M. (1986). Taxonomic studies on encyrtid parasitoids (Hymenoptera: Chalcidoidea) of India. Aligarh Muslim Univ. Publication. *Zoological Series on Indian Insect types*, **13**: 36-39.

*Kaul, Angali (1987). *Studies on Protozoan parasites of some Invertebrates in Kashmir*. M.Phil. Dissertation. P.G.Deptt. of Zool. Faculty of Science, University of Kashmir, Srg.

Kaul, H. N., Kulkarni, S. M., Dhanda, V. and Geevarghese, G. (1978). *Hunterellus sagarensis*, parasitizing *Haemaphysalis bispinosa* (Acari: Ixodida) in Jammu and Kashmir, India. *Acrology Newsl.*, **6**: 4-5.

Kaul, H.N. and Shetty, P. S. (1983). Sandflies of Jammu Region, with a review in their records in Jammu and Kashmir, India. *Indian J. Medical Res.*, **78**: 643-646.

Kaul, H. N., Shetty, P. S., Ghalasi, G. R. and Dhanda, V. (1990). Survey of ticks (Acarina: Ixodidae) for Crimean haemorrhagic fever virus activity in Jammu and Kashmir State, India. *Indian J. Medical Res.*, Section A, *Infections Diseases*, **91** (Jan.): 5-8.

*Kaul, K and Agarwal, M. M. (1986). Taxonimic studies on encyrtid parasitoids (Hymenoptera: Chalcidoidea) of India. *Aligarh Muslim University Publication-Zoological Studies on Indian types*, **13**: 36-39.

Kaul, N., Gupta, V., Bhardwaj, S., Dogra, D. and Dogra, N. (2016). A new focus of cutaneous leishmaniasis in Jammu Division of Jammu and Kashmir State, India. Indian *J. Dermatol., Venereol. and Leprology*, **82** (2): 145-150.

*Kaul, Ramesh (1985). *Studies on the nematode parasites of Brassicaceous plants of Kashmir*, Ph. D. Thesis P. G. Deptt. of Zool., Faculty of Sci., University of Kashmir, Sgr., Kmr. (J&K). 338 pp.

Kaul, P. and Qadri, M. S. (2010). Acute superficial lymphangitis following pigeon mite bite. *BHJ/Case Reports*, 1-3 DOI: 10.1136/bcr.09.2010.3310.

Kaul, S. (1949). Relapsing fever tick-borne on account of an outbreak in Jammu and Kashmir forces, India. *Indian Med. Gaz.*, **84**: 433-440.

*Kaul, S. M., Jain, D. C. (1995). Distribution of Phlebotomine sandflies (Diptera: Psychodidae), according to physiographic divisions of India. *J. Commun. Dis.*, **27**: 155-163.

Kaul, S. M., Gupta, M. R., Jain, D. C. Gupta, R. S. and Sharma, R. S. (1995). Malaria control in Jammu City, Jammu and Kashmir State. *J. Communicable Diseases*, Sept., **27**(3): 191-192.

Kaul, S. N. (1944). *Report of typhus outbreaks at Banihal submitted to D. M. S.*, J&K (Govt.).

Kaul, V. (1984). Seasonal fluctuation of phytonematodes in two fruit nurseries in Kashmir. *Indian J. Helminthol.*, **37**(1): 4-11.

*Kaul, V. (1984 *a*). *Studies on the Plant-parastic Nematodes in Kashmir*. Ph. D. Thesis, P. G. Deptt. of Zool., Faculty of Science, Univ. of Kamr, Srinagar, Kmr. (J&K), 247 pp.

Kaul, V. (1987). Two new species in the super-family Hopolaimoidea (Nematoda) from Kashmir. *Geobios new Reports*, **6**: 162-164.

Kaul, V. (1998). Plant-Nematology in Jammu and Kashmir: Aspects and Prospects, pp.18-22. *In:* Trivedi, P. C. (Ed.), *Phytonematology in India*. CBS Publishers and Distributors, Daryaganj, New Delhi, 205 pp.

Kaul, Vijay; Masoodi, M. A., Bhagat, K. C. and Bhat, O. K. (2000). Biology, behaviour and bio-control agents of *Drosicha dalbergia* (Grien), infesting Almonds in Kashmir. *Ann. Plant. Prot. Sci.*, **8** (1): 24-26.

Kaul, V. and Fotedar, D. N. (1987).Morphometric variability in *Helicotylenchus erythrinae*(Eimmermann) 1904 Golden 1956. *Geobios new Rept.*, 6: 138-142.

Kaul, V. and Wali-ullah, M. I. S. (1989). A new species of genus *Pratylenchus* Micotetzky, 1922 (Pratylenchidae: Nematoda) from Kashmir. *Indian J. Nematol.*, **19** (1): 66-69.

Kaul, V. and Wali-ullah, M. I. S. (1995). *Tylenchorhynchus oryzae* sp. n. (Tylenchorhynchidae: Nematoda) from Kashmir, India. *Annals. Plant Protection Sci.*, **3** (2): 155-157.

Kaur, D., Ganai, D., Parveen, S., Borkataki, S., Yadav, A., Katoch, R. and Godara, R. (2015) Occurrence of *Setaria digitata* in cow. *J. Parasitic Dis.*, **39** (3): 477-478.

Kaur, Ramindeer (1992). Mosquito types present in the collection of National Institute of Communicable Diseases (Diptera: Culicidae). *Oriental Insects*, **26**: 367-382.

Kaur, R. (2003). An update on the distribution of mosquitoes of the Tribe Aedini in India (Diptera: Culicidae). *Oriental Insects*, **37**: 439-455.

Kaw, B. L. (1941). Studies on the Helminth parasites of Kashmir. Part I. Descriptions of some new species of genus *Pomphorhynchus* Monticelli (1905). *Proc. Indian Acad. Sci.*, **13**: 369-378.

Kaw, B. L. (1943). Studies on the helminth parasites in Kashmir. Part II. On two new trematodes of the sub-family Pleurogenitinae looss, 1899, with a review of the genus *Pleurogenes* Looss, 1896. *Proc. Indian Acad. Sci.*, **18**(3), Sec B: 97-108.

Kaw, B. L. (1944). Studies on helminth parasites of Kashmir. Part III. Description of a new Allocreadid, *Crepidostomum indicum* from a fresh water fish, *Schizothorax niger* from Dal lake, Kashmir. *Proc. Ind. Acad. Sci.*, **20**: 72-77.

Kaw, B.L. (1945). On the present status of the genus. *Loxogenes. Proc. Indian Acad. Sci.*, **21** (6): 342-43.

Kaw, B. L. (1950). Studies in Helminthology: Helminth parasites of Kashmir. Part-I. Trematodes. *Indian J. Helminthol.*, **2** (2): 67-126.

Kaw, B. L. (1951). Studies on Helminthology. Helminth Parasites of Kashmir. Part II. Acanthocephala. *Indian. J. Helminthol.*, **3**(2): 117-132.

*Kaw, Lalita (1968): *Studies on Helminth Parasites of some vertebrates in Kashmir: Helminth parasites of some birds in Kashmir*, Ph. D. Thesis. P. G. Deptt. of Zool., Faculty of Sci., Univ. of Kmr., Sgr., Kmr. (J&K).

Kazmi, S. I. and Hayat, M. (1998). Revision of the Indian Copidosomatini (Hymenoptera: Chalcidoidea, Encyrtidae). *Oriental Insects*, **32**: 323.

Kettle, D. S. (1995). *Medical and Veterinary Entomology* (2nd edition): CAB International, Willingford Oxon, U.K., 725 pp.

Khajuria, A., Azim, S., Rehman, S., Sudan, N. A., Sharma. M., Sultan, Z. and Ahmad, A. (2013). Prevalence of parasitic Diseases of liver in sheep and goats in organized and unorganized slaughter houses in Jammu. *J. Vet. Parasitol.*, **27** (1): 46-50.

Khajuria, J. K. and Kapoor, P. R. (2003). Prevalence of parasites in sheep and goats at Kathua, Jammu. *J. Vet. Parasitol.*, **17**: 121-126.

Khajuria, J. K., Yadav, A. and Raina, A. K. (2005). Incidence of helminth parasites in Equines of Jammu Region. *Centaur*, **21**: 58- 61.

Khajuria, J. K., Katoch, R., Yadav, A., Arora, S., Soodan, J.S., Borkataki, S. and Singh, Ajitpal (2010). Prevalence of Gastro-intestinal parasites affecting pigs in Jammu Districts of Jammu and Kashmir. *Vet. Practioner*, **11** (2): 167-168.

Khajuria, J. K., Katoch, R., Yadav, A., Godara, R. and Singh, A. (2012). Prevalence of gastro-intestinal heminth in small ruminants of low-altitude sub-tropical zone of Jammu Province. *J. Vet. Parasitol.*, **26** (2): 151-155.

Khajuria, J. K., Katoch, R., Yadav, A., Godara, R., Gupta, S. K. and Singh, A. P. (2013). Seasonal prevalence of gastro-intestinal helminths in sheep and goats of middle agro-climatic zone of Jammu Province. *J. Parasitic Dis.*, **37**: 21-25.

Khajuria, J. K., Yadav, A. and Raina, A. K. (2005). An outbreak of Psoroptic mange in sheep and its treatment. *J. Res.*, **2**: 126-132.

*Khajuria, J. K., Yadav, A., Vohra, S., Katoch, R., Sood, S., Singh, A. and Agarwal, R. (2009). An outbreak of of parasitic gastroenteritis in sheep and goats in R. S. Pora, Jammu. *Vet Practitioner*, **10**: 168-169.

Khajuria, R. R. (1966). Incidence and control of parafilariasis in bovine in Jammu Province. *Indian Vet. J.*, **43** (8): 749-50.

*Khajuria, V., Godara, R., Yadav, A. and Katoch, R. (2014). Deltametrin Resistance in *Rhipicephalus microplus* in Jammu region. *Indian Vet. J.*, **91**: 21-24_.

Khajuria, V., Godara, R., Yadav, A. and Katoch, R. (2015). Prevalence of Ixodid ticks in dairy animals of Jammu Region. *J. Parasitic Diseases*, **39** (3): 418-421, doi: 10.1007/s12639-013-0354-v.

Khan, A. A. and Shah, M. A. (2017). Records of aphids and their natural enemies in agro-ecosystem with special reference to horticultural ecosystem of Kashmir. *J. Entomol and Zool. Studies*, **5** (4): 189-203.

*Khan, A. R. (1974). *Studies on the trematode parasites of some vertebrates of Kashmir.* Ph. D. Thesis P. G. Deptt. of Zool., Faculty of Sci., Univ. of Kmr., Sgr., Kmr. (J&K).

Khan, A.R. 1986(1987). On a new trematode, *Opisthorchis anasi* sp. nov. from Kashmir. *Indian J. Helminthol.*, **38** (2): 119-122.

*Khan, A. R., Rasheed, M., Magray, A. and Pandit, B.A. (2008). Comparative Morphology of oocytes *of Eimeria* in domestic Rabbit of Kashmir valley, *In*: M. Z. Chishti and Fayaz Ahmad (eds.), pp.481-482, *Science For Better Tomorrow*, printed at Universal Printers, Khanyar, Srinagar.

Khan, A. R. and Ahmad, D. 1991 (1992). Redescription of *Hypoderaeum conoideum* (Bloch, 1982), Dietz, 1909 from *Anas platyorhynchos domesticus* in Kashmir. *Indian J. Helm.*, **43**(2): 162-164.

Khan, A. R and Ahmad, F. (1992a). A new record of *Echinoparyphium chinense* Lu. Li et Chu, 1964 from *Anas platyrhynichos domesticus* in Kashmir. *Indian J. Helm.*, **44** (2): 146-148.

Khan, A. R. and Chishti, M.Z. (1982). On *Echinostoma revolutum* (Froelich) Loosse, 1899 and synonymy of *Neoechinostoma spinosa* Agarwal, 1963. *Kashmir Univ. Res. J.*, **2**: 22-24.

Khan, A. R. and Chishti, M. Z. (1983). Studies on *Plagiorchis* (*Plagiorchis*) *tehri* n.sp. from *Passer domesticus* from Bachpora, Kashmir, *J. Indian, Instit. of Sci.*, **64** (12): 189-192.

Khan, A. R. and Chishti, M. Z. (1984). *Euparyphium pindchi* sp. nov. from a common Grebes, *Podiceps ruficollis capensis. Indian J. Helminthol.*, **36** (2): 134-138.

Khan, A. R. and Chishti, M. Z. 1987(1989). The incidence of trematode parasites in a common Kashmiri fowl population and study of *Catatropis kashmirensis* n.sp. (Digenea: Notocotylidae Luhe, 1909). *Riv. Parassitol.*, **48** (3): 355-360.

Khan, A. R., Ahmad, F., Rashid, M. and Bakshi, S. (2004). Seasonal occurrence of helminth parasites in *Schizothorax* in Dal Lake in Kashmir. *J. Parasitic Dis.*, **28** (1): 23-28.

Khan, A. R., Perihar, A. K., Sharma, D. B. and Lahori, V. C. (1975). Kala-azar in Jammu. *J. Indian Med. Res.*, **65** (3): 85-86.

Khan, A. R. and Rasheed, Majidah 1993 (1994). Influence of season on the intensity of helminth infection in *Schizothorax esocinus* Heckel from the Wular Lake. *Indian J. Helminthol.*, **45** (42): 64-73.

Khan, A. R. Rasheed, Majidah and Kaul, J. A. (2002). Prevalence of Giardiasis in Kindergarten School Children of Urban Kashmir. *J. Parasitic Diseases,* **26** (1): 30-33.

Khan, A. R., Rasheed, Majidah and Mir, Tajamul, H. (2003). A study on Mallophagon lice and some helminthes of Whooper Swan, *Cygnus Cygnus* of Kashmir. *Proc. Smr. on Interdisciplinary Approaches* of *Zoological Research*, held at Deptt. of Zool. Univ. Kmr., Sgr. Kmr. (J&K), from 29th September – 1st October, 2003, p.15 (Abt.).

Khan, A. R., Zargar, S. A., Lateef, M. and Nazir, Muzaffar (2008). Epidemiological Survey of *Taenia saginata* infection in Kashmir, *pp.* 500-502, **In**: M. Z. Chishti and Fayaz, Ahmad (eds.), *Science For Better Tomorrow*, printed at Universal Printers, Khanyar, Srinagar, Kashmir.

Khan, E. and Siddiqi, M. R. (1963). *Criconema laterale* n. sp. (Nematoda: Criconematidae) from Srinagar, Kashmir. *Nematologica,* **9**: 584-86.

Khan, I. A., Malla, N., Wattal, C. and Agarwal, S. C. (1985). Amoebiosis in Kashmir(India)- a laboratory study. *Trans. Roy. Soc. Trop. Med. Hyg.*, **79** (3): 425.

Khan, M. D. (1985). *Studies on Amphistomes of Domestic Ruminants in Kashmir*. Ph. D. Thesis. P. G. Deptt. of Zool., Faculty of Sci., Univ. of Kmr., Sgr., Kmr. (J&K).

Khan, M. R., Ahmad, N., Khan, S. M. and Askary, T. H. (2005). Damage caused by *Meloidogyne inconita* to five cultivars of *Chrysanthemum* and its chemical control. *Intl. J Nematol.*, **15** (2): 199.

Khanday, A. L., Sureshan, P. M., Buhroo, A.A., Ranjith, P. and Tselikh, Ekaterina (2019). Pteromalid wasp (Hymenoptera: Chalcidoidea, associated with bark beetles, with description of a new species from Kashmir, India. *J. Asia-Pacific Biodiversity*, 12(2): 262-272.http doi.org./10.10.16/j.japb.2019.0101.

Khuroo, M.S. (1996). Ascariasis. *Gastroenterol.Clin.North America*, **25**: 553-57.

*Khuroo, M. S. (2002). Hydatid disease: Current Status and Recent Advances. Ann. Saudi Med, 22: 56-64 (Covers, *inter alia*, hydatid disease as endemic to Kashmir).

Khuroo, M. S., Zargar, S. A. Yattoo, G. N. Dar, M.J. Javid, G., Khan, B. A. Boda, M. I. and Mahajan, R. (1992). Sonographic findings in gall bladder ascariasis *J. Clinical Ultra Sound*, **20**(9): 287-91.

Khurshid, I., Ahmad, J.M. and Sheikh, A.A. (2018). Description of a new species of *Halticoptera spinola* (Chalcidoidea: Pteromalidae) from Kashmir. *J. Entomol. and Zool. Studies*, **6** (2): 3031-3034.

*Khurshid, Ibraq and Ahmad, F. (2014). Prevalence of Helminth parasites of *Schizothorax* from Shallabugh Wetland and River Sindh in Kashmir. *Global Veterinaria*, **12** (5) 731-735.

*Khurshid, I., Ahmad, F. and Ahmad, T. (2013). Parasitic distribution in relation to gender, season and length of fish hosts in Shallabugh wetland. *Intl. J. Scientific and Engineering Res.*, **4** (4): 1083-1091. (Results of this study have shown the prevalence of infection due to cestodes, trematodes, nematodes and acanthocephalans, increased with increase in length of hosts. Male and female hosts were equally prone to infection and season-wise distribution of parasites showed positive correlation).

Khurshid, I., Ahmad, F. and Sofi, T. A. (2013). Comparative study of parasites infecting some fishes of Shallabugh wetland, Kashmir. *New York Science J.*, **6** (1): 68-72. (Helminth parasites included for the study are: Trematodes, Cestodes and Acanthocephalans).

Kharoo, V. K. (2011). Studies on Monostomes from the winter Migratory birds in Kashmir. Indian *J. Fundamental and Applied Life Sci*, **1** (3): 209- 216.

Kharoo, V. K. and Dhar, R.L. 1981(1982). On a new species of Trematoda, *Leucochloridium gallinuli* n.sp. from the Indian Moorhen, *gellinula chloropus* from Kashmir. *Indian. J. Helminthol.*, **33** (1): 15-19.

Khuroo, M. S. (1999). Hepato-biliary and Pancreatic ascariasis in endemic areas. *JK – Practioner*, **6**(2): 189.

Khuroo, M. S., Dar, M.Y., Yattoo, G.N. Zargar, S.A. Javaid, G. Khan, B.A. and Boda, M.I. 1993. Precutaneous drainage versus albendazole theraphy in hepatic hydatidosis: A prospective randomized study. *Gasteroenterology*, **104**: 1452-1459.

Khuroo, M. S., Datta, D. V. Khoshy, A., Mitra, S. K. and Chuttani, P. N. (1980). Alveolar hydatid disease of the liver, with Budd-Chiari syndrome. *Post-graduate Medical J.*, **653**: 197-201.

*Khuroo, M. S. (1996). Ascariasis. *Gastroenterol. North America*, **25**: 553-77.

*Khuroo, M. S. Wani, N. A. and Javid, G. *et al.* (1997). Precutaneous drainage compared with surgery for hepatic hydatid cyst. *N. Engl. J. Med.* 337 -.

Khuroo, M. S. and Zargar, S. A. (1985). Biliary ascariasis: Common cause of biliary and pancreatic disease in an endemic area. *Gastroenterology*, **88**(2): 417-423.

Khuroo, M. S., Zargar, S. A., Koul, P., Khan, B. A., Dar, M. Y. and Alai and M. S. (1992 *a*). Ascaris-induced acute pancreatitis. *British J. Surgery*, **79** (12): 1335-1338.

Khuroo, M. S., Zargar, S. A. and Mahajan, R. (1991). *Echinococcus granulosus* cysts in liver: Management with preceutaneous drainage. *Radiology*, **180** (1): 141-145.

Khurma, Uma, R. (1988). Plant-parasitic nematodes from some hilly areas of North-India. *Indian J. Nematol.*, **18** (1): 132-133.

*Khurshid, I. and Ahmed, F. (2012a). A short note on heavy infection of acanthocephalan worm (*Pomphorhynchus kashmirensis*) in alimentary canal of naturally infected snow trout, *Schizothorax niger* (Heckel). *Intl. J. Adv. Life Sciences*, **5** (1): 79-83.

*Khurshid, I. and Ahmed, F. (2012). Gastro-intestinal helminth infection in fishes relative to season from Shallabugh Wetland. *Intl. J. Recent Scientific Res.*, **3** (4): 270-272.

*Khurshid, I. and Ahmed, F. (2013). Epidemiological studies of helminth infected fishes of Shallabugh Wetland. *Intl. J. Curr. Res.*, **5** (6): 1346-1347.

*Khurshid, I., Ahmed, F. and Ahmad, T. (2013). Parasitic distribution in relation to gender, season and length of fish hosts in Shallabugh Wetland. *Intl. J. Scientific and Engineering Res.*, **14** (4): 1083 -1091.

*Khurshid, I. and Ahmed, F. (2014). Parasitic Burden of some fresh water fishes of River Sindh in Kashmir in relation to season. *Intl. J. Innovative Res. in Sci., Engineering and Technol.* **3** (8): 15688-15690 (on cestodes and acanthocephalans in *Schizothorax* spp.).

*Khurshid, I. and Ahmed, F. (2014). Population dynamics of parasites as evaluation metric to assess the trophic quality of fresh water bodies – A case study showing relationship infection level of helminths in *Schizothorax* spp. of River Sindh, Kashmir. *Intl. J. Fisheries and Aquatic Studies*, 2 (2): 206-209.

*Khurshid, I. and Ahmad, F. (2014). Prevalence of Helminth Parasite of *Schizothorax* from Shallbugh wetland and River Sindh in Kashmir.*Global Veterians*, **12** (5): 731-735.

Knight, K. L. and Stone, A. (1977). *A Catalogue of the mosquitoes of the world* (2nd *Edition*). The Thomas Say Foundation Vol.VI, Entomological Society of America, Maryland, U.S.A.

Koehsler, M., Walochnik, J., Georgopoulos, M., Pruente, C., Boeckeler, W., Auer, H. and Barisani-Asenbauer, T. (2011). *Linguatula serrata*, tongue-worm in Human eye, Austria. *Emerg. Infect. Dis.*, **17**(5): 870-2: doi: 10.3201/eid1705.100790.

Kotwal, S., Abrol, D. P., Dar, S. and Gandotra, A. (2013). Mite pests of Honey bee (*Apis mellifera*) and their seasonal incidence in Jammu Division of Jammu and Kashmir, India. *Bioscan,* **8** (2): 529-531.

Koul, P. A; Waheed, A; Hayat, M. and Sofi, B.A. (2000). Praziquantel in Niclosamide resistant *Taenia saginata,* infection. *Scandinavian J. Infectious Diseases,* **31** (6): 603-604.

Koul, P.L., Raina, M.K. Bambroo, P. and Koul, U. 1991(1992) *Pallisentis jagani* sp. nov. from *Channa channa* in Jammu, India. *Indian J. Helminthol.,* **43** (2): 124-28.

Koul, Ravinder, Kumar (1977). *Studies on helminth parasites of Reptiles of Jammu and Kashmir.* Ph. D. Thesis. P. G. Deptt. of Zool., Faculty of Sci., Univ. of Kmr., Sgr., Kmr. (J&K). 315 pp. and LXXX plates.

Koul, Tej, Krishni and Raina, M. K. (1989). A new copepod parasite, *Lernaea gambusiae* n.sp. (Lernaeidae. Lernaeninae) infecting the fish, *Gambusia affinis* in Kashmir. *Mysore J. Agric, Sci.,* **23** (4): 543-548.

Kour. R., Katoch, R., Godara, R. and Yadav, A. (2015). Incidence of myiasis and bionomics of Cyclorraphan flies in small ruminants of Jammu. *Indian J. of Small Ruminants,* **21** (1): 69-70.

Krishna – Prasad, K. S. (1986). Plant-parasitic nematodes of India, Problem and progress *In:* M Swarup, G. and Das – Gupta, D. R. (Eds.), Indian. Agricultural Research Institute, New Delhi.

*Kuchai, J. A., Ahmad, F., Chishti, M. Z., Tak, H., Javid, A., Ahmad, S. and Rasool, M. (2012). A study of morphology and morphometry of *Haemonchus controtus. Pakistan J. Zool.,* **44** (6): 1737-1741.

Kuchai, J. A., Ahmad, F., Chishti, M. Z., Dar, J. A. and Tak, H. (2013). On the Morphology and Morphometry of *Trichuris ovis* Abildgaard, 1795, recorded from ruminants of Ladakh. *J. Buffalo Sci,* **2** (1): 49-52.

Kuchai, J. A., Chishti, M. Z., Ahmad, F. and Rasool, M. (2010). Prevalence of trematode parasitic infestation in Yak of Ladakh *Bos grunniens. Pakistan J. Wildlife.* **1**: 64-66.

Kuchai, J. A., Chishti, M. Z. and Dar, S. A. (2011). First Report of *Dicrocoelium dendriticum* from sheep of Ladcakh (J&K), India. *World's Vet. J.,* **1** (1): 17-19.

Kuchai, J. A., Chishti, M. Z., Bhat, A. A. and Tak, H. (2011a). A Survey of Caprine Nematodiasis in Laddakh. *World's Vet. J.,* **1** (1): 1-4.

*Kuchai, J. A., Chishti, M. Z., Ahmad, F. and Mir, J. A. (2013). Impact of health status and species of the host on prevalence of helminthiasis in sheep and goats of Ladakh. *Intl. J. Agronomy and Plant Production,* 4 (5): 869-872.

Kuchai, J. A., Chishti, M. Z., Tak, H. and Lone, B. A. (2012). Faecal examination of Pashmina Goats (*Capra siberica*) of Ladakh for Nematode Infections. *Global J. Sci. Frontier Res. Biol. Sci.*, **12** (4): 37-40.

*Kuchai, J. A., Chishti, M. Z., Zaki, M. M., Ahmad, J., Rasool, M., Dar, S. A. and Tak, H. (2011). Prevalence of Nematode parasites in sheep of Ladakh, India. *J. Agric. Ext. Rural Dev.*, **3:** 229- 231, doi. 10.5897/JAED11099.

Kuchai, J. A., Chishti, M.Z., Zaki, M. M., Dar, S. A., Rasool, M., Ahmad, J. and Tak, H. (2011a). Some epidemiological aspects of Fascioliosis among cattle of Ladakh. *Global Veterinaria*, **7** (4): 342- 346.

Kuchai, J. A., Chishti, M. Z., Zaki, M. M., Ahmad, J., Rasool, M., Dar, S. T. and Tak, H. (2011b). Epidemiology of Helminth parasites in small Ruminants o f Ladakh. *Online J. Animal Feed Res.*, **1** (5): 239- 242.

Kulkarni, S. M. and Bhat, H. R. (1972). *Neopsylla kashmirensis*, sp. nov. from Kashmir, India. (Siphonaptera: Hystrichopsyllidae). *Oriental Insects*, **6** (4): 513-519.

*Kulkarni, S. M., Bhat, H. R. and Modi, G. A. (1978). Survey of Phelbotomid sandflies from the Himalayan region, India (Diptera) (Phlebotominae). *Indian J. Med. Res.*, **67**: 583-588.

Kullandar, S. O., Fang Fang Delling, B. and Ahlander, E. (1999). *Fishes of the Kashmir Valley*. pp.99-167, *In:* Nyman, L. (Ed.). *River Jehlum, Kashmir Valley Impact on the Aquatic Environment*. Published by Swedmar. The International Consultancy Group of the National Board of Fisheries, Sweden.

Kumar, A., Gabhane, G., Gogoi, D. and Raut, B. (2008). Coccidiosis in broilers: An outbreak in cold desert region. *J. Vet. Parasitol.*, **22** (1): 61-62.

Kumar, K., Rahman, S. J., Nim, S., Chirwatkar, K. R. and Kalra, N. L. 1983 (1984). A note on sandflies of Jammu. *J. of Communicable Dis.*, **15** (4): 287-289.

Kumar, R., Kumar, S., Khurana, S. K. and Yadav, S.C. (2013). Development of an antibody- ELISA for seroprevalence of *Trypanosoma evansi* in equids of North and North-westren regions of India. *Veterinary Parasitol.*, **196** (3-4): 251-257.doi. org./10.1016/j.vetpar.2013.04.018.

Kumar, S., Kishore, K. and Sharma, M. K. (2015). Live leech in the larynx. *Indian J. Medical Specialities.* Doi: http: //dx. org/10. 7713/ijms 2013.0015.

Kurahashi, H. and Okadome, T. (1976). A new species of *Polleniopsis* from Kashmir, India (Diptera: Calliphoridae). *Bull. Japan Entomol. Academy*, **9** (2): 42-44.

Lakhshminarayana, K.V. (1979). A synoptic list of Mallophaga Sen. Lat. (Phthiraptera: Insecta) from India and adjacent countries together with Host and Regional indices. *Recd. Zool. Surv. India*, **75**: 39-201.

Lateef, M., Zargar, Showkat, A., Khan, A. A., Nazir, M. and Showkat, Abid (2008). Successful treatment of niclosamide- and praziquantel resistant beef tapeworm infection with nitazoxanide. *Intl. J. Infectious Dis.*, **12** (1): 80-82.

Lazarevic, M., Milosevic, M., Stankovic, S.S. and Zikic, V. (2017). Morphological discrimination of the genera *Binodoxys* Mackauer and *Trioxys* Haliday (Hymenoptera: Braconidae: Aphidiinae), based on the general shape of forewing. *Biologica Nyssavia* 8 Doi- 10.5281/Zenodo 96447.

Lazo, R.F., Hidalgo, E., Lazo, J.E., Bermeo, M., Llaguna, M., Murillo, J. and Teixera, V.P.A. (1999). Ocular linguatuliosis in Ecuador: Case Report and morphometric study of the larva of *Linguatula serrata*. *American J. Trop. Med. Hygiene*, **60**: 405-409.

Leherwal, M. A., Yasin, S.B. and Ahmad, S.B. (2004). Diagnosis of Cutaneous leishmaniasis by FNAC- Report of three cases. *J. Cytol.*, **21**: 103-105.

Lehrer, A. Z. (2008). *Thyrsocnema kashmirensis* n.sp.- new Indian species of the genus *Thyrsocnema* Enderlein (Diptera: Sarcophagidae). *Fragmenta Dipterologica,* **15**: 24-25.

Lewis, D. J. (1978). The Phlebotomine sandflies (Diptera: Psychodidae) of the Oriental region. *Bull. British Mus.* (Nat. Hist.), **37** (6): 219-243.

*Lewis, R. E, (1998). Resume of the Siphonaptera (Insecta) of the world. *J. Medical Entomolgy*, **35** (4, 1): 337-389. http: //doi. org./10.1093/lmedent/35.4.377.

Lichtwardt, R.W. (1909). *Deut. Ent. Zeits*, p.124.

Logar, J., Marimi Fiser and Natasa (2008). Cutaneous myiasis caused by *Hypoderma lineatum*.*Wiener Klinische Wochenschrift*, **120**: 19-20.

*Lone, A. H., Ahmad, T., Anwar, M., Habib, S., Sofi, G. and Imam, H. (2011). Leech Therapy: A holistic approach of treatment in Unanai (Greeko- Arab) medicine. *Ancient Science of Life*, **31** (1): 31-35.

Lone, B. A., Ahmad, F. and Tak, H. (2012). Impact of helminth parasites on plasma proteins in children in Kashmir. *Intl. J. Adv. Res. Biotechnol.*, **1**: 5-7.

Lone, B. A., Ahmad, F. and Tak, H. (2012 b). Soil transmitted Helminthic infections in children of Kashmir Valley. *DAV Intl. J. of Sci.*, **1** (1): 72-74.

*Lone, B. A. Ahmad, F. and Chishti, M. Z. (2016). *Handbook of Veterinary Parasitology*. IP. Innovative Publication Pvt. Ltd.

Lone, B. A., Chishti, M. Z. and Ahmad, F. (2011a). Prevalence of Coccidia and Gastro-intestinal Nematode infections in Goats of Baramulla District of Kashmir Valley. *Global Veterinaria*, **7** (1): 27-30.

Lone, B. A., Chishti, M. Z., Bhat, F. A. and Tak, H. (2011b). Serodiagnosis of *Oesophagostomum columbianum* infection in goats, using indirect ELISA. *Veterinary World*, **4** (11): 503- 506.

Lone, Bashir. A., Chishti, M.Z., Ahmad, F. and Tak, H. (2012 a). A Survey of gastro-intestinal helminth parasites of slaughtered sheep and goats in Ganderbal, Kashmir. *Global Vet.*, **8** (4): 338- 341.

*Lone, B. A., Chishti, M. Z., Ahmad, F. and Tak, H. (2012). Immunodiagnosis of *Haemonchus controtus* infection in sheep by Indirect Enzyme Linked Immunosorbent Assay (ELISA). *Iranian J. Vet. Res.*, **13** (1) (Ser. No.38): 49-53.

*Lone, B. A., Chishti, M. Z. and Ahmad, F. (2013). Seasonal occurrence of gastro-intestinal parasites in sheep in Kashmir Division. *Intl. J. Livestock Res.*, **3** (2): 135-136 (Overall prevalence of 51.21 per cent of parasites, *viz. Moniezia, Avitellina, Fasciola, Dicrocoelium, Amphistomes* and *Paramphistomum*).

Lone, B. A., Chishti, M. Z., Ahmad, F., Tak, H., Bandh, S. A. and Khan, A. (2016). A field Survey on the Status of gastro-intestinal helminth parasites in Hangul (*Cervus elaphus hangul*) in Dachigam National Park of Kashmir. *J. Parasitic Dis.*, **40** (3) 750- 755.

*Lone, G. M., Wali-ullah, M. I. S., Zaki, F. A., Ali, A. and Mir, S. A. (2012 a). Occurrence of virus vector nematodes in disturbed, undisturbed and neglected apple ecosystem of Kashmir, India. *Indian J. Nematol.*, **42** (2): 92- 94.

Lone, G. M. and Zaki, F. A. (2013). Report of dagger nematode, *Xiphinema americanus* (Cobb, 1913) (Dorylaimida: Longidoroidea) from rhizosphere on apple trees in Kashmir (India). *SKAUST J. Res.*

Lone, G. M. and Zaki, F. A. (2014). Report of *Xiphinema insigne* (Loos, 1949) and *Longidorus elongatus* (de Man, 1876, Thorne and Swanger, 1936) (Dorylaimida: Longidoroidea) from rhizosphere of neglected apple trees of Kashmir, India. *Indian J. Nematol.*, **44** (2): 232-237.

*Lone, G. M. and Zaki, F.A. (2018). Nematode Pests as vectors of Plant viruses in Fruit Ecosystem. **In**: Nazeer Ahmad, S. A. Wani and W.M. Wani (eds.), pp. 299-314, Chap. 20, *Apple: Production and Value Chain Analysis*. Daya Publishing House, New Delhi.

Lone, G. M., Zaki, F. A. and Wali-ullah, M. I. S. (2012 b). Report of *Xiphinema index* Thornie and Allen 1950 (Dorylaimida: Longidoroidea) associated with neglected apple orchards in Baramulla, Kashmir, India. *Indian J. Nematol.*, **42** (2): 71- 74.

Lone, G.M., Zaki, F. A., Pathania, S. S. and Muhi, S. (2018). Population abundance and distribution of ectoparasitic plant nematodes associated with apple (*Malvus domestica*) in apple belts of Kashmir, India. *Intl. J. Curr. Microbiol. Appl. Sci.*, **7** (6): 3817-3824.

*Lone, R., Syed, K., and Lone, A. (2019). Recent pattern and risk factors of intestinal helminth infection among school childfren Kashmir, India. *Archives of Clinical Microbiol.*, **23** (2011): 2 (Sumbal area of district Budgam showed high incidence of helminthic infection, with very heavy load of parasites).

Luc, M. and Goodey, J. B. (1963). *Hirshmanniella* nom nov. for *Hirschannia Nematologica,* **9**: 471.

Luc, M., Sikora, R. A. and Bridge, J. (Eds.) (1990). *Plant Parasitic Nematodes in sub-tropical and Tropical Agriculture.* CAB International Willingford, U. K., Oxon.

Luc, M., Sikora, R. A and Bridge, J. (Eds.) (2005). Plant Parasitic Nematodes in Sub-tropical and Tropical Agriculture (**2nd Edition**). CABI Bioscience, Egham, U. K, CAB International Wallingford, U. K, .

*Mao, T. C. (1969). A Revised Checklist and concise Host Index of Hippoboscidae (Diptera). *Pacific Insect Monograph*, **20**: 261- 299.

Magnaval, J- Francois, Glickman, L. T., Dorchies, P. and Morassin, B. (2001). Highlights of Human Toxocariasis. *The Korean J. Parasitol.*, **39** (1): 1-11, doi: 10.3347kjp 2001.39.1.1.

Magray, A. R., Tak, I. R., Tak, H., Ganai, B. A., Chishti, M. Z., Dar, J.S. and Ahmad, F. (2017). Abattoir Survey in order to ascertain the prevalent Trematode parasites of cattle in North Kashmir. *Global Veterinaria*, **18** (3): 188-196.

Mahajan, A., Jasrotia, D. S. Charak, R. S., Kumar, T. Bhagat, P.L, Sharma, N. and Gupta, B. B. (2009). Scrub Typhus Jammu outbreak-2009. *JK Science*, **12** (2), April-June, 98-101.

Mahajan, D., Bhat, M.L., Singh, J.B. and Hans Deepa (2009). Visceral leishmaniasis in a native Kashmiri Boy. *J.K. Science*, **11**: 152-153.

Mahajan, R. (1970). Effect of light on the survival of *Hopolaimus* sp., the soil nematode surrounding the roots of *Zea mays* in Kashmir. *Indian J. Helminthol.*, **22**(1): 36-40.

Mahajan, R. (1971). A preliminary report on the occurrence of plant-parasitic and soil nematodes in Jammu and Kashmir State. *Kashmir Sci.*, **8** (12): 115-119.

Mahajan, R. (1972). On two soil nematodes, *Discolaimoides indicum* sp. nov. and *Pungentus juglensi* sp. nov. from around roots of *Juglans regta* in Kashmir. *J. Helminthol.*, **46**(2): 161-165.

Mahajan, R. (1972 *a*). *Studies on the soil and plant-parasitic Nematodes in Kashmir.* Ph. D. Thesis, Deptt. of Zool., Faculty of Sci., Univ. of Kmr., Sgr., Kmr.

Mahajan, R. (1973). Two new nematodes, *Tylenchus kashmirensis* n. sp. and *Dorylaimoides pari* n. sp. from Kashmir. *Rivista Parasit.*, **34**(3): 201-204.

Mahajan, R. (1974). *Tylenchorhynchus kashmirensis* sp. n. and *Quinisulcius himalayae* sp. n. (Nematoda: Tylenchorhynchinae) from India. *Proc. Helminthol. Soc. of Washington*, **41** (1): 13-16.

Maity, A., Mukhopadhyay, M., Naskar, A., Hazara, S. *et al.* (2015). A checklist of Indian Big-headed flies (Insecta: Diptera: Pipunculidae). *ZSI, e- publication, Zoological Surv. Of India*. DOI.10.13140/RGZ.1.5155.4809 Kolkata.

Maity, A., Naskar, A., Sengupta, P. P., Homechaudhuri, S. and Banerjee, D. (2016). An annotated checklist of Horseflies (Diptera: Tabanidae) from India with remarks on Surra disease vectors. *J. Zool-Studies*, **3**(4): 53-84.

Majidah, Rasheed and Khan, A.R. (1998). Seasonal population dynamics of some helminth parasites of fishes in Wular Lake, Kashmir. pp.34-40, *In: Parasitic Diseases-New Horizon* (*Proc.* 8th Natl. *Congr. Vet. Parasitol* and *Natl. Symp. On Molecular parasitol.*, held at CCS Haryana Agr. Univ., Haryana, from 9-11 Oct. 1998.

Mackauer, M. (1972). The aphid attacking genera of Aphelinidae (Hymenoptera), including new genus. *Canadian Entomologist*, **104** (11): 1771-79.

Makhdoomi, D. M., Nasreen, S., Bandy, S. D. and Moulvi, B. A. (1995). Incidence of different ovine gastro-intestinal parasites of Kashmir. *Indian Vet. J.*, **72** (8): 898-900.

Malhotra, Y. R. and Joyti, M. K. (1972). A new copepod parasite, *Lernaea kashmirensis* n. sp. (Lernaeidae, Lernaeinae) infecting stone loach in Kashmir. *Vest.* Cs. *Spol. Zool.*, **36** (2): 119-122.

*Malik, A. A., and Bari, Shams-ul (2019). *Human abdominal Hydatidosis*. Springer, 103 pp. https: / /doi. org./10.1007/978-132152-89.

*Malik, A. A., Wani, Rauf, A. and Bar, S. (2008). Acute acalculous cholecystis due to *Taenia saginata*. *Anu Saudi Med.*, **28** (5): 388-389 (On *T. saginata* from Kashmir).

*Malik, A. A., Bari, Shams-ul, Younis, M., Wani, K. A. and Rather, A. A. (2011). Primary splenic hydatidosis. *Indian J. Gastroenterol.*, **30** (4): 175-177.

Malik, G. M., Romshoo, G. J., Hussain, Ashaq, Ahmad Zubair, Basu, J. A. and Bhat, G. H. (1997). Hydatid Disease of liver, presenting as cholangitis. *JK-Practioner*, **4** (3): 194-96.

*Malik, G. M., Kamili, Azra, N., Chishti, M. Z., Tanveer, S., Ahad, S. and Johri, R. K. (2016). Synergistic approach for treatment of chicken coccidiosis using berberine- A plant natural product. *Microbial Pathogenesis*, **93**: 56-62.

Malla, A. K. (1981). *Morphology, Histology and Histochemistry of some cestodes.* Ph. D. Thesis. P. G. Deptt. of Zool., Faculty of Sci., Univ. of Kmr., Sgr., Kmr. (J&K).

*Malla, Nancy, Aggarwal, A. K. and Mahajan, R. C. (2002). A serological study of Human toxocariasis in north India. *The National Med. J. India*, **15** (3): 145-147.

Malla, N., Wattal, C. Khan, I., Kaul, P. and Raina, V. (1989). Study of trichomoniasis in Kashmir, North India. Indian *J. Med. Microbiol.*, **7**: 121-6.

Mandal, C. K. (2004). Endemic Leech-fauna of India. *Rec. Zool. India.*, **103** (Part 1-2): 103-110.

Mani, M. S. and Kaul, B. K. (1973). On some chalcidoidae from India. *In:* Mani, M.S., Dubey, O.P. Kaul, B. K. and Saraswat, C. G. *Mem. School. Entomol.* St. John College, Agra, U.P.

Mani, M. S. and Saraswat, G.G. (1973). On some Chalcidoidea from India. *Mem. School of Entomol.* **No 1973, 2**: 78-125.

Manickavasagam, S. and Athithya, A. (2018). An updated Checklist of Mymaridae (Hymenoptera: Chalcidoidea) of India. *J. Entomol. and Zool. Studies*, **6** (4): 1654-1663.

Manickavasagam, S. and Menakadevi, C. (2014). A checklist of Aphelinidae (Hymenoptera: Chalcidoidea of India. *Rec. Zool. Surv. India*, 114 (Part I), pp 13-34.

Mantoo, M. A., Zaki, F. A. and Bhat, M. A. (2008). Occurrence of distribution of plant parasitic and other nematodes around the rhizosphere of vegetable crops in

Kashmir. *In*: M.Z. Chishti and Fayaz Amad (eds.), pp.15-19. *Science For Better Tomorrow*. Printed at Universal Printers, Khanyar, Srinagar, Kashmir.

Maqbool, A. and Ahmad, I. (2016). Haematological response of snowbelt, *Schizothorax plagiostomus* Heckel, naturally infected with a new *Trypanosoma* species. *J. Parasit. Dis.*, **40** (3); 791-800, doi: 10.1007/s12639-014-0580S.

*Maqbool, I., Wani, Z. A., Allaie, I. M., Shahardar, R. A., Baba, F. A. and Ashraf, A. (2016). Parasitic- fauna infection in Pashmina goats of cold arid region of Ladakh, India. *Indian J. Animal Sci.*, **86** (1): 42-44.

*Maqbool, I., Shahardar, R. A., Wani, Z. A., Allaie, I. M. and Shah, M. M. (2018). Efficacy of antihelmintics against G. I. helminth of cattle in Pulwama District of South Kashmir. *J. Pharmacognosy and Phytochemistry*, **7** (5): 1214-1217.

Moravec, F., Scholz, T., Ash, A. and Kar, P. K. (2010). New data on the morphology and taxonomy of three species of *Rhabdochona* (Nematoda: Rhabdochonidae), parasitizing fishes in India. *Folio Parasitologica*, **57** (4): 95-306 (*Rhabdochona* (*R. hellichi turkestanicus* Skrybin), is synonym of *Rhabdochona kashmirensis* Thapar).

Masarat, S., Ahmad, F., Chishti, M.Z., Hamid, S. and Sofi, B.A. (2012). Prevalence of *Cryptosporidium* species among HIV positive asymptomic and symptomic immigrant population in Kashmir, India. *Iranian J. Microbiol.*, **4:** 35-39.

Masood, Qazi, Majid, Imran, Khan, D., Singh, G., Sameen, F., Chisti, M. and Qureshi, W. (2009). Cutaneous Leishmaniasis- A new phenomenon. *Kashmir Medical J.*, **3** (3): 395-397.

Masoodi, M. A., Bhagat, K. C. and Sofi, M. R. (1993). Toxicity of Insecticides to *Encarsia perniciosi* Tower and *Aphytis proclia* Walker. *J. Biological Control*, **7**: 37-39.

Masoodi, M. A; Bhagat, K. C. and Sofi, M. R. (1996). Preliminary observations on the natural enemies of San Jose Scale, infesting apple trees in Kashmir. *Shashpa*, **3**: 77-78.

Masoodi, M.A., Trali, A. R., Bhat, A. M., Tickoo, R. K. and Nehru, R. K. (1986). Incidence of Parasites of *Lymentria obfuscata* (Lymantriidae: Lepidoptera) in Kashmir. *Entomophaga*, **31** (4): 401-404.

Masoodi, M. A. and Trali, A. R. (1987). Seasonal history and biological control of San Jose Scale, *Quadraspidiotus perniciosus* Comstock (Diapididae: Homoptera) on apple in Kashmir. *J. Biological Control*, **1**: 3-6.

Masoodi, M. A. and Trali, A. R. (1987*a*). New records of Scale Insects (Diaspidae: Homoptera) from Kashmir. *Entomon.*, **12** (3): 227.

Masoodi, M. A., Trali, A. R. and Bhat, A. M. (1986). Further observations on the natural enemies of *Lymantria obfuscata* Walker (Lymantriidae: Lepidoptera) in Kashmir.*Entomon*, **11** (4): 251-254.

Masoodi, M. A., Trali, A.R., Bhat, A. M., Tickoo, R. K. and Nehru, R. K. (1986a) Incidence of *Lymantaria obfuscata* (Lymantridae: Lepidoptera) in Kashmir. *Entomophaga*, **31** (4): 401-404.

Masoodi, M. A; Trali, A. R., Bhat, A. M., Tikoo, R. K. and Nehru, R. K. (1989). Establishment of *Aphytis* sp. *Proclia* group on San Jose Scale in Kashmir. *Indian J. Plant. Prot.,* **31**: 71-73.

Masoodi, M. A., Trali, A. R., Bhat, A. M., Tikoo, R. K. and Nehru, R. K. (1989*a*). Establishment of *Encarsia* (*Prospaltella perniciosi*), a specific parasite of San Jose Scale, on apple in Kashmir. *Entomophaga,* **34** (1): 39-43.

Mathur, A. C. and Sharma, M. L. (1962). *Hadronotus* sp. (Scelionidae, Proctrotrupoidae) as a parasite of eggs of *Cletus bipunctatus* West (Corecidae: Hemiptera). *Indian J. Ent.,* **23**(4): 303.

Mathur, R. N. (1956). A new species of *Anastatus* Motschulsky (Chalcidoidea, Eupelmidae) from Kashmir. *Proc. R. Ent. Soc. London* (B), **25:** 93-97.

Mathur, V. K. and Prasad, S. K. (1971). Occurrence and description of *Hirschmanniella oryzae* in Indian Union, with description of *H. mangoriensis* n. sp. *Indian J. Nematol.,* **1**(1): 220-225.

Maviaux, J., Tkach, V., Vasileva, G. I. and Georgieva, B. B. *et al.* (2017). Cyclophyllidea van Beneden in Brown, 1900. *In*: J. H. Caira and K. Jensen (eds.), *Planetary Biodiversity Inventory (2008-2017): Tapeworms from vertebrate Bowls of Earth.* No. 25, Chap. 6, pp. 77- 148. University of Kanas, Natural History Museum, Lawrence KS, USA.

Mehra, H. R. (1937). Certain new and already known distomes of the family Lepodermatidae Odhuer (Trematoda) with discussion on the classification of the family. *Zeitschr. Parasit.,* **9**: 429-469.

Mehta, U. R. and Raski, D. J. (1971). Revision of genus *Criconema* Hofmanner and Menzel, 1914 and other related genera (Criconematidae: Nematodes). *Indian J. Nematol.,* **2**: 145-198.

Mehlhorn, M. (Ed.) (2009). *Encyclopedia of Parasitology.* (2nd Edition) Springer, Berlin, Heidelberg, Germany.

Mehmood, S., Nashir-uddullaha, N., Ahmed, J. A. and Borkataki, S. (2019). Parasitic affections of domestic Pigeon (*Columba livia*) in Jammu. *Ann. Parasitol.,* **65** (1): 53-64, doi: 10.17420/ap6501.182.

Mir, A. S., Pandit, B. A., Shahardar, R. A. and Banday, M. A. (1993) Prevalence of ectoparasites of indigenous fowl of Kashmir. *Indian Vet. J.,* **70** (11): 107-172.

Mir, A. S., Shahardar, R. A., Pandit, B. A. and Ahmad, M.R. (1996). Occurrence of histomoniasis (entrohepatitis) in hen reared partridges (*Gracia eletoris*). *Indian Vet. J.,* **73** (1): 98-99.

*Mir Abdul Aziz (1992). *Studies on the Parasitoid complex of Galliform, Anseriform and Podicipediform Birds of Kashmir.* Ph. D. Thesis. Deptt. of Zool., Faculty of Sci., Univ. of Kmr., Srg, Kmr. (J&K).

Mir, I. A., Wani, N.A. Ahangar, A. G. Shah, O. J., Saleem, K. and Patnaik, Rekha (2002). Surgical ascariasis in children. *JK-Practioner,* **10** (1): 17-21.

*Mir, M. A., Chishti, M.Z., Zargar, M. A, and Tak, H. (2003).Effect of *Haemonchus contortus* on haematological Profile and eye colour in sheep. *J. Res. Dev.*, **8**: 69.

Mir, M. A., Raina, Puran, Matto, T.K. and Sethi, A. S. (1979).Incidence of Giardiasis in Kashmiri Children. *Indian. J. Pediatr.*, **46**: 358-361.

Mir, M. R., Chishti, M. Z., Rashid, M., Dar, S. A. and Katoch, R., Kuchai, J. A. and Dar, J. A. (2013). Point prevalence of gastro-intestinal helminthiasis in large ruminants of Jammu, India. *Intl. J. Scientific and Research Publ.*, **3** (3): 1-3.

*Mir, M. R., Chishti, M. Z., Dar, S. A. and Rashid, M. (2013). Bionomics of helminth parasites in goats of sub-tropical Jammu area of J&K, India. *Agric. Biol. JN Am*, **4** (3): 316-318 (Necroscopic examination revealed 72. 88 per cent of parasitic infection).

*Mir, M. R., Chishti, M. Z., Rashid, M., Dar, S. A., Katoch, J. K., Mehraj, M., Dar, M. A. and Rasool, R. (2013). Incidence of gastro-intestinal nematodosis in sheep of Jammu. *Trends Parasitol. Res.*, **2** (1): 1-4.

*Mir, M. R., Chishti, M. Z., Dar, S. A., Katoch, R., Rashid, M., Ahmad, F. and Tak, H. (2013). Incidence and histopathological study of Monieziosis in goats of Jammu (J&K), India. *Cibtech. J. of Zool.*, **2** (1): 19-23.

*Mir, M. R., Chishti, M.Z., Rashid, M., Katoch, R., Mehraj, M., Dar, M. A. and Rasool, R. (2013). The epidemiology of caprine fascioliasis in Jammu (J&K), *India. Intl. J. Food, Agri. and Vet. Sci.*, **3** (1): 233-237 (During the period 2007-2008, in Jammu region, from slaughter houses, out of a total of 284 livers, 15 showed Fasciola hepatica infection i.e 5.28 per cent prevalence).

Mir, M. S., Darzi, M. M., Kamil, S. A., Nashiruddullah, and Saleem, N. I. (2006). Pathology of *Taenia pisiformis* Bloch infestation in Angora rabbits. *J. Vet. Parasitol.*, **20** (2): 129-132.

Mir, M. S., Darzi, M.M., Hussain, I. and Wani, S. A. (2009). Concurrent occurrence of visceral linguatulosis and paratuberclosis in alpine cross goats (*Capra hirus*). *Veterinarski. Arhiv.*, **79** (3): 301-314.

Mir, M. S., Darzi, M. M, Mir, A. Q. and Malik, H.U. (2004). Report on occurrence of *Linguatula serrata* in cow in Kashmir. *Proc Natl. Symp. On Advances in Pathological Techniques in Diagnosis of animals, Birds and Fish Diseases* and XXI. Annl. Confr. Of IAVP, 23-25 Nov., West Bengal, Univ. of Animal and Fishery Sciences, Kolkata, West Bengal, p.88.

Mir, M. S., Darzi, M. M., Mir, A.Q., Malik, H. U. and Mustafa, M. (2012). Pathomorphology of bovine linguatulosis (*Linguatula serrata*). Indian *J. Vet. Pathol.* **36** (1): 10-14.

Mir, M. R., Chishti, M. Z., Dar, S. A. and Rashid, M. (2012). Bionomics of helminth parasites in goats of sub-tropical Jammu area of Jammu and Kashmir. *Agriculture and Biol. J. North. America,* **4** (3): 316-319.

Mir, M. R., Chishti, M.Z., Rashid, M., Dar, S. A., Katoch, R., Kuchai and Dar, J. A. (2013 b). Point prevalence of gastro-intestinal helminthiasis in large ruminants of Jammu, India. *Intl. J. Scientific and Res. Publ.*, **3** (3): 1-3.

*Mir, M. R., Chishti, M. Z., Rashid, M., Dar, S. A., Katoch, R., Khajuria, J. K., Mehraj, M., Dar, M. A and Rasool, R. (2013). Incidence of gastro-intestinal nematodosis in sheep of Jammu. *Trends Parasitol. Res.*, **2** (1): 1-4.

*Mir, M. R., Chishti, M.Z., Rashid, M., Dar, S.A., Katoch, R., Khajuria, J. K., Mehraj, M, Dar, M. A. and Rasool, R. (2013 d). Incidence of gastro-intestinal Nematodosis in sheep in Jammu. *Trends Parasitol. Res.*, **2** (1): 1-4. (Necroscopic examination made on four nematodes, *viz. Trichuris ovis, Haemonchus controtus, Dictyocaulus filarae* and *Chabertia ovina*, showing overall prevalence as 44.74 per cent).

Mir, M. R., Chishti, M. Z., Rashid, M., Dar, S. A., Kuchai, J. A. and Dar, J. A. (2013). Prevalence of gastro-intestinal nematodes in goats of Jammu region. *Intl. J. Sci. Res.*, **4** (3): 208-2010.

Mir, M. R., Chishti, M.Z., Zargar, M. A., Tak, H. and Dar, F. A. (2008). Seasonal prevalence of trematode parasites of sheep (Ovis aries) in Kashmir Valley, India. *Nigerian J. Parasitol.*, **29** (2): 80-83.

*Mir, R. A., Chishti, M. Z., Zargar, M. A., and Tak, H. (2008). Effect of *Haemonchus controtus* on Haematological Profile and eye colour in sheep. *Res. and Dev.*, **8** (2008): 71-75. (The study revealed a marked reduction in haematocrit, haemoglobin and RBC count, besides pale colour of eyes, in infected sheep).

*Mir, R. A., Chishti, M. Z., Zargar, M. A., Tak, H. and Ganie, S. A. (2009). Protective Immune response of Kashmiri native sheep to *Haemonchus controtus* Induced by parasite membrane associated with antigens. J. *Res. and Dev.*, **9**; 78-83. (This study has shown that *H. controtus* membrane-associated proteins can induce better protective response in sheep).

*Mir, S. A., Yatoo, M. I., Kanwar, M. S. and Dar, R. (2008). Ectoparasites prevalence in Pashmina goats in Changthang: A pastoral nomadic area of Ladakh. *The Indian J. Animal Sciences*, **88** (3). (On ectoparsites, *viz.* lice, ticks, fleas, mites. Lice showed common occurrence and ticks were present in less number, besides ectoparasite prevalence shown to be higher in young goats and females, during autumn and winter season).

Misger, M. S. and Mufti Mahmood (2000). Hydatidosis: a surgical jargon. *Surgery*, **5** (6): 37-38.

Mishra, A. C. (1981). The hoplopleurid lice of Indian sub-continent. *Rec. Zool. Survey India.* (Mis. Pub.), *Occasional Pap.* No. 21. Issued by Director Zoological Surv. of India.

Mishra, A. C., Kulkarni, S. M. and Bhat, H. R. (1973). *Hoplopleura himalayana*, sp. n. (Anoplura: Hoplopleuridae), parasitizing *Apodema flavicollis* in the Himalayan region of India. *Oriental Insects.*, **7**(4): 501-506.

Mittal, V., Illahi, I, Dhar, A. and Khan, M. A. (2011). Natural enemies of mulberry pyralid, *Glyphodes pyloalis* Walker (Lepidoptera: Pyralidae) in temperate climate of Kashmir. *J. Biol. Control*, **25** (1): 55-57.

Mitra, A. (1904). *A Report on the outbreak of Plague in Kashmir from 19th November 1903 to 31st July, 1904.* Published by the authority: Chief Medical Officer, Kashmir; Publisher, Central Jail Press.

Mitra, B., Roy, S., Haldar, S., Bhaumik, S., Biswas, O. and Mallick, K. (2015). Current status of thick headed flies (Conopidae: Diptera) in India. *J. Zool. Sudies,* **2** (4): 29-37.

Mitra, B. and Sharma, R. M. (2010). *Checklist of Indian Blackflies (Insecta: Diptera: Simuliidae).* Zoological Survey of India. www.zsi.gov.in/checklist/blackflies.

Mitra, B., Roy, Sankarsan, Roy, Souradip and Bhaumik, S. (2017). A current status on Diversity and Distribution of Blackflies (Insecta; Diptera: Simuliidae) in India. *Ambient Science,* **4**(1): online, DOI: 10.21276/ambi.2017.04.rv01.

Mitra, R. D. (1959). Notes on sandflies of Punch and Riasi Districts of Kashmir. *Z. Tropenmed Parasit.* **10**(1): 56-66.

Mitra, R. D. and Roy, D. N. (1953). Notes on sandflies Part III. *T. Tropenmed Parasit.,* **88**: 324-326.

*Mohammad, M., Wani, A. R., Khan, S. H. and Abbas, A. (2016). Pest complex of Potato (*Solanum tuberosum* L.), with special reference to management of white grubs in northern District of Kashmir Valley. *J. Agricultural Sciences,* **8** (3): 151-161. (Plant parasitic nematodes reported from different districts, Baramulla, Kupwara and Bandipore are: *T. kashmirensis, Hoplolaimus indicus* (= *Basirolaimus indicus, H. dishystera, Pratylenchus, Helicotylenchus indicus* and *X.basiria*).

Mohan- Rao, C.V.R. (1986). Prevalence of Crimean haemorrhagic fever Cango- virus in Jammu, and Kashmir State. *Indian J. Med. Res.,* **84**: 134.

*Mohi-uddin and Khan, Majeehur Rahman (2014). Root-knot nematode: ignored soil-borne plant pathogens causing root diseases in chickpea. *European J. Biotechnology and Bioscience,* **2** (1): 4-10.

*Moien, J. M., Burhan, R., Mir, M., Bhat, M. A., Wani, N. M and Bhat, M. A. (2017). Prevalence of gastro-intestinal nematodes in sheep and goats reared by nomads of Jammu region of J&K. *Intl. J. Livestock Res.,* **8** (8): 219-225. (Prevalence of nematode infection was reported to be 65.04 per cent and 70.55 per cent in sheep and goats respectively).

Moore, J. P. (1924). Notes on some Asciatic Leeches principally from China, Kashmir and British India. *Proc. Acad, nat. Sci. Philad.,* **76**: 343- 388.

Morley, Claude (1913). *The fauna of British India, including Ceylon. Hymenoptera – Vol. III. Ichneumonidae. Ichneumones* Deltoidei. Taylor and Francis, London, England.

*Munshi, N. A., Dar, G. H., Shawl, A. S., Amman-ullah and Dar, G. H. (2001). Report on occurrence of some mushroom nematodes in Jammu and Kashmir. *Appl. Biol. Res.* (CAAS, Shalimar, Kashmir), **3**: 64.

Munshi, Y. (2005). Prophylactic and Curative treatment of Zarabatul salaj (Frostbite) by Regimental Therapies. *Hippocratic J. Unani Medicine.* **1** (1): 21-25.

*Mustaq, S., Dogra, D. and Dogra, N. (2016) Clinical Response with intralesional Amphotericin B in the treatment of old world cutaneous leishmaniasis: a preliminary Report. Dermatologic Therapy, 29/6: 398-405, Doi: 10.1111/dth.12377. (Five consecutive patients diagnosed with CL at the Deptt. of Dermatology Govt. Medical college, Jammu and were treated with Amphotericin B (2.5 mg).

Mustaq. P., Naz, T. and Ahmad, W. (2007). Description of new and known species of the superfamily Tylencholaimoidea (Nematoda: Dorylaimidae) from Jammu and Kashmir, India. *J. of Nematode, Morphology and Systematics*, **10**: 11-29.

*Nabi, S., Tanveer, S., Ganaie, S. A., Ahad, S., Niyaz and Abdullah, I. (2015). Acanthocephalan infestation in fishes: A Review. *The J. Zool. Studies*, **2** (6): 32-37.

Nabi, Tauseef, Nadeema, Rafiq and Jameel, Imran (2015). Visceral leishmaniasis: very rare cases report from Kashmir Valley (J&K), with literature Review. *Worldwide J. Multidisciplinary Research and Dev.*, **1** (6): 1-5.

Nagaraja, H., Ahmad, M. J. and Gupta, A. (2007). Description of new species of *Trichogramma* (Hymenoptera: Trichogrammatidae) from Anantnag, J&K, India. *Indian J. Biol. Control*, **21** (2) 287-289.

*Naik, I. M., Wani, Z.A., Shahardar, R. A., Allaie, I. M. and Shah, M. (2018). Prevalence of gastro-intestinal helminth of cattle in Sourth Kashmir. Indian *J. Animal Sci.*, **88.** (Cattle of Pulwama showed the prevalence of Paramphistomes, followed by *Fasciola* spp. and *Dicrocoelium* spp. Among cestodes, *Monezia* spp. showed the prevalence. Nematodes included *Strongyle*, followed by *Trichuris* spp. and *Toxocara vitulorum*).

Naik, S. R., Rao, P. N., Datta, D., Mehta, S. K., Mahajan, R. C., Mehta, S.K. *et al.* (1979). Kalazar in North-westren India- A study of 23 patient.s. *Trans. Roy. Soc. Trop. Med. Hyg.*, **73**: 61-65.

Nair, C. P. (1973). Malaria in Kashmir Province of Jammu and Kashmir State – India. *J. Communicable Dis.*, **5** (1): 22-46.

Najar, I. and Bashir, A. (2016). Insect Diversity of Dodhpathri (Budgam), Jammu and Kashmir, India. *Intl. J. Fauna and Biol. Studies*, **3** (6): 28-32.

Nandi, N. C. (1986). *Leucocytozoon squamatus* sp. n from scaly-bellied green wood-peeker, *Picus squamatus squamatus* Vigora. *Acta Protozool.*, **25**(2): 223-226.

Nandi, N. C. (1989). Blood parasites of some Indian Birds. *Proc. Zool. Soc. Calcutta*, West Bengal, **40**: 49-50.

Nandi, N. C. (1999). Blood parasitism in wild Indian birds. *Rec. Zool. Surv. India* (Part I): 27-41.

Nandi, N. C. and Das, A. R. (2010). Synoptic list of Protozoans from India and their assessment of Taxonomic Diversity. Part I. Parasitic Protozoans. *Zoolgical Survey of India. Occ. Pap. No. 319*, pp 1-56, published by the Director, Zoological Survey of India, Kolkatta, India.

Nandi, N. C. and Mandal, A. K. (1978). Studies on some avian haematozoa from Jammu and Kashmir. *Bull. Zool. Surv. of India.*, **1** (2): 103-106.

*Narain, Kanwar (1993). *Protozoan and Platyhelminth parasites of some aquatic vertebrates of J&K.* Ph. D. Thesis, Faculty of Sci., P. G. Deptt. of Zool., The Univ. Kmr., Sgr., Kmr. (J&K).

Narain, Kanwar and Raina, M.K. (1992). A new report of 'White spot disease' in fresh-water fish of Jammu and Kashmir, with a note on its histopathology. pp.115-120., *In: Proc. Curr. Trends in Fish and Fisheries Biology and Aquatic Ecology (Proc. 3rd. Natl, Semr on Fish and their environment,* held at Srinagar from 19-22 September, 1989. P.G. Deptt. of Zool. The Univ. of Kashmir, Srinagar, Kashmir.

Narayanan, E. S. (1957). Phenomenon of insect parasitism and their practical utilization in Biological Control of insect pests. *Proc. 44th. Sess. of Indian Sci. Congr.* **2**: 23.

Narayanan, E. S. (1961). New records and description of new species of San Jose Scale parasites *Proc.Natl. Inst. Sci.* (B) *Silver Jublee* Suppl. 1960 (B): **26**: 20-29.

Narayanan, E. S. and Lal, Kundan (1954). Studies on Indian Ichneumonidae (Hymenoptera) Part-II, Sub-family Pimplinae, Tribe Pimplini. *Indian J. Entomol.,* **16**: 282-296.

Narayanan, E. S. and Lal, Kundan (1955). Studies on Indian. Ichneumonidae (Hymenoptera) Part-IV, Sub-family Pimplinae, Tribe Ephialtini, *Indian J. Entomol.,* **17**.

Narayanan, E. S., Rao, V. P. and Subba Rao, B. R. (1967). Advances made in Insect Parasitology and biological control of Pests in India. *In: Agricultural Entomology* ICAR, New Delhi. Pp.205-246.

Narendran, T. C. (1986). Family Chalcididae. **In:** Subba Rao, B.R. and Hayat, M. (eds), pp.11-42, The Chalcidoidea (Insecta: Hymenoptera) of India and adjacent countries. *Oriental Insects,* 20: 1-430 pp.

Narendran, T. C. (1994) *Torymidae and Eurytomidae of Indian sub-continent* (Hymenoptera: Chalcidoidea). Zoological Monograph, Deptt. of Zoology, University of Calicut, Kerala, 500 pp.

Narendran, T. C. (2007). Indian Chalcidoid parasitoids of the Tetrastichinae (Hymenoptera: Eulophidae). *Rec Zool. Surv. India, Occ.* Paper No.272: 233-250.

Narendran, T. C. (2009). A Review of the species of *Anastatus* Motschulsky (Hymenoptera: Chalcidoidea: Eupelmidae) of the Indian sub-continent. *J. Threatened Taxa,* **1**(2): 82.

Narendran, T. C. (2011). *Fauna India and adjacent countries: Eulophinae (Hymenoptera: Eulophidae),* pp. 192-7, published by Director, Zool. Surv. of India, Kolkatta.

Narendran, T. C., Buhroo, A. A. and Chishti, M. Z. (2001). Taxonomic studies on four species of Chalcidoidea (Hymenoptera) of economic importance from Kashmir. *Entomon,* **26** (2): 147-159.

Narendran, T.C., Kumar, G. and Vastrad, A.S (2010). Two new species of *Megastigmus* Delman (Hymenoptera: Torymidae) from India, with revised key to the Indian species. *Records of Zool Surv. India,* **110** (2): 5.

Narendran, T. C. and Razak, N. (2010). A new species of *Torymus* Dalman (Hymenoptera: Torymidae) from India, with checklist of *Torymus* species in India and adjacent countries. *J. Bioscience,* **18:** 148-151.

Narendran, T. C., Razak, N. and Sureshan, P. M. (2013). A taxonomic Review of *Chrysocharis* Foerster (Hymenoptera: Eulophidae) of Indian Sub-Continent. *Rec. Zool. Suv. India,* 113 (Part 2): 13- 34.

Nashir-uddulah, Ahmad, Nawab, Ara, J., Borkataki, S., Pathak, A. K, Goswami, P. and Chowdhary, A. R. (2016). Cestodiasis in Scaly-bellied Woodpecker (*Picus squamatus*) of Kashmir. *Bhartiya Krishi Anushandhan Patrika,* **31** (1): 74-77.

*Nashir-ddulah, N., Darzi, M. M., Mir, M. S., Kamil, S. A. and Shahardar, R. A. (2005). Recovery of *Dictyocaulus* species from the lungs of a Kashmir red deer (*Cervus elaphus hanglu*). *The Vet. Record,* **157**: 591.

*Nashir-uddulah, N., Darzi, M. M., Shahardar, R. A., Kamil, S. A., Mir, M. S. and Mir, M. (2007). Pathology of spontaneous *Dictyocaulus* sp. infection in hangul (*Cervus elaphus* hangul), sheep and goat. *J. Vet. Parasitol.,* **21** (1): 37-40.

Narasimhan, M. V., Panda, V., Mohanty, I., Sahni, S., Padhis, S. and Dash, M. (2013). *Dipylidium caninum* infection in a child: A rare case report. Indian *J. Med. Microbiol.,* **31** (1): 82-84.

Narasimahan, D. (1962). Field survey of insecticide susceptibility of disease vectors. II. *Pediculus humanus corporis* in Srinagar. *Armed Forces Med. J. India.* **18:** 418-419.

*Nasreen, Shugufta, Jeelani, S, and Hakeem, M. (2005). Incidence of gastro-intestinal nematodes in sheep in Kashmir valley. *J. Vet. Parasitol.,* **19**: 27-29. (Study has shown positive infection prevalence of five nematodes, *viz. Strongyles, Trichostrongylus, Haemonchus, Nematodirus* and *Marshallagia,* with highest infection for all as 67.14 per cent in summer and lowest in winter (44.31 per cent).

Naveed, S., Gupta, V., Kapoor, H., Quari, H., Altaf, A. and Para, M. (2015). Liver abscess in tropics: an experience from Jammu. *Scott. Med.J.,* **59** (3): 167-71, doi: 10.1177/0036933014543049.

*Nazir, Sumayya, Ali, M. N. and Chishti, M. Z. (2013). Purification and characterization of *Pomphorhynchus kashmirensis* somatic antigens collected from naturally infected local fish, *Schizothorax* spp. *New York Science J.,* **6** (7): 24-28. (Immunological studies have been conducted on *P. kashmirensis,* to reveal the nature of their somatic antigens by affinity chromatography and SDS PAGE and their antigenic properties were also confirmed by Ouchterlony double diffusion (ODD).

Nazir, T., Katoch, R., Godara, R., Yadav, A. and Pandey, V. (2014). Observations on Buffalo Sarcoptic mange in Jammu. India. *Buffalo Bull.,* **33** (3): 308-315.

Nazir, T., Katoch, R., Godara, R. and Yadav, A. (2013). Efficacy of eprinomectin pour on against *Rhipicephalus* (*Boophilus*) *microplus* on buffaloes. *J. Parasit. Dis.*, **37** (2): 166-167.

Nazir, T., Katoch, R., Godara, A. and Yadav, A. (2013). Comparative efficacy of pour-on preparation of eprinomectin and ivermectin against ascarids in buffalo calves. *Indian Vet. J.*, **90**: 79-80.

Nazir, T., Katoch, R., Yadav, A. and Godara, R. (2016). Comparative efficacy of pour-on eprinomectin and Ivermectin against *Sarcoptes scabiei* in buffaloes. *J. Parasitic. Dis.*, 40: 359-361.

Nehru, R. K., Bhagat, K. C. and Kaul, V. (2005). Nematode-fauna associated with citrus rhizosphere in Jammu and Kashmir. *Annals of Plant Protection Sciences*, **13** (2): 524-525.

Niewiadomska, K. (2002). Superfamily Diplostomoidea Poiner, 1886. ***In:*** Gibson, D.I., Jones, A. and Bray, R.A. (eds.). *Keys to the Trematoda*. Vol. 1, pp. 159-166, Wallingford: CABI Publishing and the Natural History Museum.

Nighat, M., Malik, M. A., Dar, H. U., Sofi, M. A. and Malik, Farooq (2002). Preliminary studies on the natural enemies of *Glyphodes pyloalis* Walker (Pyralidae: Lepidoptera), a serious pest of Mulberry trees in Kashmir. *Oriental Science*, **7**: 111-113.

Nikam, P.K. (1980). Studies on Indian species of *Enicosphilus* Stephen (Hymenoptera: Ichneumonidae). *Oriental Insects*, **14** (2): 131-23.

*Pandit, B. A., Mir, A. S., Banday, M. A.A. and Shahardar, R. A. (1991). Prevalence of Helminth parasites nin indigenous fowls of Kashmir valley. *Poultry Advisor*, **24 (10)**: 37-39.

*Pandit, B. A., Shahardar, R. A., Bhat, A. S. and Darzi, M. M. (2003). Prevalence of gastro-intestinal parasitic infection in sheep in Kashmir Valley, under different management practices. *Appl. Biol. Res.*, **5** (11): 1-5.

*Pandit, B. A., Shahardar, R. A., Banday, M. A., Darzi, M. M and Mattoo, F. A. (2004). Gastro-intestinal helminth parasites of cattle of Kashmir. *J. Veterinary Parasitol.*, **18** (1): 63-65.

Pandit, B. A., Shahardar, R. A., Darzi, M. A., and Jeyabal, L. (2007). Parasitic infestation in Fox (*Vulpus vulpus*). *J. Veterinary Parasitol.*, 21: 97-98.

Pandit, B. A., Sharardar, R. A. and Jeyabal, L. (2008). Prevalence of gastro-intestinal parasitic infestation in Equines of Kashmir. *Vet. Scan*, **3** (1): Article 22.

Papp, J. (1985). Three new *Blacus* Nees, 1818 species from India (Hymenoptera; Braconidae: Blacinae). *Annl. Historica- Naturales Musci Nationalis Hungarica*, **77**: 227-232.

Parashad, R. and Duggal, K. K. (1966). Drosophilid Survey of India III. The Drosophilidae of Kashmir. *Res. Bull. (N.S.) Panjab Univ.*, 179(Part III-IV): 277-290.

Patton, W. S. (1922). Notes on the myiasis- producing Diptera of man and animals. *Bull. Entomological Res.*, **12**: 239-261.

Pont, A.C. (1986). Family Finnidae, pp.41-54. *In*: A. Sors and L.Papp (Eds.).*Catalogue of Palaearctic Diptera 11: Sarcophagidae- Hypodermatidae*. Akademiai Kiado, Budapest, 346 pp.

Puent, S., Otranto, D., Pandero, R. *et al*. (2010). First diagnosis of an imported human myiasis, caused by *Hypoderma sinense* (Diptera: Oestridae), detected in European Traveller returning from India. *J. Travel Medicine*, **17**(6): 419-423.

Pedbidri, V. S., Thakre, J.P., Risbud, A.R., Joshi, G.D., Singh, A., Mavale, M.S. and Gupta, V.P. (1996). An outbreak of dengue haemorrahagic fever in Jammu. *Indian J. Virology*, **12**: 83-87.

Pampori, R. A., Wani, Asif, Latoo, M. and Hakeem, A. (2003). Hydatid cyst in the neck – An unusual presentation. *JK-Practioner*, **10**(2): 130-131.

Pande, B. P. (1937) *Prostocus himalayai* n.sp., a frog trematoda (Lecithodendriidae). *Proc. Nat. Acad. Sci.*, **7:** 202-204.

Pande, B. P. (1938). The trematode genus *Allocreadium* in North India fresh water fishes. *Proc. Ind. Acad. Sci.*, **7**: 54-60.

*Pandit, B. A. and Mir, A. S. (1988). Prevalence of Coccidial (Eimeria sp.) infection in sheep of Jammu and Kashmir. *Indian Vet. J.*, **65**: 669-672.

Pandit, B.A. (2006). Prevalence of Caprine coccidiosis in Kashmir valley. *Indian J. Small Ruminants*, **12** (2): 223-226.

Pandit, B. A (2009). Prevalence of coccidiosis in cattle in Kashmir. *Vet. Scan*, **4** (1), Article no.33.

Pandit, B.A., Mir, A.S. and Banday, M. A. A. and Shahardar, R.A. (1991). Prevalence of Helminth parasites in indigenous fowls of Kashmir Valley *Poultry Advisor*, **24**: 37-39.

Pandit, B. A., Shahardar, R. A., Darzi, M. M., Banday, M. M. A. and Bhat, A. S. (2003). Survey of gastro-intestinal nematodes in sheep in Kashmir. *Indian J. of Small Ruminants*, **9**(1): 39-42.

Pal, T. K. (1991). On a collection of Bombyliidae (Diptera) from India and Pakistan. *Rec. Zool. Surv. India*, **89** (1-4): 277-292.

Palaq, Langer, Seema and Noorani, I. A. (2016a). Histology of gills of *Labeo rohita* infested with crustacean parasite, *Argulus* sp. *Intl. J. Adv. Res.* **4** (7): 1387-1392.

Palaq, Langer, S. and Noorani, I. A. (2016b). Susceptibility of *Lebeo rohita* and *Tor putitora* to a crustacean parasite, *Argulus foliaceous. Intl. J. of Recent Scientific Res.*, **17** (3): 9788-9792.

Parrah, J. D., Buchoo, B. A. and Moulvi, B. A. (2004). Ocular filariosis in equines: A case Report. *Centaur*, **4**: 70-71.

*Parry, F. Q. (2017). Hydatid Disease. *J. Surgery and Emergency Medicine*, **1** (2): 1-2.

Parry, F. Q., Ahmad, S. Z., Sherwani, A. Y., Chowdri, N. A. and Wani, K. A. (2010). Paraspinal Hydatid cyst: A Rare presentation of Echinococcosis. *Intl. J. Surgery*, **8** (5): 404-406.

Parry, F. Q., Wani, S. N., Bazaz, S., Khan, S. R. and Malik, N.S. (2011). Primary Pelvic hydatid Cyst: A case Report. *Case Reports in Surgery*, **vol. 2011**, Article ID 809387.

Patheja, J. S. (1963). Investigation Report of an epidemic of louse-borne typhus among Tibetan Rufugees at Leh (Eastern Ladakh) during July, 1961. *Armed Forces Med. J.* (India), **19**(3): 120-124.

Pawar, A. D. and Parry, M. (1989). Record of natural enemies of important fruit pests in Ladakh (J&K). *Ind. J. Plant Prot.*, **17**: 291-293.

Pawar, A. D., Tuhan, N. C., Balasubramanium, S. and Parry, M. (1981). Biological control of codling moth in Ladakh. *Indian J. Plant Prot.*, **8**(2): 189-191.

*Peerzada, M. Y. (1997). *Population dynamics of helminthes of some snow-fed streams of Kashmir*. Ph. D. Thesis P. G. Deptt. of Zool., Faculty of Sci., Univ. of Kmr., Sgr., Kmr. (J&K).

Peerzada, M. Y. (2009). Population studies of *Camallanus fotedari* Raina and Dhar in its hosts, *Nemachielus kashmirenis*. **In**: pp. 175-186, Veena Tandon, A. K Yadav and Roy Bishnupada (eds.), *Current Trends in Parasitology, Proc. 20th National Cong. Parasitology (Nov. 3-5, 2008)*, Published by Panima Publishing Corporation, New Delhi, India.

Peirce, M. A. (2003). Avian medicine Samour, J. (ed.) Elsevier, Edenbirg, pp.245-252.

*Perkins, F. O., Barta, J. R., Cloton, R. E., Peirce, M. A. and Upton, S. J. (2000). Phylum Apicomplexa In: Lee, J. J., Leedale, G. F. and Bradbury, P. (eds.). *An Illustrated Guide to Protozoa: Organisms traditionally referred to protozoa* or newly discovered groups 1. (2nd Ed.), published by Society of Protozoologists, pp. 190-369.

Petrotschenko, V.I. (1956). *Acanthocephala of domestic and wild animals*. J.I., 430 pp.

Philip, C, B. (1979). Further notes on Far Eastern Tabanidae (Diptera) VI. New and little-known species from the Oriental and additional records, particularly from Malaysia. *Pacific Insects*, **21**(2-3): 179-202.

Pojmanska, T. (2002). Family Brachylaimidae Joyeux and Foley, 1930. pp.32-39. *In*. Gibson, David, Jones A.J. and Bray, R.A. (Eds.), *Keys to the trematodes* (Vol.I). CBI Publishing CAB (International Willing ford, Oxon, U.K.

Pont, A. C. (1975). Himalayan Muscidae (Diptera) II.New spcies of Hydrotaeini II. Opusscula Zoologic Munchen, 139: 1-13.

*Pont, A.C. (1986). Family Fannidae, pp.41-54, In: A. Soos and L. Papp (eds.), *Catalogue of Palaearctic Diptera II: Scathophagidae, Hypodermatidae*. Akademiai Kiado, Budapest, Hungary.

Price, M. and Graham, O. H. (1997). *Chewing and sucking lice as parasites of mammals and Birds*. United States Department of Agriculture. Agricultural Research Service. *Technical Bull.* **No. 1849**, pp. 320.

Price, R. D. Hellenthal, R., Palma, L., Johnson, K. P. and Clayton, D. H. (2003). *The chewing Lice: World Checklist and Biological overview*. Illinois Natural History Survey, Special Publication. 24. + 501 pp.

Punjabi, A. A. and Saraf, S. K. (1969). Occurrence of *Varroa jacobsonii* in Kashmir. *Indian Bee J.*, **3144**.

*Punjoo, Shazia, B., Khan, A. R. and Rashid, Majidah (2008). Ecodynamics and correlation between some ecological factors and helminth parasites of fishes of Anchar Lake, Kashmir, pp. 514-520. **In**: *Science For Better Tomorrow*, Printed at Universal fPrinters, Khanyar, Srinagar, Kashmir. (Helminths- Trematodes, Acanthocephalans, Cestodes and Nematodes, are recorded and their relation with ecological parameters- physico-chemical of the Lake, have been studied).

Pramanik, S., Soujite, Banerjee, Sampa, Banerjee, Soumyajit, Goutam, K. Saha and Gautum Aditya. (2016). Observations on the predatory potential of *Lutzia fuscana* on *Aedes* larvae: implication for biological control (Diptera: Culicidae). *Fragmeta entomologica*, **48** (2): 137-142.

Prasad, K. N., Prasad, A., Verma, A. and Singh, A. K. (2008). Human cysticercosis and Indian Scenario: A Review. *J. Biosci*, **33** (4): 571-582.

Puri, I. M. (1932). Studies on Indian Simuliidae. Part 2. Descriptions of males, females and pupae of *Simulium rufibasis* Brunetti, its variety *fasciatum* nov. var. and of three new species from the Himalayas. *Indian J. Med. Res.*, **19**: 899-915.

Puri, I. M. (1932 *a*). Studies on Indian Simuliidae. Part. B. Description of males, females and pupae of *S. griseifrons* Brunetti (1911) and of four new species, with striped thorax. *Indian J. Med. Res.*, **19**: 1125-1143.

Puri, I. M. (1948). The distribution of Anopheline mosquitoes in India, Pakistan, Ceylon and Burma. Part V, Additional Records, 1963-48. *Indian J. Malar*, **2**: 67-107.

Putatunda, B.N. and Abrol, D.P. (2002). Mites associated with Bees in Jammu and Kashmir, India. *Zoos' Print J.*, **18** (2): 1021-1024.

*Qayoom, I. and Shah, F. A. (2017). Parasitic bioload in Schithoracine Fishes of Kashmir Valley, with respect to the trophic status of Anchar Lake. *Res. J. Agricultural Sciences*, **8** (2): 364-369.

Qayoom, I., Shah, F. A., Bulkhi, M., Abubaker, A., Bhat, F., Bhat, A., Kumar, A. and Bhat, B. A. (2015). Incidence of Helminth parsites in cold water fishes of river Jehlum, Srinagar, J&K. *The Ecoscan*, **9** (1 and 2): 11-16.

Rani-Puteri, A. M., Irwin, P., Gante, M., Glen, T. C., Linda, M. M. and Traub, R. J. (2010). A survey of Canine filarial disease of veterinary and public health significance in India. *Parasites and Vectors* **3**: 30/doi.org/10.1186/1756-3305-3-30.

Rani, Puteri, A. M.A., Irwin, P.J., Coleman, G.T., Gante, M. and Traub, R. J. (2011a).A survey of canine Tick-borne diseases in India. *Parasit. Vectors*, 4: 141, http: doi: 10-1186/1756-3305-4-141.

Rani - Puteri, A. M. A. R., Glen, T.C., Irwin, P. J. and Traub, R. J. (2011 b). *Hippobosca longipennis*- a potential intermediate host of a species of *Acanthocheilonema* in dogs in Northern India, *Parasites and Vectors*, **4**: 1-7 doi: org./10.1186/1756-3305-4-143.

*Raina, A. K., Bhadwal, M. S., Raina, R. and Rashid, A. (1999). Epidemiological studies on endoparasitic infection in diary animal of Jammu. *Life Science Reporter*, **1**: 78-79.

*Raina, A. K., Rakesh, K, Rajna, V. S., Sridav and Singh, R. P. (1985. Oral transmission of *Trypanosoma evansi* infection in dogs and mice. *Vet Parasitol.*, **18** (1): 67-69.

Raina, A.K., Yadav, A., Gupta, S. K., Singh, R., Agarwal, R. and Hussain, K. (2005). Occurrence of haemo-protozoan diseases and anaplasmosis in bovines in Jammu District. *Indian J. Vet.Med.*, **25** (2): 110-111.

*Raina, M.K. (1968). *Studies on Helminth Parasites of some Vertebrates in Kashmir Monographic study* on *Avitellina* Gough, 1911 *and Studies* on *nematode parasites of shrew and Rat in Kashmir.* Ph. D. Thesis. P. G. Deptt. of Zool., Faculty of Sci., Univ. of Kmr., Sgr., Kmr. (J&K).

Raina, M. K. (1970). On morphology of *Ganguleterakis spumosa* (Schneider, 1866) Lane, 1914 from the intestine of common rat, *Rattus rattus*) in Kashmir. *Kashmir Sci.*, **7**: 95-98.

Raina, M.K. (1971).On the morphology of *Prosthogonimus putschkowskii* Skrjabin, 1912 from egg of domestic hen in Kashmir. *Zool. Anz.*, **187** (5 and 6): 324-327.

Raina, M. K. (1973). Incidence of infection of sheep and goat with *Avitellina* Gough, 1911, *Stilesia* Railliet 1983 and *Moniezia* Blanchard 1891, in Kashmir. *J. Sci. Univ. Kmr.* **1**(1-2): 59-62.

Raina, M. K. (1975). A monograph of the genus *Avitellina* Gough, 1911. (Avitellinidae: Cestoda). *Zool. Jb. Syst.*, **102**: 508-552.

Raina, M. K., Chishti, M. and Kaul, P. L., 1975 (1977). A new report of *Oochoristica tuberculata* (Rudolphi, 1819) Luhe, 1898, from the intestine of *Agama tuberculata* in Kashmir, with remarks on its synonyms. *Indian J. Helminthol.*, **29**(1): 1-4.

Raina, M. K., Chishti, M. Z. and Khan, A. R. (1973). An abnormality inii *Oochoristica tuberculata* (Rudolphi, 1812) Luhe, 1898. *J. Sci. Univ. Kashmir*, **1**: 57.

Raina, M. K. and Dhar, R. L. (1972). On *Camallanus fotedari* n. sp. (Nematoda: Spiruidea) from the intestine of *Nemacheilus kashmirensis* in Kashmir, Indian., *J. Helminthol.*, **46** (2): 157-160.

Raina, M. K. and Kaul, R. K. (1974). A new report of *Trichocephalus muris* Schrank, 1788 from *Rattus* rattus in Kashmir. *J. Sci. Univ. Kashmir*, **2** (1-2): 70-72.

Raina, M. K. and Kaul, R. K. (1982). On a new species of *Capillaria* (Nematoda) from a bat in Kashmir. *Angew Parasitol.*, **23**: 28-31.

Raina, M. K. and Khan, M. D. (1982). Prevalence of Amphistomes in Ruminants in Kashmir. *Kashmir Univ. Res. J.*, **2**: 17-21.

Raina, M. K. and Khan, M.D. (1983). Excysted Amphistome metacercariae. *Indian J. Parasitol.*, **7**(1): 115-116.

Raina, M. K. and Koul, P.L. 1983. The histopathology of *Neoechinorhynchus hutchinsoni* infection in *Nemacheilus kashmirensis*. *J. Helminthol.*, **58**: 165-168.

Raina, M. K., Malla, A. K. and Raina, Usha 1975 (1978). On *Mathevotaenia sysmmetrica* (Baylis, 1927) Akumyan, 1946, from the intestine of *Ratus rattus* in Kashmir. *J. Sci. Univ. Kmr.*, **3**(1-2): 113-117.

Raina, M. K. and Narain, K. (1990). New heterotrichid ciliates from *Rana* (*Dicroglossus*) *cyanophlyctis*. *Proc.* 11th *Intl. Congr. Parasit.*, held at Paris, France, from 20-24th August, 1990 (Abt).

Raina, M. K. and Narain, K. (1992). A new Myxosporidian from Gall Bladder of *Puntius chola* in Jammu, with review of the Indian species of *Myxidium* Butschli 1882, from fishes. pp.108-114 *In*: Yousuf, A.R., Raina, M.K. and Qadri, M.Y. (Eds.), *Curr. Trends in Fish and Fisheries Biology, and Aquatic Ecology*, P. G. Department of Zoology. The Univ. of Kashmir, Srinagar, Kashmir (J&K).

Raina, M. K. Raina, Phoola (1985). On two capillarid nematodes from shrew in Kashmir, with a note on the synonymy of *Skrjabino capillaria* Skarbovitsch, 1946. *Indian J. Helminthol.*, **37**(1): 109-115.

Raina, M. K., Sey, O. and Khan, M.D. (1987). Paramphistomes (Trematoda Amphistomida) of domestic ruminants in Kashmir, India. *Misc. Zool. Hung.*, **4:** 5-12.

Raina, R. (1966). *Longidorus reneyii* sp. nov. (Nematoda: Lengidoridae) from Srinagar, Kashmir. *Indian J. Ent.*, **28**(4): 438-441.

Raina, R., Raina, A.K. and Bhadwal, M.S. (2000). Outbreak of Surra in buffaloes and ponies. *Indian J. Vet. Med.*, **20**(1): 32.

Raina, Sunil, K., Razdan, Sushil, Pandita, K. K., Sharma, R., Gupta, V.P. and Razdan, S. (2012). Active Epilepsy as Indicator of Neurocysticercosis in Rural North-west India. *Epilepsy Research and Treatment*, **2012 (2012)**, Article ID 802747.

Rajmohan, K. (2015). A checklist of Scelionidae (Hymenoptera: Platygastroidea) of India. *Zoo Print J.*, 21(12): 2506-2513.

Ramaseshiah, G. and Dharmadhikari, P. R. (1969). Aphelinid parasites of aphids of Kashmir. *Tech. Bull. Commonw. Inst. Biol. Control.*, **11**: 156-164.

Ranga-Rao, G. S. C. Sharma, R.L. and Hemaprasantha (1994). Parasitic infection of Indian Yak, *Bos* (*Poephagus*) *grunniens- an* overview. *Vet. Parasitol.*, **53**: 75-82.

Rakhshani, E., Hossein, B., Ahmad, Z., Tomanovic, Z., Jankovic, M., Andjeljko, P., Bodlah, I. and Stary, P. (2012). New species and additional evidence of aphid parasitoids (Hymenoptera: Braconidae: Aphidiinae) from India. *Zootaxa*, **3397**: 45-54.

Rao, K. N. A. (1951). Tick typhus in Srinagar. *Indian J. Med. Res.*, **39**: 293-96.

Rao, S. N. (1953). *Indian Ichneumonidae* (*Hymenoptera*) *in the Forest Research Institute of Dehra Dun*. Indian Forest Research Institute, **8**: 159-225.

Rao, S. N. and Grover, P. (1960) Studies on Indian Ichneumonidae (Parasitic Hymenoptera). *Proc. Natl. Acad. Sciences*, India, 30: 276-288.

Rao, T. R. (1981). *Anophelines in India*. W.Q. Press, Banglore, pp.1-596.

Rao, K. N. A. and Kalra, S.L. (1949). Tick-borne relapsing fever in Kashmir. *Indian J. Med. Res.*, **37**: 385-97.

Rao, T.R., Dhanda, V., Bhat, H.R. and Kulkarni, S.M. (1973). A survey of haematophagons arthropods in Western Himalayas, Sikkim and hill districts of West Bengal. A general account. *Indian J. Medical Research*, **61**(10): 1421-1461.

Rao, V. P. (1966). Survey for natural enemies of Gypsymoth Report for period 1961-66 – U.S. P.L.480 Project. *Commw. Inst. Biol. Control. Banglore*, India, 50 pp.

*Rao, V.P (1972). Evaluation of Hymenopteran parasite of Gypsy Moth. **In:** *Final Tech. Rept. Commonwealth Institute of Biological Control*, Indian Station, Banglore, India, p. 24. [on *Pimpla laothoe* (= *Coccygomimus laothoe*), ichneumonid parasite of *Lymantria obfuscata* from Kashmir].

Rao, V. P., Ghani, M. A., Sankaran, T. and Mathur, K. C. (1971). *A review of the biological control of insects and other pests in South-east Asia and Pacific region. Tec. Communication* No. 6. CIBC, Trinibad. West Indies. Commonwealth Agricultural Bureux. England.

*Rasheed, M., Khalid, S. and Pandit, B. A. (2008). Prevalence of Nematode infection in the Calves of Kashmir valley. *In*: M. Z. Chishti and Fayaz Ahmad (eds.). *Science For Better Tomorrow*, pp. 414-416. Printed at Universal Printers, Khanyar, Srinagar, Kashmir.

Rashid, Adil, Dar, L. M., Mir, M. S. and Kamil, S. A. (2014). Prevalence of *Cysticercus tenuicollis* infection in slaughtered sheep in Kashmir. *SKAUST J. Res.*

Rashid, Adil, Darzi, M.M., Mir, M. S., Dar, Mir, A. and Allie, I. M. (2017). Prevalence of Ovine cystic echinococcosis in Kashmir, North India. *Vet. Parsitol.: Regional Studies and Reports*, **10**: 85-89.

Rashid, A., Azad, Shami, A. and Farooq, U. (2014). Community analysis of plant-parasitic nematodes associated with ornamental plants in Rajouri District (J&K), India. *Intl. J. Curr. Microbiol. And Applied Sci.*, **3** (1): 194 -197.

*Rashid, M., Godara, R., Yadav, A. and Katoch, R. (2018). Prevalence of ticks in sheep and goats of Jammu region. *Indian J. Small Ruminants*, **24** (1): 183-186.

*Rashid, S., Tanveer, S. and Abdullah, S. (2019). Histopathological studies of cestodiasis in domestic fowl, *Gallus gallus domesticus*. *SM J. of Biol.*, **5**(1): 1020.

*Rashid, Suhail, Tanveer, S. and Malik, S. (2019). SDS PAGE Profile of *Raillietina tetragona* and *Raillietina cesticillus* of domestic fowl (*Gallus gallus domesticus*) for Development of Immunodiagonosis of Cestodiasis. *EC Microbiology*, **15.2** (2019): 147-152. (Investigation on development of proper immunodiagonosis for cestode infection in domestic fowl).

Rashid, A., Dar, M., Mir, M.S., Darzi, M., Khan, H. M. and Kamil, S. A. (2014). Prevalence of *Cysticercus tenuicollis* infection in slaughtered sheep of Kashmir. *SKAUST J. Res.*, **10** (1): 62- 65.

*Rashid, M., Khalid, S. and Pandit, B. A. (2008). Prevalence of Nematode infection in Calves o f Kashmir. *In*: M. Z. Chishti and Fayaz Ahmad (eds.), pp. 414-416,

Science For Better Tomorrow, Printed at Universal Printers, Khanyar, Srinagar, Kashmir.

Rather, S. and Yaseen, A. (2017). "Tin Tack" sign in localized cutaneous leishmaniasis: A finding from a non-endemic disease focus. *Indian J. Dermatol.,* **62** (5): 535-537.

Rather, P. A. and Hassan, Iffat (2014). Human *Demodex* Mite: The versatile mite of Dermatological importance, *Indian J. Dermatol.,* **59** (1): 60-66.

Rather, Sumi-ulah, Buhroo, A. A and Khanday, L. (2017). Occurrence of eulophid parsitoids on apple leaf-miner, *Lyonetia clerkella* Linn. (Lepidoptera: Lyonetiidae) in Kashmir. *Intl. J. Entomol. Res.,* **2** (6): 39-43.

Rather, S. A., Tak, H. and Kakru, D. (2015). Seroprevalence of Haemoprotozoan diseases in ruminants of Kashmir Valley. *Agricultural Advances,* **4** (11): 122-128.

Rather, S., Bashir, S., Bhat, Y. J. and Hassan, I. (2019). Chronic Relapsing cutaneous Leishmaniasis in an elderly female: A rare clinical presentation from an non-endemic area. *Indian Dermatol. Online J.,* March- April.

Rather, S. A., Tak, H. and Kakru, D. K. (2019). Sero-prevalence of *Toxoplasma gondii* infection in healthy human population in Kashmir valley, India. *Intl. J. Biol. Med. Res.,* **10** (1): 6587-6590.

Rather, Z.A., Wachkoo, A.H., Khuroo, A.H. and Dar, Tanvir-ul- Hassan (2019). First Record of *Aspicera hartigi* (Hymenoptera, Figitidae) from India, with observation on its foraging behavior on *Ephedra* plant. *J. Asia Pacific Biodiversity,* **12** (3): 473-476.

Rathi, N. and Rathi, A. (2010). Rickettsial Infection: Indian Perspective. *Indian Pediatrics,* pp. 157-164.

Rebecca, J., Riddhi, Traub, Pediekar, P., Cuttell, Leigh, Porter, R.B., Rani-Puteri, A.B.M and Gatie, M.L. (2014). The prevalence and distribution of gastro-intestinal parasites of Stray and Refuge dogs in four locations in India. *Veterinary Parasitology,* **205 (2014)**: 233-238.

Rehana, R., Bhagat, R. C., Ahmad, F., Rukhsana and Sofi, T. S. (2016). Gregarine Parasite belonging to the family Diplocystidae Bhatia, 1938, reported from the Cockroach, *Blatta orientalis* (Dictoptera, Blattidae) from Kashmir. *J. Vet. Sci, Technol.,* **7**: 399 Doi: 10.4172/2157-7579.1000.399.

Rehana, R., Bhagat, R. C., Rukhsana, A. and Tanveer, S. (2016). New record of *Leidynema appendiculata* (Leidy, 1850) Chitwood, 1932 (Oxyuroidea: Thelastomatidae) from the cockroaches, *Blatta orientalis* (Dictyoptera: Blattidae) fom Kashmir. *J. Vet. Sci. and Tech.,* **7**: 403, doi: 10.4172/2157-7579.1000403.

Rehmat, T., Anis, S.B., Khan, M.T., Fatima, J. and Begum, S. (2011). Aphelinid parasitoids (Hymenoptera: Chalcidoidea) of armoured scale insects (Hymenoptera: Diaspididae) from India. *Biol. and Medicine,* **3** (2): 270-281.

Reshi, A. A. and Tak, Hidayat-ullah (2014). The study of some Risk factors associated with coccodiosis (*Eimeria*) in sheep of Kashmir Valley. J. *Agri. and Vet. Sciences,* **7**(1): 11-13.

Ricardo, G. (1911). A Revision of the Oriental species of the genus family Tabanidae other than *Tabanus*. *Recd. Indian Mus.*, **4**: 321-397.

Rishi, N. D. (1976). Survey and studies of aphidiid parasites of aphids, pp.47-48 In: *Symposium on Modern Trends in Zoological Research in India*. Published by Zoological Society of India, Calcutta, India.

Rishi, N. D. and Shah, K. A. (1985). Survey and bioecological studies on natural enemies complex of India. Gypsy Moth, *Lymantria obfuscata* Walker) (Lepidoptera: Lymanpiidae). *J. Entomol. Res.*, **9** (1): 82-93.

Rishi, N. D. and Lone, M. A. (1984). Description of a new species of *Protaphelinus* Mackauer (Hymenoptera: Aphelinidae) parasitic on gall-forming aphid. *Pemphigus bursarius* Linn., infesting *Populus nigra* var. *italica* in Kashmir. *J. Ent. Res.*, **8** (2): 132-136.

Rodnigues, F.M., Padbidri, Ghalashi, G.R., Gupta, N.P. Mandke, V.B., Pinto, R.S., Hoon, R.S., Bapat, M.B. and Mohan Rao, C.V. R. (1986). Prevalence of Crimean haemorrhagic fever Congo virus in Jammu and Kashmir State. *Indian J. Medical Res.*, **84**: 134-38.

Rognes, K. (1972). A new genus of Halicoboscinae from the Himalaya (Diptera: Calliphoridae), with amended genus and sub-family concept. *Entomologica Scandinavica*, **23**(4): 391-404.

Rohdendrof, T.R. (1937). *Fauna USSR, Diptera*, **19** (1): 174.

Ruggiero, M. A., Gordon, D., Orrell, T.M. Bailly, N., Bourgoin, T., Brusca, R.C., Cavalier- Smith, T., Guiry, M.D. and Kirk, P.M. (2015). A Higher Level Classification of All Living organisms. *PLOSONE*, 10(4): eo 119248.

*Rukhsana, A., Bhat, B. A. and Maqbool, S. (2011). Effect of *Bothriocephalus acheilognathi* (Cestoda: Pseudophyllidea) on health of five fishes of *Schizithorax*. *Bioinfolet*, **8** (1): 107-109.

*Sahaf, K. A. (1991). *Studies on the parasitic infection of Bombay mori and their control in J&K*. Ph. D. Thesis, Faculty of Sci., P. G. Deptt. of Zool., The Univ. Kmr., Sgr., Kmr. (J&K).

Sahaf, K. A. (2002). Incidence of Protozoan disease, Pebrine of Silkworm, *Bombyx mori* in Jammu and Kashmir. *J. Entomol. Res.*, **26**(4): 305-307.

Sajad, P., Rather, S., Hassan, I. and Qureshi, W. (2017). Primary Lip Leishmaniasis-Report of 4 cases from a non-endemic Region of Kashmir Valley of North India and Review of Literature. *J. Advances in Medicines and Medical Res.*, **24**(1): 1-5. DOI: 10.9734/JMMR/2017/36072.

Salam, S. T., Khan, A. R. and Mir, M. (2008). Prevalence and Pathology of *Amoebotaenia sphenoides* in free ranging chicken of Kashmir Valley. *The Internet J. Parasitic Dis.*, **4** (1).

Salam, S. T., Khan, A. R. and Mir, M.S. (2010). Prevalence and pathology of *Raillietina cesticillus* in the indigenous chicken (*Gallus gallus domesticus*) in temperate Himalayan region of Kashmir. *Veterinarski Archiv*, **80** (2): 323- 328.

Salam, S. T., Mir, M. S., Shahnaz, S. and Khan, A. R. (2009). Prevalence and the associated Lesions of *Cheilospirura* (*Acuaria*) *harmulosa* in indigenous chicken of Kashmir Valley. *J. Parasitol.*, **95** (6): 1436-39.

Salam, S, T., Mir, M.S. and Khan, A.R. (2009). Prevalence and seasonal variations of ectoparasite load in free range chicken of Kashmir Valley. *Trop. Animal Health Production*, **41**: 1371-1376.

Saleem, Taniya, Katoch, R., Ysav, A., Mir, I. A., Mir, R., Godara and Ahmad, I. (2017). Prevalence of ovine oestrosis in Plain and Kandi areas of Jammu, North India. *Intl. J. Sci. Environ. Technol.*, **6** (2): 1117-1122.

Sarmast, A. H., Showkat, H., Parry, F. Q., Kachroo, M. Y., Mir, S. F., Lone, Y.A. Khan, A. A. and Khan, O. M. (2012). Primary hydatid cyst in the spleen – two case Repors treated surgically, with review of literature. *Acta Medica Saliniana*, **41** (10: 55-58.

Sauchis, A., Latorre, A., Gomzalez, C.F. and Michelana, J. M. (2000). An 18Sr DNA-based Molecular Phylogeny of Aphidiinae (Hymenoptera: Braconidae). *Molecular Phylogeny and Evolution*, **2** 180-194.

Scott, H. (1925). Zoogeography and Systematic Notes on the Nycteribiidae (Diptera: Pupipara) of India, Ceylon, Burma. *Recd. Indian Mus.*, **27**: 351-384.

*Shah, F. A., Qayoom, I., Balkhi, M. H. and Kumar, A. (2015). Impact of Parasitic Diseases of Fishes of North-west Himalayan streams. *Curr. World Environment*, **10** (3): 920.

*Shah, Humaira, B., Yousuf, A. R., Chishti, M.Z. and Ahamad, Fayaz (2013a). Metacercariae of *Clinostomum schizothoraxi* Kaw (Digenea: Clinostomatidae) in *Crassius crassius* (L.) under different environmental conditions. *Folio Parasitologica*, **60** (2): 163.

*Shah, H. B., Yousuf, A. R., Chishti, M. Z. and Ahmad, F. (2013 b). Seasonal changes in infra-population of *Diplozoon kashmirensis* Kaw (Monogenea: Diplozooidae) along eutrophic gradient. *Parasitol. Res.*, 112 (9): 3347-56.

Shah, K. A. and Andrabi, S.A. (2009). A note on Ivermectin and Clorsulon treatment of cattle infested with subcutaneous parasites. *Vetscan*, **4** (1): article 37, pp. 34-35.

*Shah, M. M., Maqbool, I., Manzoor, I, Shabnam, N., Rather, T, and Ahangar, S. A. (2018). Seasonal prevalence of gastro-intestinal parasites in sheep of Kashmir. *J. Entomol. and Zool. Studies*, **6** (6): 534-536.

*Shah, M. M., Shahardar, R. A., Naik, I. M., Allaie, I. M. and Wani, Z. A. (2018). Prevalence of GI helminths of cattle in low-lying and marshy areas of Kashmir valley. *Indian J. Animal Sci.*, **89** (4).

Shah, Irfan, Sharma, H. K., Shah, M.AS., Katoch, R. and Malik, M. A. (2017). Prevalence of Various intestinal zoonotic Parasites in dogs of Jammu Region of Jammu and Kashmir. *Intl. J. Livestock Res.*, **7** (1): 2277.

Shamin, M., Ahmad, Z. and Ahmad, S. (2008). Record of the genus *Peristenus* Foerster (Hymenoptera: Braconidae) from India, with descriptions of four new specie. *Rec. Zool. Surv. India*, **108**: 81-90.

Sheela, S., Narendran, T.C. and Tiwari, R. N. (2003). Contribution to the knowledge of Chalcididae of India. *Rec. Zool. Surv. India,* **101** (3-4): 247-260.

Sheikh, G. N., Willayat, M. M. and Ashraf, H. (2007). Prevalence of Cryptosporidiosis in dairy calves of Kashmir valley. *J. Vet. Pub.Health,* **5**: 21- 24.

Sheikh, J. H., Chishti, M. Z., Rasheed, M., Tak, Irfan-ur- Rauf and Dar, S. A. (2014). Occurrence of *Helicotylenchus chishti* sp. n. (Order: Tylenchida) on *Crocus sativus* (saffron) and its generic buid-up analysis and note on its possible management strategy. *J. Zool. And Biocience Res.,* **1** (3): 10-14.

Sheikh, K. A., Baba, A. A., Ahmad, S. M., Shera, A. H., Patnaik, R. and Sherwani, A. Y. (2010). Mechanical small bowl obstruction in children at a tertiary care centre in Kashmir. *Paediartic Surgery,* **7** (2): 81-85.

*Sheikh, M. M., Tak, H., Fazili, M. F., Bhat, B. A. and Wani, I. N. (2018). *Balisascaris transfuga*- a parasite with zoonotic potential. Intl. *J. Adv. Scientific Res and Management,* **3** (12): 174 -180.

*Sheikh, M. A., Khan, S. A. and Bhat, D. M. (2020). An updated checklist of Ichneumonid parasites (Hymenoptera: Ichneumonidae), associated with insects/insect pests of Agricultural, Horticultural and other Economically important plants in Union Territory of Jammu and Kashmir. *Intl. J. Curr. Microbiol. and Applied Sci.,* **9** (4): 2266-2277 (The paper deals with 46 spp., under 29 genera, belonging to 11 sub-families, associated with host crops and plants).

Shi, M. and Cheu, X.X. (2005). Molecular phylogeny of Aphidiinae (Hymenoptera: Braconidae) based on DNA Sequence of 16Sr RNA and ATPase 6 genes. *European J. Entomology,* **102**: 133-1338.

Shu-Piung, S., Tao Li, and Mao, Ling, C. (2019). Taxonomic study of the genus *Townesia* Ozols (Hymenoptera: Ichneumonidae: Pimplinae), with description of a new species and key to the world species. *Zookeys,* 878: 23-32 doi: 10.3897 (zookeys, 878.38071.

Sen, P. (1938). A check and host-list of Ixodoidea (Ticks) occurring in India. *Indian J. Vet. Sci.,* **8**: 133.

Senior-White, R., Aubertin, D. and Smart, J. (1940). *Fauna of British India, including the remaining of Oriental Region, Diptera-IV.Calliphoridae. Indian Reprint Edition.* Today and Tomorrow Printer and Publishes, New Delhi. Originally published by Tylor and Francis, London, England.

Sewell, R. B. S. and Roonwal, M. L. (Eds.) (1965). *The Fauna of India and adjacent countries. Diptera., Muscidae,* Part I by Dr. F. I. Van Emiden Published by *Zool. Surv. of* A. (1996).

*Shafi, J., Tak, Irfan-ur- Rauf, Dar, B. A. and Chishti, M.Z. (2017). Association between epidemiology and haematophagous behaviour of *Haemonchus contortus* and *Ostertegia ostertagi,* infecting sheep of Kashmir valley. *Curr. Sci.,* **113** (9): 1776.

*Shah, A. W., Parveen, Muni, Mir, S. H., Sarwar, S.G. and Yousuf, A. R. (2009). Impact of Helminth parasitism on Fish haematology of Anchar Lake. *Pakistan J. Nutrition,* **8**: 42- 45.

Shah, F. A. and Shah, T. A. 1986 (1988). Search for *Nosema apis* in Kashmir *Apis cerana. Indian Bee J.*, **48**(1-4): 55-56.

Shah, F. A. and Shah, T. A. (1988). Varroa disease of honeybees – A Review. *Indian Bee J.*, **50**: 102-105.

Shah, F. A. and Shah. T. A. (1989). *Tropilaelaps* disease of honeybees. *Indian Bee J.*, **51**: 120-125.

Shah, F.A. and Shah, T.A. 1989 (1991). *Tropilaelaps* (*Tropilaelaps*) *clareae*) not a threat to *Apis mellifera* in temperate climate. *Indian Bee. J.*, **15**(3): 120-121.

*Shah, H. B. and Yousuf, A. R. (2007). Parasitism in Crucian carp, *Crassius crassius* (L.) inhabiting different trophic Lakes. *J. Himalayan Ecology and Sustainable Dev.*, **2**: 47-54.

* Shah, Humaira, B., Yousuf, A. R. and Chishti, M. Z. (2013). Helminth communities of fish as ecological indicator of Lake Health. *Parasitol.*, **140** (3): 352-260, doi: org/10.1017/soo31182012001679.

* Shah, H., Yousuf, A.R., Chishti, M. Z., Shahnaz, S. and Ahmad, F. (2014). Trophic status and helminth infra-communities of fish populations in Kashmir Himalayan Lakes. *J. Helminthology*, **88** (3): 264-271.

Shahardar, R. A., Mir, A. S., Mir. M. M., Pandit, B. A. and Ahmad, M. A. (1995). Efficacy of fenbendazole against gastro-intestinal nematodes of Kashmir deer. *Indian Vat. J.* **72**(7): 781-789.

Shahardar, R. A., Mir, A. S., Pandit, B. A. and Banday, M. A. A. (1993). Successful treatment of goat warbles with inermectin. *Indian Vet. J.*, **70** (7): 666-667.

Shahardar, R. A. and Pandit, B. A. (2003). A Rare Report of lungworm infection in mules and its treatment. *The Indian Vet. J.*, **80** (7).

Shahardar, R. A. and Rashid, A. (1998). Occurrence of larvae of *Oestrus ovis* in goat in Kashmir. *J. Vet. Parasitol.*, **12** (2): 147.

Shahardar, R.A., Darzi, M. M. and Pandit, B. A. (2001). Recovering of larvae of *Oestrus ovis* from trachea and bronchi of sheep. *J.Vet Parasit.*, **15**(1): 85.

Shahardar, R. A., Mir, A.S., Pandit, B. A. and Banday, M. A. (1993). Successful treatmenr of goat warbles with ivermectin. *Indian Vet. J.*, **70** (7): 666-667.

*Shahardar, R. A. and Pandit, B. A. (2003). A rare Report of lungworm infection in mules and its treatment. *The Indian Vet. J.*, **80** (7).

Shahi, N., Yousuf, A.R., Rather, M. I., Ahmad, F. and Yaseen, T. (2013). First report of blood parasite in fishes from Kashmir and their effect on the haematological profile. *Open Vet. J.*, **3** (2): 89-95.

Shahnawaz, M. Shahardar, R. A., Wani, Z. A. and Shah, S. A. (2011*a*). Prevalernce of Ovine platyhelminth parasite infection in Ganderbal area of Kashmir Valley, *Indian J. Animal Sci.*, **81:** 40-43.

Shahnawaz, M. Shahardar, R. A. and Wani, Z. A. (2011*b*). Seasonal prevalence of platyhelminthosis of sheep in Ganderbal area of Kashmir valley. *J. Vet. Parasitol.*, **25**: 59-62.

Shakeebah, K. S., Wani, S. A., Allaie, I. M., Shahardar, R. A., Hussain, H. A. and Shah, S. A. (2018). *Setaria equina* in the eye of horse. *J. Vet. Parasitol.*, **32** (2): 82-84.

Sharif, M. (1928). A revision of the Indian Ixodidae with reference to the collection in the Indian Museum. *Recd. Ind. Mus.*, **30**: 217-344.

Sharif, M. (1930). A revision of Indian Siphonaptera Part I. Family Pulicidae. *Rec. Indian Mus.*, **32**: 29-62.

Sharma, B. D. (1972). *Ornithonyssus bursa* (Mesostigmata: Dermanyssidae) on the poultry birds of Kashmir. *Poultry Guide*, **9** (9): 52.

Sharma, B. D. (1973). *Psammotettix striatus* (Linn.), a rare hemipteran ectoparasite of the flying foxes of Poonch (J&K). *Curr. Sci.*, **43** (3): 108.

Sharma, B. D. (1993 a). *Medical and Vetrinary Arthropod Disease Ecology*. APH Publishing House, New Delhi.

Sharma, B. D. (1993 b). High incidence of Trypanosomiasis in Jammu and Kashmir State and its analysis, p. 288, *Medical and Veterinary Arthropod Disease Ecology*. APH, Publishing House, New Delhi, India.

Sharma, B. D. (1993 c). Observations on Fleas (Siphanoptera), Lice (Anoplura) and Bugs (Hemiptera), collected by the author, with remarks on their Ecology, pp.125-157, Chapter VIII. *Medical and Veterinary Arthropod Disease Ecology*. A.P.H. Publishing House, New Delhi.

Sharma, B. D. (1993 d). Ticks from Humans in J&K State (India). pp. 292 -296, *Medical and Veterinary Arthropod disease Ecology*. A. P. H Publishing House, New Delhi.

Sharma, B. D. (1993e). Notes on Relapsing fever tick, *Ornithodoros tholozani* (Laboulhene and Megnin, 1882), collected from the high-altitude regions of Jammu and Kashmir State. pp. 290- 291, *Medical and Veterinary Arthropod Diseases Ecology*. A. P. H. Publishing House, New Delhi.

Sharma, B. D. (1993 f). New record of *Ctenocephalides felis orientalis* (Jordan, 1925) (Siphonaptera: Pulicidae) from J&K State. In: Sharma, B.D. (Ed.), pp. 290-291, *Medical and Veterinary Arthropod Disease Ecology*. A. P. H. Publishing House, New Delhi.

Sharma, S. K., Yadav. A., Sooden, J. S. and Gupta A. (2005). Bot infestation in equine-a case report. *Intas Polivet.*, **6** (2): 364-365.

Sharma, B. D. and Das, S. M. (1971). A new record of *Boophilus decoloratus* (Koch) (Acarina: Ixodidae) from Kashmir, India. *Indian J. Entomol.*, **32**(3): 292-293.

Sharma, B. D. and Das, S. M. (1971*a*). On the occurrence of *Melophagus ovinus* Linn., 1958 (Insecta: Hippoboscidae) on sheep and cattle in Kashmir. *Indian Vat. J.*, **48** (9): 972-973.

Sharma, B. D. and Das, S. M. (1972). A checklist of ticks and their hosts in Kashmir. *Orissa Vet. J.*, **7**113.

Sharma, B. D. and Sharma, T. (1972): A new record of two ectoparasitic mites of housefly from J&K State. *Indian J. Anim. Res.*, **6** (1): 42.

Sharma, B. D. and Sharma, T. (1973). Occurrence of *Ischnopsyllus octactenus* on some bats in Kashmir (Jammu and Kashmir State), *Indian J.Entomol.*, **35** (2): 161.

Sharma, B. D. and Sharma, T. (1974). A note on contribution of tick- fauna of Jammu Province. *Indian J. Anim. Sci.*, **44** (9): 709-712.

*Sharma, H. K., Vohra, S., Yadav, A., Katoch, R., Sood, S. and Kumar, J. (2008). Prevalence of *Cysticerus tenuicollis* in sheep of Jammu Region. *Veterinary Practitioner*, **9** (1): 67-68.

Sharma, N. R., Thapa, C. D. and Nath, A. (1981). Pathogenicity and identity of myceliphagous nematodes infesting *Agaricus bisporus* (Lan.) Singh. *Indian J. Nematol.*, **11**: 230-231.

Sharma, O. P. 2003 (2004). *Illustrated Dictionary of Biodiversity*. Published by Forest Res. Instt., Jammu and Kashmir, Jammu (J&K), 159 pp.

Sharma, R. L. (1994). Parasitic bronchitis in goats and the possible use of *Dactyocaulus filaria* vaccine for its control. *Vet. Parasitol.*, **51**(3/4): 255-262.

Sharma, R. L. and Dhar, D. N. (1978). Studies on ovine Oesophagostomoiasis I. Population kinetics following varying levels in lambs. *Proc. Asian Congr. Parasitol.*, p.300 (Abt).

Sharma, R. L. and Dhar, D. N. (1979). Population kinetics of primary *Oesophagostomum columbianum* infection in lambs. *Indian J. Anim. Sci.*, **49**: 203-208.

Sharma, R. L. and Dhar, D. N. (1987). Prevalence of *Dictyocaulus viviparous* infection in Kashmir. *Indian J. Parasitol.*, **11**(1): 101.

Sharma, R. L., Peer, F. U. and Bhat, T. K. (1996). Antiparasitic activity of invermectin against *Psoroptis communis* var. *ovis* infestation in Kashmir Merino sheep. *Indian Vet. J.*, **67** (3): 264-66.

Sharma, R. L., Raina, O. K. and Hemaprasanth, 1989 (1990). Post hibernation fate of snail (*Lymnaea auricularia*) *sensu stricto* experimentally infected with *Fasiola gigantica. Riv. Parassitol.*, **50** (1-1): 185-87.,.

Sharma, S. K., Sodan, J. S., Bal, M. S., Khajuria, J. K. and Upadhaya, S. R. (2009). Parasitic dermatitis of Canines of Jammu Region. *J. of Vet. Parasitol.*, **23** (1): 65-67.

Sharma, S., Katoch, R., Yadav, A. and Godara, R. (2014). Giardia prevalence in bovine calves in Jammu. *Vet. Practitioner*, **15** (2): 272-273.

Sharma, H. K., Vohra, S., Yadav, A., Karoch, R., Sood, S. and Kumar, J. (2008). Prevalence of *Cysaticercus tenuicollis* in sheep of Jammu Region. *Vet. Practitioner*, **9** (1): 67-68.

Sharma, R., Sharma, H. K., Sharma, R., Katoch, R., Kaur, R., Rashid, M. and Ganai, A. (2017). Epidemiological study of prevalence of *Cysticercus cellulosae* in pigs of Jammu region. *Vet. Practitioner*, **18** (1): 102- 105.

Sharma, Shikha., Katoch, R., Yadav, A. and Godara, R. (2015). Giardia prevalence in pet dogs in Jammu, India. *Vet. Practioner*, **16** (1): 1030-1031.

Sharma, Sourabh, Iqbal, Azim, S. and Shah, H. (2013). Study of poultry coccidiosis in organized and backyard farms in Jammu Region. *Vet. World.* 2013, doi: 10.5455/vetworld.2013.467-469.

Sharma, S., Azim, S., Iqbal, A., Nasirudullah, N. and Mustaq, I. (2015a). Pathomorphological alternation associations with chicken coccidiosis in Jammu Division of India, *J. Parasit. Dis.*, **39** (2): 147-51.

Sharma, Sourab, Iqbal, A., Azim, S, Mustaq, Wani, I. Ahmad, Z. and Ahmad, Showkat (2015b). Prevalence of poultry coccidiosis in Jammu region of Jammu and Kashmir. *J. Parasit. Dis,* **39** (1): 85-89, doi: 10.1007/s12639-013-0286-5.

Sharma, V. and Gupta, A. (1985). A new doryctine genus *Protatus* infesting *Sinoxylon* sp., a serious pest of *Dalbergia sissoo* in Jammu (India) (Hymenoptera: Braconidae). *Reichenbachia,* **23**(1): 37-39.

Sheikh, K. A., Shams-ul- Bari, Hussain, Z., Parry, F, Q., Lone, R., Arif, S. Khawaja, A.R. and Shera, A. (2007). Clinical profile of Liver abscess in children of Kashmir and variuoius modes of management. *JK Practioner,* **14** (1): 24-27.

Shaw, A.A. (1980). Protozoan infection in J&K State. pp.200-205. In. *Heritage of Kashmir,* Hassnain F.M. (Ed.), Chap.18. Gulshan Publishers, Srinagar, Kashmir.

Shaw, A. A. (1989). Investigations on some infections in exotic, pure and cross bred cattle of Kashmir Valley. *Indian J. Comparative Microbiol., Immnul. and Infectious Diseases,* **10** (1): 33-38.

Shaw, A. A. and Das, S.M. (1970). Bovine Theileriosis in cross-bred Jersey Cow in Kashmir. *Kashmir Sci.,* **7**(1-2): 99-102.

Sheikh, A. G. 1977 (1979). Occurrence of *Anguina tritici* (Steinbuck, 1799), Chitwood, 1935, on wheat in Kashmir. *Indian J. Plant Protection,* **5**(2): 20.

Sheikh, B. A., Ahmad, F. and Sofi, T. A. (2016 a). Morphology and Prevalence of some helminth parasites in *Gallus domesticus* from Gurez valley of Jammu and Kashmir, *India. J. Fisheries Livest. Production,* **4**: 159, doi: 10.4172/23322608-1000159.

*Sheikh, B. A., Sofi, Tanveer, A. and Ahmed, F. (2014). Ecology of the Asian Tapeworm, *Bothriocephalus acheilognathi* Yamaguti, 1934 of fishes in Dal Lake of Srinagar. *J. Fisheries and Aquatic Studies,* **2** (1): (99. 8 per cent prevalence of *B. acheilognathi* found in fishes: *Schizothorax niger* and *Cyprinus carpio* in Dal Lake of Kashmir between April, 2013 to January, 2014).

Sheikh, B. A., Ahmad, F. and Sofi, T. A. (2016 b). Seasonal prevalence of Gastro-intestinal helminth of sheep (*Ovis aries*) and goat (*Capra hircus*) with respect to Age and Gender of Gurez Valley, Kashmir. *J. Animal Sci. Advances,* **6** (3): 1609-1625.

*Sheikh, B. A., Ahmed, F. and Sofi, Tanveer, A. (2018). Morphology of Helminth parasite in *Salmo trutta fario* and *Schizothorax plagiostomus* in Gurez valley of Jammu and Kashmir, India. Bio-medical J. Sciwentific and Technical Res., 6 (1): 2018 BJSTR.M.S.I.D.001306. DOI: 10.26717/BJSTR.2018.06.001306.

Sheikh, M. M., Fazili, Mustahson, Tak, H. and Bhat, B. A. (2017). Parasitic Prevalence in Himalyan Black Bear (*Ursus thibetanus*) in Kashmir Himalaya. *Intl. J. Vet. Sci. and Animal Husbandry*, **2** (10) Part I: 10-12.

Sherman, R. A. and Pechter, E. A. (1988). Maggot therapy: a review of therapeutic application to fly larvae in human medicine, especially for treatment of Oesteomyelitis. *Med. and Vat. Entomol.*, **2**: 225-230.

Sherwani, A. Y., Sarmast, A. H., Malik, A. A., Shah, M., Parry, F. Q., Wani, M. A. and Dangroo, S. (2015). Primary Retroperitoneal Hydatid cyst. *Acta Medica Iranica*, **53** (7): 448-451.

Showkat, H. I., Sarmast, A. H., Amin, A., Mir, S. F., Kotwal, S., Parry, F.Q., Lone, R., Arshad, F. and Bakhshi, I. (2012). Hydatid cyst in a 5-year old Boy presenting with prolonged fever. *The South Asian African J. of Child Health (SAJCH)*, **6** (1): 23.

Shu- Ping Sum, Tao Li, Mao, Ling Cheng, Jing-Xian and Nhi Thi Pham (2019). Taxonomic Study of the genus Townesia Ozols (Hymenoptera: Ichneumonidae: Pimplinae), with description of new spcies from China and key to world species. *Zookeys*, **878**: 23-32, doi: 10. 3897/zookeys 878.38071.

Shuja-uddin (1975). year Two new species of Aphidiidae (Hymenoptera) from India. *Rec. Zool. Surv. India*, **68**: 415-420.

Shuja-uddin (1974). Three species of *Trioxys* (Haliday) (Hymenoptera: Aphidiidae) recorded from India. *Indian J. Ent.*, **35**: 9-14.

Shuja-uddin (1976). Two new species of the genus. *Toxares* Westwood (Aphidiidae: Hymenoptera) from India. *Indian J. Entomol.*, **36**: 261-274.

Shuja-uddin (1978). New records of *Adialytus* Forster (Hymenoptera: Aphidiidae) from India, with comments on the validity of the genus. *J. Ent.* Res., **2** (2): 160 -62.

Shuja-uddin (1978*a*). New Records of *Ephedrus campestris* Stary (Aphidiidae: Hymenoptera). *Indian J. Zool.*, **6** (1): 60-61.

Shuja-uddin (1982). Description of a new species of the genus *Trioxys* Haliday (Hymenoptera: Aphidiidae) and a new record of *Trioxys* (*Trioxys*) *pallidus* (Hal) from Kashmir (India). *J. Entomol. Res*, **6** (2): 146-149.

Shuja-uddin (1983). A new species of the genus *Trioxys*, Haliday (Hymenoptera: Aphidiidae) from Kashmir, India. *J. Ent. Res.*, **7**(1): 36-38.

Siddiqi, A. U. and Khan, E. (1983). Taxonomic studies of Tylenchidae (Nematoda) of India. 3. Two new species of *Cephalenchus*, along with description of *Filenchus conicephalus* sp. n. (Nematoda: Tylenchidae) from India. *Indian J. Nematol.*, **13**(1): 84-90.

Siddiqi, M. R. (1963). Four new species in the subfamily Tylenchinae (Nematoda) from North India. *Z. fur. Parasitkde.*, **23**: 239-244.

Siddiqi, M.R. (2000). *Tylenchida: Parasites of Plants and Insects* (2nd Edition). CABI Science, CABI, Publishing, UK.

Sidhu, M., Akhter, N. and Meenia, R. (2015). Analysis of role of platelets transfusion support in the first ever dengue epidemic in Jammu Province. PARIPEX- *Indian J. Research*, **4**(6): 4-6.

Sig, A. K., Guney, M., Guclu, A. U. and Ozmen, Erkan (2017). Medicinal Leech Therapy- an overall prespective. *Integr. Med. Res.*, 6 (4)S: 337-343, doi: 10.1016 ijmr.2017.08.00.

Simmonds, F. J. and Rao, V. P. (1960). Record of *Plutella maculipennis* Curt. and some of its parasites in Kashmir, India. *Canad. Ent.*, **92** (4): 274.

Singh, A. R., Katoch, A., Yadav, A., Borkataki, S. and Khajuria, J. K. (2009). Prevalence of gastro-intestinal parasites of buffalo calves in R.S. Pora, Jammu. *Vet Practioner*, **101**: 71-72.

Singh, C., Zargar, S. A., Masoodi, I. Shoukat, A. Ahmad, B. (2012). Predictors of intestinal parasitosis in school childrens of Kashmir. A Prospective Study. *Tropical Gastroenterology*, **31**: 105-107.

Singh, D. and Sidhu, I. S. (2004). A Checklist of blowflies (Calliphoridae) from North-west India. *Uttar Pradesh J. Zool.*, **24** (1): 63-71.

*Singh, D. K. (1991). Theileriosis in India. **In**: *Orientation and Co-ordination of Research on Tropical Theileriosis*, pp. 23-28. *Proc. of Second International Workshop*, sponsored by the European Communities, Science and Technology for Development Programme, Animal Disease Research Lab., National Dairy Dev. Board, Anand, India, March 18-22 March, 1991.

Singh, G., Singh, R., Katoch, R., Wazir, V. S., Khajuria, J. K., Tikoo, A. and Anand, A. (2014). Prevalence of haemonchosis in sheep and goats of Jammu. *The Indian Vet. J.*, **91** (1): 52-53.

Singh, H. N., Udaya Lakshmi and Singh, H. K. (1984). Natural parasitization in eggs of *Lymantria obfuscata* at Shrinagar. *Indian J. Entomol.*, **46**(3)365-366.

Singh, H. N. and Udaya Lakshmi, A. (1987). Natural parasitism in eggs of *Lymantria obfuscata* Walker, at Shrinagar, J&K. *Bull. Entomol.*, **28**(1): 70-72.

Singh, I., Sharma, N. and Chohan, J. S. (1976). Incidence of cereal cyst nematode, *Heterodera avenae* Wall. in Jammu and Kashmir State. *Food Farming and Agriculture*, **8**: 20.

Singh, J., Gupta, S.K., Singh, R. and Hassan, S.A. (2014). Etiology and haemo-biochemical alteration in cattle of Jammu, suffering from anaemia. *Veterinary World*, **7** (2): 49-41.

Singh, N. P., Maqbool, N. and Rizvi, S. K. (2005). New record of parasitic mite, *Podapolipus diander* on grasshopper. *Bionotes*, **74** (4): 137.

Singh, Ranvijay, Singh., Gurdarshan and Duda, P. L. 1999 (2001). Community structure of enteric trematode of two sympatric frogs in Jammu. *Oriental Sci.*, **4**(2): 57-70.

Singh, S. (2007) Two new species of genus *Cerchysiella* Giraull (Hymenoptera: Chalcidoidea: Encyrtidae) from Northern India. *Annals of Entomol.*, **25** (1-2): 17-20.

Singh, V. K. (2009). Root-knot nematode, *Meloidogyne javanica* on citrus in Jammu. *Indian J. Nematol.*, **39**: 240.

Singh, V. K. (2009). Occurrence of Root-knot nematode, *Meloidogyne incognita* on carrot in Jammu. *Indian J. of Nematol.*, **39** (2): 245.

Singh, V. K. (2012). Roor-knot nematode, *Meloidogyne javanica* of Strawberry- a new host record from India. *Indian Phytopathology*, **65** (2): 211.

Singh, V. K., Dwivedi, M. C. and Wali, P. (2011). Occurrence of seed gall nematode, *Anguina tritici* in Jammu. *Indian J. Nematol.*, **41**(1): 109-110.

Singh, V. K. and Gupta, V. K. (2011). First report of *Meloidogyne javanica* on ginger and *Meloidogyne incognita* on coriander in Jammu and Kashmir (India). *J. Horti. Sci.*, **6** (1): 74-75.

Singh, V. K., Kala, C. S. and Kaul, V. (2007). New record of root-knot nematode, *Meloidogyne graminicola* infecting rice in Jammu. *Indian J. Nematol.*, **37** (1): 94.

Singh, V. K. and Satyaparya (2008). New record of Root knot nematode, *Meloidogyne incognita* infecting Bean and Pea in Jammu. *Indian J. Nematol*, **38** (2): 257-258.

Singh, V. K., Singh, V. B., Verma, V.S. and Singh, Permendra (2012): Record of Root-knot nematode, *Meloidogyne javanica* on Papaya and *Meloidogyne incognita* on Banana in Jammu. *Indian J. Nematol.*, **42** (1): 83-102.

Singh, V.K., Singh, V. B. and Zalpuri, L. (2012). Survey, Pathogenic effect and management of root-knot nematode, *Meloidogyne incognita* on Okara. *Indian J. of Nematol.*, **42** (2): 161-168.

Sinton. J. (1932). Notes on some Indian species of the genus *Phelbotomus*. xxx. Diagnostic table for the females of the species recorded from India. *Indian. J. Med. Res.*, **20**: 55-74.

Sinton. J. (1933). Notes on some Indian species of genus *Phlebotomus*. xxxv. Additions and alterations to the diagnostic table of females. *Indian J. Med. Res.*, **21**: 225-228.

Sinton. J. (1933 *a*). Notes on some Indian species of genus *Phlebotomus*. Diagnostic table for the males of the species recorded from India. *Indian J. Med. Res.*, **21**: 417-428.

Smit, F. G. A. M. (1977). Some Siphonaptera of small mammals in Kashmir (Insecta). *Senckenbergiana biol.*, **58** (1-2): 45-53.

Smith, K. G. V. (1973). *A Manual of Forensic Entomology*. The Trustees of the British Museum (Natural History) and Cornell University Press.

Sofi, B. A., Yattoo, G. N., Dar, P. R., Tanveer, S., Wani, S.A. and Dar, Z. A. (2008). Seroprevalence of Toxocariasis in children in Kashmir (J&K), India. *Iranian J. Parasitol.*, **3** (1-2).

*Sofi, Muneer, A, Zaki, F. A., Raina, R. H. and Khan, Z. H. (2008). Parasitization Record of San Jose Scale, *Quadraspidiotus pernicious* in apple orchards of Kashmir. *Indian J. Plant Protection*, **36** (1): 123-124.

*Sofi, Tanveer, A., Ahmed, F. and Sheikh, B. A. (2016). Occurrence, spread and control measures of *Bothriocephalus acheilognathi* (Bothriocephalidae: Cestoda) *J. Parasitic Dis.*, **40** (3): 1096-1108, published online 2014 Oct.30, doi: 10.1007/12639014-0568 (This paper has been retacted fom the journal, refer to *J. Parasit. Dis.*, 2018, December 1, **43** (1): 165, vide also PMC Retraction Policy).

Sood, S., A. Yadav, S., Vohra, Katoch, R., Ahmad, D.B. and Borkataki, S. (2009). Prevalence of coccidiosis in Poultry birds of R. S. Pora, Jammu. *Vet. Pract.*, **10**: 68-70.

Soodan, J. S., Sharma, R.K. and Bhadwal, M. S. (2000). Demodicosis in leopards in Manda wildlife Sanctuary, Jammu. *J. Vet. Parasitol.*, **14** (2): 177-178.

Soota, J. D. (1959). Fauna of the Kashmir Valley, Leeches. *Recd. Ind. Mus.*, **54**(1-2): 1-4.

*Sosha, T. A. and Wani, S.A. (2014). Aphid parasitoids (Hymenoptera: Braconidae: Aphidiinae) of Kashmir, India: A checklist. *Acta Zoologica Mexico* (*N.S*), **30** (539-552). Doi: 10.21829/azm.2014.30376.

*Sosha, T. A. and Wani, S. A. (2014). Natural enemies of Walnut aphid, *Chromaphis juglandicola* and *Panaphis juglandis* (Hemiptera: Aphididae) of Kashmir, India. *Researcher*, **6** (6): 35-39.

Srivastava, S. P. and Wattal, B.L. (1981). Parasitic arthropods of some vertebrates in Gulmarg and Tangmarg areas of Jammu and Kashmir State (India). *J. entomol. Res.*, **5**(2): 121-126.

Stary, P. and Bhagat, R. C. (1978). Records and descriptions of aphid parasitoids from Kashmir, India. (Hymenoptera: Aphidiidae). *Act. Ent. Bohemosl.*, **75**: 387-393.

Stary, P. and Ghosh, A. K. (1983). *Aphid parasitoids of Indian and adjacent countries* (Hymenoptera: Aphidiidae). *Tec. Monograph* No.7, Edited and published by Director, Zoological Survey of India, Calcutta, West-Bengal, India.

Stone, A. (1973). Superfamily Anisopodoidea, Family Anisopodidae, pp. 8 In: Catalogue of the Diptera of the Oriental Region Vol. I, Sub-order Nematocera, Delfinado, M.D. and Hardy, D.E. (eds.). The University Press of Hawaii, Honolulu, USA.

Stone, A. and Philip, B. C. (1974). The Oriental species of the Tribe Haematopotini (Diptera, Tabanidae). *Tech. Bull. Agri. Res. Serv.United State Department of Agriculture, Washington D.C.* 248 pp.

Subba-Rao, B. R. and Hayat, M. (Eds.) (1985). The Chalcidoidea (Insecta: Hymenoptera) of India and adjacent countries. Part-I. Review of families and keys to families and genera. *Oriental Insects*, **19**: 163-310.

Subba-Rao, B. R. and Hayat, M. (Eds.) (1986). The Chalcidoidea (Insecta: Hymenoptera) of India and adjacent countries, Part-II. *Oriental Insects*, **20**: 1-43.

Subba-Rao, B.R. and Hayat, M. (1986). Family: Mymaridae. In: Subba Rao, B.R. and Hayat, M. (eds.), pp. 179-191, The Chalcidoidea (Insecta: Hymenoptera) of India and adjacent countries. *Oriental Insects*, **20**: 1-430 pp.

Sudan, S., Sharma, S., Gupta, M., Kumar, R. and Sambyal, S. S. (2016). Sero-epidemilogical trends of Dengue fever in Jammu Province of J&K State. *International J. Medical Research and Health Sciences*, **5**: 81-86.

Sudhan, N. A. and Shahardar, R. A. (1999). Occurrence of *Coenurus* cyst in sub-cutaneous tissue of ear of a goat. *J. Vet. Parasitol.*, **13**(1): 75- 76.

Sureshan, P.M. (2009). Two species of *Megastigmus* Dalman, associated with wild Rose, *Rosa webbiana* (Rosaceae) from Ladakh region, India, with key to the Oriental species (Hymenoptera: Chalcidoidea: Torymidae). *J. Threatened Taxa*, **1** (5): 305-308.

Sureshan, P. M. (2010). A new species of *Ecdamua* Walker from cold desert of Ladakh, India (Hymenoptera, Chalcidoidea: Torymidae). *Rec. Zool. Surv. India*, Part I, **110** (1): 103-106.

Sureshan, P. M. (2012). New Distributional records of Pteromalidae (Hymenoptera: Chalcidoidea) from Ladakh, India. *Rec. Zool. Surv. India*, **112** (Part 4): 39-45.

Sureshan, P. M. and Narendran, T. C. (2003). A checklist of Pteromalidae (Hymenoptera: Chalcidoidea) from Indian sub-continent. *Zoos' Print Journal*, **18** (5): 1099-1110.

*Sureshan, P. M. and Narendran, T. C. (2004). Key to the genera of Pteromalidae of India and Adjacent countries. *Zoological Survey of India*, Calcutta, India. 56 pp.

Swarup, G. and Das-Gupta, D. R. (1986). *Plant parasitic nematodes in India– Problem and Progress*. Indian Agriculture Research Institute, New Delhi, India. 497 pp.

Swarup, G., Prasad, S. K. and Raski, D. J. (1964). Some *Heterodera* species from India. *Plant Diseases Reporter*. **48** (3): 235.

Tak, Hidayat-ullah, Chishti, M. Z. and Ahangar, A. G. (2003). Prevalence of hydatid disease in slaughtered animals in Kashmir Valley, pp.225-230. *In*: Gupta, Neelan and Gupta, D.K. (Eds.) *Proc. Natl. Congr. Parasitol.*, held at Barielly, U. P. in Feb., 2003.

Tak, I. R., Chishti, M. Z. and Ahmad, F. (2013). Epidemiological studies of abomasal nematodes of sheep of Kashmir valley, with particular reference to *Haemonchus controtus*. *Nature and Sci.*, **11**: 34-39.

Tak, Irfan-ur-Rauf, Dar, S. A., Chishti, M. Z., Kaur, H. and Dar, G. A. (2014). Parasites of some fishes (*Labeo rohita* and *Schizothorax niger*) of Jammu and Kashmir in India. *Intl. J. Fisheries and Aquaculture*, **6** (9): 104-107. Doi: 10.5897/IJFA14.0435.

*Tak, I. R., Ali, S., Dar, S. A., Ganai, B. A., Chishti, M. Z. and Ahmad, F. (2014). A brief study of morphology of *Haemonchus controtus* and its Haematophagous Behaviour. *Global Veterinaria*, **13** (6): 960-965. Doi: 10.5829/idosi.gv.2014.13.06.9118.

Tak, Irfan-ur- Rauf, Dar, S.A., Chishti, M. Z., Kaur, H. and Dar, G. A. (2014). Parasites of some fishes (*Labeo rohita* and *Schizothorax niger*) of Jammu and Kashmir, India. *Intl. J. Fisheries and Aquaculture*, **6** (9): 104-109.

*Tak, I., R., Ganai, B. A., Chishti, M. Z. and Ahmad, F. (2013). Epidemiological studies of ostertagiosis infecting sheep of Kashmir valley. *DAV Intl J. Sci.*, **2**: 68-73.

Takada, Hajimu and Rishi, N. D. (1980). Records of fifteen species of Aphidiidae (Hymenoptera) from Kashmir, India, with descriptions of three new species. *Kontyu*, **48** (2): 234-240.

Talwar, P.K. and Jhingran, A.G. (1991). *Inland Fishes of India and Adjacent. Countries.* Vol. I-II. Oxford and IBH Publishing House New Delhi. Vol.I, 1-541 pp., Vol. II, 542-1158 pp.

Tantary, T. A., Nzir, R. Chishti, M.Z., Ahmad, F., Dar, G.H. and Dar, S. (2016). A report on the incidence of *Trichodina heterodentata* from fishes of Jammu (J&K), India. *J. Parasit. Diseases*, **40** (2): 524, doi: 10.1007/s 12639-014-0538-ZEpub 2014, August.

*Tanveer, S. (1996). *Biology and pathogenicity of parasites of some poultry and economically important birds of Kashmir.* Ph. D. Thesis. P. G. Deptt. of Zool., Faculty of Sci., Univ. of Kmr., Sgr., Kmr. (J&K).).

*Tanveer, S. (2012). *Cestode Parasites of Birds: Morphological Pecularities*. LAP Lambert Academic Publishing (2012-04-07), Saabruchen, Germany.

* Tanveer, S. and Ahmed, F. (2012). Prevalence of parasites infecting some selective fishes from the water bodies of Kashmir valley. *Trends in Fisheries Res.*, **1** (2): 7-10. (On helminths of *Cyprinus carpio* and *Schizothorax* of fishes, in Dal Lake and River Jehlum).

Tanveer, S. and Chishti, M. Z. (1999). Studies on some strongylid and trichostrongylid nematodes in avian hosts of Kashmir. *Oriental Sci.*, **4**(2): 93-100.

Tanveer, Syed and Chishti, M.Z. (2001). Studies on Notocotylid trematode genus *Paramonostomum* (Leuhe) 1909 in domestic fowl and common coot in Kashmir, with description of a new species. *J. Parasitic Diseases*, **25** (2): 95-99.

*Tanveer, S., Ahad, Shazia and Chishti, M. Z. (2015). Morphological characterization of nematodes of genera: *Capillaria, Acuaria, Amidostomum, Streptocara, Heterakis* and *Ascardida*, isolated from intestine and gizzard of domestic birds from different regions of the temperate Kashmir Valley. *J. Parasitic Dis.*, **39**: 745-760.

Tarello, W. (2005).*Trypanosoma avium* incidence, pathogenicity and response to Melarsomine in Falcons from Kuwait. Parasite, 12: 85-87.http: / /www.parasite-journal.org., doiorg/10.1051/parasite/2005121085.

* Tariq, K. A., Chishti, M. Z., Ahmmed, F. and Shawl, A. S. (2006). Epidemiology of paramphistomosis in small ruminants in Kashmir Valley. *Oriental Science*, **11**: 11-16.

*Tariq, K. A., Chishti, M. Z. and Ahmed, F. (2008). Epidemiological study on *Paramphistomum* infection in goats- Kashmir Valley. *World J. Agric. Sci.*, **4**: 61-66.

Tariq, K. A., Ahmed, F., Chishti, M. Z., Tak, H. and Shawl, A. S. (2008 a). An Abattoir Survey of Gastro-intestinal Helminth parasites of sheep in Hazratbal region of Kashmir. **In**: M. Z. Chishti and Fayaz Ahmad (eds.), pp. 409-413, *Science For Better Tomorrow*, Printed at Universal Printers, Khanyar, Srinagar, Kashmir.

*Tariq, K. A., Chishti, M. Z., Ahmed, F. and Shawl, A. S. (2008 b). Antihelminthic efficacy of *Achilea milifolium* against nematodes in sheep: *In-vitro* and *in-vivo* studies. *J. Helminthol.*, **82**: 227-233.

Tariq, K. A., Chishti, M. Z., Ahmad, F. and Shawl, A. S. (2008 c). Epidemiology of gastro-intestinal nematodes of sheep managed under traditional husbandry system in Kashmir valley. *Vet. Parasitol.*, **158**: 138-143.

*Tariq, K. A., Chishti, M. Z., Shawl, A. S., Ahmad, F. and Tak, H. (2008 d). Seasonal Prevalence of Gastro-intestinal nematodes according to faecal examination of sheep. **In**: M. Z. Chishti and Fayaz Ahmad (eds.), *Science For Better Tomorrow*, Printed at Universal Printers, Khanyar, Srinagar, Kashmir.

*Tariq, T. A., Chishti, M. Z., Ahmad, F. and Shawl, A. S. (2009). Antihelmintic activity of extracts of *Artemisia absinthium* against ovine nematodes. *Vet. Parasitol.*, **160** (1-2): 83-87.

Tariq, K. A., Chishti, M. Z. and Ahmad, F. (2010). Gastro-intestinal nematode infection of goats, relative to season, host sex and age from Kashmir valley. *Indian J. Helminthol.*, **84**: 93-97.

Tavassoli, M., Pourseyed, S.H., Ownagh, A., Bernousi, Iraj and Mardani, Karim (2011). Biocontrol of Pigeon tick, *Argas reflexus* (Acari: Argasidae) by entomopathogenic fungus, *Metarhizium anisopliae* (Ascomycota: Hypocreales). *Brazilan J. Microbiol.*, **42** (4): Oct./Dec.2011. http: dx.doi.org./10.1590/s15178382201.

Thapar, G.S. (1950). Two new species of the genus *Rhabdochona* Railliet from Indian Fishes *Indian J. Helminthol.*, **2** (1): 35-40.

Theodor, O. (1948). Classification of the old world species of the sub-family Phlebotominae (Diptera: Psychodidae) (illustrated). *Bull. Entomol. Res.*, 85-115.

Thokar, Manzoor, A. (1986). *Comparative evaluation of enzyme linked immunosorbent Assay for detection of Toxoplasma gondii antibody in clinically suspected toxoplasmosis.* M. D. Thesis Department of Microbiology, Shir-e-Kashmir Institute of Medical Sciences, Srinagar, Kashmir. pp.69.

Thokar, M. A., Malla, N. and Wattal, C. (1988). Serological study of patients clinically suspected to have toxoplasmosis in Kashmir. *Indian J. Medical Res.*, **88**: 29-34.

Tickoo, Ratni (1970). On the morphology of *Haematoloechus almorai* (Pande, 1937) from the lungs of *Rana cyanophlyctis* in Kashmir. Part I: General Morphology cuticle and musculature. *Kashmir Sci.*, **7**(1-2): 156-157.

Tiku, J. L. (1980). Incidence trypanosomiasis in some breeding bulls of Jammu Province. *Livestock Adviser*, **5**(9): 55-5.

Townes, H., Townes, M. and Gupta, V.K (1961). A Catalogue and re-classification of Indo- Australian Ichneumonidae. *Mem. American Ent. Inst.* No. 1.

Tramboo, S. R., Shahardar, R. A., Allaie, M., Wani, Z. A. and Bushra, M. S. (2015). Prevalence of gastro-intestinal helminth infection in ovine population of Kashmir valley. *Vet. World*, 8 (10): 1119- 1204, doi: 10.14202/vetworld 2015-1119-1204.

Tramboo, S. R., Shahardar, R. A., Allaie, I. M. and Wani, Z. A. (2018). Prevalence of ticks infesting livestock of Kashmir valley. *J. Entomol. Zool. Studies*, **6** (6): 877-879.

Tramboo, S. R., Shahardar, R. A., Allaie, I. M., Bulbul, K. H., Khan, A., Mir, M. S., Hussain, I., Rather, M. A. and Ahanger, A. R. (2019). Prevalence of ectoparasites in equines of Kashmir valley. *J. Entomol. and Zool. Studies*, **7** (6): 528- 534.

Tripathi, Y. R. (1956). Studies on parasites of Indian fishes.*V. Acanthocephala. Rec. Indian Mus.*, **54**: 61-99.

Triapitsym, S.V. (2013). Review of *Gonatocerus* (Hymenoptera: Mymaridae) in the Palaearctic Region, with notes on extralimital distribution. *Zootaxa*, **3644** (1): 1- 178. http: dx. dx doi. Org./10.11646/zootaxa, 3644.11.

Tritscher, M., Retschnig, G., Yanez, O. Williams, G., Geoffrey, R. and Nezemann, P. (2016). Host-sharing by honeybee parasites, *Lotmania passim* and *Nosema ceranae*. Wiley *Ecology and Evolution, Doi: 10.1002/ece.3.2796.*

Trivedi, P. C. 1998 (Ed.). *Phytonematology in India*. CBS Publishers and Distributors, Daryaganj. New Delhi, India, 205 pp.

Tufani, N. A., Fazili, M. R., Malik, H.U., Beigh, S. A and Dar, K. H. (2015). Clinico-Haematological Profile and Therapeutic Management of acute babesiosis in Holstein- Friesian Cross bred cow. *Vet. Clinical Sciences*, **3**(3): 11-14.

Tufani, N. A., Hafiz, A., Malik, H. U., Peer, F.U. and Makhdomi, D. M. (2009). Clinico-therapeutic management of acute babesiosis in bovine. *Intas Polivet.*, **10** (1): 49-50.

Tufani, N. A., Malik, H.U. and Fazili, M.R. (2017). Clinico-haematological Profile and Therapeutic management of acute Babesiosis in sheep and goats. *J. Anim. Res.*, **7** (5): 857-863.

Upadhyay, S. R., Bhardwaj, R. K., Singh, R. and Sharma, R. (2013). Concurrent infection of *Mycoplasma haemocanis* and *Hepatozoon* in German Shepherd dog. Indian Vet. J., 90(6): 99.

Van Cleave, H.J. (1948). Expanding horizon in the recognition of Phylum. *J. Parasitol.*, **34**: 1-20.

Van-Emden, F.I. (1965). *The Fauna of India and adjacent countries, Diptera Vol. VII, Muscidae, Part 1.* Issued by the Zoological Survey of India, Calcutta, Published by Baptist Mission Press, Calcutta, 647 pp.

Vavilova, Valeriya, Luzyanin, S., Woyciechowski, M. and Blinov, A. (2017).Parasites of the genus *Nosema*, *Crithidia* and *Lotmania* in honeybee and bumblebee population. A case study in India. *J. Genetics Breeding*, **21**(8): 943-951. Doi: 10.18699/VJ17.317.

Verma, K. C. and Magotra, M. L. (1976). Vivax cerebral malaria in Jammu. *Indian Pediatrics*, **13**(3): 229-31.

*Verner, Michelsen (1983).*Thyreophora anthropophaga* Robineau- Desvoidy, an ' extinct' bone skipper rediscovered in Kashmir (Diptera: Piophilidae, Thyreophorinae). *Insect Systematics and Evolution*, **14** (4): 411- 414.

Vohra, S., Katoch, R., Yadav, A., Khajuria, J. K. and Gupta, S. (2016). An outbreak of gastro-intestinal parasitic infection in sheep of R. S. Pora region of Jammu. *Intl. J. Sci., Env. And Technol.*, **5** (4): 2057-2061.

Wachkoo, A., and Akbar, S. A. (2019). *Protichneumon pisoricus* (Hymenoptera: Ichneumonidae), the first record from India. *National Acad. Science Letters* Doi: 10.1007/s 40009.018-07752.

Wali, Asifa, Balkhi, M.H., Maqbool, R., Darzi, M. M., Shah, F. A., Bhat, F. A. and Bhat, B. A. (2016 a). Distribution of Helminth Parasites in intestine and their seasonal rate of infestation in three fresh water fishes of Kashmir. *J. Parasitol. Res.*, **Vol. 2016**, *Article ID 8901518, doi: 10.1155/2016/8901518.*

Wali, Asiafa, Balkhi, M. H., Maqbool, Feroz, A. S., Farooz, A. B., Darzi, A.K and Bhat, A.A. (2016 b). Histopathological alteration and distribution of *Pomphorhynchus kashmirenisis* in intestines and their seasonal rate of infestation in three fresh water fishes of Kashmir. *J. Ent. Zool. Studies*, **4** (5): 22-28.

*Wali, A., Balkhi, M. H., Maqbool, R. and Shah, F. A (2017). Seasonal Histopathological alternation caused by *Adenoscolex oreini* in *Schizothorax niger* of Kashmir Valley. *Indian J. Animal Res.*, Doi: 10.18805/ijar.voiof.7815.

Wali-ullah, M. I. S. (1992). Nematodes associated with vegetable crops in Kashmir Valley, India. *Nematol. Medit.*, **20**: 47-48.

Wali-ullah, M. I. S. (2005). Nematodes associated with Kiwi (*Actinidia delicious* Clev) in Kashmir Valley, India. *Indian J. of Nematol.*, **35** (2): 227.

Wali-ullah, M. I. S. and Bhat, M. A. (1990). Occurrence of mixed population of *Pratylenchus* Filepjer, 1936 on potato (*Solanum tuberosum*) in Srinagar, Kashmir. *Indian J. Nematol.*, **20** (2): 239-241.

Wali-ullah, M. I. S., Masoodi, M. A. and Kaul, V. (1993). Population behaviour of nematode associated with strawberry (*Fragaria vesca*) in Srinagar, Kashmir. *Curr. Nematol.*, **4**: 41-44.

Wali-ullah, M. I. S. and Kaul, V. (1997). Nematodes associated with cherry plant in Kashmir Valley. *Indian J. Nematol.*, **27** (2): 237.

Wali-ullah, M. I. S., Kaul, V. and Zaki, F. A. (1991): The Rice-root nematode, *Hirschmanniella* spp. in Jammu and Kashmir. *Geobios new Report*, **10**: 175-176.

Wali-ullah, M. I. S., Wanchoo, R. N. and Mantoo (2008). Diversity of nematodes of Brinjal (*Solanum melongena*) in Jammu District of J&K State, **In:** pp. 9-14, M. Z. Chishti and Fayaz Ahmad (eds.), *Science For Better Tomorrow*, printed at Universal Printers, Khanyar, Srinagar, Kashmir.

Walter, T. C. and Boxhall, G. (2018). World of Copepods database. Accessed at http: //www.marinespecies.org./ copepod on 2018.02.08.

Wani, G., M., Ahmad, S. M., Khursheed, B. (2015). Clinical study of cutaneous leishmaniasis in Kashmir Valley, India. *Dermatology Online J*. 6: 387-92, DOI: 4103/2229-5178.169732.

Wani, B. A. (1970). *Incidence and Symptomatology of ascariasis in Kashmir*. M. D. Thesis (Medicine). Govt. Medical College, Srinagar. The Univ. of Kashmir, Srinagar, Kashmir (J&K).

Wani, M. A., Babu, K. M., Ahangar, A. G., Kirmani, A. R. and Saxena, R. K. (1999). Hydatid cyst of Brain – A case Report and Review of literature. *JK-Practitioner*, **6** (2): 145-146.

Wani, N. A. (1993). Hydatid Disease of of the Spleen. *Indian J. Surgery*, **55** (3): 155-160.

Wani, N. A., Shah, O. J., Zargar, J. I., *et al.* (2000). Hydatid cyst of the Pancreas. *Digestive Surgery*, **17**: 188-190, doi: org/10.1159/000018829.

Wani, Rauf, Ahmad, Malik, Chowdri, N. A., Wani, K. A. and Naqash, S. A. (2005). Primary extrahepatic abdominal hydatidosis. *Intl. J. Surgery*, **3** (2): 125-127.

*Wani, R. A., Malik, A. A., Parray, F. Q., Wani, A. A. and Dar, A. H. (2012). Hydatid Disease at Unusual Sites. *Intl. J. Case Repts. and Images*, **3** (6): 1-6.

Wani, S. A., Ahmad, F., Zargar, S. A., Fomda, B. A., Ahmad, Z. and Ahmad, P. (2007a). Helminth infestation in children of Kupwara District: A Prospective Study. *Indian J. Med. Microbiol.*, **25** (4): 398-400.

Wami, S. A., Ahmad, F., Zargar, S. A., Ahmad, Z., A. (2007 b). Prevalence of Intestinal Parasites and associated Risk factors among school children in Srinagar city, Kashmir, India. *J. Parasitol.*, **93** (9) 15541-1543.

*Wani, S. A. and Ahmad, F. *et al.* (2007c). Epidemiology of Gastro-intestinal helminths in school going children of District Budgam, India. Pp. 219-225, **In**: B. L. Koul (ed.), *Advances in Fish and wildlife Ecology and Biology*. **4** (2007). Published by Daya Publishing House, Darya Ganj, New Delhi.

Wani, S. A., Ahmad, F., Zargar, S. A., Fomda, B. A., Ahmad, Z. and Ahmad, P. (2007 a). Helminth infestation in children of Kupwara Disrtrict: A Prospective Study. *Indian J. Med. Microbiol.*, **25** (4): 398-400.

Wani, S. A., Ahmad, F., Zargar, S. A., Ahmad, Z., Ahmad, P. and Tak, H. (2007 b). Prevalence of intestinal Parasites and associated risk factors among school children in Srinagar City, Kashmir, India. *J. Parasitol.* **93** (9): 1541-1543.

Wani, S. A., Ahmad, F., Zargar, S. A., Dar, P. A., Dar, Z. A. and Jan, T. R. (2008). Intestinal helminths in a population of children from Kashmir Valley, India. *J. Helminthol.*, **82** (4): 313-317, Doi: 10.1017/S0022149x08019792.

Wani, S. A., Ahmad, F., Zargar, S. A., Amin, Ayesha, Dar, Z., Dar, P. A. (2010). Intestinal helminthiasis in children in Gurez of Jammu and Kashmir State, India. *J. Global Infect. Dis.*, **2** (2): 91-94 doi: 10.4103/0974-777x62872.

Wani, S. A. and Ahmed, F. (2009). Intestinal helminths and associated Risk Factors in children of District Pulwama, Kashmir, India. *Indian J. Medical Microbiol.,* **27.1** (2009): 81-82.

Wani, S. A., Ahmad, F. and Amin, A. (2017). Burden of intestinal Helminth infection among children of the Kashmir Valley, India, **In**: *3rd world Conference on Parasitology and Pathogenesis*, July 12-13, 2017, Chicago, U.S.A, *J. of Bacteriol. Parasitol.*, vol. **8** (3), Suppl., p. 57.

Wani, S. A. and Amin, Ayshia (2016). Intestinal helminth infections among children of District Shopian of Kashmir Valley. *J. Parasitic Dis.*, **40** (4): 1422- 1425.

*Wani, S. A. and Salam, S. T. (2020). Intestinal Helminthiasis in school age children of Kashmir valley and the need for intervention: A systematic Review. *Acta Scientific Microbiol.*, **3** (5): 37-44.

Wani, W. A. and Chrungoo, R. K. (1992). Biliary Ascariasis: Surgical aspect. *World J. of Surgery,* **16**: 976-79.

Wani, Z. A., Shahardar, R. A., and Shahnawaz, M. (2011). Prevalence of nemathelminth parasites in sheep of Ganderbal District of Kashmir Valley. *J. Vet. Parasitol.*, 2**5:** 26-29.

Wani, Z. A., Shahardar, R. A., Shahnawaz, M. and Mir, M. S. (2012). Pathological changes due to Nematode helminth parasites in sheep of Kashmir Valley. *Intl. J. of Livestock Res.*, **2** (1): 173-178.

Wani, Z. A., Allaie, I. M., Bushra, M. S., Hussain, H. A., Malik, Asloob, A., Tramboo, S. R. and Sofi, K. A. (2013). Occurrence of parasitic infections in the livestock of Zanskar regfon of Kargil, India. *Intl. J. Livestock Res.*, **3**(1): 88-91.

Wani, Z. A., Allaie, I. M., Shah, B. M., Raies, A., H. Athar and Junaid, S. 2015. *Dipylidium caninum* infection in dogs infested with fleas. *J.Parasitic Diseases,* **39** (1): 73-75.

Wan-Qixue, Vettchrves, Y.G. and Du, J. (2011). A Review of tribe Boettcheriscina Verves, 1990(Diptera: Sarcophagidae), with description of 7 new species and genus from China. *Annl. Of Soc. of Entomol.*, **47** (3-4): 303-329.

*Wattal, Chand, Malla, N., Khan, I. A. and Agarwal, S. C. (1986). Comparative evaluation of enzyme- linked immunosorbent assay for diagnosis of pulmonary echinococcosis. *J. Clin. Microbiol.*, **21**: 41-46.

Wattal, Chand, Malla, Nancy, Khan, Imtiyaz and Agarwal, S.C. (1986). *In vitro* development and hatching of *Ascaris lumbricoides* var. *hominis* egg. *Indian J. Med. Res.*, **83**: 257-264.

White, S. R., Aubertin, D. and Smart, J. (1940). *The Fauna of British India, including the remainder of the Oriental region,* Vol.VI. Calliphoridae. Taylor and Francies, London, England.

Weckesser, S., Hilger, C., Lent, Z.D., Jakob.T. (2010). Anaphylactic reactions to bites of pigeon tick, *Argas reflexus*. *Eur. J. Dermatol.*, **20** (2): 244-245.

WoRM (2008). *Linguatula serrata* Froilich, 1789. Accessed at http: www. Marinespecies. Org/aphia-php ?=taxdetails and id=730130 on 2018.02.11.

Yadav, A., Khajuria, J. K. and Devi, J. (2006). Cat fleas infestations in goats. *Indian Vet., J.,* **83**: 439-440.

Yadav, A. Khajuria, J. K. and Raina, A. K. (2004/2005). Gastro-intestinal parasitic infestation profile of bovines in R. S. Pura. Jammu. *J. Vet. Parasitol.,* **18** (2): 167-169.

Yadav, A., Katoch, R. Godara, R. and Borah, M.K. (2014). Occurrence and histopathology of *Przhevalskiana silensus* infestation in sheep of Jammu region. *J. Parasitic Diseases,* **38**: 324-327.

*Yadav, A., Katoch, R., Khajuria, J. K., Katoch, M. and Rastogi, A. (2012). Economic impact of *Przhevaskiana silensus* infestation in native goats of Northern India. *Trop. Anim. Health Prod.,* **44**: 580-582.

Yadav, A., Katoch, R., Khajuria, J.K., Godara, R. and Agarwal, R. (2013). Prevalence of *Hypoderma lineatum* in cattle of Jammu Region. *J. Parasitic Diseases,* **37** (2): 196-198.

*Yadav, A., Katoch, R., Godara, R. and Borah, M. K. (2014). Occurrence and histopathology of *Przhevalskiana silensus* larvae in sheep. *J. Parasit. Dis.,* **38**: 324 – 327.

*Yadav, A., Katoch, R., Khajuria, J. K., Katoch, M. and Rastogi, A. (2011). Economic impact of *Przhevalskiana silensus* infestation in native goats of Northern India. *Tropical Animal Health and Production,* **44** (3): 581-587.

Yadav, A., Katoch, R., Khajuria, J. K., Katoch, M. and Agarwal, R. (2011). Prevalence and biology of warble fly infestation by *Przhevalskiana silenus* (Brauer) in Jammu Province, India. *Tropical Animal Health and Production,* **43** (8): 1487-92.

Yadav, A., Katoch, R., Katoch, M., Aggarwal, R., Khajuria, J., Godara, R. and Kalha, R. (2012). Cross sectional study and analysis of potential risk factors for *Cryptosporidium* spp. infection in buffalo calves in Jammu. *Vet. Pract.,* 13 (2): 278-281.

Yadav, A., Katoch, R. Singh, P., Khajuria, J. K. and Vohra, S. (2009). Occurrence of *Melophagus ovinus* in sheep of Poonch in Jammu Region. *J. Vet. Parasitol.,* **23**: 91-92.

Yadav, A., Khajuria, J. K. and Soodan, J. S. (2006).Warblefly infestation in goats in Jammu. *J. Vet. Parasitol.,* **20**: 149- 152.

Yadav, A., Khajuria, J. K. and Raina, A. K. (2006). Seasonal prevalence of gastro-intestinal parasites in sheep and goats of Jammu, India. *J. Parasitol.,* **20**: 65-68.

Yadav, A., Khajuria, J. K., Sharma, R. K., Samanta, A.K. and Mandal, G. P. (2007). Acute coccidiosis and its treatment in kids. *J. Vet. Parasitol.,* **21** (1): 71-72.

Yadav, A., Kumar, A., Bhadwal, M. S., Khajuria, J. K. and Gupta, A. (2006). Ocular sekariasis in horses: A case study. *J. Vet. Parasitol.,* **20**: 183-184.

Yamaguti, S. (1958). *Systema Helminthum* Vol.I (Part I and II). *The Digenetic Trematodes of FVertebrates,* Vol. IV. *Monogenea and Aspidocotylea* Interscience Publishers Inc. Indian Edition (Reprint) 1985, International Books and Periodical Supply Co. New Delhi, India.

Yamaguti, S. (1959). *Systema Helminthum* Vol.II. The *Cestodes of Vertebrates.* Johan Wiley and Sons, Japan.

Yamaguti, S. (1963). *Systema Helminthum,* Vol. V. *Acanthocephala.* John Wiley and Sons. Inc. Indian Reprint Edition, 1985, International Books and Periodical Supply Service. New Delhi, 423 pp.

Yamaguti, S. (1971). *Synopsis of Digenetic trematodes of Vertebraters,* Vol. I. Keigaku. Publishing Co. Tokyo- Japan.

Yamaguti, Satyu (1985). *Systema Helminthum* Vol, . III, Part I and II, *Nematodes of Vertebrates,* Indian Reprint Edition, Part I, 1-679 pp. and Part II 681-126 pp. International Books and Periodical Supply service, New Delhi, India.

Yasmeen, M. (1979). *Intestinal functions of Ascariasis.* M.D. Thesis (Medicine). Govt. Medical College, Srinagar. The Univ. of Kashmir, Srinagar, Kashmir (J&K), 55 pp. (unpublished).

Yattoo, G. N., Khuroo, M.S. Zargar, S. A., Bhat, F. A. and Sofi, B. A. (1999). Case Report: Precutaneous drainage of the pancreatic head hydatid cyst with obstructive jaundice. *J. Gastroenterology* and *Hepatology,* **14** (9): 931-34.

Yousuf, M., Tak, Hidayat-ullah and Gul, Nazima (2013). Prevalence of Eimerosis in domestic free range chickens of south Kashmir, J&K (India). *Intl. J. Curr. Res.,* **51** (06): 1606-1608.

*Yousuf, M. and Shafee, S. A. (1987). Taxonomy of Indian Trichogrammatidae (Hymenoptera: Chalcidoidea). *Indian J. Syst. Ent.,* **4** (2): 200.

Yousuf, R. Mir, S.M., Mir, S., Shah, A.W. and Ganaie, G.A. (2011). A comparative study of parasites infecting some selected fishes from water bodies of Kashmir Valley. Indian. *J. Fundamental and Applied Life Sciences,* **1** (2); 150-153.

Zaitzev, V.F. (1988). New species of Bombyliidae from India and Sri Lanka. *Entomol. Review,* 1-9.

Zaka-ur-Rab 1972(1974). Record of *Syrphus confrator* Wiedmann as predaceous on woolly aphid, *Eriosoma lanigerum* (Husum). *Indian J. Ent.,* **34**(4): 348-49.

Zaki, F. A. (1999). A note in Ladakh (J&K). *Appl. Biol. Res.,* **1**(1): 175-177.

Zaki, F. A. and Bhatti, D. S. (1989). Effect of Castor (*Ricinus communis*) leaves in combination with different fertilizer doses on *Meloidogyne javanica* infesting tomato. *Indian J. Nematol.,* **19** (2): 171-176.

*Zaki, F. A, and Bhatti, D.S. (1989 b). Phytotherapeutic effect of some plant leaves on *Meloidogyne javanica,* infecting tomato plants. *Nematologia Mediterranean,* **17** (2): (Effective results were found in control of *M. javanica* with leaves of plants, *viz. Calotropis procera, Ricinus communis, Azadirachta indica*).

Zaki, F. A., Kaul, V. and Wali-ullah, M. I. S. (1991). Plant-parasitic nematodes associated with potato-turnip rotation and other vegetable crops in Kashmir. *Indian J. Nematol.,* **21**(2): 174-75.

Zaki, F. A., Kaul, V. and Wali-ullah (1993). Crop losses estimation in tomato and potato caused by plant-parasitic nematodes in Kashmir. *Indian J. Nematol.,* **23**(1): 101-104.

Zaki, F. A. and Mantoo, M. A. (2003). Plant -parasitic nematodes associated with temperate fruits in Kashmir. *Pest Management and Economic Zoology,* **11** (1): 97-101.

Zaki, F. A. and Mantoo, M. A. (2008). Integrated pest management in Saffron. *In:* pp. 209-222, F. A. Nehvi and S. A. Wani (eds.). *Saffron Production in Jammu and Kashmir*. Directorate of Extension Education, SKAUST – K, India.

Zaki, F. A. and Mir, A. A. (1994). First record of ear- cockle nematode (*Anguina tritici*) on wheat in Ladkah. *Indian J. Nematol.,* **24**: 228-29.

Zalpuri, L., Tara, J. S. and Singh, V. K. (2013 a). Plant and soil nematodes: Diversity and Interactions. *Global J. Biol. Agric. And Health Sci.,* **2** (2): 123- 124.

Zalpuri, L., Tara, J. S. and Singh, V. K. (2013 b). Plant parasitic nematodes associated with citrus plantations in Jammu Region. *Indian J. Nematol.,* **43** (2): 217-219.

Zalpuri, L., Tara, J. S. and Singh, V. K. (2013 c). Prevalence of Plant-parasitic nematodes (*Citrus* species) in various villages of Jammu region. *Intl. J. Scientific and Res. Publ.,* **3** (6): 1-3.

Zargar, A. H., Masoodi, S. R., Laway, B. A., Sofi, B.A. and Wani, A. I (1998). Seroprevalence of toxoplasmosis in women with repeated abortion in Kashmir. *J. Epidemiol. Community Health,* 52: 135-136.

*Zargar, U. R., Yousuf, A. R, Chishti, M. Z., Ahmed, F. and Bashir, H. (2012). Effect of water quality and trophic status on helminth infection in the cyprinid fish, *Schizothorax niger* Heckel, 1838 from three lakes in the Kashmir Himalaya. *J. Helminthol.,* 86: 70- 76.

Zargar, U. R., Chishti, M Z., Yousuf, A. R. and Ahmad, F. (2012). Infection level of Monogenean parasite, *Diplozoon kashmirensis* (Monogenea, Polyopisthocotylea) in Crusian carp, *Crassius crassius* from Lake ecosysatem of altered water quality: What factors do have an impact on the Diplozoon infection. *Vet. Parasitol.,* **189** (2-4): 218-226.

*Zargar, U. R., Chishti, M. Z., Yousuf, A. R. and Ahmed, F. (2012). Infection level of the Asian tapeworm (*Bothriocephalus acheilognathi* in cyprinid fish, *Schizothorax niger* from Anchar Lake, relative to season, sex, length and condition factor. *J. Parasitol. Res.,* **110** (1): 427-435 [(Infection was found higher in summer (39. 5 per cent) lesser in winter season (8.1 per cent)].

*Zargar, U. R., Chishti, M. Z., Yousuf, A. R. and Ahmed, F. (2013). Species spectrum, diversity profile and infection indices of helminth parasite-fauna of Chirruh snow trout, *Schizothorax esocinus* (Heckel) in Lake ecosystems of Kashmir

Himalaya. Do similarity and host parasite association arise. *Veterinary Research Communication*, **37**(3): 197-207.

Zeya, S. B. and Hayat, M. (1995). A revision of the Indian species of Gonatocerus Nees. (Hymenoptera: Chalcidoidea: Mymaridea). *Oriental Inst.*, **29**: 47-160.

Zutshi, M. K. (1967). *Lymantria obfuscata* Walker and its natural enemies in Kashmir (Lymantridae: Lepidoplera). *J. Bombay Nat. Hist. Soc.*, **64**(1): 126-129.

Zumpt, F. (1965) Myiasis in Man and Animals in old World. Butterworth, London, England, 267.

Subject Index

B

C

D

E

F

R

S

T

V

W

I

K

L

M

9 789354 615108

Printed at Repro India Ltd.